Modern Analysis of
Customer Surveys

Statistics in Practice

Series Advisors

Human and Biological Sciences
Stephen Senn
University of Glasgow, UK

Earth and Environmental Sciences
Marian Scott
University of Glasgow, UK

Industry, Commerce and Finance
Wolfgang Jank
University of Maryland, USA

Statistics in Practice is an important international series of texts which provide detailed coverage of statistical concepts, methods and worked case studies in specific fields of investigation and study.

With sound motivation and many worked practical examples, the books show in down-to-earth terms how to select and use an appropriate range of statistical techniques in a particular practical field within each title's special topic area.

The books provide statistical support for professionals and research workers across a range of employment fields and research environments. Subject areas covered include medicine and pharmaceutics; industry, finance and commerce; public services; the earth and environmental sciences, and so on.

The books also provide support to students studying statistical courses applied to the above areas. The demand for graduates to be equipped for the work environment has led to such courses becoming increasingly prevalent at universities and colleges.

It is our aim to present judiciously chosen and well-written workbooks to meet everyday practical needs. Feedback of views from readers will be most valuable to monitor the success of this aim.

A complete list of titles in this series can be found at http://eu.wiley.com/wileyCDA/Section/id-300612.html

Modern Analysis of Customer Surveys

with applications using R

Edited by

Ron S. Kenett

KPA Ltd., Raanana, Israel, University of Turin, Italy, and
NYU-Poly, Center for Risk Engineering, New York, USA

Silvia Salini

Department of Economics, Business and Statistics,
University of Milan, Italy

A John Wiley & Sons, Ltd., Publication

This edition first published 2012
© 2012 John Wiley & Sons, Ltd

Registered office
John Wiley & Sons, Ltd, The Atrium, Southern Gate, Chichester, West Sussex, PO19 8SQ, United Kingdom

For details of our global editorial offices, for customer services and for information about how to apply for permission to reuse the copyright material in this book please see our website at www.wiley.com.

Wiley also publishes its books in a variety of electronic formats. Some content that appears in print may not be available in electronic books.

Designations used by companies to distinguish their products are often claimed as trademarks. All brand names and product names used in this book are trade names, service marks, trademarks or registered trademarks of their respective owners. The publisher is not associated with any product or vendor mentioned in this book. This publication is designed to provide accurate and authoritative information in regard to the subject matter covered. It is sold on the understanding that the publisher is not engaged in rendering professional services. If professional advice or other expert assistance is required, the services of a competent professional should be sought.

Library of Congress Cataloging-in-Publication Data

Modern analysis of customer surveys : with applications using R / edited by Ron S. Kenett, Silvia Salini.
　　p. cm.
　Includes bibliographical references and index.
　ISBN 978-0-470-97128-4 (cloth)
　1. Consumer satisfaction – Research – Statistical methods.　2. Consumer satisfaction – Evaluation.
3. Consumers – Research – Statistical methods.　4. Consumers – Research – Data processing.　5. Sampling
(Statistics) – Evaluation.　6. Surveys – Statistical methods.　7. Surveys – Data processing.　I. Kenett, Ron.
II. Salini, Silvia.
　　HF5415.335.M63 2011
　　658.8′3402855282–dc23

　　　　　　　　　　　　　　　　　　　　　　　　　　　　　　　　　　　　　　2011030182

A catalogue record for this book is available from the British Library.

Print ISBN: 978-0-470-97128-4
ePDF ISBN: 978-1-119-96116-1
oBook ISBN: 978-1-119-96115-4
ePub ISBN: 978-1-119-96138-3
Mobi ISBN: 978-1-119-96139-0

Set in 10/12pt Times by Aptara Inc., New Delhi, India
Printed and bound in Singapore by Markono Print Media Pte Ltd

To Sima, our children, their families and children;
Jonathan, Alma, Tomer and Yadin, they are my sources
of inspiration and motivation.
Ron

To Riccardo, Marco and Angela: my cornerstones.
Silvia

In memory of Ennio Isaia

Contents

21 Fuzzy Methods and Satisfaction Indices   **439**
Sergio Zani, Maria Adele Milioli and Isabella Morlini

Appendix   An introduction to R   **457**
Stefano Maria Iacus

Foreword

The key to a successful business is understanding one's customers – knowing what they want and, more generally, how satisfied they are with one's product or service. This book describes how to construct such a key, and how to use it to unlock the door to understanding one's customers. It describes insights, introduces methodology, and outlines the cutting-edge statistical tools necessary for effective customer management.

Customer satisfaction is complementary to actual purchase patterns. The latter will show what the customers *buy*, but the former can show what they *would like to buy*. And a company which does not think ahead, to discover and provide what the customers would like, is a company destined for oblivion. Moreover, while traditionally considerable emphasis has been placed on customer recruitment, recent years have seen a shift towards putting more emphasis on the importance of customer retention. Retaining customers, reinforcing loyalty by providing them with what they want, is cheaper than recruiting new customers. One also has more information on existing customers than on those who 'might' become customers, which means that one can construct much more accurate models of how the customers behave and of what they want.

'Satisfaction' is a perfect example, like 'happiness' and 'well-being', of the sort of concept which triggers debates about the feasibility of measurement in the social sciences. Such attributes defy the traditional hard science notions of 'representational' measurement, in which a formal mapping is established from the system being studied to a mathematical model. Instead, one adopts so-called 'pragmatic' measurement, in which the precise property being measured is defined simultaneously with the procedure for measuring it. Since such concepts can be defined in multiple ways, one has to be very explicit about what exactly one means by 'satisfaction'. The measurement of satisfaction also illustrates some of the other complications with which measurement in the social sciences has to contend, including such things as the dependence on context, how questions are worded, the reactive nature of human beings, and even the time of day. Worse, intrinsic in its very definition is that fact that satisfaction is subjective. There is no concrete property, existing out there in the 'real world', which we are trying to measure. All of these complications add up to make measuring customer satisfaction a difficult and challenging problem – which this book ably and comprehensively tackles.

Moreover, we are now faced with multiple ways in which data can be – and indeed is – collected: the traditional face-to-face interview and postal surveys have been joined by telephone surveys and web surveys of various kinds, as well as comprehensive monitoring of customer behaviour via such things as loyalty cards. These new technologies present exciting new opportunities, but also new challenges: selection bias can be a particular challenge. All of this makes this book very timely. The breadth of statistical tools and models described in this book illustrates perfectly what a rich area this has become.

One of the particularly attractive features of this book is that it is about *practical* statistics. It does not simply describe statistical tools which can be applied in an idealised setting, but confronts the issues which all practicing statisticians have to face – of missing data, outliers, measurement error, selection bias, differing definitions, outdated data, and so on. Its treatment of such issues is cutting-edge. A second particularly attractive feature is its comprehensiveness: it covers all aspects, from basic data sources to high-level statistical modelling. Put together, these two attractive features make it a genuinely valuable book for all researchers concerned with measuring customer satisfaction, be they academic or commercial. Indeed, one might even say, more broadly, that it could serve as a valuable resource for statisticians who work in customer-driven applications in general, and who intend their work to make a difference.

David J. Hand
Professor of Statistics, Imperial College, London
Past President, Royal Statistical Society

Preface

All over the world, organizations are focusing on retaining existing customers while targeting potential customers. Measuring customer satisfaction provides critical information on how an organization is providing products or services to the marketplace. In order to measure customer satisfaction, survey questionnaires are used, in which respondents are asked to express their degree of satisfaction with regard to multiple aspects of the product or service. Statistical analysis of data from these surveys is carried out and measures of various aspects and overall satisfaction are computed. This data is, however, non-trivial to handle because of the subjective nature of the observed variables.

First of all, as described by Ferrari and Manzi (*Quality Technology and Quantitative Management*, Vol. 7, No. 2, pp. 117–133, 2010), the relevance correct weighing of the variables that determine the level of satisfaction are unknown. In addition, these variables often have an ordinal measurement scale which needs to be suitably dealt with. Moreover, the level of satisfaction is generally dependent on both expectations and individual characteristics of respondents as well as on contextual variables. Surveys also contain measurement errors caused by the subjective nature of the variables and by cognitive dissonance that can affect data, with undesired consequences on the reliability of the results. With the objective of handling, or at least to controlling, some of these problems, many different methods to assess customer satisfaction have been proposed in the literature. *Modern Analysis of Customer Satisfaction Surveys* presents, in two parts and an appendix, basic and advanced methods for the analysis of customer satisfaction survey data. We have gathered together contributions from world-class authors who are leading the field with new techniques and innovative ideas.

Most of the chapters include an application of various techniques to a standard set of data collected from 266 companies (customers) participating in the ABC Annual Customer Satisfaction Survey (ACSS). The data refers to a questionnaire consisting of 81 questions. The ABC Company is a typical global supplier of integrated software, hardware, and service solutions to media and telecommunications service providers. The ACSS questionnaire was designed to provide feedback on all company touch points and interactions with customers. It covers topics such as equipment, sales support, technical support, training, supplies, software solutions, customer website, purchasing support, contracts and pricing and system installation. Information on customers includes: country, industry segment, age of ABC's equipment, profitability, customer seniority and position of respondent.

The first part of the book consists of nine chapters. The first two chapters introduce the basics of customer satisfaction survey data analysis. The remaining chapters cover topics such as: sampling, surveys and census; measurement scales; integrated analysis; web surveys;

customer satisfaction and customer loyalty; missing data and imputation methods; and outliers and robustness for ordinal data.

The second part of the book consists of 12 chapters with new and innovative techniques and models. They include: causality models; Bayesian networks; log-linear models; CUB models; Rasch models; decision trees; partial least squares models; nonlinear principal component analysis; multidimensional scaling; multilevel models for ordinal data; control charts applied to customer surveys; and fuzzy methods.

Our interest in statistical analysis of customer satisfaction data has followed different paths. Ron has worked on real problems of customer satisfaction for several decades, doing consulting and academic research on methods designed to generate high-impact conclusions and recommendations. Silvia first dealt with the topic during her studies at the Department of Statistics of the Catholic University of Milan. She then dealt with the evaluation of the quality of education services on multiple projects in her current department at the Faculty of Political Science of the University of Milan. Many of her teachers and supervisors, from the Catholic University of Milan and the University of Milan, have contributed to this book as authors or as reviewers.

The seed for this book was planted on 22 July 2005 when a project to investigate the application of non-standard techniques to survey data was launched at the Department of Statistics and Applied Mathematics of the University of Torino under the auspices of the Diego de Castro Foundation and the strong support of Professor Roberto Corradetti, the head of the department and the foundation general secretary. The late Professor Diego de Castro combined an interest in the development of statistical methodology and its application to business and industrial problems. We believe that this work is well in line with his many contributions to statistics, demography, economics and jurisprudence. Professor Corradetti showed vision and constancy of purpose in making sure that the project we started in 2005 matured and produced important publications. Some of these publications appeared in two special issues of the international journal *Quality Technology and Quantitative Management*.

One of the key participants in the first meetings of this project was Professor Ennio Isaia from the Department of Statistics and Applied Mathematics of the University of Torino, who unfortunately passed away in February 2010. Ennio was a gifted statistician with a unique capability of combining theoretical knowledge, interest in applied problems and a great ability to program in R and other software languages. In a sense, this book is designed to have similar characteristics.

In preparing this book we have enlisted the participation of external reviewers who gave us invaluable feedback. They all deserve sincere thanks. The reviewers were: Anthony Atkinson, Mojca Bavdaž, Maurizio Carpita, Enrico Ciavolino, Benito Vittorio Frosini, Salvatore Ingrassia, Patrik Mair, Monica Pratesi, Matteo Pessione, Susanne Raessler, Christersen Rune Humbo, Marco Scrutari, Roberta Siciliano, Pierpaolo d'Urso, Richard Paul Waterman and Silvia Figini. We are also extremely grateful to Professor David Hand for taking the time to write a foreword that ably summarizes the aims of this book and the challenges of the topics covered. Readers should begin with this foreword which is where the book really starts.

The book provides background for both basic and advanced methods for analysing customer satisfaction survey data. The book chapters are supported by R applications so that practitioners can actually implement the methods covered in the book. The book's website (`www.wiley.com/go/modern_analysis`) provides the data sets and R scripts used in the book.

The authors who contributed have helped us present the state of the art in the analysis of customer satisfaction surveys. We thank them for their work and professionalism. While we retain responsibility for any typos and mistakes that may have sneaked in, the chapter authors deserve full credit.

Ron S. Kenett
KPA Ltd., Israel and University of Turin, Italy

Silvia Salini
University of Milan, Italy

Contributors

Alessandro Barbiero   Post-doctoral fellow in Statistics, Department of Economics, Business and Statistics, University of Milan, Italy.

Silvia Biffignandi   Professor of Business and Economic Statistics, University of Bergamo, Italy.

Giuseppe Boari   Professor of Statistics, Faculty of Economics, Università Cattolica del Sacro Cuore, Milan, Italy.

Andrea Bonanomi   Assistant Professor of Statistics, Department of Statistics Sciences, Università Cattolica del Sacro Cuore, Milan, Italy.

Gabriele Cantaluppi   Associate Professor of Statistics, Faculty of Economics, Università Cattolica del Sacro Cuore, Milan, Italy.

Luciana Dalla Valle   Lecturer in Statistics, School of Computing and Mathematics, University of Plymouth, UK.

Francesca De Battisti   Assistant Professor of Statistics, Department of Economics, Business and Statistics, University of Milan, Italy.

Laura Deldossi   Associate Professor of Statistics, Department of Statistical Sciences, Catholic University of Milan, Italy.

Pier Alda Ferrari   Professor of Statistics, University of Milan, Italy.

Stephen E. Fienberg   Maurice Falk University Professor of Statistics and Social Science, Carnegie Mellon University, Pittsburgh, USA.

Roberto Furlan   Senior methodologist, Kantar Health, Epsom, UK.

Iddo Gal   Senior Lecturer, Department of Human Services, University of Haifa, Israel.

Giuliano Galimberti   Assistant Professor of Statistics, Department of Statistical Sciences, University of Bologna, Italy.

Leonardo Grilli   Associate Professor of Statistics, School of Economics, University of Florence, Italy.

Stefano Maria Iacus   Associate Professor, Department of Economics, Business and Statistics, University of Milan, Italy.

Maria Iannario   Assistant Professor of Statistics, Department of Theory and Methods for Human and Social Sciences – Statistical Sciences Unit, University of Naples Federico II, Italy.

Ron S. Kenett Chairman and Chief Executive Officer, KPA Ltd., Raanana, Israel; Research Professor, University of Turin, Italy; and international Research Professor at the Center for Risk Engineering of NYU Poly, New York.

Daniel Manrique-Vallier Post-doctoral fellow, Social Science Research Institute and Department of Statistical Science, Duke University, Durham, North Carolina, USA.

Diego Martone President and chief executive, Demia Studio Associato, Trieste, Italy; and contract professor, Department of Psychology 'Gaetano Kanizsa', University of Trieste, Italy.

Alessandra Mattei Assistant Professor of Statistics, Department of Statistics 'G. Parenti', University of Florence, Italy.

Fabrizia Mealli Professor of Statistics, Department of Statistics 'G. Parenti', University of Florence, Italy.

Maria Adele Milioli Associate Professor of Statistics, Department of Economics, University of Parma, Italy.

Isabella Morlini Associate Professor of Statistics, Department of Economics, University of Modena and Reggio Emilia, Italy.

Giovanna Nicolini Professor of Statistics, University of Milan, Italy.

Irena Ograjenšek Associate Professor of Statistics, Faculty of Economics, University of Ljubljana, Slovenia.

Barbara Pacini Associate Professor of Statistics, Department of Statistics and Applied Mathematics, University of Pisa, Italy.

Giovanni Perucca Research assistant, Polytechnic University of Milan, Italy.

Domenico Piccolo Professor of Statistics, Department of Theory and Methods of Human and Social Sciences, Statistical Sciences Unit, University of Naples Federico II, Italy.

Carla Rampichini Professor of Statistics, Department of Statistics, University of Florence, Italy.

Marco Riani Professor of Statistics, University of Parma, Italy.

Donald B. Rubin John L. Loeb Professor of Statistics, Department of Statistics, Harvard University, USA.

Silvia Salini Assistant Professor of Statistics, Department of Economics, Business and Statistics, University of Milan, Italy.

Gabriele Soffritti Associate Professor of Statistics, Department of Statistical Sciences, University of Bologna, Italy.

Nadia Solaro Assistant Professor of Statistics, Department of Statistics, University of Milano-Bicocca, Milan, Italy.

Francesca Torti Division of Statistics, Faculty of Economics, University of Parma, Italy.

Sergio Zani Professor of Statistics, University of Parma, Parma, Italy.

Diego Zappa Associate Professor of Statistics, Department of Statistical Sciences, Catholic University, Milan, Italy.

Part I

BASIC ASPECTS OF CUSTOMER SATISFACTION SURVEY DATA ANALYSIS

1

Standards and classical techniques in data analysis of customer satisfaction surveys

Silvia Salini and Ron S. Kenett

Customer satisfaction studies are concerned with the level of satisfaction of customers, consumers and users with a product or service. Customer satisfaction is defined as 'The degree of satisfaction provided by the goods or services of a company as measured by the number of repeat customers' (www.businessdictionary.com). Customer satisfaction therefore seems to be an objective and easily measured quantity. However, unlike variables such as revenues, type of product purchased or customer geographical location, customer satisfaction is not necessarily observed directly. Typically, in a social science context, analysis of such measures is done indirectly by employing proxy variables. Unobserved variables are referred to as *latent variables*, whilst proxy variables are known as *observed variables*. In many cases, the latent variables are very complex and the choice of suitable proxy variables is not immediately obvious. For example, in order to assess customer satisfaction with an airline service, it is necessary to identify attributes that characterize this type of service. A general framework for assessing airlines includes attributes such as on-board service, timeliness, responsiveness of personnel, seating and other tangible service characteristics. In general, some attributes are objective, related to the service's technical characteristics, and others are subjective, dealing with behaviours, feelings and psychological benefits. In order to design a survey questionnaire, a set of observed variables must be identified.

In practice, many of the customer satisfaction surveys conducted by business and industry are analysed in a very simple way, without using models or statistical methods. Typical reports include descriptive statistics and basic graphical displays. As shown in this book, integrating a basic analysis with more advanced tools, provides insights into non-obvious patterns and

Modern Analysis of Customer Surveys: with applications using R, First Edition. Edited by Ron S. Kenett and Silvia Salini.

important relationships between the survey variables. This knowledge can significantly affect findings and recommendations derived from a survey.

After presenting classical customer satisfaction methodologies, this chapter provides a general introduction to customer satisfaction surveys, within an organization's business cycle. It then presents standards used in the analysis of survey data. Next it gives an overview on the techniques commonly used to measure customer satisfaction, along with their problems and limitations. Finally, it gives a preview and general introduction to the rest of the chapters in this book.

1.1   Literature on customer satisfaction surveys

Survey questionnaire design, data collection approaches, validation of questionnaires, sampling problems, descriptive statistics and classical statistical inference techniques are covered in many books and papers. This book presents such topics, but also provides a large range of modern and non-standard techniques for customer satisfaction data analysis. Moreover, these various techniques are compared by applications to a common benchmark data set, the ABC 2010 annual customer satisfaction survey (ACSS). For details on the benchmark data set and the ABC company, see Chapter 2.

A non-exhaustive list of relatively advanced books dealing with customer satisfaction data analysis includes Grigoroudis and Siskos (2010), Jacka and Keller (2009), Hayes (2008), Allen and Rao (2000), Johnson and Gustafsson (2000), Vavra (1997) and Biemer and Lyberg (2003). Grigoroudis and Siskos (2010) describe service quality models and the Multicriteria Satisfaction Analysis (MUSA), with examples of satisfaction barometers. Hayes (2008) gives special attention to reliability and validity of questionnaires with a link to customer loyalty. The book by Allen and Rao (2000) is most comprehensive in terms of statistical methods. Although not written by statisticians, it provides a useful and well-written description of techniques of descriptive analysis of univariate, bivariate and multivariate data; it also describes dependent models (linear and logistic regression), explanatory techniques (factor analysis, principal component analysis), causal models (path analysis), and structural equation models. Appendix C of Johnson and Gustafsson (2000) presents an interesting comparison of alternative data analysis methods, in particular considering (1) gap analysis, (2) multiple regression, (3) correlation, (4) principal component regression and (5) partial least squares (PLS). Vavra (1997) covers theories of customer satisfaction and loyalty with several examples of scales, analytic procedures and best practices. Biemer and Lyberg (2003) provide a comprehensive treatment of classical design and analysis of sample surveys.

This book is focused on statistical models for modern customer satisfaction survey data analysis. It addresses modern topics such as web surveys and state-of-the-art statistical models such as the CUB model and Bayesian networks (BN). The book chapters, written by leading researchers in the field, use practical examples in order to make their content also accessible to non-statisticians. Our ultimate goal is to advance the application of best practices in the analysis of customer satisfaction survey data analysis and stimulate new research in this area. As stated in the book foreword by Professor David Hand, we aim to make a difference.

1.2   Customer satisfaction surveys and the business cycle

Statistical analysis is a science that relies on a transformation of reality into dimensions that lend themselves to quantitative analysis. Self-administered surveys use structured questioning designed to map out perceptions and satisfaction level, using a sample of observations from a population frame, into data that can be statistically analysed. Some surveys target all customers; they are in fact a type of census. In others, a sample is drawn and only customers in the sample receive a questionnaire. In drawing a sample, several sampling schemes can be applied. They range from probability samples such as cluster, stratified, systematic or simple random sampling, to non-probability samples such as quota, convenience, judgement or snowball sampling. For more on the different types of surveys, see Chapters 3 and 7.

The survey process consists of four main stages: planning, collection, analysis and presentation. Modern surveys are conducted with a wide variety of techniques, including phone interviews, self-reported paper questionnaires, email questionnaires, internet-based surveys, SMS-based surveys, face-to-face interviews, and video conferencing.

In evaluating the results of a customer satisfaction survey three questions need to be asked:

1. Is the questionnaire properly designed?

2. Has the survey been properly conducted?

3. Has the data been properly analysed?

More generally, we ask ourselves what is the quality of the data, what is the quality of the data analysis, and what is the quality of the information derived from the data (for more on *information quality*, see Kenett and Shmueli, 2011). Addressing these questions requires an understanding of the survey process, the organizational context and statistical methods.

Customer satisfaction surveys can be part of an overall integrated approach. *Integrated models* are gaining much attention from both researchers and practitioners (Rucci *et al.*, 1998; MacDonald *et al.*; 2003; Godfrey and Kenett, 2007). Kenett (2004) presents a generic integrated model that has been implemented in a variety of industries and businesses. The basic building blocks of the model are data sets representing the voice of the customer, the voice of the process and the voice of the workforce. The integration of these, through BN or other statistical methods, provides links between the variables measured on these three dimensions. These links can show, for example, the extent to which satisfied employees generate happy customers and improved financial performance. As an example, the integration at Sears Roebuck has shown that a 5-point increase (out of 100) in employee satisfaction results in an increase of 1.5 units (out of 5) in customer satisfaction, which resulted in a 0.5% increase in revenue growth (Rucci *et al.*, 1998). For more on integrated models, see Chapter 5.

In handling customer satisfaction, several statements are commonly made on the impact of increased customer loyalty and satisfaction. These are based on practical experience and research (see, for example, http://tarp.com/home.html). Some of the more popular statements are:

1. Growth from retention

   • A very satisfied customer is 6 times more likely to repurchase your product than a customer who is just satisfied.

   • Loyal customers spend 5–6% more of their budget with you than customers who are not loyal.

2. Profit boost from retention

   • An increase in customer retention of just 5% can boost profits by 25–85%.

   • Loyal customers are not as price sensitive.

3. Reducing the cost of acquisition

- Acquiring a customer costs 5–7 times more than retaining one.

- Satisfied customers, on average, tell 5 other people about their good experience.

4. The cost of defection

- The average customer who experiences a problem eventually tells 9 other people about it.

- 91% of unsatisfied customers will never buy from you again.

Annual customer satisfaction surveys (ACSS) are conducted in order to:

- identify key drivers of satisfaction and prioritize actions;

- compare data over time to identify patterns in customers' experiences;

- disseminate the results throughout the appropriate audiences within the company to drive change within the organization.

In Chapter 2 we present a fictitious but realistic company called ABC, and its ACSS questionnaire. Data collected with this questionnaire will serve as a benchmark for the models presented in the book. By analysing a benchmark data set we are able to compare insights and added value provided by different models.

A typical internet-based ACSS plan, and it deliverables, is presented in Table 1.1. Typical technical service level agreements (SLAs), when conducting an internet-based ACSS, are presented in Table 1.2.

The ACSS is usually part of a larger plan that is designed and approved at the beginning of the financial year. At that point, decisions are made with strategic and budgetary impact. If the financial year starts in January, the kickoff of the ACSS cycle is usually planned in August. In this context, a general framework for conducting ACSS consists of the following activities:

Month	Activity
August	Survey plan and design
September	Survey communication and launch
October	Survey execution
November	Data analysis and report presentation
December	Annual budget process
January	Launch of annual strategic initiatives
February	Monitoring of improvement areas and key performance indicators
March	Review of progress and mid-course corrections
April	Detailed plans and execution
May	Execution
June	Execution
July	Progress review

To operate this annual cycle, one needs an effective steering committee and improvement methodology. For details on such organizational capabilities, see Kenett and Baker (2010).

Table 1.1   Main deliverables in an internet-based ACSS project

	Category	Deliverables

1	Infrastructure	• Questionnaire evaluation (if relevant)
		  ○ An evaluation of effectiveness of previously used questionnaires
		• Questionnaire design and development
		  ○ (Re)design of questionnaire
		  ○ Setting up of a survey website
		  ○ Testing and validation
		• Contact list management

2	Data collection	• Data collection (e-survey and/or phone or one to one interviews) • Open-ended responses through survey or open lines

3	Data analysis	• Data clean-up phase
		• Reporting and analysis
		  ○ Full report with insights and trend analysis
		  ○ Executive summary and management presentations
		  ○ Database for drill-down tools

4	Support and maintenance	• Project manager
		• Technical support for:
		  ○ Monitoring real-time data (during the survey)
		  ○ Resolving problems operating the questionnaire by customers (via email or phone).
		• Conducting the phone surveys (where relevant)
		• Quality management – a function that is responsible for key performance indicators and quality metrics.

Tables 1.1 and 1.2, and the annual plan sketched above, provide the flavour of a typical ACSS, within an overall strategic initiative for achieving operational excellence. When applying an integrated approach, the ACSS initiative is complemented by other initiatives such as employee surveys, dashboards that reflect the voice of the process, and event-driven surveys that are triggered by specific events. Examples of events followed by a satisfaction survey questionnaire include calls to a service centre or acquisition of a new product. Chapters 3, Chapter 5 and 7 present a range of surveys conducted by modern organizations. This book is focused on analysis aspects of an ACSS. The next section presents standards used in planning and conducting such customer surveys.

1.3   Standards used in the analysis of survey data

In the United States, the National Center for Science and Engineering Statistics (NCSES), formerly the Division of Science Resources Statistics, was established within the National Science Foundation with general responsibility for statistical data. Part of its mandate is

Table 1.2   Service level agreements for internet-based customer satisfaction surveys

Subject	Metric
SLA for maintenance Maintenance includes incidents and problems such as: 1. Customer cannot access survey site 2. Customer cannot enter a specific answer/s 3. Survey is not responsive 4. Response time is poor 5. ABC personnel cannot see progress reports	• MTTR (mean time to repair) – 3 hours (working hours, on working days) • MTBF (mean time between failures) – 3 days • MTBCF (mean time between critical failures) – 2 weeks
SLA for system availability **SLA for performance**	• % Availability – 95% • Time until web page is loaded (initially) – 4 seconds • Time until page is refreshed according to user answers – 2 seconds

to provide information that is useful to practitioners, researchers, policy-makers, and the public. NCSES prepares about 30 reports a year based on surveys. In Europe, Eurostat uses several quality dimensions for evaluating the quality of a survey. These are: relevance of statistical concept, accuracy of estimates, timeliness and punctuality in disseminating results, accessibility and clarity of the information, comparability, coherence, and completeness. For more on criteria for assessing information quality see Kenett and Shmueli (2011).

This section draws on standards and guidelines found at NCSES and other government agencies such as Eurostat. The purpose of survey standards is to set a framework for ensuring data and reporting quality. Guidance documents are meant to help increase the reliability and validity of data, promote common understanding of desired methodology and processes, avoid duplication and promote the efficient transfer of ideas, and remove ambiguities and inconsistencies. The goal is to provide the clearest possible presentation of data and its analysis. Guidelines typically focus on technical issues involved in the work rather than on issues of contract management or publication formats.

Specifically, NCSES aims to adhere to the ideals set out by Citro *et al.* (2009). As a US federal statistical agency, NCSES surveys must follow guidelines and policies as set forth in the Paperwork Reduction Act and other legislation related to surveys. For example, NCSES surveys must follow the implementation guidance, survey clearance policies, response rate requirements, and related orders prepared by the Office of Management and Budget (OMB). The following standards are based on US government standards for statistical surveys (see http://www.nsf.gov/statistics/). We partially list them below:

SECTION 1. DEVELOPMENT OF CONCEPTS, METHODS, AND DESIGN

Survey Planning

Standard 1.1: Agencies initiating a new survey or major revision of an existing survey must develop a written plan that sets forth a justification, including: goals and objectives;

potential users; the decisions the survey is designed to inform; key survey estimates; the precision required of the estimates (e.g., the size of differences that need to be detected); the tabulations and analytic results that will inform decisions and other uses; related and previous surveys; steps taken to prevent unnecessary duplication with other sources of information; when and how frequently users need the data; and the level of detail needed in tabulations, confidential microdata, and public-use data files.

Survey Design

Standard 1.2: Agencies must develop a survey design, including defining the target population, designing the sampling plan, specifying the data collection instrument and methods, developing a realistic timetable and cost estimate, and selecting samples using generally accepted statistical methods (e.g., probabilistic methods that can provide estimates of sampling error). Any use of nonprobability sampling methods (e.g., cut-off or model-based samples) must be justified statistically and be able to measure estimation error. The size and design of the sample must reflect the level of detail needed in tabulations and other data products, and the precision required of key estimates. Documentation of each of these activities and resulting decisions must be maintained in the project files for use in documentation (see Standards 7.3 and 7.4).

Survey Response Rates

Standard 1.3: Agencies must design the survey to achieve the highest practical rates of response, commensurate with the importance of survey uses, respondent burden, and data collection costs, to ensure that survey results are representative of the target population so that they can be used with confidence to inform decisions. Nonresponse bias analyses must be conducted when unit or item response rates or other factors suggest the potential for bias to occur.

Pretesting Survey Systems

Standard 1.4: Agencies must ensure that all components of a survey function as intended when implemented in the full-scale survey and that measurement error is controlled by conducting a pretest of the survey components or by having successfully fielded the survey components on a previous occasion.

SECTION 2. COLLECTION OF DATA

Developing Sampling Frames

Standard 2.1: Agencies must ensure that the frames for the planned sample survey or census are appropriate for the study design and are evaluated against the target population for quality.

Required Notifications to Potential Survey Respondents

Standard 2.2: Agencies must ensure that each collection of information instrument clearly states the reasons the information is planned to be collected; the way such information is planned to be used to further the proper performance of the functions of the agency; whether responses to the collection of information are voluntary or mandatory (citing authority); the nature and extent of confidentiality to be provided, if any, citing authority; an estimate of the average respondent burden together with a request that the public direct to the agency any comments concerning the accuracy of this burden estimate and any suggestions for reducing this burden; the OMB control number; and a statement that an agency may not conduct and a person is not required to respond to an information collection request unless it displays a currently valid OMB control number.

Data Collection Methodology

Standard 2.3: Agencies must design and administer their data collection instruments and methods in a manner that achieves the best balance between maximizing data quality and controlling measurement error while minimizing respondent burden and cost.

SECTION 3. PROCESSING AND EDITING OF DATA

Data Editing

Standard 3.1: Agencies must edit data appropriately, based on available information, to mitigate or correct detectable errors.

Nonresponse Analysis and Response Rate Calculation

Standard 3.2: Agencies must appropriately measure, adjust for, report, and analyze unit and item nonresponse to assess their effects on data quality and to inform users. Response rates must be computed using standard formulas to measure the proportion of the eligible sample that is represented by the responding units in each study, as an indicator of potential nonresponse bias.

Coding

Standard 3.3: Agencies must add codes to collected data to identify aspects of data quality from the collection (e.g., missing data) in order to allow users to appropriately analyze the data. Codes added to convert information collected as text into a form that permits immediate analysis must use standardized codes, when available, to enhance comparability.

Data Protection

Standard 3.4: Agencies must implement safeguards throughout the production process to ensure that survey data are handled to avoid disclosure.

Evaluation

Standard 3.5: Agencies must evaluate the quality of the data and make the evaluation public (through technical notes and documentation included in reports of results or through a separate report) to allow users to interpret results of analyses, and to help designers of recurring surveys focus improvement efforts.

SECTION 4. PRODUCTION OF ESTIMATES AND PROJECTIONS

Developing Estimates and Projections

Standard 4.1: Agencies must use accepted theory and methods when deriving direct survey-based estimates, as well as model-based estimates and projections that use survey data. Error estimates must be calculated and disseminated to support assessment of the appropriateness of the uses of the estimates or projections. Agencies must plan and implement evaluations to assess the quality of the estimates and projections.

SECTION 5. DATA ANALYSIS

Analysis and Report Planning

Standard 5.1: Agencies must develop a plan for the analysis of survey data prior to the start of a specific analysis to ensure that statistical tests are used appropriately and that adequate resources are available to complete the analysis.

Inference and Comparisons

Standard 5.2: Agencies must base statements of comparisons and other statistical conclusions derived from survey data on acceptable statistical practice.

SECTION 6. REVIEW PROCEDURES

Review of Information Products

Standard 6.1: Agencies are responsible for the quality of information that they disseminate and must institute appropriate content/subject matter, statistical, and methodological review procedures to comply with OMB and agency Information Quality Guidelines.

SECTION 7. DISSEMINATION OF INFORMATION PRODUCTS

Releasing Information

Standard 7.1: Agencies must release information intended for the general public according to a dissemination plan that provides for equivalent, timely access to all users and provides information to the public about the agencies' dissemination policies and procedures including those related to any planned or unanticipated data revisions.

Data Protection and Disclosure Avoidance for Dissemination

Standard 7.2: When releasing information products, agencies must ensure strict compliance with any confidentiality pledge to the respondents and all applicable Federal legislation and regulations.

Survey Documentation

Standard 7.3: Agencies must produce survey documentation that includes those materials necessary to understand how to properly analyze data from each survey, as well as the information necessary to replicate and evaluate each survey's results (See also Standard 1.2). Survey documentation must be readily accessible to users, unless it is necessary to restrict access to protect confidentiality.

Documentation and Release of Public-Use Microdata

Standard 7.4: Agencies that release microdata to the public must include documentation clearly describing how the information is constructed and provide the metadata necessary for users to access and manipulate the data (See also Standard 1.2). Public-use microdata documentation and metadata must be readily accessible to users.

These standards provide a comprehensive framework for the various activities involved in planning and implementing a survey in general. Our focus here is on customer satisfaction surveys and, as such, these standards apply equally well. In the next section we present some special aspects of customer satisfaction surveys.

1.4   Measures and models of customer satisfaction

Models for analysing data from customer satisfaction surveys address two components: the conceptual construct and the measurement process.

1.4.1   The conceptual construct

The concept of customer satisfaction is related to the concept of quality. Quality, however, is different in the case of a product and of a service. Evaluations of service quality are not exclusively linked to the observed output, as for products, but, and to a greater extent, to the whole process through which the service is provided. Consumer evaluation of the service received is determined by factors affected by psychological interactions that are established during the exchange transaction, and by factors connected to technical-specific characteristics of the service. The former factors concern behaviour, sensations and psychological benefits, which are difficult to measure, whilst the latter factors can be evaluated by objective indicators similar to those utilized for product quality.

Knowledge and interpretation of how consumers perceive and evaluate product or service quality are essential for the orientation of the company management and strategy. As a result, the availability of measurements and interpretation models for *customer satisfaction* linked to subjective evaluations of product/service quality is becoming as important as the corresponding objective technological measurements. This subjective input is therefore an important ingredient in the statistical control of a quality system. In general, customer satisfaction, from the relevant measurements and interpretation models, is one of the essential components of modern management since it is the basis of a company's success (Kanji and Wallace, 2000; Kenett, 2004; Kenett and Salini, 2009). Customer satisfaction indicators, at the national level or by economic sectors, are becoming essential in the evaluation of the overall performance of economic systems (see Anderson and Fornell, 2000).

An important paradigm for evaluating service quality is the gap model developed by Parasuraman *et al.* (1985). Following executive interviews and focus groups in four different service business areas, the authors proposed a conceptual model of service quality where consumers' perceptions on service quality depend on gaps between the service provider organization and the consumer environment. Later, they developed in-depth measurement scales for service quality.

Perceived service quality is defined, according to the gap model, as the difference between consumers' expectation and perceptions. This depends on the size and the direction of four gaps in a company's delivery of service. These gaps determine the difference between customer expectations and perceptions – the service quality gap. The key points for each gap can be summarized as follows:

1. the difference between what customers expected and what management perceived about the expectation of customers;

2. the difference between management's perceptions of customer expectations and the translation of those perceptions into service quality specifications and designs;

3. the difference between specifications or standards of service quality and the actual service delivered to customers;

4. the difference between the service delivered to customers and the promise of the firm to customers about its service quality

The gap model clearly determines the two different types of gaps in service marketing, namely the customer gap and the provider gap. The latter is considered an internal gap, within a service firm. This model views services as a structured, integrated model which connects external customers to internal services provided by different functions in a service organization. Important characteristics of the model are the following:

1. The gap model of service quality gives insights into customers' perceptions of service quality.

2. Customers always use dimensions to form the expectation and perceptions of service quality.

3. The model helps predict, generate and identify key factors that cause the gap to be unfavourable to the service provider, in meeting customer expectations.

For more on these models, see Chapters 3 and 7.

1.4.2   The measurement process

Zanella (2001) provides an overview of measures and models of customer satisfaction. He classifies the main features of models and related techniques used for describing customer satisfaction. We provide below a brief description based on Zanella's work.

Composition or formative models

In composition models, customer satisfaction is considered a 'multidimensional attribute', where each component corresponds to a dimension of the conceptual construct, i.e. to an aspect of a product or service considered essential in determining customer satisfaction. The synthesis of the evaluations of the single 'marginal' satisfaction attributes has a defining, and therefore conventional, nature. In fact, there is a lack of explicit research into the functional links of the latent variables that correspond to the various dimensions, and the latent one-dimensional variables associated with the concept under investigation, i.e. customer satisfaction. The latter is turned into a target variable by giving it a value obtained through *composition*, i.e. addition of values of target variables corresponding to the various dimensions. These models thus came to be called *composition models* or *formative models*.

Starting from the fundamental work by Parasumaran *et al.* (1988, 1991), the well-known composition model known as SERVQUAL was developed. The most serious criticism of the original gap model approach was expressed by Cronin and Taylor (1992), who raised doubts about the SERVQUAL indicator being appropriate to describe service quality. This criticism gave rise to another improved model, SERVPERF. A description of these models can be founded in Chapter 7.

Explanatory or decomposition models: Regression models

A self-declared 'questionnaire' provides an overall assessment of customer satisfaction with a specific product or service. The response variables of the underlying customer satisfaction model are typically expressed on a semantic differential scale, with corresponding conventional scores such as a five-point or seven-point scale. However, this scale could also be dichotomous or made so by summarizing judgements in two categories. In Chapters 2 and 20 we use two dichotomizing schemes. The first of these identifies customers who responded '5' on a five-point scale. Their percentage yields a satisfaction index labelled 'TOP5'. At the other end of the scale, customers who responded '1' or '2' are aggregated to form an index labelled 'BOT1+2'. TOP5 represents a measure of excellence. BOT 1+2 is very effective in identifying pockets of dissatisfaction. Some organizations combine the labels 'satisfied' and 'very satisfied' which produce indices with higher values but much reduced resolution. The common use of average response often produces uninformative statistics. Below we present three ways to model data on an ordinal or nominal scale.

Ordinary linear regression model   Explanatory variables describe dimensions related to specific aspects of a product or service. These could be, for example, age of equipment or geographical location. This is the case with composition models, that produce data that can be expressed on conventional ordered rating scales. Such data can, however, refer to respondents' personal characteristics, such as age or the number of purchases in a previous period, that are measured on metric scales.

The usual statistical analysis techniques for such data apply the least squares criteria for deriving estimates of the unknown parameters and for determining the goodness of fit.

Regression models and techniques accounting for the ordinal character of the response and of explanatory variables   In this context, monotonic regression analysis plays an important role (see Kruskal, 1965). In Zanella (1998) a non-linear regression model with latent variables is presented for deriving a ratio scale representation of the response.

Logistic regression model   If one can assume a probability distribution for the response portraying overall satisfaction, the expected value of the response can be presented, with conditioning on the different situations described by the values of the explanatory variables. The logistic regression approach allows us to take into consideration the fact that the values of the response variable are on an ordinal scale. Rasch models, presented in Chapter 14, are a particular case of logistic models.

Linear structural models with latent variables (LISREL)

The LISREL models allow us to establish links between latent variables, which are related to dimensions describing customer satisfaction (Bollen, 1989). This is at the core of the conceptual construct. Thus LISREL provides a complete determination of the construct under study. The model is composed of two systems of equations: *structural equations* and *measurement model*. Baumgartner and Homburg (1996) give comments and recommendations on the basis of cases of complete structural model application in marketing. In particular, they recommend an accurate assessment of the identifiability conditions and the use of a suitable set of indicators for checking model adequacy, such as chi-square, root-mean-square residual, goodness-of-fit index, and determination coefficients for measurement equations.

Another approach to estimation for this type of models is the PLS method presented in Chapter 16 of this book. The LISREL method is used for calculating the American Customer Satisfaction Index; for more details see Anderson and Fornell (2000) and Kenett (2007). A similar approach is used to compute the European Customer Satisfaction Index. The main problem of the LISREL approach is that metric scales are assumed. In general, however, the variables are measured with ordinal scales. A transformation to obtain metric scales can be used with caution (see Zanella, 2001). The approach here is that we wish to deal with the more realistic assumption that the variables are ordinal. Chapter 4 is devoted to measurement scales.

1.5   Organization of the book

The book is organized into two parts. Part I consists of nine chapters on general topics in customer satisfaction surveys. Part II consists of 12 chapters with new and innovative techniques and models.

Chapter 2 introduces the ABC company and its annual customer satisfaction survey. It provides the background for the case study used throughout the book, including a basic analysis of the ABC 2010 ACSS data.

Chapter 3 discusses the sampling problem. The two main types of surveys, census and sample, are compared and non-sampling errors are listed, focusing on their presence in customer surveys and on their effects on estimates. Data collection methods are also described, linking every method with one or more non-sampling errors. Finally, methods to correct these errors are proposed.

Chapter 4 considers the problem of measurement scales and the problem of scale construction.

Chapter 5 is about integrating data in a business data system, with emphasis on marketing integrated systems.

Chapter 6 introduces different types of web surveys. The main economic and non-economic benefits of web surveys and the main drawbacks associated with online research are presented. The final part of the chapter discusses the application of web surveys to customer and employee satisfaction research projects.

Chapter 7 addresses the interrelated concepts of customer satisfaction, perceived service quality and customer loyalty. The chapter deals with methodological issues relevant in survey collection of customer satisfaction data and evaluates the ABC ACSS questionnaire from the conceptual and methodological points of view.

Chapter 8 is devoted to missing data and their imputation, with a special focus on customer satisfaction data. Imputation, multiple imputation, and other strategies to handle missing data, together with their theoretical background, are discussed. Some examples of and advice on computation are provided using the ACSS example.

Chapter 9 tackles the topics of robustness and multivariate outlier detection in presence of ordinal data. A review of outlier detection methods in regression for continuous data is presented. The second part of the chapter concentrates on ordinal data and illustrates how to detect atypical measurements in customer satisfaction surveys, with application to the ACSS data.

Chapter 10 provides an overview of the approach to the estimation of causal effects based on the concept of potential outcomes, stemming from the work on randomized experiments and then extended in the 1970s to non-randomized studies and different modes of inference.

Chapter 11 introduces BN and their application to customer satisfaction surveys. A theoretical introduction to BN is given, and BN are applied to the ABC 2010 ACSS data set and to the Eurobarometer transportation survey.

Chapter 12 introduces the structuring of categorical data in the form of contingency tables, and then provides a brief introduction to log-linear models and methods for their analysis, followed by an application to customer satisfaction surveys. The focus is on methods designed primarily for nominal data like the data gathered in the ABC 2010 ACSS.

Chapter 13 introduces a class of statistical models, CUB models, based on the psychological mechanism which induces customers to choose a definite item or to manifest an expressed preference towards some object/brand. The approach has been applied to the ABC 2010 ACSS data set and to students' satisfaction towards their university orientation service.

Chapter 14 describes the Rasch model and its use in the context of customer satisfaction surveys. A detailed application based on the ABC 2010 ACSS data is given.

Chapter 15 presents a selection of decision tree methodologies, useful for evaluating customer satisfaction. An illustrative example where tree-based methods are applied to the ACSS data set is provided.

Chapter 16 introduces the partial-least squares estimation algorithm and its use in the context of structural equation models with latent variables (SEM-LV). After a short description of the general structure of SEM-LV models, the PLS algorithm is introduced; then, statistical and geometrical interpretations of PLS are given. A detailed application example based on the ACSS data concludes the chapter.

Chapter 17 describes homogeneity analysis and non-linear principal component analysis (NLPCA), which allow us to set up a synthetic numerical indicator of the level of satisfaction, starting from ordinal responses. Problems connected to the presence of missing data and their treatment are presented. NLPCA is applied to the ABC data and the main aspects and findings of this application are described.

Chapter 18 reviews the basic concepts and methodological results of multidimensional scaling (MDS) methods. It also offers several applications of MDS to customer satisfaction studies. Certain metric MDS models are applied to data collected in the ABC 2010 ACSS.

Chapter 19 is devoted to regression models for ordinal responses, with special emphasis on random effects models for multilevel or clustered data. The last part of the chapter presents an application of random effects cumulative models to the analysis of student ratings on university courses.

Chapter 20 presents an application of methods and standards used in quality management and quality control to the analysis of customer satisfaction surveys. The chapter covers in detail the ISO 10004 standard that provides guidelines for monitoring and measuring customer satisfaction, introduces control charts and describes the corresponding ISO 7870 guidelines. It also discusses how standard control charts (p, c, and u charts) can be used to analyse customer satisfaction surveys to monitor, over time, the number or proportion of satisfied or unsatisfied customers.

Chapter 21 develops a framework that uses fuzzy set theory in order to measure customer satisfaction, starting from a survey with several questions. The basic concepts of the theory of fuzzy numbers are described. A criterion based on the sampling cumulative function, which assigns values to the membership function with reference to each quantitative, ordinal and binary variable, is suggested. Weighting and aggregation operators for the synthesis of the variables are considered and applied to the 2010 ABC survey data.

1.6 Summary

This book is about modern and applied methods for analysing customer satisfaction survey data. It can be used by practitioners who wish to provide their customers with state-of-the-art methods, and by researchers who wish to dedicate their efforts to this challenging domain. The challenge is inherent in such multidisciplinary fields where quantitative skills need to be combined with methods drawing on psychology and cognitive science, and eventually management theory techniques, so that the surveys generate a measurable and effective impact. Customer satisfaction surveys are a prime example where we explicitly put the customer at the centre of our attention, so that we can improve and achieve high effectiveness and high efficiencies.

References

Allen, D.R. and Rao, T.R.N. (2000) *Analysis of Customer Satisfaction Data*. Milwaukee, WI: ASQ Quality Press.

Anderson, E.W. and Fornell, C. (2000) Foundations of the American Customer Satisfaction Index, *Total Quality Management*, 11, 869–882.

Baumgartner, H. and Homburg, C. (1996) Applications of structural equation modelling in marketing and consumer research: A review. *International Journal of Research in Marketing*, 13, 139–161.

Biemer, P. and Lyberg, L. (2003) *Introduction to Survey Quality*. Hoboken, NJ: Wiley.

Bollen, K.A. (1989) *Structural Equations with Latent Variables*. New York: Wiley.

Citro, C.F., Martin, M.E. and Straf, M.L. (eds) (2009) *Principles and Practices for a Federal Statistical Agency*, 4th edn. Washington, DC: National Academies Press.

Cronin, J.J., Jr. and Taylor, S.A. (1992) Measuring service quality: A re-examination and extension, *Journal of Marketing*, 56, 55–68.

Godfrey, A.B. and Kenett, R.S. (2007) Joseph M. Juran, a perspective on past contributions and future impact, *Quality and Reliability Engineering International*, 23, 653–663.

Grigoroudis, E. and Siskos, Y. (2010) *Customer Satisfaction Evaluation: Methods for Measuring and Implementing Service Quality*. New York: Springer.

Hayes, B.E. (2008) *Measuring Customer Satisfaction and Loyalty: Survey Design, Use, and Statistical Analysis Methods*. Milwaukee, WI: ASQ Quality Press.

Jacka, J.M. and Keller, P.J. (2009) *Business Process Mapping: Improving Customer Satisfaction*. Hoboken, NJ: Wiley.

M. D. Johnson and A. Gustafsson, (2000) *Improving Customer Satisfaction, Loyalty, and Profit: An Integrated Measurement and Management System*. San Francisco: Jossey-Bass, 2000.

Kanji, G.K. and Wallace, W. (2000) Business excellence through customer satisfaction. *Total Quality Management*, 11, 979–998.

Kenett, R.S. (2004) The integrated model, customer satisfaction surveys and Six Sigma. In *Proceedings of the First International Six Sigma Conference*, Center for Advanced Manufacturing Technologies, Wrocław University of Technology, Wrocław, Poland.

Kenett, R.S. (2007) Cause and effect diagrams. In F. Ruggeri, R.S. Kenett and F. Faltin (eds), *Encyclopedia of Statistics in Quality and Reliability*. Chichester: Wiley & Sons, Ltd.

Kenett, R.S. and Salini, S. (2009) New frontiers: Bayesian networks give insight into survey-data analysis. *Quality Progress*, August, 31–36.

Kenett, R.S. and Baker, E. (2010) *Process Improvement and CMMI for Systems and Software*. Taylor and Francis, Auerbach CRC Publications.

Kenett, R.S. and Shmueli, G., (2011) *On Information Quality*. Robert H. Smith School Research Paper No RHS 06-100. Available at SSRN: http://ssrn.com/abstract=1464444.

Kruskal J.B. (1965) Analysis of factorial experiments by estimating monotone transformations of data. *Journal of the Royal Statistical Society, Series B*, 27, 251–263.

MacDonald, M, Mors, T. and Phillips, A. (2003) Management system integration: Can it be done? *Quality Progress*, October, 67–74.

Parasuraman, A., Zeithaml, V. and Berry, L. (1985) A conceptual model of service quality and its implications for future research. *Journal of Marketing*, 49, 41–50.

Parasuraman, A., Zeithaml, V.A. and Berry, L.L. (1988) SERVQUAL: A multiple-item scale for measuring customer perceptions of service quality. *Journal of Retailing*, 64, 11–40.

Parasuraman, A., Berry, L.L. and Zeithaml, V.A. (1991) Refinement and reassessment of the SERVQUAL scale. *Journal of Retailing*, 67, 420–450.

Rucci, A., Kim, S. and Quinn, R. (1998) The employee–customer–profit chain at Sears. *Harvard Business Review*, 76(1), 83–97.

Vavra, T.G. (1997) *Improving Your Measurement of Customer Satisfaction: A Guide to Creating, Conducting, Analyzing, and Reporting Customer Satisfaction Measurement Programs*. Milwaukee, WI: ASQ Quality Press.

Zanella, A. (1998) A statistical model for the analysis of customer satisfaction: Some theoretical and simulation results. *Total Quality Management*, 9, 599–609.

Zanella, A. (2001) Measures and models of customer satisfaction: The underlying conceptual construct and a comparison of different approaches. In *The 6th World Congress for Total Quality Management, Business Excellence – What Is to be Done, Proceedings, Volume 1*, The Stockholm School of Economics in St Petersburg, pp. 427–441.

2

The ABC annual customer satisfaction survey

Ron S. Kenett and Silvia Salini

This chapter introduces the ABC company and its 2010 annual customer satisfaction survey (ACSS). This provides the background for the case study used throughout the book, including a basic analysis of the survey data.

2.1 The ABC company

Media, telecommunications, and software companies are now competing on each other's traditional turf. Cable TV companies are offering phone services, telephone companies provide satellite TV, and software developers put it all together. In this competitive environment, service providers need to control operating costs and excel at customer satisfaction as a driver to retention and loyalty, while continuously enhancing their products and services.

ABC (a fictitious but realistic company) is a typical global supplier of integrated software, hardware, and service solutions to media and telecommunications service providers. The company develops, installs and maintains systems combining hardware, software and advanced electronics. These enabling systems support various lines of business, including video on demand, cable, and satellite TV, as well as a range of communications services, such as voice, video, data, internet protocol, broadband, content, electronic, and mobile commerce

The company also supports companies that offer bundled or convergent service packages. In addition, the company's information technology services consist of system implementation, hardware and software integration, hardware supplies, training, maintenance and version upgrades. Its customers include media and communications providers, network operators and service providers. The company was founded in 2005 in Canada. Today, the company's workforce consists of more than 5000 professionals located in 10 countries and serves customers in Europe and elsewhere.

Modern Analysis of Customer Surveys: with applications using R, First Edition. Edited by Ron S. Kenett and Silvia Salini.
© 2012 John Wiley & Sons, Ltd. Published 2012 by John Wiley & Sons, Ltd.

ABC is therefore a typical business-to-business supplier operating on the global scene with state-of-the-art products combining electronics, software and mechanical components. In this book we will consider the ABC 2010 annual customer satisfaction survey (ACSS).

The ABC 2010 ACSS questionnaire has been designed to provide feedback on all company touch points and interactions with customers. It covers topics such as equipment, sales support, technical support, training, supplies, software solutions, customer website, purchasing support, contracts and pricing and system installation. Descriptive variables for each customer include: country, industry segment, age of ABC's equipment, profitability, customer seniority and position of respondent

The first part of the questionnaire consists of an assessment of overall satisfaction, with two specific variables evaluated on a five-point anchored scale: a variable assessing repurchasing intentions and a variable assessing whether the customer will recommend ABC to others. Finally, a binary variable indicates whether ABC is considered the best supplier. In the second part of the questionnaire, there are almost 50 statements grouped according to the various touch point topics. For each statement, there are two types of scores: the item evaluation score, based on a five-point semantic differential, and a measure of item importance based on a three-point anchored scale. Each topic is covered by specific items that set a context and an overall satisfaction question from the topic. The 2010 ACSS return rate was 45%, with all survey regions well represented (see Section 20.6). The questionnaire itself is presented in the Appendix to this chapter.

2.2   ABC 2010 ACSS: Demographics of respondents

Data for the ABC 2010 ACSS was collected from 266 ABC customers (companies) with a questionnaire consisting of 81 questions, one questionnaire per customer. For each customer, we have various descriptive variables such as: country, segment, age of ABC's equipment, profitability, customer seniority (years) and position of respondent. A segment represents the value of a customer to ABC. The platinum segment represents key strategic customers. Only customers above a certain value to ABC are classified into segments.

An analysis of respondents by country of origin and other demographics proved the respondents to be representative of the install base of ABC in these countries. For techniques to perform this test, see Section 20.6.

A basic frequency analysis of the responses is presented in Tables 2.1–2.6. Appendix A shows how to generate tables and graphs in this chapter using R. This basic analysis

Table 2.1   Respondents by country (no missing values).

Country	Frequency	(%)
Benelux	26	9.8
France	15	5.6
Germany	112	42.1
Italy	39	14.7
UK	51	19.2
Israel	23	8.6
Total	266	100.0

Table 2.2   Respondents by segment (43 missing values).

	Frequency	(%)
Other	112	50.2
Silver	42	18.8
Gold	43	19.3
Platinum	26	11.7
Total	223	100.0

Table 2.3   Respondents by age (in years) of ABC's equipment (one missing value).

	Frequency	(%)
Less than 1	46	17.4
1–2	65	24.5
2–3	44	16.6
3–4	35	13.2
More than 4	75	28.3
Total	265	100.0

Table 2.4   Companies by profitability (25 missing values).

Profitability level	Frequency	(%)
Profitable	78	32.4
Break-even	105	43.6
Below break-even	58	24.1
Total	241	100.0

Table 2.5   Customer seniority in years (one missing value).

Seniority	Frequency	(%)
1	46	17.4
2	65	24.5
3	44	16.6
4	35	13.2
5	75	28.3
Total	265	100.0

Table 2.6   Respondents by position (no missing values).

Position of respondent	Frequency	(%)
Owner	142	53.4
Management	49	18.4
Technical management	30	11.3
Technical staff	15	5.6
Operator	20	7.5
Administrator	6	2.3
Other	4	1.5
Total	266	100.0

shows that the majority of customers came from Germany (42%), do not belong to a specific segment (Segmentation = Other), have a break-even profitability, and the respondent was most likely to be the company owner. The distribution of the age of ABC's equipment and of the customer seniority was more or less uniform.

2.3   ABC 2010 ACSS: Overall satisfaction

The first part of the questionnaire assesses *overall satisfaction*, with two specific variables (throughout we use the terms 'variable', 'question' or 'item' interchangeably) evaluated by a score ranging from '1' (very low satisfaction) to '5' (very high satisfaction), a variable on repurchasing intentions and another on willingness to recommend ABC to others both measured by a score ranging from '1' (very unlikely) to '5' (very likely). Part of this section is a binary variable that indicates whether ABC is the best supplier. Table 2.7 shows the overall satisfaction with ABC and Table 2.8 shows the overall satisfaction with ABC's improvements during 2010. As we can see, 8.6% of the customers are highly satisfied with ABC improvements. Table 2.9 shows that slightly less than 40% of customers consider ABC to be their best supplier. Figures 2.1a and 2.1b respectively show that 27.2% of customers are very likely to recommend ABC and 32.8% of customers are very likely to repurchase an ABC product.

Table 2.7   Overall satisfaction level with ABC (4 missing values), from '1' (very low) to '5' (very high).

Level	Frequency	(%)
1	11	4.2
2	25	9.5
3	70	26.7
4	118	45.0
5	38	14.5
Total	262	100.0

Table 2.8   Overall satisfaction level with ABC's improvements during 2010 (9 missing values), from '1' (very low) to '5' (very high).

Level	Frequency	(%)
1	23	8.9
2	38	14.8
3	93	36.2
4	81	31.5
5	22	8.6
Total	257	100.0

Table 2.9   Is ABC your best supplier? (15 missing values).

	Frequency	(%)
Yes	97	38.6
No	154	61.4
Total	251	100.0

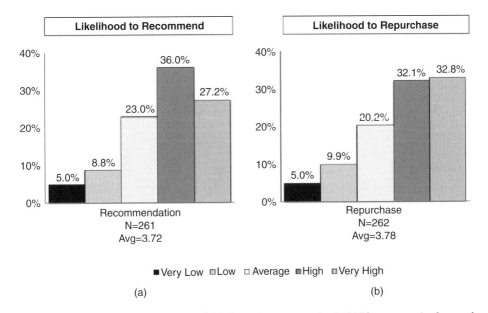

Figure 2.1   (a) Would you recommend ABC to other companies? (b) If you were in the market to buy a product, how likely would it be for you to purchase an ABC product again?

Table 2.10   Net customer loyalty index from ABC 2010 ACSS, by country.

Country	Net CLI (%)	CLI (%)	CRI (%)	N
ABC	2.3	24.1	21.8	261
Benelux	−56.0	0.0	56.0	25
France	−21.4	14.3	35.7	14
Germany	31.5	42.3	10.8	111
Italy	−10.3	128	23.1	39
UK	−24.5	6.1	30.6	49
Israel	17.4	26.1	8.7	23

From the responses to overall satisfaction, recommendation and repurchasing intention one can generate three indices that assess customer loyalty. The customer risk index (CRI) is the percentage of customers who gave a rating of '1' or '2' to any of these three questions. Such customers are considered detractors who, in a sense, work for the competition. The customer loyalty index (CLI) is the percentage of customers who gave a rating of '5' or '4' to overall satisfaction, '5' to recommendation and '5' to repurchasing intentions. These customers are considered promoters and help the sales force with references and success stories. The net CLI is the difference between the CLI and CRI. Net CLI has been found to predict growth and success (Reichheld, 2003). Table 2.10 presents these indices for ABC customers, by geographical area. Based on this we can predict growth in Germany and significant problems ahead in Benelux. Moreover, the success in Germany is reflected by 42.3% of customers being loyal, and the problems in Benelux are due to very dissatisfied customers with a CRI of 56.0%.

2.4   ABC 2010 ACSS: Analysis of topics

In the second part of the questionnaire there are questions grouped by different topics: *equipment and system, sales support, technical support, training, supplies and orders, software add-on solutions, customer website, purchasing support, contracts and pricing, system installation and overall satisfaction with other suppliers.* For each variable in these dimensions we have two types of scores: the item evaluation score (from '1' to '5') and the item importance level (low = '1', medium = '2', high = '3', and N/A). For each topic there is also an evaluation of overall satisfaction.

In analysing the data we use a methodology and tools developed by KPA Ltd. to emphasize areas for improvement and areas of excellence (for more on the tools, see www.kpa-group.com/en/our-expertise/products). A basic element in this analysis is the computation of the proportion of '1' and '2' ratings, labelled BOTI+2, and the proportion of '5' ratings, labelled TOP5. For more on this type of analysis, including statistical tests of significance, see Chapter 20 in the present volume, Kenett and Zacks (1998) and Kenett (2002, 2004).

Figures 2.2a and 2.2b show the TOP5 and BOTI+2 values for the overall satisfaction questions from the various topics. The topics with highest levels of satisfaction (TOP5) are

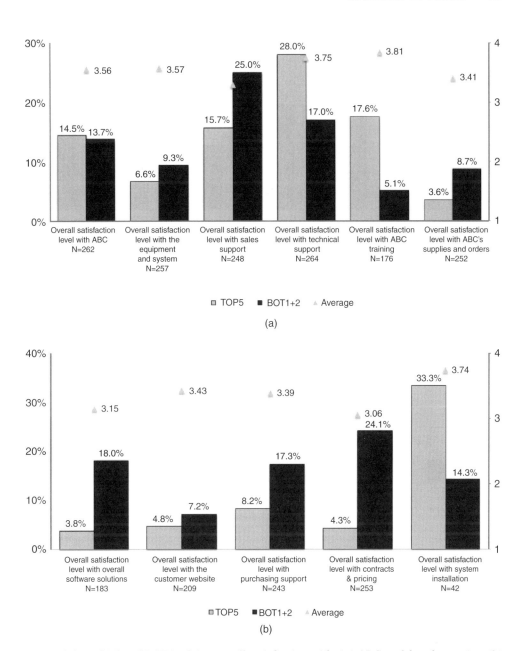

Figure 2.2 TOP5 *and* BOT1+2 *in overall satisfaction with: (a) ABC and first five topics; (b)* *remaining five topics.*

installation (33.3%) and technical support (28.0%). The topics with highest levels of dissatisfaction (BOT*1+2*) are sales support (25.0%) and contracts and pricing (24.1%).

Figures 2.3a and 2.3b zoom in on the specific items in the topics with highest levels of dissatisfaction: sales support and contracts and pricing. The specific items with highest

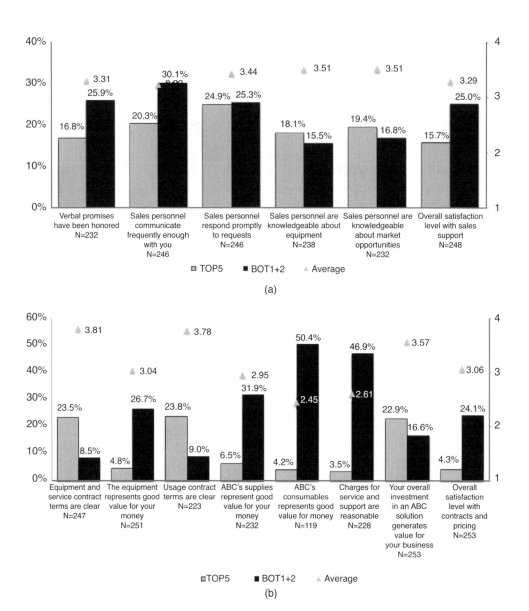

Figure 2.3   TOP5 *and* BOT1+2 *in items of: (a) sales support; (b) contracts and pricing.*

dissatisfaction levels in these topics (BOT*1+2*) are:

12. Verbal promises have been honoured.

13. Sales personnel communicate frequently enough with you.

62. ABC's consumables represent good value for money.

63. Charges for service and support are reasonable.

2.5 ABC 2010 ACSS: Strengths and weaknesses and decision drivers

Strengths in customer satisfaction are identified from a survey as items with high level of satisfaction (TOP5) and high importance rating. Weaknesses in customer satisfaction are identified from a survey as items with high levels of dissatisfaction (BOT$1+2$) and high importance (high percentage of importance rating '3'). Figure 2.4 presents all items on a scatter plot of importance and dissatisfaction. Questions 12 and 63 are identified as weaknesses. Questions 13 and 62 also had high BOT$1+2$, but do not represent items with high importance.

Declared importance, as reflected by the importance rating on each item, can be compared to generated importance derived from a partial least squares regression or principal components analysis of the item's response on overall satisfaction with ABC. For more on these topics, see Chapters 16 and 17. Figure 2.5 presents a scatter plot of declared importance against normalized coefficients of generated importance.

A *key to success* (quadrant III in Figure 2.5) is an item with high declared and high generated importance. An example of such an item is question 22, 'The remote support care center is valuable and meets your expectations'. ABC needs to achieve excellence in such areas since customers consider them essential.

A *future opportunity* (quadrant II in Figure 2.5) is an item with relatively low declared importance but high generated importance. An example of such an item is question 56, 'When you have an administrative problem, you know who to contact'. Clarifying the ABC focal point handling customer administrative problems can have a significant effect on customer satisfaction. This has not been explicitly stated by customers but has, *de facto*, an impact on overall satisfaction. Improvements in such items have the potential of creating a 'wow effect', with the customer being happily surprised at the company's initiative.

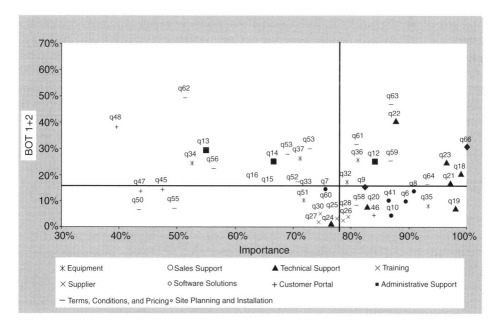

Figure 2.4   ABC weaknesses from 2010 ACSS survey.

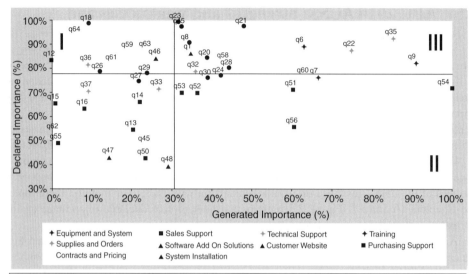

I-Price of Entry – Issues that, statistically, do not affect overall satisfaction but are declared 'important' by customers, are issues which must be addressed. These issues can cause dissatisfaction but do not generate significant advantage.
II - Future Oportunities – Issues that statistically affect overall satisfaction but which are not declared 'important' by customers. Such issues are our opportunity to create a WOW effect by anticiating customer needs.
III - Keys to Success - Issues that statistically affect overall satisfaction and are also declared as 'important' by customers. These issues should be addressed since they are key to generating customer satisfaction.

Figure 2.5   Declared importance plotted against generated importance.

2.6   Summary

As introduced in Chapter 1, the ACSS is a process and should be managed as such. Its annual cycle requires a focal point, an adequate budget and dedication. Proper management of the process involves technical and organizational skills. To handle it properly one needs to ensure good data quality, a questionnaire that has been validated organizationally, i.e. with the involvement of key stakeholders, and a management team that understands the need to take a snapshot of the voice of the customer for planning and management purposes. Companies that direct design and improvement efforts on the basis of ACSS data, integrating it with information from internal processes (voice of the process) and the workers (voice of the workforce), have achieved competitive positions and business growth (Kenett, 2004). The following chapters provide state-of-the-art models and methods for companies that want to get the most from such survey efforts.

References

Kenett, R.S. (2002) Issues in customer satisfaction surveys. DEINDE 2002, Torino, Italy.

Kenett, R.S. (2004) The integrated model, customer satisfaction surveys and Six Sigma. In *Proceedings of the First International Six Sigma Conference*, Center for Advanced Manufacturing Technologies, Wrocław University of Technology, Wrocław, Poland.

Kenett, R.S. and Zacks, S. (1998) *Modern Industrial Statistics: Design and Control of Quality and Reliability*. Spanish edition 2002, 2nd paperback edition 2002, Chinese edition 2004. Pacific Grove, CA: Duxbury Press.

Reichheld, F.F. (2003) The one number you need to grow. *Harvard Business Review*, December.

Appendix

The ABC 2010 Annual Customer Satisfaction Survey

Company: _____

Completed by: _____

Title/Position: 1. Owner   2. Management   3. Technical Management   4. Technical Staff
                   5. Operator   6. Administrator   7. Other, please specify: _____

Dear Customer,

For each of the following statements, please select a number indicating the extent of your agreement with the statement concerning your experience with ABC during 2010. Then, under 'Importance Level', select another number indicating the importance of the statement to you. If a certain statement is not relevant or not applicable, please select N/A.

Overall Satisfaction with ABC

	Evaluation				
	Very low				**Very high**
1. Overall satisfaction level with ABC:	1	2	3	4	5
2. Overall satisfaction level with ABC's improvements during 2010:	1	2	3	4	5
3. Is ABC your best supplier?	a.	Yes			
	b.	No			
	Very unlikely				**Very likely**
4. Would you recommend ABC to other companies?	1	2	3	4	5
5. If you were in the market to buy a product, how likely would it be for you to purchase an ABC product again?	1	2	3	4	5

Equipment and System

	Evaluation					Importance Level			
	Strongly disagree				**Strongly agree**	**Low**		**High**	
6. The equipment's features and capabilities meet your needs.	1	2	3	4	5	1	2	3	N/A

	Evaluation					Importance Level			
7. Improvements and upgrades provide value.	1	2	3	4	5	1	2	3	N/A
8. Output quality meets or exceeds expectations.	1	2	3	4	5	1	2	3	N/A
9. Uptime is acceptable.	1	2	3	4	5	1	2	3	N/A
10. For customers who purchased a system during 2010: ABC's equipment met your requirements.	1	2	3	4	5	1	2	3	N/A

	Very low			Very high	
11. Overall satisfaction level with the **equipment**:	1	2	3	4	5

Sales Support

	Evaluation					Importance Level			
	Strongly disagree				Strongly agree	Low		High	
12. Verbal promises have been honored.	1	2	3	4	5	1	2	3	N/A
13. Sales personnel communicate frequently enough with you.	1	2	3	4	5	1	2	3	N/A
14. Sales personnel respond promptly to requests.	1	2	3	4	5	1	2	3	N/A
15. Sales personnel are knowledgeable about equipment.	1	2	3	4	5	1	2	3	N/A
16. Sales personnel are knowledgeable about market opportunities.	1	2	3	4	5	1	2	3	N/A

	Very low			Very high	
17. Overall satisfaction level with **sales support**:	1	2	3	4	5

Technical Support

	Evaluation					Importance Level			
	Strongly disagree				Strongly agree	Low		High	
18. Technical support is available when needed.	1	2	3	4	5	1	2	3	N/A
19. The technical staff is knowledgeable.	1	2	3	4	5	1	2	3	N/A

20. The technical staff is well informed about the latest equipment updates/enhancements.	1	2	3	4	5	1	2	3	N/A
21. Parts are available when needed.	1	2	3	4	5	1	2	3	N/A
22. The remote support care center is valuable and meets your expectations.	1	2	3	4	5	1	2	3	N/A
23. Problems are resolved within the required time frame.	1	2	3	4	5	1	2	3	N/A
24. The technical staff is courteous and helpful.	1	2	3	4	5	1	2	3	N/A

	Very low				**Very high**
25. Overall satisfaction level with **technical support**:	1	2	3	4	5

Training

	Evaluation					Importance Level			
	Strongly disagree				**Strongly agree**	**Low**		**High**	
26. The trainers are knowledgeable about the equipment.	1	2	3	4	5	1	2	3	N/A
27. The trainers are effective communicators.	1	2	3	4	5	1	2	3	N/A
28. The initial operation training was productive and met your needs.	1	2	3	4	5	1	2	3	N/A
29. On-site training was beneficial for your business.	1	2	3	4	5	1	2	3	N/A
30. Operator training was effective.	1	2	3	4	5	1	2	3	N/A

	Very low				**Very high**
31. Overall satisfaction level with ABC **training**:	1	2	3	4	5

Supplies and Orders

	Evaluation					Importance Level			
	Strongly disagree				**Strongly agree**	**Low**		**High**	
32. Performance of supplies has consistently improved.	1	2	3	4	5	1	2	3	N/A
33. ABC branded performance meets your expectations.	1	2	3	4	5	1	2	3	N/A
34. The web-based service meets your expectations.	1	2	3	4	5	1	2	3	N/A
35. Orders placed are delivered when promised and are delivered complete.	1	2	3	4	5	1	2	3	N/A
36. The range of commercial consumables is sufficient.	1	2	3	4	5	1	2	3	N/A
37. The range of specialty consumables is sufficient.	1	2	3	4	5	1	2	3	N/A
	Very low				**Very high**				
38. Overall satisfaction level with ABC's **supplies and orders**:	1	2	3	4	5				

Software Add-On Solutions

39. Do you use ABC add-on software
    a.   Yes
    b.   No (Please skip to question 43.)

40. Do you use third-party software?
    a.   Yes
    b.   No

	Evaluation					Importance Level			
	Strongly disagree				**Strongly agree**	**Low**		**High**	
41. Capabilities and features of tools meet your needs.	1	2	3	4	5	1	2	3	N/A
	Very low				**Very high**				
42. Overall satisfaction level with **add-on software solutions**:	1	2	3	4	5				
43. Overall satisfaction level with overall **software solutions**:	1	2	3	4	5				

Customer Website

44. Did you register at ABC's customer website?	a.	Yes	
	b.	No (Please skip to question 50.)	

	Evaluation					Importance Level			
	Strongly disagree				**Strongly agree**	**Low**		**High**	
45. The website resources are helpful.	1	2	3	4	5	1	2	3	N/A
46. The online ordering process is effective.	1	2	3	4	5	1	2	3	N/A
47. ABC reports provide valuable information for your operations.	1	2	3	4	5	1	2	3	N/A
48. The special application has helped you select consumables.	1	2	3	4	5	1	2	3	N/A
	Very low				**Very high**				
49. Overall satisfaction level with the customer website:	1	2	3	4	5				

Purchasing Support

	Evaluation					Importance Level			
	Strongly disagree				**Strongly agree**	**Low**		**High**	
50. Invoices are provided on time.	1	2	3	4	5	1	2	3	N/A
51. Invoices are correct when first received.	1	2	3	4	5	1	2	3	N/A
52. Invoices are clear and easy to understand.	1	2	3	4	5	1	2	3	N/A
53. Credits are issued promptly.	1	2	3	4	5	1	2	3	N/A
54. Complaints are handled promptly.	1	2	3	4	5	1	2	3	N/A
55. Administrative personnel are friendly and courteous.	1	2	3	4	5	1	2	3	N/A
56. When you have an administrative problem, you know who to contact.	1	2	3	4	5	1	2	3	N/A
	Very low				**Very high**				
57. Overall satisfaction level with **purchasing support**:	1	2	3	4	5				

Contracts and Pricing

	Evaluation					Importance Level			
	Strongly disagree				Strongly agree	Low		High	
58. Equipment and service contract terms are clear.	1	2	3	4	5	1	2	3	N/A
59. The equipment represents good value for your money.	1	2	3	4	5	1	2	3	N/A
60. Usage contract terms are clear.	1	2	3	4	5	1	2	3	N/A
61. ABC's supplies represent good value for your money.	1	2	3	4	5	1	2	3	N/A
62. ABC's consumables represent good value for money.									
63. Charges for service and support are reasonable.	1	2	3	4	5	1	2	3	N/A
64. Your overall investment in an ABC solution generates value for your business.	1	2	3	4	5	1	2	3	N/A
	Very low				Very high				
65. Overall satisfaction level with **contracts and pricing**:	1	2	3	4	5				

System Installation

Please complete this section only if your equipment was installed during 2010.

	Evaluation					Importance Level			
	Strongly disagree				Strongly agree	Low		High	
66. Equipment worked properly after installation.	1	2	3	4	5	1	2	3	N/A
	Very low				Very high				
67. Overall satisfaction level with **site installation**:	1	2	3	4	5				

Overall Satisfaction with other suppliers

68. Of all the suppliers you are working with in the capital equipment market, please indicate your **best supplier**, excluding ABC. (Please select only one 'other supplier'.)

1. Company 1
2. Company 2
3. Company 3
4. Other, please specify:

_____

5. None
6. Don't know

69. Which ABC product do you use?

_____

Based on your personal experience with the supplier you selected in question 68, please rate your satisfaction with each of the following:

	Evaluation				
	Very low				**Very high**
70. Overall satisfaction level with *other supplier*:	1	2	3	4	5
	Very unlikely				**Very likely**
71. Would you recommend *other supplier* to other companies?	1	2	3	4	5
72. If you were in the market to buy a product, how likely would it be for you to purchase the other brand's product again?	1	2	3	4	5

	Evaluation				
	Very low				**Very high**
73. Overall satisfaction level with the other brand's **equipment and system**:	1	2	3	4	5
74. Overall satisfaction level with the other brand's **sales support**:	1	2	3	4	5
75. Overall satisfaction level with the other brand's **technical support**:	1	2	3	4	5
76. Overall satisfaction level with the other brand's **training**:	1	2	3	4	5
77. Overall satisfaction level with the other brand's **supplies and orders**:	1	2	3	4	5

78.	Overall satisfaction level with the other brand's **purchasing support**:	1	2	3	4	5
79.	Overall satisfaction level with the other brand's **contracts and pricing**:	1	2	3	4	5
80.	Overall satisfaction level with the other brand's **customer business development**:	1	2	3	4	5
81.	Overall satisfaction level with the other brand's **system installation** (only if your equipment was installed during 2010):	1	2	3	4	5

Thank you for your cooperation!

3

Census and sample surveys

Giovanna Nicolini and Luciana Dalla Valle

This chapter is about the type of surveys used by firms who want to find out about the satisfaction level of their customers. We begin with a brief introduction on the importance of customer satisfaction surveys (CSSs) for a firm, and proceed to present the two main types of surveys, census and sample. The choice between the two depends on costs, timeliness, goals of the survey, and mainly on the type of firm, its dimension and its customers. For each survey, the quality of the data collected depends on different types of errors. We focus our attention on non-sampling errors in CSSs and discuss their effect on statistical estimates. Then we show various data collection methods applied in CSSs, linking every method to one or more non-sampling errors. Finally, some methods to correct these errors are suggested.

3.1 Introduction

The measurement of customer satisfaction can be direct or indirect. In the first case, customers are directly asked how satisfied they are; in the second case, satisfaction is measured as the difference between the perceived and expected satisfaction for each dimension. In the evaluation of customer satisfaction, information has to be collected directly from customers, since the performance of the service and its quality are perceived subjectively and may differ from an objective study of performance and quality. This information is usually gathered on a regular basis, in order to monitor the success of process improvement activities.

Every competitive firm should be interested in assessing the satisfaction level of its customers. Some of the reasons are that customer loyalty guarantees the continuity of the firm, loyal customers are promoters of the firm, and satisfied customers are more tolerant of the firm's mistakes and of choices such as changes in service levels or price increases.

Customer satisfaction surveys (CSSs) need complex procedures requiring financial commitment as well as organizational effort. A key success factor is the support of top management as well as the capability to evaluate the results of the survey and to monitor its follow-up

Modern Analysis of Customer Surveys: with applications using R, First Edition. Edited by Ron S. Kenett and Silvia Salini.
© 2012 John Wiley & Sons, Ltd. Published 2012 by John Wiley & Sons, Ltd.

actions. Most firms dedicate to CSSs suitable resources and consider customer satisfaction indicators on a par with economic and production performance metrics. Indeed, CSSs are able to deliver representative and reliable information that can be very important in understanding the successes and difficulties of a company.

CSSs can be conducted by the company itself or can be outsourced. Surveys conducted with internal resources have the advantage of being less expensive, but when the company lacks the required competence, external suppliers are to be preferred.

Customers can be asked about their satisfaction level periodically (as a sequence of cross-surveys) or continuously (longitudinal surveys). Data collection is done periodically more often than continuously, since periodical surveys have a relatively low organizational complexity. Continuous (or event-driven) surveys are spread throughout the year. They require a greater organizational effort, but provide an ongoing focused view of customer satisfaction. Examples include surveys conducted with customers who called the help desk in the past week, had a new product installed or experienced a technical problem. These surveys allow us to recognize trends in customers satisfaction and provide early warning signs of developing problems. Periodical surveys have the advantage of better reflecting the long-term impression of the customer. In assessing loyalty, long-term impressions appear more important than short-term impressions. The characteristics of the sample in periodical and continuous surveys will be discussed in the following sections.

Different levels of satisfaction measurement may be used to obtain detailed information from a CSS. Examples of such survey questions are listed below (for more details, see Griffiths and Linacre, 1995):

- First, the global aspects of customer satisfaction are measured with questions concerning the service and overall satisfaction: 'How satisfied are you on the whole?', 'Would you recommend this service to friends/relatives?' and 'Would you like to use this service again?'.

- At a second level the dimensions of the service are introduced. For example, in a CSS of an airline company, with its dimensions being 'booking', 'cabin' and so on, the questions would be: 'How satisfied are you with the ticket booking process on the whole?' and 'How satisfied are you with the cabin environment on the whole?', and so on.

- The third level consists of specific and detailed questions about satisfaction with single performance dimensions. In the 'cabin' dimension example, the questions could be: 'How satisfied are you with the level of cleaning of the cabin?' and 'How satisfied are you with the comfort of the cabin?', and so on.

We can distinguish between types of CSS: census surveys, when the whole population is interviewed; and sample surveys, where a subset of individuals is selected as survey participant.

An important issue in CSSs is the quality of the data. In periodical surveys, data quality and response rates should be assessed throughout the questionnaire completion process and not just at the end of the data collection phase. Similarly, the quality of continuous surveys has to be evaluated on an ongoing basis (Delmas and Levy, 1977; Hothum and Spinting, 1977). In the following sections we illustrate the main types of error that may affect survey quality, and show the principal methods to correct them.

3.2   Types of surveys

The first phase of a statistical survey involves the definition of its main purpose, the definition of the target population, the generalization of the survey to the whole population or only to a part of it, the nomination of a focal point that is responsible for the survey (internal or external to the firm), and the organization and deployment of the periodical or continuous surveys. These features are analysed below in some detail.

3.2.1   Census and sample surveys

The definition of the target population and of the primary units (in this case the customers) to be surveyed is critical for the effectiveness and the usefulness of the survey. At this stage it is necessary to determine, for example, if the target population is generalizable to:

- all the customers of the firm that used the service at least once;

- only customers that frequently use the service;

- customers with a decreasing demand for the service.

The target population size, together with the planned survey costs and the schedule deadlines, can determine the choice of running a census survey or a sample survey. The size of the target population and the availability of a complete list of customers as a survey frame are typically related to the firm size and to the type of customers.

The type of firms considered here mainly depends on their size and the markets they operate in. Hence, we have three types of firms:

- *Large firms* operate globally or nationally in an oligopolistic regime in 'imperfect competition'. Such firms operate in areas such as telecommunications, electricity, transportation, banking, finance, insurance and health care.

- *Medium firms* operate in a smaller context. If they operate nationwide, they do so in an environment of 'perfect competition' (local transport, tour operators, private clinics, etc.).

- *Small firms* typically develop locally a wide range of services in a climate of 'perfect competition'.

Large firms can carry out sample surveys with large samples, whilst medium or small firms tend to carry out census surveys. If we focus our attention on the types of customers, we can distinguish between firms that offer services to other firms (business to business – B2B) and firms that offer services to the final consumer (business to consumer – B2C). In the first case, the list of customers (generally not large) is available and the survey is quite often a census survey. In the second case, the list of customers is not necessarily known and the survey can only be a sample survey. Table 3.1 summarizes the classification based on these two factors.

Obviously, the choice of the type of survey, based on a census or sample, has an effect on the survey errors. This requires an understanding of the errors that influence the quality of survey research.

Table 3.1   Types of surveys for types of firms and customers.

Firm's dimension	B2B	B2C
Small	C	C/S
Medium	C	C/S
Large	C	S

C = census survey; S = sample survey

3.2.2   Sampling design

The survey sampling design and the process of drawing respondents (units) has a major role in sample surveys. When the sampling frame (the list of customers in the target population) is known, a probabilistic sampling design can be used, either 'simple' or 'stratified'. The units are usually drawn with equal probability. Sampling with probability proportional to size is also possible. If the population of customers is not very large, co-located in a small area and quite homogeneous, the simple sample design is to be preferred; otherwise, the stratified sampling design has to be preferred using auxiliary variables for the strata definition. These variables typically characterize the type of customer. For individual respondents they can be job function, seniority, gender or location. If the units are firms, the variables can be the sector, the number of employees, the type of market etc. It is well known that stratified sampling generally guarantees a better representativeness of the sample and a greater efficiency of the estimates than simple sampling, besides giving specific estimates for the stratum populations. This allows the firm to accurately analyse special groups of customers, called *domains of study*. If the domain coincides with the population of one or more strata, we have a *stratified domain*. In this case the estimates of the level of satisfaction of the customers are unbiased and generally more accurate than similar estimates obtained through simple random sampling. Conversely, if the domain concerns segments of populations that cross several strata, we have a *cross domain*. These domains are usually generated by post-stratification. The estimates of their levels of satisfaction are unbiased, but less accurate, because of the variability of the dimension of the post-stratum sample. In order to increase the efficiency of the estimate, a post-stratum ratio estimator can be employed, although it may be biased if the post-stratum sample size is not sufficiently large. Such ratio estimators are asymptotically unbiased (Särndal *et al.*, 1992).

When the frame population is unknown, we can apply quota sampling, which is similar to a stratified sample but is not a probabilistic sample. In such a case we cannot estimate the sampling error unless we assume a model generating the study variable. Moreover, for certain types of services, cluster or area sampling is used. For example, a mobility services firm can interview users at its bus stops or a railway service company can interview users travelling by train. In both cases, samples of areas (bus stops) or clusters (train carriages) are extracted through a probability sample.

3.2.3   Managing a survey

Whatever the type of survey, the interviewees need to respond to the survey and provide representative answers. In order to increase the number of respondents, the questionnaire should not be too long and should be engaging, with questions concerning issues that really matter to the customer. The interviewee should also be convinced that the firm will give his/her responses careful consideration in order to improve the service. There are many ways to collect data. The researcher should pay attention to possible errors in the survey plan. For example, if face-to-face interviews are adopted, in order to guarantee truthful responses, data collection is usually not conducted by internal firm employees; otherwise interviewers may influence the interviewees' responses. Employing properly trained people from external companies to carry out face-to-face interviews can minimize this type of error.

3.2.4 Frequency of surveys

There is no rule suggesting how often a survey should be repeated. The service could be offered to the same customer many times or only once. In the former case, although the standard of the service is always the same, it could be delivered by different people, and we might ask the customer to repeat the survey. However, as the service varies over time, the quality and satisfaction level need to be monitored in a timely and accurate manner. The replication of the survey and its timing should be defined by top management and are influenced by the budget of the firm. However, for census as well as for sample surveys, running a survey once is not enough if we want to monitor customer satisfaction over time. In continuous or event-driven surveys, a questionnaire is forwarded to a customer who has received a service. This could, however, result in high level of unit non-response from customers who frequently use the service. Monitoring service on a regular basis, with different customer samples, provides information on satisfaction level from a wide range of customers. If the service is improved, repeatedly interviewing the same customer sample over time allows us to assess the shift in customer satisfaction over time. For more on tracking customer satisfaction over time, see Chapter 20.

3.3 Non-sampling errors

The usefulness of a survey (be it a census or a sample survey) depends on the quality of the data collected and the accuracy of the results. If the quality is poor, the results are inevitably biased and the entire survey may prove worthless. Poor quality may have many causes, some due to the organization of the survey and the firm itself; others can be due to the willingness of the customer to cooperate. A great part of these very causes is at the core of many remarkable errors. Errors may be *sampling* or *non-sampling errors*. Properties of the sample are evaluated by statistical sample theory. A measure of sampling error is the standard error of the estimator. Non-sampling errors include a set of all the likely errors we could encounter at each step of the survey process and their possible interactions, sampling errors excluded. As they are not affected by sampling errors, census surveys should in theory generate the true value of the parameter of the population. This, however, does not happen because of non-respondents. The incidence of these two distinct types of errors – sampling and non-sampling – is very different. Sampling errors can be reduced by modifying the dimensions of the sample and using more accurate sampling designs. Non-sampling errors are due to different causes and tend to grow with the number of units considered. Non-sampling errors are related to survey procedures and are present in both census and sample surveys; they are often more relevant than sampling errors.

A common classification of non-sampling errors is provided by Groves (1989):

- *Measurement error* is measured by the difference between the real value of an item, related to a surveying unit, and the value observed.

- *Coverage error* is observed when the number of units of the target population (customers of a firm involved in the survey) and the number of units of the frame population (the list of customers) do not coincide.

- *Unit non-response and non-self-selection error* occurs when the unit does not respond to the questionnaire. This may be caused by many factors; however, in a CSS, we consider it as generated by the behaviour of the customer.

- *Item non-response and non-self-selection error* occurs when the unit does not respond to specific items in the questionnaire. This error may be caused by many factors; however, as in the previous case, we consider it as generated by the behaviour of the customer.

We now analyse these errors and their influence on a CSS.

3.3.1   Measurement error

Measurement error occurs in the data imputation phase and when the questionnaires are filled in. However, considering that customer satisfaction questionnaires are made by simple questions with default answers on an ordinal scale and that in many circumstances the questionnaires are filled in by the customer himself, we can assert that amongst the enlisted errors the measurement one may appear, in CSSs, the least relevant. The methods suggested in the literature to find and correct these types of errors are complex, usually increasing survey costs and extending survey schedules (Lessler and Kalsbeek, 1992).

3.3.2   Coverage error

Coverage error, along with measurement error, can be attributed to the firm or the way the survey is carried out. Coverage error occurs when there is a frame (the list of the target population units) which is incorrect or incomplete because it does not reflect the target population. This happens when the frame is not up to date – for example, if it contains subjects that are no longer customers (*over-coverage* error) or is missing new customers that have not yet been added (*under-coverage* error). In the former case there are units that are no longer suitable for the survey; in the latter case there are missing units that will not been involved in the survey. Over-coverage error is treated as less serious than under-coverage error since it is easier to identify in a CSS, and thus easier to correct. Under-coverage is, however, more difficult to identify and potentially generates bias in estimates or in parameter values. Assume that the target population size is the sum of the units included in the frame, N_F, and of the missing units, $N_{\bar{F}}$, that is, $N = N_F + N_{\bar{F}}$, with an under-coverage rate of $W_{\bar{F}} = N_{\bar{F}}/N$. The mean value of the target variable in the population is formed of two parts, the mean value of the units included in the frame and the mean value of the missing units: $\bar{Y} = \bar{Y}_F + \bar{Y}_{\bar{F}}$. Consequently, considering the mean value $\bar{Y}_F$ instead of $\bar{Y}$, we have an absolute bias of

$$B(\bar{Y}_F) = \bar{Y} - \bar{Y}_F = W_{\bar{F}}(\bar{Y}_F - \bar{Y}_{\bar{F}}), \tag{3.1}$$

and a relative bias of

$$RB(\bar{Y}_F) = \frac{W_{\bar{F}}(\bar{Y}_F - \bar{Y}_{\bar{F}})}{\bar{Y}} = \frac{W_{\bar{F}}(1-r)}{W_{\bar{F}}r + (1 - W_{\bar{F}})}, \qquad (3.2)$$

where $r = \bar{Y}_{\bar{F}}/\bar{Y}_F$. In particular, note from equation (3.2) that when $r = 1$ the bias is zero, when $r > 1$ the bias is negative, while when $r < 1$ the bias is positive. Therefore, a large gap between the target and the frame population can significantly affect survey results. These may be correct for the frame population but incorrect for the target population. The frame usually contains a set of variables, called *auxiliary variables*, which are useful in the selection of units (sampling with probabilities proportional to size), in the construction of the sampling design (stratified or post-stratified sampling) and in the specification of the domains of interest. If the frame is not up to date, the auxiliary variables will not be updated as well and consequently generate unsatisfactory results. The existence of the frame is the basis of a census survey or a sample survey with a probability sample. On the other hand, if we do not know the frame we cannot carry out a census survey, but will be able to conduct a sample survey with sampling designs such as quota, area or cluster sampling.

3.3.3   Unit non-response and non-self-selection errors

This kind of error is attributed to the interviewee/customer. We define *unit non-response* as the choice not to participate in the survey by a population unit that has been selected for the survey. It is thus the consequence of a selection, generating a potential error in a sample survey. Similarly, *unit non-self-selection* is the choice not to participate by a population unit that has not been selected, inasmuch as the survey involves the whole population. Usually people discuss self-selection errors. We define non-self-selection error in line with the definition of non-response error. These two concepts each have a complement in *unit response* and *unit self-selection*, in sample surveys and censuses, respectively. In any case, the presence of respondents and non-respondents, or self-selected and non-self-selected, divides the target population into two subpopulations. It is necessary to understand whether these two subpopulations are similar to or different from the target variable in terms of satisfaction levels. In the first case the non-response/non-self-selection can be considered random and does not generate errors in the parameters, while in the second case it can bias the results. A high level of non-responses reduces the dimension of the sample and increases the variance of the estimators. A small number of self-selected subjects suggests that the collected data is not representative of the population but of one of its samples that is certainly not random. If we suppose that the probability of a unit non-response is equal to the probability of a non-self-selection, the effects of these two errors can be considered equal. However, we cannot exclude surveys where these two probabilities are not equal. In this case, the effect of the two errors should be assessed on a case-by-case basis. Below, we consider these two errors as equal. Therefore, let R be the response rate and $\bar{R} = 1 - R$ the non-response rate. Then the mean value of the target variable $\bar{Y}$ has a non-response bias

$$B(\bar{Y}_{\bar{R}}) = \bar{Y}_R - \bar{Y} = \bar{Y}_R - (\bar{R}\bar{Y}_{\bar{R}} + R\bar{Y}_R) = \bar{R}(\bar{Y}_R - \bar{Y}_{\bar{R}}), \qquad (3.3)$$

which is the product of the non-response rate and the difference between the mean of the respondents $\bar{Y}_R$ and of the non-respondents $\bar{Y}_{\bar{R}}$. Therefore, if the difference between the two means is zero, the bias is zero, although the percentage of the non-respondents can be high. If the percentage of non-respondents is small, but the difference between the two means is high, the bias will be high. Therefore, the non-response bias is not always due to a high percentage of non-respondents.

In order to quantify the error, understanding the causes of participation/non-participation in the survey will be useful. The researcher knows the features of the people who responded and did not respond by considering the frame. Various weighting methods have been suggested in the presence of non-response or non-self-selection that are based on models in which auxiliary variables are employed. What the researcher does not know are the reasons behind the non-responses or the non-self-selection. The reasons in CSSs are different from those in surveys about sensitive matters such as diseases or private life habits, that may make some interviewees feel uneasy. Indeed, causes of non-response or non-self-selection in CSS may be a good level of satisfaction, sheer boredom or the reluctance of people to fill in the questionnaire due to the number of CSSs one is asked to fill in such as in hotels, organized tours or training sessions. Poorly designed questionnaires and a lack of obvious improvement in services following the implementation of CSSs can also be demotivators. For more on evaluating the effect of unit non-response and a test for determining sample representativeness, see Chapter 20 and, in particular, Section 20.6.

3.3.4   Item non-response and non-self-selection error

The literature defines an *item non-response* as a missing answer to one or more questions in the questionnaire. Like the definitions in Section 3.3.3, we associate this error with a sample survey. The missing answer to one or more questions in the questionnaire in a census survey is called *item non-self-selection*. As was defined for the unit, item non-response typically has the same probability as item non-self-selection. Responses to an item can may be missing because: (a) the subject did not want to respond; (b) the subject did respond but the response is useless because of incoherence with other responses; (c) the subject did respond, but information has been lost due to errors in the data processing phase. Case (a), called in the literature *missing not at random*, is the most severe, since the missing response depends on the non-observed value of the item and creates systematic differences between respondents and non-respondents. This can generate bias in the calculation of the parameters of interest. This error could be considered less serious than the unit error, since its consequences involve only one item. This suggests that just for that item, the dimension of the sample is smaller, with an increased variance of the estimate. However, partial non-response creates an incomplete data set and can affect the application of certain analytical techniques. For example, the application of the Rasch model for the evaluation of customer satisfaction requires the data set to be complete. Deletion of incomplete units can greatly reduce the data set. Therefore, if the missing data mechanism is random, simple imputation methods are available, and special imputation procedures are needed (see Chapter 8).

3.4   Data collection methods

Firms may use various methods for data collection, according to the various types of customers, the amount of time and the costs involved. In this section we highlight the features and problems of a CSS with regard to the types of firms providing services, the types of customers, the data collection methods carried out throughout the surveys and, consequently, the potential errors connected with them.

The most common data collection methods and the types of errors associated with them are as follows:

- *Face-to-face interviewing* is used when the complete list of customers is not known. It is frequently used by mobility services in interviews of the customers who actually use the service. It is generally carried out in a sample survey context (see Section 3.3.2). This is a very expensive method, and the presence of the interviewer may produce a bias caused by measurement errors. However, through this type of interview, open answers are more feasible, giving the customer the chance to express his opinion about some features of the service, as well as to give useful suggestions for improvements.

- *Computer-assisted telephone interviewing* (CATI) is frequently used and not very expensive. In CATI, the response rate may be quite high. Furthermore, this method has the advantage of immediate availability of results that can be quickly analysed and interpreted. It is possible to employ CATI if a list of customers already exists. It can be used in census surveys and in sample surveys with random samples. With this approach it is possible to keep a close watch on measurement errors, while coverage errors (if the list of customer is not updated) and non-response errors are more frequent.

- *Computer-assisted web interviewing* (CAWI) is the modern version of the now rarely used mail interview. This method is correctly applied if extended to all the customers who have an email address known to the firm. Since such surveys typically target all customers, it is possible to verify the occurrence of self-selection error. Conversely, if not every customer has internet access, the choice of this method can involve various coverage errors. The main advantage of CAWI is certainly the low cost of data collection. Moreover, the absence of an interviewer enables the respondent to freely decide when to answer the questions. However, this method has the drawback of a limited response rate. In order to prompt subjects to participate, the questionnaire should be kept as short as possible, requiring not more than 15–20 minutes to complete. Moreover, emails should clearly point out the importance of the interviewees' contribution and reminders should be sent 2–3 weeks after sending out the first emails.

- *Web surveys* (WS) can be of various types (Couper, 2000; Bethlehem and Biffignandi, 2011). Their use is becoming more widespread every day, in part because the number of people with internet access is increasing every day. The web approach most frequently employed in a CSS is an online questionnaire published on a dedicated website or a questionnaire that pops up for every xth visitor on the firm's website. As far as online questionnaires on the firm's website are concerned (WS1), the filling in of a questionnaire that is open to everyone depends on the customer's cooperation. Hence, we have self-selection and statistical inference is not always possible. In the second case (WS2), a systematic sample is used but there may be unit non-response errors. If every customer of the firm has internet access (e.g., online banking), or if the survey is only directed at the visitors of the site, coverage error is not present. On the other hand, if not every customer of the firm is an internet user, the coverage error adds up to the errors we have already described.

- *Open Surveys* involve giving the customer a paper questionnaire on completion of the service. The filling in and submission of the questionnaire depend on the customer. This survey is very easy, it is not supported by any kind of organization, but self-selection error is quite frequent.

For more on web-based surveys see Chapter 6, for other types of surveys see Chapter 7. In the next section we suggest some methods to correct the most important non-sampling errors in a CSS. We do not discuss measurement errors, since they are not frequent, or coverage errors, since they can be easily removed by updating the frame. We focus on suggestions to correct the other two types of errors in a CSS.

3.5 Methods to correct non-sampling errors

In this section we discuss methods to correct (unit and item) non-response errors in a CSS. These methods can also potentially be applied to non-self-selection errors.

3.5.1 Methods to correct unit non-response errors

Unit non-responses can be reduced through special procedures activated during the data collection phase. If, on completion of a survey, a high percentage of unit non-responses is observed, one should apply methods to correct the potential bias in CSS estimates. Such methods directly or indirectly use auxiliary variables which are known for all the population or for the selected sample (for respondents and non-respondents). Some methods are based on the idea that the bias is generated by a gap between the response model, which represents the real distribution of the responses, and the observed responses. It is then necessary to estimate the response model to correct the observed response distribution and, indirectly, correct the estimates of the target variables (Särndal *et al.*, 1992). Conversely, other methods act directly on the estimates of the target variables, with appropriate changes of weights. However, before applying any estimate correction procedure, it is necessary to assess whether the observed sample is representative of the design sample or if unit non-responses have altered the design sample. If the observed sample is not representative of the selected sample, it is necessary to investigate correction methods.

R-indicators

The representativeness of a sample was widely discussed in the literature (Kruskal and Mosteller, 1979) and was often linked to the response rate. The latter is traditionally thought of as an indicator of the quality of a survey: if it is low, the sample of respondents can misrepresent the total sample and the population. However, more recently, this indicator has also been criticized, since the bias of the estimates caused by non-responses is believed to depend on whether the respondents represent the selected sample or not. When the group of respondents represents the sample well, the difference between the estimate of the expected value of the target variable, computed from respondents, and the estimate computed from the whole sample is quite close to zero. Therefore, it is necessary to identify indicators, called *representativity indicators* or *R-indicators* (Schouten *et al.*, 2009), measuring how much the response data set is different from the sample data set. These indicators do not provide inputs to the estimation process, but indicate the need to correct the estimates when, due to unit non-response, the sample is no longer representative. These indicators are constructed with the probabilities of response (in the following called *response propensities*) that are estimated, for example, using a logit or probit regression model, through a set of demographic and socioeconomic auxiliary variables, Ξ, able to describe the response behaviour of an individual or firm (Schlomo *et al.*, 2009). Let $\rho(i)$ for $i = 1, 2, \ldots, N$, the estimate of the response propensity; then the response data set is called *strongly representative* if $\rho(i) = \rho$ for all *i*. This says that the non-responses are *missing completely at random* and the probability of response does not depend on the value of the target variable; and this has to be applied to every target variable. However, this case is not feasible in everyday practice, because criteria to estimate response propensities do not exist. More realistic is the case of a weakly representative data set, where the response propensity is estimated by the set Ξ. A data set is *weakly representative* of the sample and of the set Ξ if the average response propensity is the same for each auxiliary variable. For example, if X is the variable 'age', the average response propensity of youth $(X = x)$ is $\rho_X(x)$: this means that the auxiliary variable and the response behaviour are independent. The response propensities can be used for the construction of indicators that measure the divergence from representativity,

$$R(X) = 1 - 2S(\rho_X), \tag{3.4}$$

where $R \in [0, 1]$ and $S(\rho_X)$ is the standard deviation of response propensities. A data set is representative if $R = 1$ and not representative if $R = 0$. In the first case $S = 0$, so all people have the same response propensity; in the second case $S = 0.5$, where there is maximum variability in response propensities. The R-indicators method is mainly applied in the data collection phase, in order to preserve the representativeness of the selected sample. For a related method based on standardized residuals, see Section 20.3.

Adjustment classes

This method is founded on the post-stratification of the selected sample through auxiliary variables, allowing us to group the sampled units into C classes, called *adjustment classes*, with different response probabilities but with the same answers for respondents and non-respondents within the same class. In other words, if the average satisfaction values for the target variables of the two groups – respondents and non-respondents – are equal within each class, the bias can be lowered to zero. The construction of adjustment classes is not easy, because, if socio-demographic variables are used (e.g., for an individual, gender, age, profession; for a firm, sector, number of employees), classes composed of subjects with similar characteristics may give different responses. In order to guarantee this equality, it is necessary to use variables of different types, for example associated with satisfaction levels recorded in a previous CSS. Moreover, the subsample size for each class must be large enough to guarantee at least the asymptotic accuracy of estimators, if the Horvitz–Thompson estimator is not used.

   If the information about auxiliary variables is known for the population, we know the size of each class N_c, with $c = 1, \ldots, C$ and $\sum_c N_c = N$, and the class weights $W_c = N_c/N$. It follows that the estimator is similar to the post-stratified one, where each class contributes with its weight to the average satisfaction of the target population:

$$\hat{\bar{Y}}_{ps} = \sum_c W_c \bar{y}_{Rc}, \tag{3.5}$$

where $\bar{y}_{Rc}$ is the sample mean for class c calculated over the class respondents. On the other hand, if the average satisfaction of the respondents in the adjustment class is equal to the average satisfaction of the non-respondents, then (3.5) is an unbiased estimator of the average satisfaction of the target population. When the class weights W_c are unknown, since the values of the auxiliary variables are known only for the selected sample, they are estimated through the class sample weights $w_c = n_c/n$. In this case we have the estimator

$$\hat{\bar{Y}}_{ps}^{*} = \sum_c w_c \bar{y}_{Rc}, \qquad (3.6)$$

which is still unbiased if the difference in average satisfaction between respondents and non-respondents within the same class is zero. However, estimator (3.6) is less efficient than (3.5) for the estimation of W_c. An alternative to (3.6) is the ratio estimator

$$\hat{\bar{Y}}_{rt} = \frac{\sum_c y_{Rc}}{\sum_c n_{Rc}}, \qquad (3.7)$$

with $y_{Rc} = \sum_{i_R \in c} y_{i_R}$ being the total of the target variable observed on the respondents in each class c and n_{Rc} being the number of respondents in the class. If we adopt a reweighting approach for the basis weights, $w_{Rc} = n_{Rc}/n_c$ denoting the response rate in the class c, we have the weighted estimator (Little, 1986)

$$\hat{\bar{Y}}_{wrt} = \frac{\sum_c w_{Rc}^{-1} y_{Rc}}{\sum_c n_c}. \qquad (3.8)$$

Detailed studies of the bias and variance of the estimators presented can be found in Thomsen (1973, 1978) and Oh and Scheuren (1983).

We have already mentioned the complexity of building adjustment classes. This complexity can be overcome if response propensities are estimated through two-phase sampling and then used to define inclusion probabilities. In the first phase of this design, we draw a simple sample s of size n from the customer population of size N, with inclusion probability for the ith population elements equal to π_i. In the second phase the sample of respondents r is assumed to be drawn from s with a Poisson sampling design, which associates with the ith population element a response probability θ_i, for $i \in s$, under the hypothesis that it depends on the subject i and not on sample s. This means that in two-phase sampling, the probability that one population element belongs to the respondent sample is equal to $\pi_i \theta_i$. People with the same value θ_i belong to the same adjustment classes. Therefore, assuming that the will to respond is an independent event with different probability among subjects, the adjustment classes consist of all subjects with the same probabilities and are no longer based on equality of answers between respondents and non-respondents. In general, we can state that the non-response bias is negligible if in each class the correlation between response probabilities and the target variable goes to zero.

Calibration

Calibration is a method which allows, through one or more auxiliary variables, the basic weights of an estimator to be modified. Let s be a simple sample, without replacement, of size n. Then the unbiased estimator of the total is

$$\hat{Y} = \sum_{i \in s} a_i y_i, \qquad (3.9)$$

where the basis weights $a_i = \pi_i^{-1}$ are the reciprocal of the first-order inclusion probabilities. Let $\mathbf{X}_i = (X_{1i}, X_{2i}, \ldots, X_{Ji})$ be the vector of the auxiliary information of J variables, known for unit i (individual or firm) of the population. When we have unit non-responses, the weights a_i cannot be applied to a sample of n_R units ($n_R < n$), since the estimator (3.9) would be biased. Alternatively, the calibrated estimator

$$\hat{Y}_{cal} = \sum_{i \in r} \omega_i y_i \qquad (3.10)$$

is employed, where the weights ω_i of the respondent sample r are defined such that they satisfy two conditions:

(a) being as close as possible to the basis weights a_i;

(b) $\sum_r \omega_i \mathbf{x}_i = X_i$.

For condition (a) we can define different calibrated estimators according to the distance measures adopted. On the other hand, for condition (b), called the *calibration equation*, the *alignment* property is satisfied.

This property requires the use of weights applied to the sample data of the auxiliary variables in order to obtain estimates without errors, that is, with values equal to the real parameters X_i. The new weight is the product of the old weight and the adjusted weight: $\omega_i = a_i v_i$, where $v_i = 1 + \lambda' \mathbf{x}_i$, with $\lambda' = (\mathbf{X} - \sum_r d_i \mathbf{x}_i)' (\sum_r d_i \mathbf{x}_i \mathbf{x}_i')^{-1}$ (Särndal and Lundström, 2005). If the value X_i is not known, but the auxiliary variable total is known only for the planned sample s, in the *calibration equation* X_i is replaced by its unbiased estimator (Horvitz–Thompson estimator).

In a CSS the use of the calibrated estimator is always suggested when: there is a high percentage of missing responses; adjustment classes are not feasible; and missing responses are classified as *not applicable*. In customer satisfaction questionnaires, a set of questions may be marked as not applicable. In this case the usual item imputation methods are not feasible. This happens in services with k dimensions, where, for whatever reasons, h of them have not been completed, and the service refers only to $k - h$ dimensions. Consider, for example, the customers of an airline company with online ticketing and check-in. In this case customers cannot answer questions about 'counter service'. The sample size n can be thought as split into two groups of customers: n_1 who filled in the questionnaire with k dimensions and n_2 who filled in the questionnaire with $k - h$ dimensions. Therefore, the weights a_i must be modified with two different values v_{1i} and v_{2i}

3.5.2   Methods to correct item non-response

The methods that will be shown here have a specific end: to differently correct the bias caused by item non-response or non-self-selection. The first method is to evaluate the item satisfaction of the non-respondents in probabilistic terms. The second considers two equations tied together by a latent factor that allows the missing data associated with the non-respondents to be correctly estimated. Finally, the third method proposes a linear model which is supposed to be able to show the relations between the target variable and a group of variables thought to have generated the observed data. If the model is easily adaptable to the observed data, it will be possible to predict the values of the target variable for non-responding customers.

Propensity score matching

Propensity score matching was first proposed by Rosenbaum and Rubin (1983), building on earlier work by Rubin (1974), in the context of an experiment to evaluate the effect of a health treatment (see also Chapter 8).

This method has been applied in different contexts to correct the bias due to self-selection error in web surveys (Biffignandi and Pratesi, 2003; Biffignandi *et al.*, 2003; Lee, 2006). In the context of a CSS, the method aims to compare the outcomes of the target variable observed on a group of individuals subjected to a treatment (those who answered all questions) to the outcomes that should have been observed if the treatment had not been applied in a control group (those who did not answer all questions).

Having calculated the propensity scores as far as the subjects of the two groups are concerned, the same level of satisfaction of an individual who has filled in the questionnaire and has got the same propensity scores will be attributed to an individual from the control group who has not filled in the questionnaire (through the 'matching' phase). The propensity score matching method allows us to correct the item non-response and self-selection error and to estimate the level of satisfaction of those who have not filled in the questionnaire. It is based on the assumption that item non-response does not depend on the target variable (conditional independence assumption). This implies that the pre-intervention variables, on which the matching is influenced, must not affect the non-responses or self-selection (Rosenbaum and Rubin, 1983). In the context of CSSs, pre-intervention variables could differ according to the type of service received.

Heckman two-step procedure

Heckman (1979) proposed a two-step procedure which allows the values of the target variable for non-responding individuals to be estimated. Heckman's idea is based on a model consisting of two related equations: a substantial equation and a selection equation.

The *substantial equation* for individual i $(i = 1, \ldots, n)$ is

$$Y_{1i} = X_{1i}\beta_1 + U_{1i}, \tag{3.11}$$

where Y_{1i} represents the continuous target variable, which in this case represents the satisfaction level. In case of non-self-selection n is replaced by N (the population size). Suppose that one wishes to estimate equation (3.11) but that data are missing on Y_1 for $n - m$ observations (where m is the number of item responses). The *selection equation* for individual i is

$$W_i = X_{2i}\beta_2 + U_{2i}, \tag{3.12}$$

where W_i is an unobserved latent random variable such that $W_i \geq 0$ corresponds to individuals responding to the item, while $W_i < 0$ corresponds to non-responding individuals.

For both equations (3.11) and (3.12) we make the following assumptions: X_{1i} and X_{2i} are ith vectors of auxiliary variables known for all n subjects; β_1 and β_2 are vectors of parameters; U_{1i} and U_{2i} are the vectors of residuals; $E(U_{ji}) = 0$, $E(U_{ji}U_{j'i'}) = \sigma_{jj'}$ for $i = i'$, and $E(U_{ji}U_{j'i'}) = 0$ for $i \neq i'$, where $i = 1, \ldots, n$ and $j = 1, 2$; and the joint density of U_{1i} and U_{2i} is bivariate normal (Nicolini and Dalla Valle, 2009, 2011).

The method allows us to use equation (3.11) to estimate the level of satisfaction of non-responding customers, taking into account, via equation (3.12), the existence of a correlation between the residuals of the two equations.

Hierarchical Bayesian approach

In the hierarchical Bayesian approach the value of the variable Y on the ith unit of the population is not fixed but an observed value of a random variable Y_i $(i = 1, \ldots, n)$.

The technique was traditionally developed and employed in the small area estimation field, as explained by Rao (2003). Indeed, this is a very powerful tool when dealing with lack of data (Trevisani and Torelli, 2006). Here we employ it successfully in a CSS, where we estimate the parameters of interest for the individuals who did not respond to some items of the questionnaire through the information provided by covariates known for every subject.

The hierarchical Bayesian model is derived from a linking model and a sampling model as formulated by Fay and Herriot (1979). Suppose one is interested in estimating the characteristic Y_i for every subject, and that the auxiliary data are known for each individual i (for $i = 1, \ldots, N$). The *linking model* is

$$Y_i = X_i \beta + U_i, \tag{3.13}$$

where $U_i \sim N(0, \sigma^2)$, X_i is a vector of auxiliary variables and β is a vector of parameters. Equation (3.13) is merely a mixed linear model where β are fixed effect coefficients, accounting for the effects of the auxiliary variables X_i, valid for the entire population, while the U_i are random individual specific effects (Rossi *et al.*, 2005). The *sampling model* is

$$\hat{Y}_i = Y_i + e_i, \tag{3.14}$$

where $\hat{Y}_i$ is the estimate of Y_i and $e_i | Y_i \sim N(0, \psi_i)$, ψ_i being the sampling variance, which is typically assumed to be known.

The hierarchical Bayesian model is described by the equations

$$\hat{Y}_i | Y_i, \psi_i \sim N(Y_i, \psi_i), \tag{3.15}$$
$$Y_i | \beta, \sigma^2 \sim N(X_i \beta, \sigma^2). \tag{3.16}$$

The parameters are estimated by means of MCMC methods in the hierarchical Bayesian model for $i = 1, \ldots, m$ and then, for $i = m + 1, \ldots, n$, Y_i is estimated by $\hat{Y}_i$ with the posterior predictive distribution.

3.6 Summary

Typically, much attention is given to the results of a survey, while the methodology used is neglected. In particular, the choice of a CSS methodology is influenced primarily by cost and implementation schedules, which have an impact on the quality of the survey.

A weak method, together with lack of information, decreases the quality of the survey for the firms that commissioned it and that must draw conclusions that are relevant for future planning.

In this chapter we consider the initial phases of a statistical survey, such as the objectives, the target population, sample design, data collection methods. Each phase is analysed from the CSS point of view and in relation to the type of service supplier company and its customers.

For each type of non-sampling error, the links among types of company, customers and data collection methods are highlighted. Non-response (or non-self-selection) errors are the more evident non-sampling errors in a CSS. Some methods of correction are suggested.

The (item or unit) non-response errors in a CSS may have characteristics that are different from similar errors in other fields (social, demographic or economic). Item errors always refer to ordinal data (see Chapter 4) and unit errors are usually not random. However, the methods we suggest for unit errors are those usually proposed by the literature (adjustment classes and calibration). For item errors, on the other hand, the usual methods (such as the deductive methods of imputation, mean imputation, hot deck and cold deck) are not recommended. For errors of this type, we propose probability methods such as propensity score matching, the Heckman procedure and the hierarchical Bayesian approach.

The use of R-indicators, introduced in Section 3.5.1, allows us to determine whether, in presence of unit non-responses, the observed sample is still representative of the planned sample. Indeed, the representativeness of the observed sample affects the use of unit non-response correction methods.

References

Bethlehem, J. and Biffignandi, S. (2011) *Handbook of Web Surveys*. Hoboken, NJ: Wiley.

Biffignandi, S. and Pratesi, M. (2003) Potentiality of propensity score matching in inference from web-surveys: a simulation study. Working Paper no. 1, Dipartimento di Statistica, Informatica e Applicazioni.

Biffignandi, S., Pratesi, M. and Toninelli, D. (2003) Potentiality of propensity scores methods in weighting for Web surveys: A simulation study based on a statistical register. In *Bulletin of the International Statistical Institute: Proceedings of the 54th Session (Berlin)*. ISI.

Couper, M.P. (2000) Web surveys: A review of issues and approaches. *Public Opinion Quarterly*, 64, 464–494.

Delmas, D. and Levy, D. (1977) Consumer panels. In *Seminar on Marketing Management Information Systems. Organising Market Data for Decision Making*, pp. 273–317. Amsterdam: European Society for Opinion and Marketing Research.

Fay, R.E. and Herriot, R.A. (1979) Estimates of Income for small places: An application of James-Stein procedures to census data. *Journal of the American Statistical Association*, 74, 269–277.

Griffiths, G. and Linacre, S. (1995) Quality assurance for business surveys. In B.G. Cox, D. Binder, B. Chinnappa, A. Christianson, M. Colledge and P. Kott (eds), *Business Survey Methods*, pp. 673–690. New York: Wiley.

Groves, R.M. (1989) *Survey Errors and Survey Costs*. New York: Wiley.

Heckman, J.J. (1979), Sample selection bias as a specification error. *Econometrica*, 47, 153–161.

Hothum, C. and Spinting, S. (1977) Customer satisfaction research. In *Seminar on Marketing Management Information Systems. Organising Market Data for Decision Making*, pp. 853–890. Amsterdam: European Society for Opinion and Marketing Research.

Lee, S. (2006) Propensity score adjustment as a weighting scheme for volunteer panel web surveys. Journal of Official Statistics, 22, 329–349.

Kruskal, W. and Mosteller, F. (1979) Representative sampling, III: The current statistical literature. *International Statistical Review*, 47, 245–265.

Lessler, J.T. and Kalsbeek, W.D. (1992) *Nonsampling Error in Surveys*. New York: John Wiley & Sons, Inc.

Little, R.J.A. (1986) Survey nonresponse adjustments. International Statistical Review, 54, 139–157.

Nicolini, G. and Dalla Valle, L. (2009), Self-selected samples in customer satisfaction surveys. In *Book of Abstracts*, ESRA 2009 Conference, Warsaw.

Nicolini, G. and Dalla Valle, L. (2011) Errors in customer satisfaction surveys and methods to correct self-selection bias. *Quality Technology & Quantitative Management*, 8(2), 167–181.

Oh, H.L. and Scheuren, F.S. (1983) Weighting adjustments for unit nonresponse. In W.G. Madow *et al.* (eds), *Incomplete Data in Sample Surveys*, Vol. 2. New York: Academic Press.

Rao, J.N.K. (2003) *Small Area Estimation*. Hoboken, NJ: John Wiley and Sons, Inc.

Rosenbaum, P. and Rubin, D. (1983) The central role of the propensity score in observational studies for causal effects. *Biometrika*, 70(1), 41–55.

Rossi, P.E., Allenby, G.M. and McCulloch, R. (2005) *Bayesian Statistics and Marketing*. Hoboken, NJ: John Wiley and Sons, Inc.

Rubin, D.B. (1974) Estimating causal effects of treatments in randomized and nonrandomized studies. *Journal of Educational Psychology*, 66, 688–701.

Särndal, C.-E. and Lundström, S. (2005) *Estimation in Surveys with Nonresponse*. Chichester: John Wiley and Sons, Ltd.

Särndal, C.-E., Swensson, B. and Wretman, J. (1992) *Model Assisted Survey Sampling*. New York: Springer.

Schlomo, N., Skinner, C., Schouten, B., Bethlehem, J. and Zhang, L.C. (2009) *Statistical Properties of Representativity Indicators*. RISQ Deliverable, http://www.risq-project.eu.

Schouten, B., Cobben, F. and Bethlehem, J. (2009) Indicators for the representativeness of survey response. *Survey Methodology*, 35(1), 101–113.

Thomsen, I. (1973) A note on the efficiency of weighting subclass means to reduce the effects of nonresponse when analysing survey data. *Statistik Tidskrift*, 11, 278–283.

Thomsen, I. (1978) A second note on the efficiency of weighting subclass means to reduce the effects of nonresponse when analysing survey data. *Statistik Tidskrift*, 16, 191–196.

Trevisani, M. and Torelli, N. (2006) Comparing hierarchical Bayesian models for small area estimation. In B. Liseo, G. E. Montanari and N. Torelli (eds), *Metodi Statistici per l'Integrazione di Basi di Dati da Fonti Diverse*, pp. 17–36. Milan: Franco Angeli.

4

Measurement scales

Andrea Bonanomi and Gabriele Cantaluppi

This chapter considers the problem of measurement scales, which is an essential issue in every statistical analysis, in particular also in the analysis of customer satisfaction surveys. We begin by presenting the problem of scale construction and the different types of scales commonly used in customer satisfaction surveys, that is, the nominal, ordinal, interval and ratio scales. Then we consider the problem of obtaining an interval scale from an ordinal one; the resulting scale has a more informative characterization. In this context scaling methods are taken into account to obtain a unifying scale for a set of homogeneous items in a questionnaire.

4.1 Scale construction

Let $x_1, x_2, \ldots, x_n$ be the responses given by n individuals to a generic item in a questionnaire. Each response is chosen from set $\{s_1, s_2, \ldots, s_r\}$ of alternatives – work positions, gender, the date on which a service relationship began, the number of contacts that have taken place over a specified period of time, or degrees of agreement or satisfaction, the latter commonly measured on five- or seven-point Likert scales, where the minimum integer value represents 'complete dissatisfaction' or 'completely disagree' while the maximum value stands for 'complete satisfaction' or 'completely agree'. Such responses are very different in nature: some are qualitative statements of subjects' inner feelings or opinions, others are metric measures of an objective characteristic.

Observe that responses on a metric scale may also be affected by the presence of an inner evaluation on the part of the subject when they concern sensitive matters, e.g. when an individual is asked to state his income.

Responses may be classified according to the nature of the item they refer to. Stevens (1946) gives a four-way classification into nominal, ordinal, interval and ratio scales. We illustrate the characteristics of these types of scales, recalling that, usually, variables measured

Modern Analysis of Customer Surveys: with applications using R, First Edition. Edited by Ron S. Kenett and Silvia Salini.
© 2012 John Wiley & Sons, Ltd. Published 2012 by John Wiley & Sons, Ltd.

on a nominal or an ordinal scale are identified as categorical variables, while those measured on an interval scale or a ratio scale are said to be quantitative or metric variables.

Sometimes it is advisable to include multiple items, related to a single matter, in a survey in order to have a more reliable measure of a latent, non-observable concept. The reliability level of a conceptual construct, giving the strength of the relationship between the proposed items and the latent feature, may be assessed by means of the Cronbach's α coefficient. See Van Zyl *et al.* (2000) for the construction of a confidence interval for α and Zanella and Cantaluppi (2006) for the assumptions behind the Cronbach's α coefficient and for the construction of a test procedure for establishing the existence and the validity of a latent construct.

We observe that measurement scales are an important subject in the theory of measurement. Advances on the matter can be found in Hand (1996, 2009), which contain a comprehensive survey on the field and the relationship between the theory of measurement and statistics; the major approaches and their relationships are presented and different kinds of models within each theory described. Representational measurement theory, operational measurement theory and the classical and other theories of measurement are discussed. In particular, the first approach makes reference to theoretical work by Krantz *et al.* (1989), which defines a system of axioms and theorems intended to explain how some attributes of objects can reasonably be represented numerically. We will briefly refer to this approach in describing the measurement scales related to ordinal variables (see Section 4.1.2).

4.1.1   Nominal scale

The *nominal scale* represents the most unrestricted assignment of objects characteristics, in order to identify and cluster objects having the same category of the observed variable (Kinnear and Taylor, 1996).

Given two generic objects, a_1, a_2, we are only able to state a relationship of equivalence or similarity, in the sense that we can only recognize if a_1 is equal to a_2 ($a_1 \sim a_2$) or is different ($a_1 \not\sim a_2$).

The scale is defined through a set of r 'mutually exclusive categories', s_i, $i = 1, \ldots, r$, which may be identified according to the available responses.

Integers are sometimes conventionally associated to the categories of the scale; we observe that in this case the numbers only have identification and equivalence properties, thus losing their mathematical properties.

From a methodological point of view only the permutation transformation may be performed on a nominal scale (Krantz *et al.*, 1989). Sometimes a transformation is performed to aggregate two or more categories in order to avoid the occurrence/presence of categories with too small a frequency.

From a statistical point of view, the only summary statistics which may be applied to data measured on a nominal scale are the mode as a central tendency measure, the heterogeneity as a variability measure, and χ^2 to assess the conformity between two scales or of one scale to a particular distribution.

From an applications point of view the categorical attributes (e.g. gender, geographical areas, social classes) are typical examples of nominal scales in customer satisfaction surveys, but also situations may arise where numbers are used to describe the categories of a nominal scale like the code identifying a customer in a data warehouse.

Some authors, such as Krantz *et al.* (1989), do not consider nominal scales as measurement scales in the strict sense, since these do not satisfy an ordering relationship.

4.1.2 Ordinal scale

When data are measured on an *ordinal scale* it is possible to identify the objects considered by also taking into account their ranking, thus clustering the ordered objects which have in common the same category of the observed variable of interest.

Given three generic objects, a_1, a_2, a_3, we are able to state a relationship of equivalence, like for nominal scales, but also a relationship of ordering, in the sense that we can state if a_1 precedes a_2 ($a_1 \preceq a_2$) or a_1 follows a_2 ($a_1 \succeq a_2$), and a relationship of transitivity: if $a_1 \preceq a_2$ and $a_2 \preceq a_3$ then $a_1 \preceq a_3$.

Integer numbers are sometimes conventionally associated to the categories of the scale. Observe that in this case the numbers possess only identification, equivalence and ordering properties, and it is not possible to establish a distance between two categories.

Since objects are ordered we can establish a transformation $\varphi(\cdot)$ from the set $\mathcal{A}$, consisting of the objects $a_1, \ldots, a_n$, to a set of the ordered codes $\mathcal{S}$, which corresponds to a set of arbitrary, but ordered, numerical values defining the measurement scale:

$$\mathcal{A} \to \mathcal{S} = \{s_i, \ i = 1, 2 \ldots, r\} \subset \mathcal{Z}. \tag{4.1}$$

The values s_i are in a bijective relationship with a subset of $\mathcal{Z} = \{\ldots, -1, 0, 1, 2, \ldots\}$ satisfying the empirical ordering, which already exists among the responses. The ordering direction is completely arbitrary.

The representation and uniqueness result [see Krantz *et al.*, 1989, Theorem 1, p. 15] states the existence of a function $\varphi(\cdot)$ establishing relationship (4.1):

$$a_i \succeq a_j \quad \text{iff} \quad \varphi(a_i) \geq \varphi(a_j), \quad i, j = 1, \ldots, r.$$

The assignment of numbers through $\varphi(\cdot)$ is indeterminate, in the sense that an ordinal scale is obtained even if the numbers s_i are transformed by means of a monotone increasing function $f(\cdot)$, that is,

$$\varphi'(a_i) = f[\varphi(a_i)], \quad i = 1, \ldots, r;$$

the transformation $\varphi'(\cdot)$ establishes a mapping from $\mathcal{A}$ to $\mathcal{S}'$. Observe that the two sets $\mathcal{S}$ and $\mathcal{S}'$ are related by means of a bijective relationship only when $f(\cdot)$ is a strictly monotone function; the same ordering in the two sets is preserved in a weak sense.

A set of ordered codes, $s_1, \ldots, s_r$, thus stems from the r categories that characterize the item. The set may be constructed, e.g. for an item on education, by means of the following procedure defining an ordinal scale.

An arbitrary code, say a numerical value s_1, is assigned to the first object. If the category of the second object succeeds (precedes) that of the first, the second object is assigned a numerical value greater (lower) than s_1. This step is repeated until a value is assigned to all objects. When two objects have identical categories, they are considered equivalent and are assigned the same numerical value on the scale.

From a methodological point of view the permutation transformation cannot be applied. Sometimes a transformation is made which aggregates two or more adjacent categories. This can justify the unique direction of the relationships between $\mathcal{A}$ and $\mathcal{S}$ in (4.1).

An ordinal scale may be characterized by an even or an odd number r of categories; an indifference or neutral point is present when r is odd, corresponding to the median position on the scale.

Summary statistics which may be applied to an ordinal scale are the mode and the percentage points as central tendency indices; the heterogeneity as a variability index, and χ^2 to measure the conformity between two scales or between a scale and a distribution; non-parametric statistics like the *Spearman rank-order correlation coefficient* or the sign test may be used to measure or verify the presence of association between two scales.

A proper definition of the percentage point like the one proposed by David (1981) has to be applied, whenever the observed categories of the scale are not numeric measures; that is

$$
x_p = \begin{cases} x_{\min} & \text{if } p = 0 \\ x_{([np]+1)} & \text{if } 0 < p < 1 \\ x_{\max} & \text{if } p = 1, \end{cases}
$$

where p is the percentage order, $x_{(k)}$ is the element at position k in the ordered series of the n observed results and $[\cdot]$ denotes the integer part function, and $x_{\min} = x_{(1)}$ and $x_{\max} = x_{(n)}$ are respectively the smallest and the largest observed values.

The scales commonly used in customer satisfaction surveys are typical examples of ordinal scales. Ordered categorical attributes find widespread application in problems of measurement stemming in the social sciences. Evaluations stated in those fields have a qualitative and *intangible* nature and a measurement problem ensues (Bonanomi 2004): how can we measure something that is latent, not directly measurable, like the level of satisfaction with some dimension of a service? An adequate and appropriate construct for a qualitative issue, based on empirical relationships, can be difficult to establish. A relationship between the measure and the corresponding construct has to be stated to define a measurement scale for a qualitative issue.

The practical procedure to construct an ordinal measurement scale to evaluate social and behavioral judgments takes usually into consideration responses and evaluations from subjects. Each subject is required to judge if he perceives two or more stimuli in a equal or a different manner and he has possibly to order his perceptions. A different transformation function $\varphi(\cdot)$ may exist for each subject and thus the statistical formulation of a measurement model is difficult if not impossible. Usually the following assumption is made [see Zanella 2001, pp. 7–8] which overcomes this problem: the evaluations expressed by each subject can be considered properly ordered; we observe that this assumption resembles the 'pragmatic' measurement approach described by Hand (2009).

The subjective nature of the evaluations clearly represents a major drawback in the scale definition. As previously stated, only relationships of equivalence and ordering may be defined for subjective statements; the ordinal nature of the scale follows.

Observe that since in customer satisfaction surveys most items refer to subjective evaluations, the resulting scales are of an ordinal type. Since some techniques of analysis require metric scales, appropriate transformations should be applied in order to make possible the analysis of the available scales; see Section 4.2.

4.1.3 Interval scale

We can use an interval scale whenever we have something that is empirically measurable. The objects of interest are identified, clustered and assigned a metric value; the elements, so built, have in common the same outcome of the observed variable; moreover, it is possible to define a distance between the objects, the points on the scale usually being equally spaced (Kinnear and Taylor 1996). However, ratios on values defined by an interval scale cannot be defined.

The zero point on an interval scale is a matter of convention or convenience, as is shown by the fact that the scale form remains invariant when a constant is added. (Stevens 1946)

Given two objects, a_1, a_2, it is possible to state relationships of equivalence, as for the preceding scales, ordering, transitivity and distance $d(a_1, a_2) = |a_1 - a_2|$. Distances between the objects on an interval scale are defined on a ratio scale (see Section 4.1.4), since in this case a distance of zero means equivalence of the objects and consequently the zero point is not a matter of convention.

From a methodological point of view the permutation transformation cannot be applied. All other transformations can be applied only if they do not alter the internal distances among objects. So affine and linear transformations may be applied, while power and monotone transformations may not.

The most common summary statistics, like the percentage points and the arithmetic mean, may be applied to an interval scale as central tendency indices; the standard deviation and the heterogeneity as variability indices, χ^2 and the Kolmogorov–Smirnov statistic to measure the conformity between two scales or between a scale and a distribution; the Pearson correlation statistic, which is defined on standardized variables, may be used to measure the association between two scales.

Physical measurements, which obviously are of capital importance in modern society, are based on the definition of a pertinent measurement unit for the quantity of interest and a corresponding measurement procedure allowing to assess of how many units is composed the quantity in the examined case and leading to a number that expresses its measure. Typical examples of interval scales are temperature (Celsius and Fahrenheit) and, in customer satisfaction surveys, dates, e.g. the beginning of a customer relationship.

Original interval scales are not commonly encountered in surveys; however, proper transformations may be adopted to convert ordinal scales into interval ones (see Section 4.2).

Most psychological measurement aspires to create interval scales, and it sometimes succeeds. The problem usually is to devise operations for equalizing the units of the scales – a problem not always easy of solution but one for which there are several possible modes of attack. Only occasionally is there concern for the location of a 'true' zero point, because the human attributes measured by psychologists usually exist in a positive degree that is large compared with the range of its variation. (Stevens 1946)

Observe that ordinal scales are sometimes treated as if they were of interval type by invoking the 'pragmatic' measurement approach, according to which 'the precise property being measured is defined simultaneously with the procedure for measuring it', under the assumption of explicitly defining the meaning of the concept one is measuring (see Hand 2009).

4.1.4   Ratio scale

We can apply a ratio scale to an attribute that is empirically measurable. The objects of interest are identified and assigned a metric value corresponding to the possessed level of the observed variable. It is possible to define a distance between the objects. The main difference with respect to the interval scale is that in the case of a ratio scale the zero point is not a matter of convention or convenience (Kinnear and Taylor 1996). It is possible to uniquely define the ratio between the measures of two objects which are (mathematically) similar.

In presence of two objects, a_1, a_2, it is possible to state relationships of equivalence, as for the preceding scales, ordering, transitivity and distance $d(a_1, a_2) = |a_1 - a_2|$. Moreover, ratios on values defined as ratio scales are defined.

From a methodological point of view the permutation transformation cannot be applied. With regard to the other transformations (affine, power, linear and monotone) only the linear one can be applied since it does not alter the zero and the internal distances among objects. When the scale has only positive values the logarithm transformation may also be applied, allowing a different comparison of object measurements in relative terms.

All statistical methods may be applied to values defined on a ratio scale, bearing in mind, of course, their meaning and their utility to interpret the problem under study.

Physical measures are typical examples of ratio scales and, in customer satisfaction surveys, all measures of counts and the distances between two dates, e.g. the length of a customer relationship, are defined on a ratio scale; but not some economic variables (e.g. the utility, which is measured on an ordinal scale, since it is defined according to the Krantz assumptions regarding the qualitative ordering of objects (Varian 1992)).

> It is conventional in physics to distinguish between two types of ratio scale: *fundamental* and *derived*. Fundamental scales are represented by length, weight, and electrical resistance, whereas derived scales are represented by force, density and elasticity. (Stevens 1946)

We observe that the derived scales density and elasticity are commonly also used in statistics and economics.

4.2   Scale transformations

In Section 4.1 the properties of the measurement scales have been reviewed within the axiomatic theory and the scales have been characterized as mutually exclusive. Measurement scales may be ordered according to their informative power: the nominal scale is the least informative, the ratio scale the most informative. All kinds of statistical indices are allowed only for the ratio scale. Most in customer satisfaction measurements have only an ordinal nature, as we have already recalled.

We will now consider the scale transformation methods that have been proposed in the psychometric, sociological, econometric and statistical literature in order to obtain an interval scale from a given ordinal scale. The resulting measures can be treated in a more complete and exhaustive way from a statistical point of view.

These transformations may refer only to each individual item, with the primary aim of studying the association among them (Bartholomew 1996; Bollen 1989; Maddala 1983), or to the whole set of items (Jones 1986; Thurstone 1959), where the aim is to obtain a common scale transformation. The following subsections are devoted to the illustration of these two approaches.

The framework common to the two approaches is to suppose that we observe a K-dimensional random categorical variable $\mathbf{X} = (X_1, X_2, \ldots, X_K)'$, whose components assume the same number of ordered categories I, denoted for simplicity by the conventional integer values $x_{ki} = 1, 2, \ldots, I, k = 1, 2, \ldots, K$.

Let $P(X_k = i) = p_{ki}$, with $\sum_{i=1}^{I} p_{ki} = 1$, for all k, be the corresponding marginal probabilities and let

$$F_k(i) = \sum_{j \leq i} p_{kj} \tag{4.2}$$

be the cumulative probability of observing a conventional value x_k for X_k not larger than i.

Furthermore, assume that to each categorical variable X_k there corresponds an unobservable latent variable Z_k, which is represented on an interval or a ratio scale, with a continuous distribution function:

$$P\left(\frac{Z_k - \mu_k}{\sigma_k} \leq \frac{\xi_k - \mu_k}{\sigma_k}\right) \equiv \Psi_k\left(\frac{\xi_k - \mu_k}{\sigma_k}, \boldsymbol{\alpha}\right), \tag{4.3}$$

where ξ_k is a generic quantile of Z_k and μ_k and $\sigma_k > 0$ are respectively a location and a scale parameter, which may depend on some other real positive parameters $\alpha_1, \ldots, \alpha_K$, summarized by vector $\boldsymbol{\alpha}$: $\mu_k = \mu_k(\boldsymbol{\alpha})$, $\sigma_k = \sigma_k(\boldsymbol{\alpha})$; that is, for any but fixed $\boldsymbol{\alpha}$ a location–scale family of distributions is assumed; when no ambiguity arises we shall use the simplified notation μ_k, σ_k.

4.2.1   Scale transformations referred to single items

In the case of transformations related to every single item, the continuous K-dimensional latent random variable $(Z_1, Z_2, \ldots, Z_K)$ is usually assumed multinormally distributed. A non-linear monotone function is defined to relate the observed ordinal indicators $(X_1, X_2, \ldots, X_K)$ to the latent continuous ones; for example, with reference to the generic X_k, $k = 1, 2, \ldots, K$, we have [see Bollen 1989]:

$$X_k = \begin{cases} 1 & \text{if } Z_k \leq a_{k,1} \\ 2 & \text{if } a_{k,1} < Z_k \leq a_{k,2} \\ \vdots & \\ I_k - 1 & \text{if } a_{k,I_k-2} < Z_k \leq a_{k,I_k-1} \\ I_k & \text{if } a_{k,I_k-1} < Z_k, \end{cases} \tag{4.4}$$

where $a_{k,1}, \ldots, a_{k,I-1}$ are marginal threshold values defined as

$$a_{k,i} = \Psi^{-1}(F_k(i)), \quad i = 1, \ldots, I_k - 1,$$

$\Psi(\cdot)$ being the cumulative distribution function of a random variable.

Observe that the thresholds $a_{k,1}, \ldots, a_{k,I_k-1}$ are particular values of the quantiles ξ_k, defined in relationship (4.3).

It is usually assumed that Z_k is distributed according to a standard normal random variable (Thurstone 1959), for which conventionally $\Phi(\cdot) \equiv \Psi(\cdot)$ and the $a_{k,i}$ are assumed to satisfy the following condition:

$$\max(-3, \Psi^{-1}(F_k(i))) \leq a_{k,i} \leq \min(3, \Psi^{-1}(F_k(i)))$$

possibly assuming that $a_{k,i} = -3$ or $a_{k,i} = 3$; $I_k \leq I$ depending on the categories effectively used by the respondents ($I_k = I$ when each category has been chosen by at least one respondent).

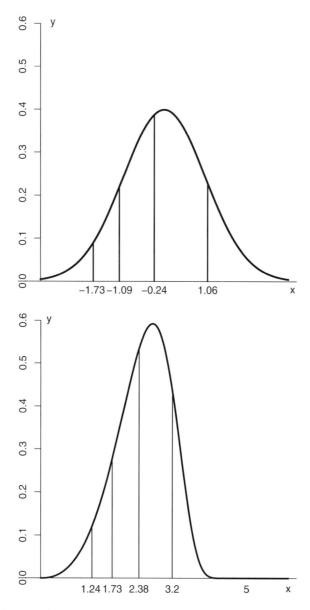

Figure 4.1   Definition of the threshold values for the item 'Overall satisfaction level with ABC' in the ABC questionnaire, corresponding to a normal latent variable (top) and to a logistic Weibull latent variable (bottom)

Figure 4.1 shows the definition of the threshold values for the item 'Overall satisfaction level with ABC' in the ABC questionnaire that was proposed by the editors for the customer satisfaction analysis. The item is characterized by 262 valid cases over the categories 1, ..., 5. In Table 4.1 the categories i, the counts n_i and the relative and cumulative frequencies, f_i and F_i, with the corresponding quantiles $a_{k,i}$ for the normal distribution are reported.

Table 4.1   Computation of the normal and logistic Weibull distribution quantiles, $_N a_{k,i}$ and $_{LW} a_{k,i}$, for the item 'Overall satisfaction level with ABC'

i	$n_k(i)$	$f_k(i)$	$F_k(i)$	$_N a_{k,i} = \Phi^{-1}(F_k(i))$	$_{LW} a_{k,i} = \Psi_k^{-1}(F_k(i))$
1	11	0.042	0.042	−1.728	1.239
2	25	0.095	0.137	−1.092	1.732
3	70	0.267	0.405	−0.242	2.384
4	118	0.450	0.855	1.058	3.198
5	38	0.145	1.000		

Jöreskog (2005) and Bartholomew (1996) justify the use of the normal distribution since any continuous variable with a density and a distribution function can be transformed by a monotone transformation to a normal distribution.

Zanella (1998, 1999) proposed using a latent variable with a logistic Weibull distribution:

$$\Psi_k(x) = 1 - \exp\left\{-\left[\frac{\gamma_k}{\alpha_k}\ln\left(\frac{\tau_0 + x}{\tau_0 - x}\right)\right]^{\alpha_k}\right\} = F_k(i), \tag{4.5}$$

$\alpha_k > 0$, $\gamma_k > 0$, with conventional values 0 if $x \leq 0$ and 1 if $x \geq \tau_0$, based on the limits for $x \to 0^+$, $x \to \tau_0^-$. This distribution has a finite support and allows one also to take into account the asymmetry which usually is present in customer satisfaction survey responses. Figures 4.2 and 4.3 give examples of plots of relationship (4.5) and of the corresponding probability densities for $\alpha_k = 1.8, 2.72$, $\gamma_k = 2$, $\tau_0 = 5$.

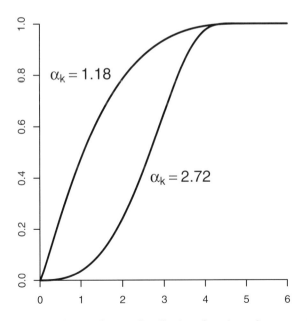

Figure 4.2   Logistic Weibull cumulative distribution functions for $\mu_k = 0$, $\sigma_k = 1$, $\alpha_k = 1.8, 2.72$

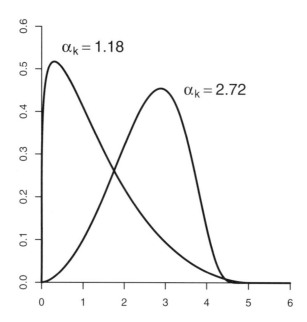

Figure 4.3   Logistic Weibull cumulative density functions for $\mu_k = 0$, $\sigma_k = 1$, $\alpha_k = 1.8, 2.72$

With reference to the 'Overall satisfaction level with ABC' item in the ABC questionnaire, in the last column of Table 4.1 the quantiles for the logistic Weibull distribution are reported, which may be obtained by minimizing, with respect to the thresholds and to the parameters α_k and γ_k, the statistic

$$D = \sum_{i=1}^{I} \frac{\{f_k(i) - [\Psi_k(i) - \Psi_k(i-1)]\}^2}{\Psi_k(i) - \Psi_k(i-1)},$$

where $\Psi_k(0) = 0$; the statistic is based on the χ^2 distance. We observe that other criteria such those reported in D'Agostino and Stephens (1986, Ch. 4) (see also Frosini (1978)) may be followed. The corresponding estimated density function is represented in Figure 4.1 for $\hat{\alpha} = 3.47$ and $\hat{\gamma} = 2.77$.

For two generic ordinal categorical variables X_h and X_k, $h, k \in 1, 2, \ldots, K$, it is possible to calculate the *polychoric* correlation, which is defined to be the value of ρ maximizing the log-likelihood typically conditional on the marginal threshold estimates

$$\sum_{i=1}^{I} \sum_{j=1}^{I} n_{ij} \ln(\pi_{ij}),$$

where n_{ij} is the number of observations for the ith category of X_h and jth category of X_k, and

$$\pi_{ij} = \Phi_2(a_{h,i}, a_{k,j}) - \Phi_2(a_{h,i-1}, a_{k,j}) - \Phi_2(a_{h,i}, a_{k,j-1}) + \Phi_2(a_{h,i-1}, a_{k,j-1}),$$

$\Phi_2(\cdot)$ being the standard bivariate normal distribution function with correlation ρ, and $a_{h,i}$, $a_{k,j}$ the threshold values for X_h and X_k, respectively defined with respect to the two marginal latent standard normal variates, $a_0 = b_0 = -\infty$ and $a_I = b_I = +\infty$. The procedure may be performed using the function `tetrachoric` available in the R library `psych` (Revelle 2011) or the function `polychor` in the library `polycor` (Fox 2010).

Sometimes it is useful to reduce the expressed scores into only two categories, e.g. 'low' and 'high' levels with respect to the items of interest; the corresponding transformation (see (4.4)) is characterized by only one threshold value and the polychoric correlation, which is now referred to two dichotomous variables, reduces to the *tetrachoric* correlation.

Given a K-dimensional random ordinal categorical variable $\mathbf{X} = (X_1, \ldots, X_K)'$, both the polychoric and the tetrachoric correlation matrices, whose generic elements are the proper correlations between the generic ordinal variables X_h and X_k, $h, k \in \{1, 2, \ldots, K\}$, may be a suitable starting point for implementing a structural equation model with latent variables of the LISREL type (Bollen 1989; Jöreskog 2005).

The threshold values for the generic ordinal categorical variable X_h may be also determined, according to an ordered logistic or an ordered probit regression model, with the parameters of a generalized linear model explaining the behaviour of X_h as a function of a set of P exogenous observed variables $(W_1, \ldots, W_P)$, which can be of a metric type or dummy variables recoding the categories of a variable measured on a nominal scale (Bartholomew 1996; Verbeek 2008).

We wish to drawn attention to considerations concerning the expression of possibly different scores stated by different subjects with regard to ordinal scales.

The above assumptions refer to a common probability distribution for all the subjects and, as we have seen, the definition of transformation techniques considers the empirical distribution function, which takes into account the behaviour of all the subjects: in presence of a group of subjects, it is thus assumed that they all have the same way of expressing their judgements; this assumption may theoretically refer to the ideal subject of the group.

We have to observe that intrinsically personal features may be present that characterize the tangible formulation of a latent judgement. Two subjects may have different functions $\varphi_1(\cdot), \varphi_2(\cdot)$ to assess their evaluations. This fact is consistent with the indeterminacy of the scale transformations stated in the second part of the Krantz theorem, see Section 4.1.2 and Krantz *et al.* (1989, p. 15).

Figure 4.4 shows the transfer functions of two subjects: one (A) with a propensity to give scores higher than the other (B). On the abscissa the theoretical values of the latent variable defining their internal mental judgements, from 'very unsatisfied' to 'very satisfied', are reported; this latent variable, we recall, is of a metric type. Their expressed scores, discrete in nature, are on the ordinates. We note that for subject B the observed scores $s_1 = 3$ and $s_2 = 4$ correspond to the two mental states $m_1 < m_2$ (referred to two ordered objects $a_1 \prec a_2$); while for subject A a discrete score of 7 corresponds to the same $m_1 < m_2$ state. That is to say, the latent variables for the two subjects are associated with different cumulative distribution functions – the latent variables are of different types or may belong to the same family but with different parametrizations. As previously observed, the statistical analysis is based on the evaluations of the subjects and not on the ordering of the objects they evaluate (Zanella 2001); but it is worthwhile to take into account the different behaviour of the subjects in a statistical model in order to allow the segmentation of statistical units and to study the heterogeneity of their response functions.

In order better to take into account the response behaviour characterizing different groups of subjects, we suggest the use of an ordered logistic or probit, or a logistic Weibull, regression model in the context of a multilevel model, thus establishing not only different intercepts and coefficient values, but also different threshold values for the different groups. This suggestion

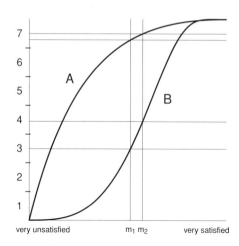

Figure 4.4   Transformation functions of two subjects with different behaviour

might go some way towards addressing the previous observation, in the sense that a unique latent family is used to link the latent scores with the observed ones but with possibly different parametrizations for different groups of subjects.

4.2.2   Scale transformations to obtain scores on a unique interval scale

To each categorical variable[1] X_k, $k = 1, \ldots, K$, describing the generic element in a set of items measuring a common latent trait (see Section 4.1) there corresponds an unobservable latent variable Z_k, represented on an interval or ratio scale. In particular, with reference to relationship (4.3), the standardized quantiles ζ_k may be defined:

$$\zeta_k = \frac{\xi_k - \mu_k}{\sigma_k}, \tag{4.6}$$

$k = 1, 2, \ldots, K$, which are real variables describing the possible values on the right-hand side, which for a fixed α are assumed to be invariant for any scale–location change corresponding to a linear transformation of the type

$$\tilde{\xi}_k = m\xi_k + q, \tag{4.7}$$

with q and m real positive values, where m expresses the change of unit measurement of ξ_k and q a displacement of the origin of measurements. Observe that the new measurement system implies that

$$\tilde{\mu}_k = m\mu_k + q, \quad \tilde{\sigma}_k = m\sigma_k, \tag{4.8}$$

where we assume that μ_k, σ_k are expressed in the same unit as ξ_k and, in addition, the positive quantities σ_k do not depend on the origin of measurements. Thus, according to (4.7), it follows from (4.6) that

$$\frac{(\tilde{\xi}_k - q)/m - \mu_k}{\sigma_k} = \frac{\tilde{\xi}_k - (m\mu_k + q)}{m\sigma_k} = \frac{\tilde{\xi}_k - \tilde{\mu}_k}{\tilde{\sigma}_k} = \frac{\xi_k - \mu_k}{\sigma_k}$$

---

[1] The methodological presentation in this section is based on Zanella and Cantaluppi (2004).

as a consequence of (4.8), which proves the invariance to linear transformations, such as the scale–location one. Correspondingly, for fixed $\boldsymbol{\alpha}$, we can identify $\Psi_k(\zeta_k, \boldsymbol{\alpha})$, in (4.3), with the standard element of the family defined by putting $\tilde{\mu}_k = 0$, $\tilde{\sigma}_k = 1$, i.e. $q = -\mu_k/\sigma_k$, $m = 1/\sigma_k$, in (4.8).

By inversion of (4.3) and with regard to (4.2) by equating the second member to $F_k(i)$, define

$$\frac{\xi_{ki}(\boldsymbol{\alpha}) - \mu_k(\boldsymbol{\alpha})}{\sigma_k} = \Psi_k^{-1}[F_k(i), \boldsymbol{\alpha}] = \zeta_{ki}, \tag{4.9}$$

with a suitable conventional or approximate definition of ζ_{ki} for $F_k(i) = 0$ or 1 where necessary. Recall that the ζ_{ki} are related, for each k, to the threshold values defined in (4.4).

Later on we shall examine the case where $\xi_{ki} = \xi_i$, for all k, that is, there exists a unique set of $I - 1$ threshold values ξ_i, $i = 1, \ldots, I - 1$, corresponding to the categories $i = 1, \ldots, I$, regardless of the random variable X_k. With this assumption (4.9) becomes

$$\frac{\xi_i(\boldsymbol{\alpha}) - \mu_k(\boldsymbol{\alpha})}{\sigma_k} = \Psi^{-1}[F_k(i), \boldsymbol{\alpha}] = \zeta_{ki}, \tag{4.10}$$

$k = 1, 2, \ldots, K$, $i = 1, 2, \ldots, I - 1$. Figure 4.5 illustrates, for the logistic Weibull case, the interpolation of the cumulative probabilities $F_k(i)$ – which can be estimated by means of the observed cumulative proportions – obtained from the distribution function of the standard element of the assumed family (4.3) of continuous latent distributions and the corresponding thresholds ξ_i representing the categories i, $i = 1, 2, \ldots, I$, on an interval scale given by the inversion formula (4.10). As an example we consider the real case presented by Jones (1986), who examines the conventional preference scores, ranging from 1 to 9, which we might as well interpret as degree of customer satisfaction, obtained from a sample of 255 army enlisted men with respect to 12 food items, or categories, $k = 1, 2, \ldots, 12$, $i = 1, 2, \ldots, 9$,

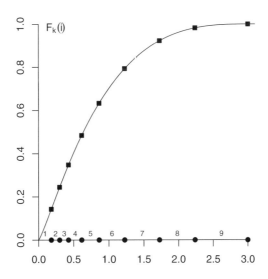

Figure 4.5   Transforming the conventional scores i into the threshold values ξ_i on a unique interval scale by inverting the cumulative logistic Weibull distribution function of the unitary element of the latent family of distributions – equation (4.5) for $\alpha_k = 1.18$, $\gamma_k = 2$ and $\tau_0 = 3$

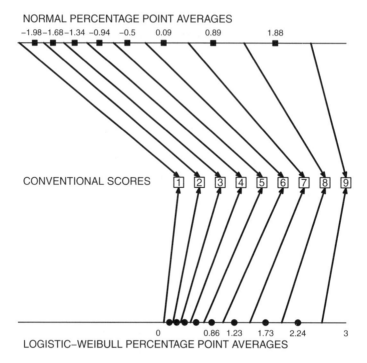

Figure 4.6 Relationship between the threshold values ξ_i on an interval scale and the conventional scores i

and treated by the author assuming 12 underlying normal distributions. Theorem 1 in Zanella and Cantaluppi (2004) shows that if (4.10) is valid, the unique thresholds ξ_i can be defined as $\xi_i = \sum_{k=1}^{K} \zeta_{ki}/K = \bar{\zeta}_i$, $i = 1, 2, \ldots, I = 8$, which are the arithmetic means respectively obtained for each category i by taking the average of the theoretical values ζ_{ki}, $k = 1, 2, \ldots, K$, over the $K = 12$ categorical variables.

The quoted Theorem establishes that a necessary condition so that the former structural relationships among percentiles are satisfied is that $\sum_k 1/(K\sigma_k) = 1$ and $\sum_k \mu_k/(K\sigma_k) = 0$. Figure 4.6 illustrates graphically the relationship between the eight threshold values $\xi_i = \bar{\zeta}_i$ and the conventional scores i, $i = 1, 2, \ldots, 9$, with reference to Jones's example (see also Table 4.2); it is supposed that the relationship holds for any categorical variables X_k, $k = 1, 2, \ldots, K$.

The unifying thresholds were computed by assuming both that the underlying latent model was a normal distribution and that it belonged to logistic Weibull family of distributions (see

Table 4.2 Percentage point (threshold) averages related to class i, $i = 1, 2, \ldots, 8$, for the latent variables referring to the normal and logistic Weibull distribution functions

	1	2	3	4	5	6	7	8
Normal	−1.979	−1.683	−1.335	−0.936	−0.495	0.089	0.886	1.876
Logistic Weibull	0.181	0.300	0.428	0.615	0.862	1.225	1.725	2.236

(4.5)) with $\gamma = 2$ and $\tau_0 = 3$ for $x = (\xi_{ki} - \mu_k)/\sigma_k$ (Zanella and Cantaluppi 2004). Observe that in both cases the inner distances between the consecutive thresholds are not constant. The Logistic Weibull distribution takes also into consideration the possible asymmetry behaviour present in the data.

Acknowledgements

We warmly thank Prof. A. Zanella for having introduced us to this topic and for his continuous valuable and stimulating encouragement.

References

Bartholomew, D.J. (1996) *The Statistical Approach to Social Measurement*. San Diego, CA: Academic Press.

Bollen, K.A. (1989) *Structural Equations with Latent Variables*. New York: John Wiley & Sons, Inc.

Bonanomi, A. (2004) Variabili ordinali e trasformazioni di scala, con particolare riferimento alla stima dei parametri dei modelli interpretativi con variabili latenti. PhD thesis, Università degli Studi di Milano Bicocca, Milan.

D'Agostino, R.B. and Stephens, M.A. (1986) *Goodness-of-Fit Statistics*. New York: Marcel Dekker.

David, H.A. (1981) *Order Statistics*, 2nd edn. New York: John Wiley & Sons, Inc.

Fox, J. (2010) polycor: Polychoric and Polyserial Correlations. R package version 0.7-8. Available at http://CRAN.R-project.org/package=polycor.

Frosini, B.V. (1978) A survey of a class of goodness-of-fit statistics. *Metron*, 36, 3–49.

Hand, D.J. (1996) Statistics and the theory of measurement. *Journal of the Royal Statistical Society, Series A*, 159(3), 445–492.

Hand, D.J. (2009) *Measurement Theory and Practice: The World through Quantification*. Chichester: Wiley.

Jones, L.V. (1986) Psychological scaling. In S. Kotz and N.L. Johnson (eds), *Encyclopedia of Statistical Sciences*, Vol. 7, pp. 340–343. New York: John Wiley & Sons, Inc.

Jöreskog, K.G. (2005) Structural equation modeling with ordinal variables using LISREL. Available at http://www.ssicentral.com/lisrel/techdocs/ordinal.pdf.

Kinnear, T.C. and Taylor, J.R. (1996) *Marketing Research: An Applied Approach*, 5th edn. New York: McGraw-Hill.

Krantz, D., Luce, R.D., Suppes, P. and Tversky, A. (1989) *Foundations of Measurement, Vol. I: Additive and Polynomial Representations*. New York: Academic Press.

Maddala, G.S. (1983) *Limited-Dependent and Qualitative Variables in Econometrics*. Cambridge: Cambridge University Press.

Revelle, W. (2011) Package 'psych': Procedures for personality and psychological research. R package version 1.0-95. Available at http://personality-project.org/r/psych.manual.pdf.

Stevens, S.S. (1946) On the theory of scales of measurement. *Science*, 103, 677–680.

Thurstone, L.L. (1959) *The Measurement of Values*. Chicago: University of Chicago Press.

Van Zyl, J.M., Neudecker, H. and Nel, D.G. (2000) On the distribution of the maximum likelihood estimator of Cronbach's alpha. *Psychometrika*, 65, 271–280.

Varian, H.R. (1992) *Microeconomic Analysis*, 3rd edn. New York: Norton.

Verbeek, M. (2008) *A Guide to Modern Econometrics*, 3rd edn. Chichester: John Wiley & Sons, Ltd.

Zanella, A. (1998) A statistical model for the analysis of customer satisfaction: Some theoretical and simulation results. *Total Quality Management*, 9, 599–609.

Zanella, A. (1999) A stochastic model for the analysis of customer satisfaction: Some theoretical aspects. *Statistica*, 98(LIX), 3–40.

Zanella, A. (2001) *Valutazione e modelli interpretativi di customer satisfaction: una presentazione di insieme*. Serie E.P. N. 105, Istituto di Statistica. Università Cattolica del Sacro Cuore, Milano.

Zanella, A. and Cantaluppi, G. (2004) Simultaneous transformation into interval scales for a set of categorical variables. *Statistica*, 103(LXIV), 401–426.

Zanella, A. and Cantaluppi, G. (2006) Some remarks on a test for assessing whether Cronbach's coefficient α exceeds a given theoretical value. *Statistica Applicata*, 18, 251–275.

5

Integrated analysis

Silvia Biffignandi

This chapter is about integrating data in a business data system, especially with regard to a marketing integrated system. Different data source typologies are discussed. Secondary sources, broken down into internal and external sources, are considered and advantages and problems in data integration are illustrated. Basic principles concerning root cause analysis as a methodology for processing integrated data is introduced. In this context, root cause analysis plays an important role in studying customer satisfaction within a marketing integrated system. Among various customer satisfaction problems, it is shown that root cause analysis can be also useful in providing an integrated analysis of customer and employee satisfaction.

5.1   Introduction

As organizations become increasingly information driven, their boundaries have broadened. Recent trends reveal that most competitive businesses require external data in their enterprise data warehouse to position themselves strategically in the market. The integration of data from external sources, especially transactional data, is full of challenges. Such integrated models have been proposed in Rucci *et al.* (1998), Kenett (2004) and Godfrey and Kenett (2007).

The universal use of XML, as the common syntax for message exchange, and the use of Hypertext Transfer Protocol as a common transport protocol, provides enterprises with the ability to commodify application technology integration within an enterprise. Businesses and financial data are communicated electronically using an extended business reporting language (XBRL) which is revolutionizing business reporting. These new technologies permit organizations to collaborate more easily with internal as well as external business partners and data sources (Kenett and Raanan, 2010).

Modern Analysis of Customer Surveys: with applications using R, First Edition. Edited by Ron S. Kenett and Silvia Salini.
© 2012 John Wiley & Sons, Ltd. Published 2012 by John Wiley & Sons, Ltd.

These technical abilities enable business institutions to both tactically optimize their margins and strategically evolve their business models.

This chapter will show that technological expertise is not sufficient for achieving competitive advantage. In order to be competitive, businesses need to create a well-formed business model, integrate sources into a coherent and high-quality data environment, and undertake relevant statistical analyses. The integrated database generated should then be kept up to date and maintained by checking the obsolescence of concepts and definitions, and comparability over time.

On this basis, we focus on the data sources, data and analyses that characterize an integrated business system. We do not consider the entire integrated system, which requires a more extended study. We restrict our attention to the marketing information system (MIS), which is related to customer satisfaction issues and, indirectly, to financial performance. The impact of integrated data on information quality is evaluated in Kenett and Shmueli (2011). For more on integrating data sources, including extract–transform–load and semantic analysis, see Kenett and Raanan (2010).

An MIS is a structure, within an organization, designed to gather, process and store data from the organization's external and internal environment and to disseminate it in the form of information to the marketing decision-makers. Updated state-of-the-art marketing data management roadmaps include new roles and processes to sustain progress with increasing demand for data.

The understanding behind such a roadmap is that if one can measure a situation, one can manage it.

Better data means that focused attention is given to the need for enhanced data quality, data integration, and measurement and analysis systems.

Data integration is a central and key issue that involves data source selection as a preliminary step for integrated database construction. Once data integration has been achieved, further steps involve data processing and statistical analysis within the given information framework.

Data-driven decisions and evidence-based management are becoming essential because of:

- the intensification of competition;

- problems with data, systems and processes that arise in post-merger systems integration;

- reduction in customer satisfaction, customer loyalty, and profit growth;

- a growth in client expectations.

The activities performed by an MIS include: information discovery, data collection and integration, statistical analysis and interpretation.

Section 5.2 focuses on issues related to sources of information for data integration. Source classification type and related problems are briefly discussed, and special attention is paid to secondary source problems. Section 5.3 deals with the question of how to analyse integrated data for marketing and customer satisfaction purposes using root cause analysis methodology.

5.2   Information sources and related problems

5.2.1   Types of data sources

Sources of information are classified as primary and secondary sources (Green *et al.*, 1988; Kenett and Shmueli, 2011). *Primary data* is collected for specific research purposes. *Secondary data* is data originally collected by individuals or agencies for purposes other than those of the research study. For example, if a government unit conducted a survey on markets of various shoe brands, then a shoe manufacturer might use this data to evaluate the potential market for a new product. The assembly and analysis of secondary data almost invariably improves the researcher's understanding of the marketing problem. Such a study helps elicit the various lines of inquiry that could or should be followed, and the alternative courses of action that might be pursued.

Effective market research typically begins with a thorough search through secondary data sources. Secondary sources of information can be divided into two categories (Dillon *et al.*, 1994; Kenett and Shmueli, 2011): internal and external. Internal secondary sources rely on statistical data a business has collected for some purpose, typically different form the current research. Such data might provide interesting information even if collected in a different decision context. External sources are, for example, official statistics as well as data provided from marketing research societies and other organizations. Secondary data is collected and published independently of the specific problem the firm is facing.

An integrated data system generates competitive advantage and benefits since information related to customers (such as customer satisfaction) can be evaluated and analysed taking into account market research studies and, conversely, market research studies can be set in the context of information on customer satisfaction. Often a search through secondary sources (also termed *desk research*) is done.

Many businesses have low expectations on the usefulness of secondary sources. There are two main reasons for this. The first is that the topic under study is often considered so unique, or specialized, that research involving secondary sources, especially external, seems futile. This is based on the assumption that no one else has made the effort to collect relevant data and publish it. Market researchers who seek out useful secondary data are often surprised by its abundance. Thus, the problem usually becomes how to properly select secondary data.

The second reason is that in most businesses the people involved in building the integrated system (and thus selecting the data to be included and its sources) and people who might be interested in using such information are different.

These two factors are a barrier to effective use for external data sources. It is therefore very important to assign a significant role to secondary source data collection and related problems. A search through external data (Kenett and Shmueli, 2011) is recommended. The market manager who selects data for analysis, and the person who collects data, should be conscious of available external data.

5.2.2   Advantages of using secondary source data

The use of secondary data has many advantages. First of all, it is usually cheaper to collect than primary data. Extensive research into secondary sources can yield information that cannot be obtained by primary data collection (for instance, demographic information collected in census or national surveys). If the desired information is available in secondary sources, searching such data is much less time-consuming and costly than primary data collection. In addition, secondary sources can yield more accurate data than that obtained through primary research. This is not always true, but when a government or international agency conducts a large-scale survey, or even a census, it is likely to yield more accurate results than custom-designed surveys with relatively smaller sample sizes. Secondary external sources play a substantial role in the exploratory phase of a research project. Furthermore, they can be extremely useful both in defining the population and in structuring the sample to be taken. For instance, government statistics on agriculture in a country will help decide how to stratify a sample. When agriculture sector firms are sampled (stratified sample) a business producing silos might interview users about their satisfaction with the product brand they are using. This data collection is useful to compute estimates of satisfaction with the silos product.

5.2.3   Problems with secondary source data

Whilst the benefits of secondary sources are considerable, their shortcomings must be acknowledged. The quality of both the data source and the data itself must be evaluated.

Many statistical problems arise when integrating the external and internal sources of a business. A good rule to follow is that besides archives containing data, a so-called *metadata database* (i.e. data collection description and quality indicators) should also be available, for both internal and external data. *Metadata information* is the basic support for integrating data so as to maintain high quality standards.

The main problems with secondary data sources may be categorized as follows: definitions, errors, bias, reliability and coherence.

When making use of secondary data, attention should be paid to the *definitions* used in the data collection. Suppose a researcher is interested in studying family size in a local community. If published statistics are used, then the definition of 'family size' needs to be clarified. It may refer only to the family nucleus or include the extended family. Another example is the term 'farm size' which needs careful handling. Suppose that a firm producing innovative tractors wishes to evaluate customer satisfaction. The firm wants to set up a stratified sample by firm size. This is an important variable since customers with large farms have significant experience in using tractors. Statistics may refer to the official statistical definitions of the land an individual owns, the land an individual owns plus any additional land he/she rents, the land an individual owns minus any land he/she rents out, all of his/her land, or only that part which he/she actually cultivates. Thus, for stratification purposes the most appropriate definition should be adopted (see Kenett and Shmueli, 2011).

When a researcher conducts fieldwork, two types of *errors* occur: sampling error and non-sampling errors. Sampling errors are related to statistical sampling theory and sample representativeness. Sampling errors are measured by standard errors which are usually computed but not always published in secondary sources. Non-sampling errors are due to various survey activities or steps (coverage, measurement). These errors are difficult to measure and are affected, for example, by the quality of the data collection procedures. Due to the peculiar nature of these errors, they are rarely published in secondary sources. One possible solution is to contact the researchers involved in the data collection in order to obtain guidance on the level of accuracy and quality of their data. For more on such errors, see Chapter 3. For techniques to determine sample representativeness, see Chapters 3 and 20.

The problem is sometimes not so much 'error' but differences in levels of accuracy required by decision-makers. When the research has to do with large investments, for example in food manufacturing, management will want to set very tight margins of error in making market demand estimates. In other cases, having a high level of accuracy is not so critical. For instance, if a food manufacturer is merely assessing the prospects for one more flavour for a snack food already produced by the company, then there is no need for highly accurate estimates in order to make the investment decision.

Bias in the source is not related to statistical concepts; it means that researchers have to be aware of vested interests when consulting secondary sources. Those responsible for their compilation may have reasons for wishing to present a more optimistic or pessimistic set of results. For example, officials responsible for estimating food shortages could exaggerate figures before sending aid requests to potential donors. Similarly, commercial organizations could inflate estimates of their market shares. As a secondary external source, official statistics are not affected by this kind of bias. This is one of the great advantages of official statistics data: it is, and should be, trusted.

Several aspects of research methodology affect the *reliability* of secondary data, like the sample size, response rate, questionnaire design and modes of analysis.

Coherence over time is highly important, too. Changes in data collection procedures (target population or definitions, for example) affect data comparability over time. An example refers to geographical or administrative boundaries: they may be changed by government, and new statistics rely on the new boundaries. Definitions should be monitored since they may change. Not only geographical areas may have their boundaries revised, but units of measurement may change and goods and products may be reclassified from time to time.

Whenever possible, market researchers should use multiple sources of secondary data. In this way, these different sources can be cross-checked as confirmation for each another. Where differences occur they must be explained or the data should be set aside. Due to the problems related to the use of secondary data, some caution must be applied in using such data. The researcher should conduct critical evaluation steps as shown in Figure 5.1. In practice, early stages of the evaluation process relate to the relevance of the data to the research objectives. Later stages are concerned with the accuracy of secondary data.

5.2.4   Internal sources of secondary information

In their day-to-day activities, businesses collect a variety of data. In most organizations, data is collected and stored by separate groups with individual databases. Isolated standalone data environments – as well as different methods for recording and archiving this data – make it extremely difficult to come to grips with all customer data and issues. Integration of internal databases is needed. Moreover, proper integration of internal and external sources is a crucial competitive point.

Sales, operations, after-sales support and marketing units typically have different databases with applications dedicated to specialized functions that relate to specific business requirements in each unit. Data integration combines, and provides access to, information on customers and markets coming from all sections of the organization, as well as other business-related data. The whole information set becomes more valuable than the sum of the parts. Such an integration is part of implementation efforts in enterprise resource planning (ERP) and customer relationship management (CRM) systems (see Wognum *et al.*, 2004; Kenett and Lombardo, 2007). Even if

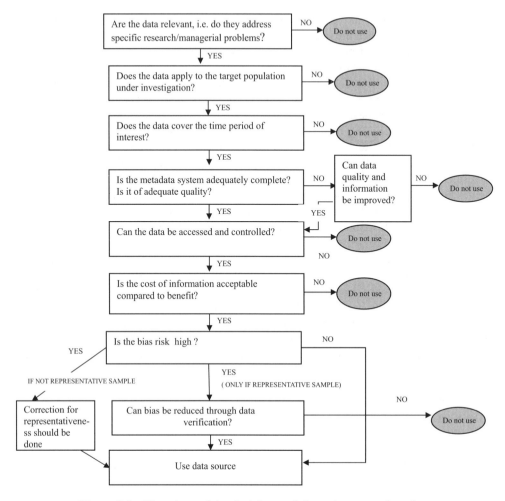

Figure 5.1   Flowchart of the decision path for using secondary data

ERP and CRM systems have specific characteristics and tasks, full integration is a synergistic bridge between them.

Take, for example, a business with customer information systems that include billing systems, marketing systems, shipping systems, and data warehouses. A short description of the internal data usually available within such an organization is given below.

Data collection in everyday operations. All organizations collect data in the course of their everyday operations. Orders are received and delivered, costs are recorded, sales personnel submit site-visit reports, invoices are sent out, returned goods are recorded, and so on. Much of this data is of potential use in marketing research, but a surprising amount of it is not actually used. Organizations frequently overlook this valuable resource by not beginning their search of secondary sources with an internal audit of sales invoices, orders, inquiries about products not stocked, returns from customers and sales force customer call sheets. For example, a great deal of information can be obtained from sales orders and invoices, such as sales by territory, sales by customer type, prices and discounts, average size of customer order, customer type, geographical area, average sales by sales person, sales by pack size and pack type, etc. This type of data is useful for identifying an organization's most profitable product and customers. It can also serve to track trends within the enterprise's existing customer group.

Financial data. An organization has a great deal of data within its files on the cost of producing, storing, transporting and marketing each of its products and product lines. This is of particular interest in determining the point at which an organization's unit costs begin to fall. Cost data has many uses in marketing research, including allowing measurement of the efficiency of marketing operations. With integrated databases, costs of new products under consideration can be better estimated.

Logistical data. Companies that keep good records relating to their transport operations are well placed to establish which are the most profitable routes, and loads, as well as the most cost-effective routeing patterns. Good data on transport operations enables the enterprise to carry out trade-off analysis. It thus becomes possible to establish whether it makes economic sense to own or hire vehicles, or the break- point at which a balance between the two gives the best financial outcome.

Storage data. The physical space occupied by individual products and the time period over which they occupy the space have an economic value in the accounting system. Therefore the stock turnover rate, stock handling costs, the efficiency of certain marketing operations, and the efficiency of the marketing system as a whole are important information if it is desired to measure in a more sophisticated way the profitability per unit, rate of sale, and direct product profitability.

This overview of the business internal information shows how crucial is adequately statistically integrated data. The methodological issues are related to the fact that applications are typically performed in a standalone environment and are optimized for particular needs. When the data is optimized in a global operational system designed to fully understand the customer, there is a need to consolidate that data, by customer, into a single customer-centric database. This gives rise to many integration problems. The main economic statistics steps in the consolidation process are: data profiling, data quality, data integration, date adding value, data monitoring.

Data profiling. This step is devoted to discovering, understanding and documenting the sources within the organization that maintain customer information. This analysis phase encompasses an inventory of the data source as well as the collection of information on data characteristics. This includes data definition problems and is supported by internal metadata databases (if available). This step is crucial in proceeding with the integration of the data.

Data quality. This step relates to the process of correcting the data. After profiling, you can begin to make the data useful. Data is often invalid, out of range, incompatible, or inconsistent with current business rules. The data quality process brings the business customer data to a standard that meets your business requirements. Thousands, even millions, of records are stored in multiple, enterprise-wide systems, with new data added daily. The proliferation of data sources leads to ambiguous, duplicate or inaccurate customer representation. For example, names and addresses can be depicted in various ways, leading to duplicate or misaligned names within and across databases.

Data integration. This phase involves matching data archives. For example, a researcher will want to identify and locate individual customers within and across the data sources. A full understanding of the firm's customer requires full collection of all of the data from all sources. To gain complete view of the customer base, the researcher must remove duplicates and consolidate customer information across data sources. Links between data sources are, therefore, provided to gain an aggregate understanding of interrelationships between customers. Linking (also known as clustering) can occur at different levels depending on need: at the customer level, at the household level (all customers at the same address, for example), at the business or corporate level, or some other combination of attributes.

Data adding value by creating a consolidated view of the customer. Here, one can enhance customer relationships by understanding more about the individual customers, their preferences and characteristics. There are many external data sources that provide geographical, demographic, financial, and lifestyle information on businesses or consumers. Augmenting the customer data hub with this type of information gives the researcher additional insights into the market and helps to provide segmentation and predictive analysis.

With the emerging paradigm of linked open data, existing information infrastructures for processing research data can be opened up for the use of external data published by sources on the web. The recent linked open data approach (Bizer *et al.*, 2009) is a method to expose, share and connect freely available data on the web using Semantic Web standards. Linked data is based on the unique identification of metadata elements and of certain entities. Linked open data resources provides the possibility to include (integrate) in business databases other freely available data sources from the web. The process of data integration which determines the basis for the statistical calculations needed can be easier to implement with standardized formats and interfaces, such as linked open data.

Data monitoring. This is a maintenance phase. A monitoring approach is undertaken: the researcher continuously identifies and addresses new problems arising in the data sources and checks coherence.

5.3   Root cause analysis

5.3.1   General concepts

Root cause analysis (RCA) has been defined and used in various different fields. In accident analysis and occupational safety and health, RCA is called *safety-based RCA*. In quality control for industrial manufacturing it is called *production-based RCA. Process-based RCA* is an expansion of production-based RCA to include business processes. *Failure-based RCA* is the practice of failure analysis as employed in engineering and maintenance, and *systems-based RCA* is the amalgamation of the preceding schools, along with ideas taken from fields such as change management, risk management, and systems analysis.

The focus in this chapter is on systems-based RCA. In the following we define a general process for performing RCA and some general principles. Note that despite the apparent disparity in purpose and definition among the various schools, processes and principles in this chapter hold for every RCA approach.

The purpose of RCA is to find effective solutions to the problems under study so as to avoid their reoccurrence (Andersen and Fagerhaug, 2000). The RCA technique is used to understand factors that affect a (usually negative) event. The analysis aims not to identify errors, but to analyse the entire process leading up to the event. The task is to understand what happened, why it happened, and how to stop it happening again. Accordingly, an effective RCA process should provide a clear understanding of exactly how the proposed solutions meet this goal.

To give such an assurance, an effective process should:

- define the problem clearly and its significance to the problem owners;

- describe precisely the known causal relationships that determined the problem;

- establish causal relationships between the root cause(s) and the problem under investigation;

- provide the documentation used to support the existence of identified causes;

- explain how the solutions will prevent a reoccurrence of the problem;

- provide a final RCA report describing the above quoted steps and analysis.

This last action is important if others are to be able easily to follow the logic of the analysis, which may be complicated since many phases, methods, tools and statistical analyses are involved. RCA is the application of a series of well-known common-sense and statistical techniques which, when used in different combinations, can produce a systematic, quantified and documented approach to the identification, understanding and resolution of the underlying causes of under-achieved quality in organizations.

The key process steps (see Figure 5.2) are general problem and causal factor identification (i.e. *Define problem and Determine causes*), hypothesis development and hypothesis testing (i.e. *Select solutions*). At each step several methods and tools are applied. Some of these methods and tools are described in Section 5.3.2. Notice that at each step, different data is needed. An integrated database consists of internal and external secondary data source integration and internal data integration. The recent availability of linked open data facilitates the use and integration of external statistical data sources. Thus, statistical analyses on a broader range of available data sets can easily be done. They can also save time-consuming data conversion when using established standards and interfaces.

The first two steps, *general problem identification* and *causal factor identification*, include gathering data or creating samples, failure investigation and experimentation, examination/dissection/physical testing, brainstorming, use of tools (such as fishbone and causal factor charts).

The third stage is *hypothesis development* (i.e. *Identify possible solutions*). At this stage a link analysis between the data and causal factors is undertaken.

The last step is *hypothesis testing* (i.e. *Select solutions*). This may involve the use of simple tools such as root cause identification and root cause mapping, as well as the application of several statistical techniques such as design of experiments, analysis of variance, Bayesian networks and the classical hypothesis testing approach. For literature on these technique, see Portwood and Reising (2007). For a general treatment of statistical techniques, see Kenett and Zacks (1998). For an introduction to Bayesian networks and causality models, see Salini and Kenett (2009) and Chapters 10 and 11 in the present volume.

Implementing solutions and tracking effectiveness is a necessary final step in order to activate actions, and to fully understand the processes and analyses and evaluate the effect of the solutions undertaken.

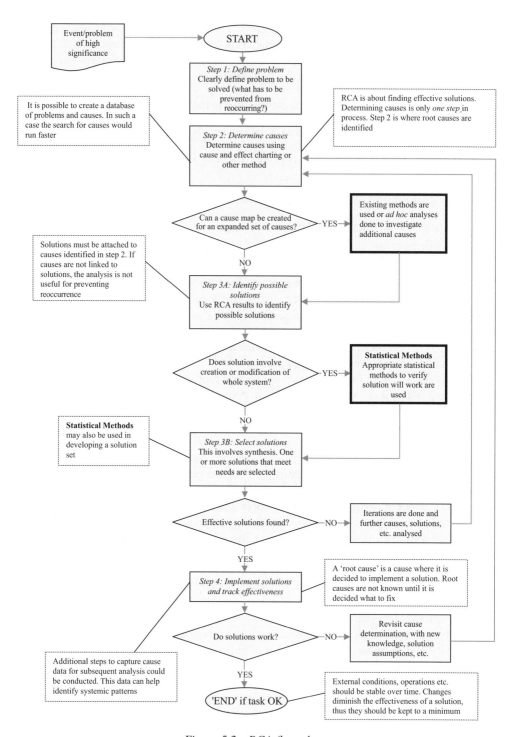

Figure 5.2   RCA flow chart

5.3.2 Methods and tools in RCA

An integrated approach is characterized by many steps. This is especially important in the case of RCA, which is involving a wide range of statistical methods, from the simplest to the most complex. First the problem should be clearly defined, and then causes should be determined. At this stage a cause–effect diagram is especially useful. Statistical methods could then be used to investigate causes and find out whether additional causes can be identified. Solutions should thus be identified and linked to causes; statistical methods can be used to develop a set of possible solutions. Solutions are iteratively selected and implemented. Checks on the effect of the chosen solutions and reassessment of causes is finally done. Figure 5.2 shows the different steps in which quantitative methods should be used in RCA. Methods and tools for conducting RCA are described below.

Charting events and causal factors

This method is a process that first seeks to identify a sequence of events and align them with the conditions that caused them. These events and respective conditions are aligned in a time-line to help determine the causal factors. Events and conditions for which there is evidence are shown as solid lines, but evidence is not listed; all others are shown as dashed lines. After completing the representation of the problem, an assessment is made by 'walking' the chart and asking whether the problem would be different if the events or conditions were changed. This leads to a list of causal factors that may then be evaluated in a further step of the analysis.

Sometimes the representation of the problem is based on the *storytelling method*, also known as the *fill-out-a-form method*. This is not a real RCA method, but is often presented as such. It is the single most common incident investigation method and is used by nearly all businesses and government entities. It typically uses predefined forms that include problem definition, a description of the event, who made a mistake, and what is going to be done to prevent reoccurrence. There is often a short list of root causes to choose from. A Pareto chart can be created to show where most problems originate. The storytelling technique is often used to communicate and predict the thoughts, emotions and behaviours of customers as they experience and interact with a company.

The primary difficulty with this approach is that it relies completely on the experience and judgement of the report authors in ensuring that the recommended solutions connect to the causes of the problems. The precise mapping between the problem and the recommended solutions is not provided. The primary purpose of this method is to document the investigative and corrective actions. These forms usually do a good job of capturing the what, when and where of the event, but little or no analysis occurs. Consequently, the corrective actions fail to prevent reoccurrence 70–80% of the time. With such poor results, one might wonder why organizations continue to use this method. There are two reasons. First, most organizations do not measure the effectiveness of their corrective actions, so they do not know when they are ineffective. Second, there is a false belief that everyone is a good problem-solver, and all they need to do is document it on a form. In any case, the use of forms is not necessarily good practice since by standardizing the thinking process only a predefined set of causes and solutions is considered.

An example of this approach can be found in the cause–event–action–outcome chains described in Wognum *et al.* (2004) and Kenett and Lombardo (2007) in the context of implementation of enterprise systems.

Change analysis

This is a tool based on a six-step process: (1) describe the event or problem; (2) describe the same situation without the problem; (3) compare the two situations; (4) note all the differences; (5) analyse the differences; (6) identify the consequences of the differences. Change analysis identifies the cause of the change. Being frequently tied to the passage of time, it therefore fits easily into an events and causal factors chart, showing what existed before, during and after the change.

Change analysis is nearly always used in conjunction with an RCA method to provide a specific cause, not necessarily a root cause. It is a very good tool to help determine specific causes or causal elements, but does not provide a clear understanding of the causal relationships of a given event. Unfortunately, many people who use this method simply ask why the change occurred and fail to complete the comprehensive analysis.

Barrier analysis

Barrier analysis can provide an excellent tool for determining where to start your RCA, but it is not a method for finding effective solutions because it does not identify why a barrier failed or was missing. This is beyond the scope of the analysis. To determine root causes, barrier analysis findings must be fed into another process to discover why the barrier failed. It is an analysis that identifies barriers used to protect a target from harm and analyses the event to see if the barriers held, failed, or were compromised in some way.

Categorization schemes

There are several categorization methods, all of which use the same basic logic. The assumption is that every problem has causes that lie within a set of categories that can also be predefined. These methods do not show all the causal relationships between the primary effect and the root causes.

Tree diagrams and Ishikawa diagrams are two examples of categorization schemes. An Ishikawa diagram is created by examining an event by asking people involved what causes may have been acting. The suggested causes are classified within broad categories. Causes are placed on the fishbone diagram according to the categories they belong to – not how they are causally connected. The original structure of an Ishikawa diagram uses manpower, methods, machinery and environment as the top-level categories. The definition of these top-level categories can then be adapted according to the specific event under investigation. Top-level categories (primary causes) contain subcategories (secondary causes) and sub-subcategories (tertiary and further causes). For example, within the category of manpower, we may find management systems; within management systems we may find training; and within training we may find training less than adequate.

Figure 5.3 gives an example of an Ishikawa diagram where primary categories were adapted to the event studied. The problem (effect) is a failure in computer hardware, primary categories are memory problems, process problems and hard drive problems. Ishikawa diagrams are also known as fishbone diagrams because of their shape: on the right-hand side the event/problem is indicated (in our case a hardware failure), and on the left-hand side a list of primary and secondary sources is given.

Ishikawa diagrams, as with all categorization schemes, are constructed by organizing brainstorming sessions, asking people to focus on one of the primary categories, reviewing

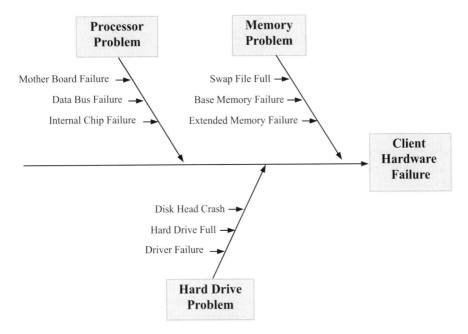

Figure 5.3   Ishikawa diagram: client hardware failure effect

what they know of the event, or inviting them to choose some causal factors from the predefined list provided. These schemes do not provide a specific actionable cause, but rather they act as a checklist of possible causes for a given effect.

The experience of the investigator who is actually thinking causally may help to suggest some kind of relationship, but these methods do not provide a means of showing how we know that a cause exists. There is no evidence provided to support the causal factors in the list, so it is not uncommon for causal factors to be included with no factual basis.

A correct application of categorization schemes restricts thinking by causing the investigator to stop at the categorical cause.

Because these methods do not identify complete causal relationships, it is not obvious which causes can be controlled to prevent reoccurrence; therefore, people are asked to guess, and vote on, which causal factors are the root causes. Statistical data and models are then applied to these 'root causes' to understand necessary solutions (for more on Ishikawa diagrams and statistical techniques. see Kenett and Zacks, 1998; Breyfogle, 2003). Actions are to be undertaken to prevent reoccurrence of the root cause.

Fault tree analysis

Fault tree analysis is a method based on a quantitative causal diagram used to identify possible failures in a system. It is a common engineering tool used at the design stages of a project and works well in identifying possible causal relationships. It requires the use of specific data regarding known failure rates of components. Causal relationships can be identified with 'and' and 'or' relationships or various combinations thereof. Fault tree analysis does not work well as a RCA method, but is often used to support an RCA. More articulated structures of similar diagrams (reality charts) have been proposed in the recent literature (Gano, 2008).

Failure modes and effect analysis

Failure modes and effects analysis (FMEA) is a tool sometimes used to find the cause of component failure. As the name implies, it identifies a component, subjectively lists all possible failures (modes) that might occur, and then makes an assessment of the consequences (effect) of each failure. It does not work well with systems or complex problems because it cannot show evidence-based causal relationships beyond the specific failure mode being analysed. FMEA is similar to fault tree analysis in that it is primarily used in the design of engineered systems rather than root cause analysis.

Why-why chart

This is one of many brainstorming methods also known as the 'five whys' method. This is the most simplistic RCA process and involves asking 'why' at least five times or until you can no longer answer the question (five is an arbitrary number). The theory is that after asking 'why' five times you will probably arrive at the root cause. The root cause that has been identified when asking 'why' does not provide any further useful information. This method produces a linear set of causal relationships and uses the experience of the problem owner to determine the root cause and corresponding solutions.

This method is quite useful when used on minor problems that require nothing more than some basic discussion of the event. Unlike most of the other methods, it identifies causal relationships, but still subscribes to the root cause approach of first finding the root cause and then assigning solutions.

It should never be used in formal incident investigations. It is perfectly acceptable for informal discussions of cause. A popular graphical representation of the 'five whys' approach is the 'why staircase', which, if used improperly, leads to a linear set of causal relationships.

Pareto analysis

Pareto analysis uses a categorical variable database to trend the frequency of categorical characteristics (for instance, failure characteristics database). A Pareto chart will only reflect causes from the predefined list provided and will focus on the frequency. For example, experience shows that the cause 'procedures not followed' is frequently caused by 'procedures not accurate'. In the Pareto analysis, this causal connection is lost. Instead, the diagram puts evidence that both 'procedures not followed' and 'procedures not accurate' are in the top causes, so we end up working on solving both problems, when in reality we may only need to solve the 'procedures not accurate' problem. This example highlights the fact that Pareto analysis provides an incomplete view of reality provided – in particular, it does not focus on cause–effect relationships. Therefore it might sometimes lead to greater use of analysis resources than necessary. This analysis can mask larger, more systemic issues. For example, if quality management has transitioned into a state of dysfunction, it can cause symptoms in many different areas, such as poor procedures, inadequate resources, outdated methods, high failure rates and low morale. Pareto analysis allows for capturing all these symptoms of a larger problem as causes, with a consequent waste of time identifying solutions for what are only symptoms. In practice, Pareto analysis is useful only when the 80–20 rule holds, i.e. 80% of the causes of the problem are in a few (two or three) categories and no special causal relationship acts among these categories. For more on Pareto analysis, see Kenett and Zacks (1998) and Godfrey and Kenett (2007).

5.3.3   Root cause analysis and customer satisfaction

General framework

As global competition increases, maintaining customer loyalty is more important than ever. Dissatisfied customers now have many options, with dozens of companies from around the world competing for their business. It is crucial for every organization to retain loyal customers by maintaining a high level of customer satisfaction. A strong understanding of the data surrounding customer satisfaction surveys is therefore most important.

The analysis of customer satisfaction data focuses on this issue and clearly demonstrates how to interpret the data gathered in customer surveys while explaining how to use this information to improve overall customer satisfaction.

Most organizations recognize the impact that both customer and employee satisfaction has on overall financial performance. The focus is on linking customer and employee satisfaction to the relationship between customer satisfaction and tangible business outcomes, such as market share, revenue and profitability (Wiley, 1996; Rucci et al., 1998; Wiley and Brooks, 2000; Derek, 2002; Kenett, 2004; Godfrey and Kenett, 2007).

In the context of an MIS an RCA can play a special role in handling data derived from employee and customer survey data in an integrated way. This approach is most suitable for advanced service quality managers and marketing researchers with more than a modest exposure to statistical data analysis.

It is well known that high employee satisfaction contributes significantly to high customer satisfaction, which drives repurchasing intentions and therefore, financial results. High employee satisfaction expresses itself as enthusiasm for one's work, which directly impacts the experience of the customer. Likewise, high customer satisfaction expresses itself as enthusiasm towards a particular organization, its products or services, which directly impacts the intent-to-return rate. It is a short leap, then, to understanding how a high intent-to-return rate among customers impacts on financial results. But with so many variables affecting employee and customer satisfaction, how does one determine those of greatest importance, so that interventions aimed at increasing satisfaction are of maximum effectiveness? In this kind of application of RCA, data are collected by surveys (Kenett, 2004).

A customer RCA can assess the perceptions of customers and identify those that drive customer behaviour. These root causes are almost always different from those driving employee satisfaction. Employee surveys are related to customer satisfaction surveys. The survey content for both employee and customer surveys should be driven by research, facts, and hard data uncovered through confidential one-to-one content interviews of a stratified random sample of the target respondents.

An employee RCA is designed to identify those perceptions in the employee population that drive the greatest number of other perceptions to the greatest extents, because it is those core, or root, perceptions that drive employee behaviour. With appropriate interventions in the root perceptions, or root causes of employee behaviour, the business can change perceptions and therefore, behaviour. Root causes of *employee perceptions* will increase *employee satisfaction* (and indirectly increase customer satisfaction), root causes of *customer perceptions* will increase customer satisfaction and intent to return, and, to close the circle, root causes of employee perceptions will increase customer intent to return.

Moreover, if we identify *employee attitudes*, opinions, or beliefs (perceptions) that drive employee behaviour with direct impact on customer repurchasing intentions and loyalty, this leads to a further gain in information, profits and in quality.

Methodology

A customer satisfaction survey serves to identify where, and quantify to what degree, each specific aspect of each issue exists throughout the organization or customer base. This is a crucial enabler of precise, targeted action, saving time and money. Interviews are a qualitative tool that is excellent for gathering textual (rather than numerical) information, such as important issues concerning different factors. Interviews may collect quantitative information, too, if necessary. Qualitative interviews will show whether 'communication', for example, is an issue. In addition, interviews can provide evidence whether communication is too fast or too slow (speed) or too much or too little (volume). Other aspects of communication and how they are affecting the business can also be determined. This way, highly specific survey questions are selected to get information either from general databases or from specific customer experience data.

Usually a stratified random sample is selected. Interviews can be conducted face to face, via the internet or by telephone (see Chapters 6 and 7). By using this data collection tool, the richest information with respect to costs is usually gained. These kinds of surveys collect significant information. An executive summary of the information obtained through the interviews is usually prepared with the main highlights. Response time, generally 20–30 minutes, is greatly minimized compared to a focus group, which may take 1–2 hours of each respondent's time. There is also the potential for an unusually deep level of understanding of the issues.

To conduct RCAs correlations between each survey item and all other survey items must be evaluated. The correlations are used to identify the items to be employed in the regression analyses to model behaviours. Usually stepwise regression is applied to the data using a linear model. Stepwise regression analysis eliminate those items that exert lesser amounts of influence over the data, and linear regression analysis provides input to the path analysis, the final 'line-up' of the primary, causal factors, and the levels of significance of each (See Chapter 11). Regression analysis is crucial to identify dominant, primary causal factors. These results will quickly and effectively turn massive amounts of information into action. Note that RCA of customer data is a process similar to that of employee data.

Application

We now focus on a problem and present an RCA approach to it. A customer-centric company addresses the problem of customer dissatisfaction with its service. The interest is in understanding the specific drivers of customer dissatisfaction – questions like why customers are dissatisfied, why they are resquesting support, and what the company can do to increase satisfaction in a cost-efficient way.

The company is registering an increase in call volume in customer service operations (service provided using a call centre). Furthermore, customer satisfaction is declining. The high cost of call centre activities forces the company to focus on this problem with a key strategic analysis for cost containment. A primary data collection effort is organized and a survey is carried out to determine why customers call.

In the first step of the data collection, a random representative sample of calls is selected each month and frequencies of the various reasons are recorded. Thus major reasons are compared over time and patterns in call reasons are identified, i.e. a 'basic view' is set up. Segments of customers by areas of concern are identified. In this way, the general problem identification and causal factor identification phases of RCA analysis are performed. In the second step, the study moves on testing hypotheses. A call monitoring team conducts an in-depth analysis of call reason. The team validates the issues and hypotheses identified. Following that, the RCA builds up more complex, secondary analyses using customer information databases.

Insights into root cause reasons for customer dissatisfaction are accurately developed using statistical models as well as stratified Pareto diagrams. For instance, a secondary analysis reveals that less than 5% of callers were responsible for more than 20% of calls. The call monitoring survey reveals that these customers were calling for the same reasons as everyone else. In addition, the findings suggest that many of these frequent callers call out of personal habit rather than for a particular reason. The financial impact of the call drivers is estimated, too. Having an integrated analysis in mind, management decide to find a solution by looking at additional data sources and appropriate operational actions are activated.

5.4   Summary

Data integration is important to support marketing systems as well as management strategies and decision throughout the business. Data integration merges internal and external data into a unified and consistent framework. To create such data integration, technological problems are only one aspect of this issue. A key topic is the definition of economic statistical concepts and rules that allow for coherence and comparability in different data sources. The recent spread of linked open data can facilitate the use of the external data sources for statistical analyses.

Once data integration is achieved, integrated statistical analyses can be processed. In this chapter, we have focused on data and analyses integrated in a marketing system with a focus on customer satisfaction. Root cause analysis is a useful approach for setting out statistical methods in an operational integrated context. In some sense, RCA is an approach which includes both the need for integrated databases and integrate analyses for customer satisfaction studies. In principle, RCA is part of every statistical technique and graphical analysis.

Acknowledgement

The author gratefully acknowledges the support of reseach grant 'University of Bergamo, BIF2008 and 2009 ex-MIUR'.

References

Andersen, B. and Fagenhaug, T. (2000) *RCA: Simplified Tool and Techniques*. Milwaukee, WI: ASQ Quality Press.

Bizer, C., Heath, T. and Berners-Lee, T. (2009) Linked data – the story so far. *International Journal on Semantic Web and Information Systems*, 5(3), 1–22.

Breyfogle, F.W. (2003) *Implementing Six Sigma: Smarter Solutions Using Statistical Methods*, 2nd edn. Hoboken, NJ: John Wiley & Sons, Inc.

Derek, M. (2002) *Linking Customer and Employee Satisfaction to the Bottom Line*. Milwaukee, WI: ASQ Quality Press.

Dillon, W.R., Madden, T., and Firtle, N. H. (1994) *Marketing Research in a Research Environment*. 3rd edn. Burr Ridge, IL: Irwin.

Gano, D.L. (ed.) (2008) *Apollo Root Cause Analysis: A New Way of Thinking*, 3rd edn. Richland, WA: Apollonian Publications.

Godfrey, A.B. and Kenett, R.S. (2007) Joseph M. Juran: A perspective on past contributions and future impact. *Quality and Reliability Engineering International*, 23, 653–663.

Green, P.E., Tull, D.S. and Albaum, G. (1988) *Research Methods for marketing decision*, 5th edn. Englewood Cliffs, NJ: Prentice Hall.

Kenett, R.S. (2004) The integrated model, customer satisfaction surveys and Six Sigma. *First International Six Sigma Conference, CAMT*, Wroclaw, Poland.

Kenett, R.S. and Lombardo, S. (2007) The role of change management in IT systems implementation. In P. Saha (ed.), *Handbook of Enterprise Systems Architecture in Practice*. Hershey, PA, and London: Information Science Reference.

Kenett, R.S. and Raanan, Y. (2010) *Operational Risk Management: A Practical Approach to Intelligent Data Analysis*. Chichester: John Wiley & Sons, Ltd.

Kenett, R.S. and Shmueli, G. (2011) *On Information Quality*. http:ssrn.com/abstract=1464444.

Kenett, R.S. and Zacks, S. (1998) *Modern Industrial Statistics: Design and Control of Quality and Reliability*. Pacific Grove, CA: Duxbury Press.

Portwood, B. and Reising, L. (2007) Root cause analysis and quantitative methods – yin and yang? Paper presented to the *25th International System Safety Conference*, Baltimore, MD, 13–17 August.

Rucci, A., Kim, S., and Quinn, R. (1998) The employee–customer–profit chain at Sears. *Harvard Business Review*, 76(1), 83–97.

Salini, S. and Kenett, R.S. (2009) Bayesian networks of customer satisfaction survey data. *Journal of Applied Statistics*, 36(11), 1177–1189.

Wiley, J.W. (1996) Linking survey results to customer satisfaction and business performance. In A.I. Kraut (ed.), *Organizational Surveys: Tools for Assessment and Change*. San Francisco: Jossey-Bass.

Wiley, J.W. and Brooks, S.M. (2000) The high performance organizational climate: How workers describe top-performing units. In N.M. Ashkanasy, C.M. Widerom and M.F. Peterson (eds), *Handbook of Organizational Culture and Climate*. Thousand Oaks, CA: Sage.

Wognum, P., Krabbendam, J., Buhl, H., Ma, X., and Kenett, R.S., (2004) Improving enterprise system support–a case-based approach. *Advanced Engineering Informatics*, 18(4), 241–253.

6

Web surveys

Roberto Furlan and Diego Martone

Since the late 1990s, market research agencies have been increasingly using web surveys to investigate market, political, or social aspects. In fact, the growing spread of the internet and the availability of fast internet access have made web surveys an attractive means of collecting survey data, as the web provides simple, fast, and, above all, cheap access to a large group of people. This chapter introduces main types of web surveys, presents their main economic and non-economic benefits, and outlines the main drawbacks associated with online research. It also discusses the application of web surveys for customer and employee satisfaction research projects.

6.1 Introduction

Since the late 1990s, market research agencies have been increasingly using web surveys to investigate market, political, or social phenomena, owing to the spread of the internet and the availability of fast internet access (Cambiar, 2006; Epstein *et al.*, 2001; Fricker and Schonlau, 2002; Schonlau *et al.*, 2002). Sinking response rates of mail and phone surveys and the increased web literacy of the general population have been the major push factors, along with key benefits offered by web surveys such as lower cost, faster data collection, and fewer problems with data editing. A major result has been the optimization of fieldwork processes and the development of what we refer to as *e-research*.

The development of new methodologies and tools for collecting information through this new mode has benefited from experience of traditional data collection methodologies, such as paper-and-pencil interviewing (PAPI), computer-assisted personal interviewing (CAPI), and computer-assisted telephone interviewing (CATI). On the one hand, the web approach has created interesting opportunities; on the other hand, it has created new issues and challenges.

Nowadays, most researchers and practitioners agree that the web mode cannot and should not be regarded as a replacement data collection mode, but as an integrative one. In fact,

Modern Analysis of Customer Surveys: with applications using R, First Edition. Edited by Ron S. Kenett and Silvia Salini.
© 2012 John Wiley & Sons, Ltd. Published 2012 by John Wiley & Sons, Ltd.

several comparative studies have shown that web surveys (computer-assisted web interviewing – CAWI) provide clear advantages over traditional modes only in some specific situations, for instance when investigating low-incidence segments or conducting advertising research (Couper, 2001).

The advantages and opportunities provided by web surveys are presented in this chapter (Sections 6.3 and 6.4), along with the main methodological aspects and possible concerns and issues faced by the researcher (Section 6.5). The chapter also discusses the application of web surveys for customer and employee satisfaction research projects (Section 6.6), as satisfaction surveys are a critical input to process improvement initiatives.

6.2   Main types of web surveys

Self-selection web surveys are probably the least interesting group of surveys, as it is not possible to compute proper estimates of the population parameters. Usually in such surveys, the outcomes cannot be used for statistical inference on the target population (Bethlehem, 2010; Couper, 2000). The questionnaire is available through a webpage and the research agency does not perform any selection process. Respondents are intercepted when they visit the website that hosts the link to the CAWI questionnaire, usually through a start or exit page popup, a banner, or a button placed in a prominent position on the webpage. When prompted, respondents may or may not decide to participate in the survey. Therefore, it is safe to assume that surfers choosing to reply will not necessarily conform to the scientific requirements of randomness and representativeness (Kenett *et al.*, 2006).

List-based web surveys are more interesting. The researcher agency sends survey invitations out based on a closed list. Depending on the country's specific privacy laws, only people who have given prior consent might be invited to participate in the survey. These surveys are particularly appropriate for research among the company's employees (e.g., employee satisfaction) or clients (e.g., customer satisfaction), members of an association, students at a university, registered users of a website, etc. Often this type of survey is a sort of census, since, typically, all customers or all employees are asked to complete the questionnaire (Kenett *et al.*, 2006). In these surveys, coverage is not a problem and the population frame is well known (Kenett *et al.*, 2006). However, non-response might be a concern (Couper, 2000).

Web panel surveys are those surveys where a group of people is periodically interviewed on the same topic. At the recruitment stage, respondents are typically asked a number of profiling questions before their inclusion in the panel, so that the distribution of these variables in the panel survey can approximate their distribution in the target population. The main benefit of a panel survey is that repeated measurements on a single sample reduce the sampling variation in the measurement of change over time and it allows variations to be analysed at the individual respondent level (Sharot, 1991). A panel survey might be affected by two types of bias: systematic attrition may produce a panel that is not representative of the target population; and repeated interviewing may change the opinion or the behaviour of the panellists (Clinton, 2001).

The most advanced type of survey consists of *web community surveys*. Recruitment for a community is somehow similar to that for web panel surveys, but profiling is an important step in the community recruitment. Web communities usually are very large, often containing over 50 000 members; potential respondents of a specific research project are selected from the community list based on the target population required for the project, and thus invited to participate. Web communities can be generalist or thematic. Generalist communities is probably the most common type and are appropriate for investigating the opinion or behaviour of large populations (e.g., adults in the UK, mobile phone users). In contrast, a thematic community is more appropriate for studying specific population targets (e.g., IT managers, oncologists). Hagel and Armstrong (1997) were among the first to highlight the opportunities arising from aggregating seemingly disparate individuals with particular interests into communities. Thematic communities gathering members sharing an interest in specific products, lifestyles, or brands are sometimes built and maintained in collaboration with the R&D departments of companies to contribute to the development process of new products or services. In fact, it has been shown (Kozinets, 2002) that the lead users of such communities tend to have the ability to anticipate new trends and identify elements of success for a new product or service, thus providing a key competitive advantage. An approach to integrate the community users within the product development process was suggested by Fuller *et al.* (2004) who introduced the concept of community-based innovation (CBI) founded on groundwork of social exchange and interaction theory. Web community surveys are probably the way forward, although they tend to be affected by representativeness issues as online respondents differ in some ways from potential offline ones. This topic is discussed in more detail in Section 6.5.

When conducting market research via the web, there are several approaches to administering the interviews. Firstly, depending on the objectives of the research, it is possible to interview a single individual at a time (typical of quantitative research) or a group of individuals (typical of qualitative research: focus groups or bulletin boards). Secondly, the researcher might decide that the information should be collected through self-administered questionnaires, with or without assistance (e.g., online help provided through text or video chat), or to use the web platform to allow an interviewer to conduct a remote interview. Finally, respondents might be invited for an interview at a specific date and time (synchronous research – in real time), a popular approach for online focus groups, or might be asked to provide the information at their most convenient time (asynchronous research – with a potential lag between the event and the measurement), as is usually the case for a CAWI questionnaire.

6.3   Economic benefits of web survey research

Web surveys, although benefiting from relatively new technologies, allow data collection activities similar to those of traditional modes. However, the internet enables significant economies of scale, an important reduction in fieldwork time, and also represents a cost-efficient means to interview low-incidence population segments that would be otherwise very expensive to contact through traditional data collection modes (McDaniel and Gates, 2002).

Web surveys are well known for being an attractive means of collecting cheap and fast information from a potentially large group of people (Clayton and Werking, 1998; Schleyer and Forrest, 2000; Shannon and Bradshaw, 2002). Unfortunately, this benefit is so important that it tends to cast a shadow over the other benefits of web surveys and on their – sometimes important – drawbacks. As a consequence, nowadays researchers and companies often use the web mode inappropriately. It is important to highlight that researchers should choose the most appropriate data collection approach based on the research objectives; this means that the cost element should not attract excessive attention, as other elements are as important as cost to successfully achieve the research objectives (Sturgeon and Winter, 1999).

In some circumstances, the researcher may decide to place web surveys and traditional modes side by side, thus working with the so called *mixed-mode* approaches to data collection. These hybrid surveys allow remarkable results to be achieved, from both a statistical and an economic point of view, that would not otherwise be possible through a pure traditional approach based on a single data collection mode (De Leeuw, 2005; De Leeuw *et al.*, 2008; Dillman, 2007; Janssen, 2006). It is quite common to adopt a mixed-mode approach that begins with the less expensive mode (i.e., a web survey) to complete as many interviews as possible at the lowest possible cost, and carry on with a more expensive mode (e.g., CATI) to improve completion rates (Groves, 2004).

6.3.1   Fixed and variable costs

Generally speaking, the web mode is considered to be cheaper than traditional modes. However, it is difficult to quantify the cost of a web survey and make appropriate comparisons with alternative modes because different authors define the cost in different ways. Moreover, academics often consider only variable costs and fail to consider, and economically quantify, the time spent by the researcher on designing and managing the survey (Fricker and Schonlau, 2002). Both fixed and variable costs strictly depend on the approach chosen by the research agency, and it is extremely hard to estimate the total cost without knowing the research approach in detail. As a consequence, in this section we present only a few general considerations about the cost structure characterizing web research.

In web surveys, the strategic advantage related to the *fixed costs* lies in the fact that most of the data collection platform is owned and managed by the interviewees, who are required to have a computer, an appropriate web browser, and an internet connection (Solomon, 2001). It is their terminals that connect to the server where the electronic questionnaire is hosted. Therefore, web surveys avoid the expensive fixed costs affecting other data collection modes (call centre management for CATI, network of interviewers for PAPI/ CAPI, etc.). More precisely, the fixed costs are only related to (Couper *et al.*, 2001):

- hardware (servers and researchers' terminals);

- data collection and management software;

- internet connection and bandwidth;

- scripting and analysis team;

- web community or panel management (when available).

In web surveys, as well as in all other types of surveys where data collection is conducted through personal computers, digitalization of the data is immediate. Moreover, since web interviews are also self-administered, many potential *variable costs* are absent (Schmidt, 1997):

- typical mail questionnaire costs such as questionnaire printing, delivery and return of paper questionnaire, and data entry;

- typical PAPI questionnaire costs such as questionnaire printing and colour images or photographs;

- typical CATI costs such as call-centre employees' salaries and phone call charges;

- typical costs of a CAPI structure such as network management costs;

- interviewers' remuneration and expenses, room rents (e.g., test centres), stationery.

It is important to highlight that with traditional research modes, a significant portion of the cost is budgeted for the interviewers. In contrast, this cost element is absent in web surveys, while a significant portion of the budget is allocated to the interviewees as they are usually compensated for their time. More precisely, while for self-selection and list-based web surveys interviewees are usually not compensated for their time, and they might be simply offered some of the survey results, for web panel and community surveys respondents usually expect some remuneration. Depending on possible legal restrictions in the country where the survey is run, respondents could be given cash or retail vouchers for each completed survey (Göritz, 2006). Lotteries are also widely used to motivate respondents, as well as points-based systems allowing them to accumulate points for each completed survey and to redeem them once enough points have been collected. Some web communities also offer the opportunity to make donations to charity. Whatever the remuneration system, this budgetary shift from interviewer to interviewee is probably the key economic feature of web surveys.

It is also important to highlight that, while with traditional data collection modes the overall cost of a study tends to significantly increase as the sample size increases, with the online mode the sample size only slightly affects the overall cost (Figure 6.1). This is due to the different variable/fixed cost ratio: this is high for studies run through traditional modes for which the interviewers' remuneration becomes the main cost element, but low for those projects run via the web for which cost increases are almost exclusively associated with respondent incentives.

The possibility of significantly increasing the sample size represents a strategic opportunity for those studies where it is important to perform a detailed and robust analysis of specific segments of the population. However, it is important to remember that the sample size alone does not guarantee the quality of the data; in fact, there are several other elements that

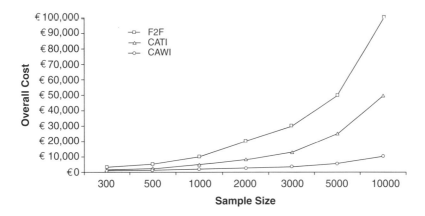

Figure 6.1   Data collection costs for different modes (face to face, CATI and CAWI) using the same questionnaire. The figures are rough estimates based on the authors' personal experience and refer to a typical 20-minute interview survey conducted in Italy and a nationally representative sample.

affect the quality, for instance an appropriate recruitment process on the basis of the research needs, the level of respondent incentives and motivators, and, last but not least, a well-written questionnaire of appropriate length.

6.4   Non-economic benefits of web survey research

Although economic considerations are very important in the choice of the research mode and tend to favour web surveys, there are other significant benefits provided by this mode. In fact, since researchers first tested data collection through the internet, some elements appeared to be quite beneficial to the overall survey management and reliability of the outcomes (Bailey, 1994; Dillman *et al.*, 1998).

The most relevant non-economic benefit is probably the *fast data capture turnaround*, thanks to the automatic questionnaire delivery, to the ready availability of the respondent's data in an electronic format, and above all to the absence of interviewers which allows quick and direct communication between interviewees and the management software. In theory, all potential interviewees could access an online questionnaire at the same time and complete it within minutes of each other. However, in reality, there are practical limitations to the number of respondents that can concurrently access the survey, due to limited bandwidth, server speed, and software features. Usually, these technical elements are properly sized based on the maximum number of respondents that are expected to access the survey at the same time. Excessively high specifications represent an inappropriate allocation of resources with likely negative consequences on the level of competitiveness of the data collection agency. The speed of data collection is such that it has been estimated that, for a web community survey, most of the interviews are completed within 48–72 hours (Harris, 1997). However, it is fair to highlight that such a high speed is usually achieved only when the community is large enough with respect to the number of interviews required and community members' historical response rates are stored in a database and taken into account when invitations are emailed out.

Another important feature of web surveys is that the *response rate* is generally higher than one using traditional data collection modes, with the sole exception of self-selection web surveys. For instance, the response rate for a web panel or community survey usually lies between 7% and 40%, depending on the recruitment and sampling framework, the research topic, the number of reminders sent out, and the type of incentives available (Couper, 2001; Martone and Furlan, 2007). In case of thematic communities, for instance a political-oriented community, the response rate is even higher, between 20% and 65%. It is worth highlighting that the response rate strongly depends on the relationship the community managers have with the community members; this is the single most important factor in driving the response rate. Statistics on the response rate achievable with list-based web surveys can be found in Fricker and Schonlau (2002).

The quick data collection turnaround achievable thanks to both the intrinsic characteristics of web surveys and the high response rate is thus a distinctive element of the online mode, allowing an impressive achievement in terms of number of interviews within a short time period. For instance, during the 2000 US presidential elections, Harris Interactive interviewed over 300 000 voters in just a week – the largest online election study in history. During the peak of the interviewing, more than 40 000 web interviews were processed per hour (Berrens *et al.*, 2003; Business Wire, 2000).

Web survey samples are generally not only large and affordable, but also *geographically well distributed*. This positively affects the representativeness of the sample, as rural respondents have theoretically the same probability of being selected for the interview as urban respondents. This is generally not true for other survey modes such as PAPI or CAPI, which provide highly skewed samples from a geographical point of view. Similar considerations apply for certain specific lifestyles (e.g., frequent flyers), for which adopting a web approach allows more representative samples to be collected.

Collecting data from different countries or from *multi-language* territories is not an issue with web surveys (Satmetrix, 2001). Data collection platforms not only can administer the questionnaire in the respondent's preferred language, but also usually offer some editing and analysis features (e.g., frequency distributions) to allow the researcher to check the progress of fieldwork and to promptly identify any low response rate or possible issues related to the logical flow of the questionnaire.

As anticipated earlier in this chapter, another key advantage of web surveys is that they represent a cost-efficient means to reach *low-incidence or hard-to-reach population segments* that would otherwise be very expensive to contact via traditional data collection modes (McDaniel and Gates, 2002). There are some population segments that are systematically excluded from traditional market research studies. For instance, people living in remote rural areas tend to be excluded from CAPI surveys, while night shift workers and managers are usually excluded from both CAPI and CATI surveys. The web approach is also fully compatible with building web communities oriented to specific themes or research areas, and appropriate to capture the interest and attention of specific population targets (e.g., credit card users, nurses, etc.). These thematic communities allow specific respondent segments to be interviewed without incurring prohibitive recruitment costs.

Convenience and anonymity are two other remarkable benefits of web surveys. Respondents are free to fill in the questionnaire at their convenience, for example during office lunch breaks or in the evening, rather than at the moment the phone interviewer happens to call (CATI) or the face-to-face interviewer knocks at the door (CAPI, PAPI). In addition, most online data collection platforms allow respondents to interrupt the interview and resume at a more convenient time, a facility particularly appreciated when interviewing population segments with little spare time (e.g., managers). Having this flexibility and freedom of choice not only helps respondents to find time to devote to the survey, but also contributes to removing the perception that the interview is invasive. Respondents' lack of availability and their perception of the interview as invasive are the two single elements that contribute most to the general trend of increasing non-response and refusal rates in traditional data collection modes.

Web surveys based on a structured or semi-structured questionnaire also benefit from the fact that there is no interviewer. Answers are not influenced by the desire to please or impress the interviewer or withdrawn to hide sensitive information (e.g., political, religious, or sexual orientation; medical conditions; illegal substances or alcohol consumption) as in traditional modes, where the interviewer is a prominent figure with a significant impact on the answers provided (de Leeuw, 1992; Levine *et al.*, 1999; Tourangeau and Smith, 1996; Turner *et al.*, 1998). Web respondents benefit from anonymity perception (Satmetrix, 2001), not only thanks to the absence of an interviewer, but also because they are well aware that research companies and academic institutions are fully committed to following rigid ethical guidelines and stricter and stricter privacy laws.

The absence of an interviewer not only provides clear economic and interview anonymity benefits, but also *reduces the probability of potential input errors* or misinterpretations occurring during the data input process. In web surveys, it is the respondent who feeds the answers directly into the data collection software, by selecting the appropriate option in the case of a pre-coded question, or by providing their answers in their own words, in the case of an open-ended question. Although the probability of errors is lower than in traditional surveys, it remains important to run consistency checks to remove possible records affected by quality problems (e.g., response set).

Web surveys offer the opportunity to establish a *direct relationship* between interviewee and researcher, as the distance between these two figures is much shorter than in other modes. In fact, in phone or face-to-face studies there are always some filters: the interviewer and, possibly, supervisors or fieldwork managers. Without these barriers, it becomes possible for the research agency to obtain feedback from the survey respondents, which is very useful in improving the quality of the questionnaires in terms of clarity of the wording and of logical flow, and in building trust. This is usually achieved by including in all questionnaires a final open-ended question asking for suggestions and comments, as well as by providing a toll-free telephone number and/or an email address for general queries or comments. Survey respondents really appreciate the opportunity to provide feedback as they perceive they are listened and not merely 'interrogated' (Conrad *et al.*, 2005).

As with CAPI questionnaires, web questionnaires allow easy inclusion of *multimedia material*, mainly images and video clips, but also audio clips, 3D content, and interactive exercises (Couper, 2000; Ray and Tabor, 2003). Multimedia has a key role when text alone is not sufficient to represent the context required in a survey or when the survey objectives are directly related to multimedia elements. A computer-based interview, with the respondent having direct access to the questionnaire, allows a full range of options for a realistic product concept administration and a more reliable data collection process (Couper *et al.*, 2001). For instance, it might be possible to improve the clarity of a rating scale and, therefore, to make the task of the respondent easier by adding some graphical emoticons to the scale, or by adopting a thermometer scale.

Last, but not least, the data collection software used in web surveys is usually extremely flexible and powerful, allowing a complex questionnaire structure (e.g., conditional and unconditional branching, showing/hiding choices, item randomization, text piping, etc.) as well as very personalized questionnaires (Solomon, 2001). A high degree of personalization is key to keeping the respondent's attention high, to optimizing the amount of information that can be collected within a specified time frame, and to adopting some highly interactive methodologies, such as an adaptive conjoint analysis (ACA) exercises (Johnson, 1987; Metegrano, 1996) or the flexible framework required by the paired comparisons conjoint exercise suggested by Furlan and Corradetti (2006). Finally, it is worth highlighting that flexible and advanced data collection and management software usually makes it possible to draw upon information collected in past surveys and stored in a database (usually socio-demographics, but also attitudinal or behavioural information) and to include or use it in the new questionnaire (the respondent might be prompted to confirm it).

6.5   Main drawbacks of web survey research

Although web surveys present undeniable advantages over traditional data collection modes, it is important to note that they present evident limitations too. Since the initial shift to web surveys, there has been intense debate about their shortcomings, with some researchers

(Mitofsky, 1999; Rivers, 2000) arguing that in some situations they might not produce trustworthy information, and internet researchers defending themselves from accusations of being under-cautious and using a non-validated approach.

As with other data collection approaches, web surveys cannot be employed indiscriminately in all projects: there are some projects that it is more appropriate to run via the web, while others should be conducted with a traditional approach. However, over the last decade web surveys have had the opportunity to mature. Both technological and statistical improvements, as well as infrastructural development and social changes (Kenett, 2004; Taylor, 1999), have made web surveys more feasible and better suited for a larger and larger variety of projects. The single most important innovation, from a statistical point of view, has probably been the adoption of propensity scoring (Schonlau *et al.*, 2003). This is a technique capable of compensating for lack of random selection, which has been adapted from other fields (Rosenbaum and Rubin, 1983, 1984).

One of the most common criticisms in the early years of web survey experimentation concerned the *low internet penetration* in the target population. Low levels of internet adoption pose inevitable representativeness problems due to the biases that arise from the differences between the target population and the sample collected. However, thanks to the growing spread of the internet and of high-speed connections, the online approach is now considered to be valid, at least for key target populations, in most developed countries (e.g., primary care physicians in the US). For instance, in the Netherlands the proportion of people with internet connection at home increased from 16% to 83% in the period from 1998 to 2005 (Bethlehem, 2007).

Unfortunately, even in developed countries, there are still some population segments characterized by low internet usage, typically those who are less likely to have internet access or are not comfortable with computers (Crawford *et al.*, 2001; Ray and Tabor, 2003). One important segment for market research with relatively low internet penetration is the elderly (Bethlehem, 2007). The perceived risks of the technology and a number of impediments related to income and education mean that a large proportion of elderly do not have the opportunity to use the internet, and thus to participate in web surveys.

Another criticism comes from researchers who accuse web surveys of *lack of representativeness* (Bandilla, *et al.*, 2001; Faas, 2003). With regard to this fundamental aspect of surveys, for the sake of clarity, it is necessary to distinguish between three types of internet surveys:

  (i) surveys naturally related to internet users;
 (ii) surveys focusing on a more general statistical population; and
(iii) surveys using closed lists of emails such as lists of customers or employees.

Online surveys of the first type are a pure product of the internet age (e.g., measurement of opinions of the surfers of a particular website). Those of the second type represent a migration of more traditional research studies (e.g., in concept testing). The third type are in fact a form of census since, typically, all customers or all employees are asked to complete the questionnaire (Kenett *et al.*, 2006).

In the first case, it is possible to collect a representative sample, and online surveys should definitely be considered the best methodological approach. However, because there are no good lists of internet users, even when the population of interest is made up of internet users, it is difficult to draw a probability sample. In census type surveys, a procedure such as the *M*-test can be used to assess the representativeness of the respondents (see Chapter 20 of this volume). For the second type of web surveys, however, it is not possible to collect a probability sample (Ray and Tabor, 2003) and, as a consequence, there are justified concerns about bias. As an example, consider the potential bias from using a banner advertisement to gather surfers, where respondents are affected by a strong and undesirable self-selection bias.

A better approach to conducting web surveys is to create a web community or panel of potential respondents (among others, All Global, Harris Interactive, SWG, Toluna). With this approach, the selection bias is strongly reduced, but not completely eliminated. Statistically speaking, communities and panels are considered to be better than simple lists of individuals, because surveys participants are recruited from a large variety of sources and their demographic and attitudinal profiles are known to the community or panel managers. Therefore, it is possible to extract from the members' list a sample of respondents potentially representative of the target population. The extraction of the contact sample from the web community or panel list should be done by taking into consideration demographic, behavioural, and attitudinal variables according to their distribution in the target population.

It is worth noting that such a sample should not be considered a probability sample, but only a raw approximation. Even with meticulous selection, the sample collected from a web community or panel is still affected by some bias. This residual bias is not only due to the different response rate of the community members, but also to the fact that online respondents differ in some ways from offline potential respondents. Generally speaking, people with online access tend to be younger, wealthier, better educated, heavy consumers of technology products, urban dwellers in dual-parent families, and of white or Asian/Pacific descent (Bethlehem, 2007; Couper, 2000; European Research into Consumer Affairs; 2001; Livraghi, 2006; Ministro per l'Innovazione e le Tecnologie, 2004; Palmquist and Stueve, 1996; Prandelli *et al.*, 2000; US Department of Commerce, 2000; White, 2000). An additional source of bias in web surveys is the profile of the participants. Apparently, among those with online access, participants in online surveys tend to be more involved with technology gadgets and tend to have more free time.

Web community and panel survey results are affected to a greater or lesser extent by these biases because of at least three decisions that people make in order to have their responses registered in a web survey:

1. The participant has to be connected to the internet. This depends on the ability to use computers, and the availability and cost of a computer with internet access. In general, a dial-up connection would suffice, but for some multimedia questionnaires a broadband connection is required.

2. The participant has to be a member of the specified community or panel. This depends on the visibility of the community or panel and the expected benefits from the registration.

3. The participant has to agree to respond to the survey. When community or panel members receive an invitation to participate in a survey, they must respond positively and complete the questionnaire in full (interrupted questionnaires are usually not considered for the analysis). This decision depends on the interest of the participant in the topic of the survey, on the expected remuneration, on the length of the questionnaire, and on spare time and availability.

Each of these decisions leads to a different type of selection problem: decision 1 leads to internet-users selection, decision 2 leads to a selection of specific web community or panel membership, and decision 3 leads to a full response to the online survey selection.

The presence of these biases has been discussed in the literature (Bradley, 1999; Taylor, 2000). One of the main criticisms of web community and panel surveys is that, since they are not based on probability sampling, they lack reliability. Although it might not be possible to completely correct the sample data for such biases, the adoption of the propensity scoring technique mentioned above provides some correction and helps to obtain more representative results from the web survey. It is important to stress that, in order to achieve valid and reliable results, the web mode should be chosen only when there is a reasonably high proportion of internet users among the target population, and the research agency should strictly follow best practices for the recruitment, sampling and analysis processes.

In web surveys, as in traditional self-administered interviews, there is also concern about the *identity and profile* of respondents. It is not always convenient or even possible to check whether respondents have provided accurate information about their identity and personal profile. Accurate identity checks are generally convenient or feasible only for high-profile community or panel members, such as healthcare professionals on whom checks are run against the list of registered medical practitioners. In order to run profile checks it is necessary to contact the respondent and ask the same profiling questions more than once and thus run a consistency analysis to spot possible discrepancies.

Web surveys also tend to be affected by a high proportion of *cheaters*, *speeders*, and *repeaters*. Cheaters are those respondents who tend to provide inaccurate or false answers because they want to complete the task in the shortest time or because they are not interested in the topic. They can be identified by consistency checks around one or more questions posed twice but in a slightly different way. Repeaters are those respondents who tend to provide the same or similar answers regardless of the question. Repeaters can be identified either at the data collection stage by the software (a warning is usually issued to the respondent) or at the analysis stage by checking the distribution of responses. Speeders are those respondents who complete the interview too quickly. Since the length of the interview is usually recorded, it is possible to identify speeders as soon as the interview is completed. These problematic respondents tend to be more common in web surveys than in traditional ones because of a number of factors:

- respondents might be interested in completing the survey only to be awarded the incentive;

- the absence of the interviewer gives the opportunity to be careless;

- the questionnaire might be complex or repetitive.

Web community and panel surveys, as well as traditional panel-based surveys, are also affected by the so-called *professional respondents*. These are community or panel members who usually belong to more than one community or panel. They join the community or panel and they agree to participate in the different surveys because they are lured by the incentive and not because they have an opinion and want to be heard. Professional respondents are often also cheaters, speeders and/or repeaters.

Although web surveys target respondents with internet access who are reasonably familiar with computer usage, it is plausible to expect that respondents have different levels of computer literacy or different experience with questionnaires (Dillman *et al.*, 1998). Therefore, the complexity of the questionnaire might be a problem for at least part of the respondent sample and lead to poor-quality results. In fact, if the web survey is self-administered, respondents cannot easily ask for clarification and support. The interviewer is absent, and thus *assistance* can only be provided via a toll-free telephone number and/or an online help desk. Careful and time-consuming design and programming of the questionnaire layout, structure and flow, clear and very detailed guidelines, and an appropriate support system are all key elements in avoiding high drop-out rates and low-quality interviews (Couper, 2000). Guidelines can be made interactive, thus fostering the social elements which are otherwise missing in a web survey.

Technical problems causing disruption to the data collection process might be serious enough to jeopardize the whole research project. On the one hand, questionnaires rich in graphics or multimedia content might cause computer freezes if the respondent's browser is not up to date or fully compatible, causing anger and frustration. On the other hand, a server breakdown or too many simultaneous accesses to the data collection agency's servers might cause serious delays and damage the reputation of the agency, possibly leading to future recruitment problems. In traditional research (CAPI or CATI), technical problems affecting the data collection process, although a cause of concern, are unlikely to lead to serious problems as it is generally possible to temporarily adopt an alternative solution (e.g., the interviewer could record the answers on a paper questionnaire). In addition, due to variations in computer specifications (e.g., screen resolution, internet browser), what is seen on the researcher's computer monitor may not be necessarily what is shown on the respondent's monitor (Dillman and Bowker, 2001; Ray and Tabor, 2003).

The researcher might also experience some problems when analysing data collected via a mixed-mode approach (Section 6.3) or when comparing web survey data against historical data collected via a traditional mode such as CAPI or CATI. In fact, *measurement differences* are quite likely to arise and are attributed to different recruitment processes, different environmental conditions of the interview (e.g., presence of the interviewer, explanations and support available), different questionnaire administration and channel of communication (e.g., questions read by the interviewer, visual scales), and different exercises (e.g., visual ranking exercises through drag-and-drop, card sorting exercise); see de Leeuw (2005), Groves (2004) and Parks *et al.* (2006).

6.6   Web surveys for customer and employee satisfaction projects

Customer satisfaction refers to the extent to which a business's customers are 'happy' with the products and services received. In the modern global village, customer satisfaction surveys are a critical input to process improvement initiatives. Both product and service providers typically conduct annual and ongoing customer satisfaction surveys to identify areas for improvement, to determine the effectiveness of improvement efforts, and to track trends over time (Kenett *et al.*, 2006).

Similarly, employee satisfaction is a measure of how 'happy' workers are with their job and working environment, providing the company with the opportunity to 'look into the mirror'. Letting the workers know that their voice is being listened to and their concerns addressed by those in charge has a positive effect on workers' morale; this is obviously of tremendous benefit to any company as it gives the opportunity to implement an improvement plan.

Customer and employee satisfaction levels can be measured using survey techniques and questionnaires. Typically, a customer or employee satisfaction survey contains three parts:

overall measures of satisfaction, rating of key performance attributes, and demographics questions (Chisholm, 1999). Sometimes customers are also asked to rate how important each key attribute is, in order to measure the gap between performance and importance and thus provide actionable insight for the management. There are pros and cons with every data collection mode. Mail surveys are cheap but tend to have a very low overall response rate. PAPI and CAPI have the potential to collect rich data, but have an unintended influence on the results and also tend to be quite expensive. Web surveys, on the other hand, are cheap and lack interviewer bias. Their potential advantages are so important that, according to Chisholm (1999), the internet will become the most widely used method to measure customer satisfaction.

List-based web surveys are appropriate whenever a closed list of potential respondents is available. Obviously, this is always the case for employee satisfaction surveys, as companies hold details of the employees as well as their corporate email addresses (when provided). However, depending on the type of business, a company might not have a list of its own customers. While list-based web surveys might be appropriate for companies operating in a business-to-business environment (e.g., marketing and advertising services, mainframe servers), as the customers are well known and their details (including email addresses) are carefully stored on a database system, they are usually not appropriate for companies supplying, directly or indirectly, final consumers (e.g., fast-moving consumer goods). In fact, with few exceptions (e.g., automotive sector), their email contact details are generally not available. When an appropriate list is not available, customer satisfaction information is generally collected through a self-selection web survey. In this case, customers are usually intercepted when browsing the company's website or through a loyalty scheme they might belong to (e.g., Nectar).

With regard to customer satisfaction studies, it should be borne in mind that some clients might be authorized to use email but might not have easy access to the internet. In this case, the invitation email will be received, but the customer will not have the opportunity to access the internet to take part in the survey (Hill *et al.*, 2003). An important concern for customer satisfaction surveys based on invitation emails is that people are becoming increasingly suspicious of unsolicited emails and the invitation email might end up in the spam folder. To make things worse, business customers receive huge numbers of emails every day, and their main survival strategy is to immediately delete all non-essential emails, inevitably including surveys (Hill *et al.*, 2003). To improve response rates, it might be a good idea to precede or follow up the invitation email with a phone call, asking the customer to kindly complete the survey and stressing that the survey results will help to offer a better service.

With regard to employee satisfaction studies, there are two main concerns to be addressed when adopting a web approach. Firstly, some employees might not have access to a corporate computer with internet connection. Should this be the case, invitations might be sent to a home computer or a suitable computer with internet connection might be made available for this purpose on the company premises. Another solution, also appropriate for those employees unable to use a computer, would be to prepare a traditional paper questionnaire for circulation only among employees excluded from the web survey. Secondly, the researcher must ensure the highest possible level of anonymity to allow workers to respond openly, without fear of repercussions; employees will not respond candidly, if at all, to satisfaction surveys unless complete anonymity is guaranteed (Chisholm, 1999). To achieve this, company intranets should be carefully avoided and invitations to participate in web surveys have to come from independent organizations; survey data should not be stored on the employer's server and the employer should be able to get hold of aggregated data only.

6.7   Summary

There is no single perfect survey mode for every possible project requirement and research condition: traditional modes can be more appropriate in some circumstances, while web surveys can represent the best available choice in other situations. The researcher should always make every effort to identify the most suitable data collection approach based on the research objectives. A sensible choice involves considering the trade-offs between the available time for the set-up, management and data collection phases, experience with each mode, the target population characteristics, the benefits and drawbacks of each mode, and, last but not least, the budget available for the research.

Web surveys present undeniable advantages over traditional modes. The most appealing ones are probably the low cost associated with the data collection process and its fast turnaround. In addition, the capacity of the internet to deliver multimedia content in a flexible format, the convenience and anonymity guaranteed to respondents, and the ability to reach low-incidence segments make the internet an ideal platform for concept tests surveys, exploration of sensitive topics, or gathering information pertaining to a market niche. Nevertheless, web surveys are affected by important drawbacks, the main one being low internet penetration in specific target populations and the consequent lack of representativeness. Although some approaches have been developed to adjust for such drawbacks, at least in part, the scientific community is not yet fully ready to accept the appropriateness of web surveys in several research areas (e.g., estimates of unemployment, welfare payments, consumer expenditures). Yet, in areas such as these, web surveys may well serve a supplemental or supporting role, particularly in mixed-mode applications or surveys of specialized populations (Couper, 2001). However, as the internet continues to spread, it is fair to expect that the current constraints in web surveys will in due time be reduced.

References

Bailey, K.D. (1994) *Methods of Social Research*, 4th edn. New York: Free Press.

Bandilla, W., Bosnjak, M. and Altdorfer, P. (2001) Effekte des Erhebungsmodus? Ein Vergleich zwischen einer Web-basierten und einer schriftlichen Befragung zum ISSP-Modul-Umwelt. *ZUMA Nachrichten*, 49, 7–28.

Berrens, R.P., Bohara, A.K., Jenkins-Smith, H., Silva, C. and Weimer, D.L. (2003) The advent of Internet surveys for political research: A comparison of telephone and Internet samples. *Political Analysis*, 11(1), 1–22.

Bethlehem, J.G. (2007) Reducing the bias of web survey based estimates. Discussion paper 07001, Statistics Netherlands, Voorburg/Heerlen, The Netherlands.

Bethlehem, J.G. (2010) Selection bias in web surveys. *International Statistical Review*, 78, 161–188.

Bradley, N. (1999) Sampling for Internet surveys: an examination of respondent selection for internet research. *Journal of the Market Research Society*, 41(4), 387–394.

Business Wire (2000) 2000 Election winners: George W. Bush and online polling; Harris Interactive scores unprecedented 99% accuracy in predicting 2000 election outcome. *Business Wire*, 14 December.

Cambiar (2006) The online research industry. An update on current practices and trends. Technical report, http://www.dufferinresearch.com/downloads/TheOnlineResearchIndustry2006.pdf (retrieved 19 August 2010).

Chisholm, J. (1999) Using the Internet to measure customer satisfaction and loyalty. In R. Zemke and J.A. Woods (eds), *Best Practices in Customer Service*, pp. 305–317. New York: AMACOM.

Clayton, R.L. and Werking, G.S. (1998) Business surveys of the future: The World Wide Web as a data collection methodology. In M.P. Couper, R.P. Baker, J. Bethlehem, C.Z.F. Clark, J. Martin, W.L. Nicholls II and J.M. O'Reilly (eds.), *Computer-Assisted Survey Information Collection*, pp. 543–562. New York: John Wiley & Sons, Inc.

Clinton, J. D. (2001) Panel bias from attrition and conditioning: A case study of the knowledge networks panel. Unpublished manuscript, Department of Political Science, Stanford University, http://www.knowledgenetworks.com/insights/docs/Panel%20Effects.pdf (retrieved 18 August 2010).

Conrad, F.G., Couper, M.P., Tourangeau R. and Galesic M. (2005) Interactive feedback can improve the quality of responses in web surveys. Paper presented at The American Association for Public Opinion Research (AAPOR) 60th Annual Conference, Miami Beach, FL.

Couper, M.P. (2000) Web surveys: a review of issues and approaches. *Public Opinion Quarterly*, 64, 464–494.

Couper, M.P. (2001) Web survey research: challenges and opportunities. In *Proceedings of the Annual Meeting of the American Statistical Association*, 5–9 August, Atlanta, GA.

Couper, M.P., Traugott, M.W. and Lamias, M.J. (2001) Web survey design and administration. *Public Opinion Quarterly*, 65(2), 230–253.

Crawford, S.D., Couper, M.P. and Lamias, M.J. (2001) Web surveys: perception of burden. *Social Science Computer Review*, 19, 146–162.

De Leeuw, E.D. (1992) *Data Quality in Mail, Telephone, and Face-to-Face Surveys*. Amsterdam: TT Publications.

De Leeuw E.D. (2005) To mix or not to mix data collection modes in surveys. *Journal of Official Statistics*, 21(2), 233–255.

De Leeuw, E.D., Hox, J.J. and Dillman, D.A. (2008) Mixed-mode surveys: When and why. In E.D. De Leeuw, J.J. Hox and D.A. Dillman (eds), *International Handbook of Survey Methodology*. New York: Lawrence Erlbaum Associates.

Dillman, D.A. (2007) *Mail and Internet Surveys: The Tailored Design Method. 2007 Update with New Internet, Visual, and Mixed-Mode Guide*. Hoboken, NJ: John Wiley & Sons, Inc.

Dillman, D.A. and Bowker, D.K. (2001) The web questionnaire challenge to survey methodologists. In U.-D. Reips and M. Bosnjak (eds), *Dimensions of Internet Science*. Lengerich: Pabst Science Publishers.

Dillman, D.A., Tortora, R.D. and Bowker, D.K. (1998) Principles for constructing web surveys: An initial statement. Technical Report 50-98, Social and Economic Sciences Research Center, Washington State University, Pullman, WA.

Epstein, J., Klinkenberg, W.D., Wiley, D. and McKinley L. (2001) Insuring sample equivalence across internet and paper-and-pencil assessments. *Computers in Human Behavior*, 17(3), 339–346.

European Research into Consumer Affairs (2001) Preventing the digital television and technological divide. Technical report, http://www.net-consumers.org (PDF version of document retrieved 23 February 2006).

Faas, T. (2003) Offline rekrutierte Access Panels: Königsweg der Online-Forschung?, *ZUMA-Nachrichten*, 53(27), 58–76.

Fricker, R.D. Jr. and Schonlau, M. (2002) Advantages and disadvantages of Internet research surveys: Evidence from the literature. *Field Methods*, 14(4), 347–367.

Fuller, J., Bartl, M., Ernst, H. and Muhlbacher, H. (2004) Community based innovation: A method to utilize the innovative potential of online communities. In *Proceedings of the 37th Annual Hawaii International Conference on System Sciences* (HICSS 2004), pp. 195–204. Los Alamitos, CA: IEEE Press.

Furlan, R. and Corradetti, R. (2006) Reducing conjoint analysis paired comparisons tasks by a random selection procedure. *Quality and Reliability Engineering International*, 22(5), 603–612.

Göritz, A. S. (2006) Incentives in Web studies: Methodological issues and a review, *International Journal of Internet Science*, 1, 58–70.

Groves, R.M. (2004) *Survey Errors and Survey Costs*. Hoboken, NJ: Wiley-Interscience.

Hagel III, J. and Armstrong, A.G. (1997) Net gain: Expanding markets through virtual communities. Boston: Harvard Business School Press.

Harris, C. (1997) Developing online market research methods and tools. *Marketing and Research Today*, 25(4), 267–279.

Hill, N., Brierley J. and MacDougall, R. (2003) *How to Measure Customer Satisfaction*, 2nd edn. Aldershot: Gower.

Janssen, B. (2006) Web data collection in a mixed mode approach: An experiment. In *Proceedings of Q2006 European Conference on Quality in Survey Statistics,* Cardiff.

Johnson, R.M. (1987) Adaptive conjoint analysis. In *Proceedings of the Sawtooth Software Conference.* Sequim, WA: Sawtooth Software.

Kenett, R.S. (2004) The integrated model, customer satisfaction surveys and Six Sigma. In *Proceedings of the First International Six Sigma Conference,* Center for Advanced Manufacturing Technologies, Wrocław University of Technology, Wrocław, Poland.

Kenett, R.S., Kaplan, O. and Raanan, Y. (2006) Surveys with new technologies: is it the end of telephone interviews? (in Hebrew), *Kesher Haeihut*, 53–54, 6–8.

Kozinets, R. (2002) The field behind the screen: Using netnography for marketing research in online communications. *Journal of Marketing Research*, 39, 61–72.

Levine, P., Ahlhauser B. and Kulp D. (1999) Pro and con: Internet interviewing. *Marketing Research*, 11(2), 33–36.

Livraghi, G. (2006) *Dati sull'Internet in Italia*. http://www.gandalf.it/dati/index.htm (retrieved 15 September 2006).

Martone, D., and Furlan, R. (2007) *Online market research: Tecniche e metodologia delle ricerche di mercato tramite Internet*. Milan: Franco Angeli.

McDaniel, C. and Gates, R. (2002) *Marketing Research: The Impact of the Internet*, 5th edn. Cincinnati, OH: South-Western College Publishing.

Metegrano, M. (ed.) (1996) *ACA system version 4.0*. Sequim, WA: Sawtooth Software.

Ministro per l'Innovazione e le Tecnologie (2004) Rapporto Statistico sulla Società dell'Informazione in Italia. Technical report, http://www.osservatoriobandalarga.it (PDF version of document retrieved 05 August 2006).

Mitofsky, W. J. (1999) Pollsters.com. *Public Perspective*, 10(24), 24–26.

Palmquist, J. and Stueve, A. (1996) Stay plugged into new opportunities. *Marketing Research*, 8(1), 13–15.

Parks, K.A., Pardi, A.M. and Bradizza, C.M. (2006) Collecting data on alcohol use and alcohol-related victimization: A comparison of telephone and web-based survey methods. *Journal of Studies on Alcohol*, 67, 318–323.

Prandelli, E., Spreafico, C. and Pol, A. (2000) Osservatorio Internet Italia: l'utenza Internet 2000. *I-LAB Centro di Ricerca sull'Economia Digitale*, http://www.unibocconi.it (PDF version of document retrieved 08 March 2006).

Ray, N.M. and Tabor, S.W. (2003) Several issues affect e-Research validity, *Marketing News*, 15, 50–53.

Rivers, D. (2000) Fulfilling the promise of the Web. *Quirks Marketing Research Review*, 34–41.

Rosenbaum, P.R. and Rubin, D.B. (1983) The central role of propensity score in observational studies for casual effects. *Biometrika*, 70(1), 41–55.

Rosenbaum, P.R. and Rubin, D.B. (1984) Reducing bias in observational studies using subclassification on the propensity score. *Journal of the American Statistical Association*, 79, 516–524.

Satmetrix (2001) *Investigating Validity in Web Survey*. Mountain View, CA: Satmetrix Systems.

Schleyer, T.K.L. and Forrest, J.L. (2000) Methods for the design and administration of web-based surveys. *Journal of the American Medical Informatics Association*, 7, 416–425.

Schmidt, W.C. (1997) World-Wide Web survey research: Benefits, potential problems and solutions. *Behaviour Research Methods, Instruments and Computers*, 29(2), 274–279.

Schonlau, M., Fricker, Jr. R.D.,and Elliott, M.N. (2002) *Conducting Research Surveys via E-mail and the Web*. Santa Monica, CA: RAND.

Schonlau, M, Zapert, K., Payne, L.S., Sanstad, K., Marcus, S., Adams, J., Spranca, M., Kan, H. J., Turner, R. and Berry, S. (2003) A comparison between a propensity weighted web survey and an identical RDD survey. *Social Science Computer Review*, 21(10), 1–11.

Shannon, D.M. and Bradshaw, C.C. (2002) A comparison of response rate, response time, and costs of mail and electronic surveys. *Journal of Experimental Education*, 70, 179–192.

Sharot, T. (1991) Attrition and rotation in panel surveys. *The Statistician*, 40, 325–331.

Solomon, D.J. (2001) Conducting web-based surveys. *Practical Assessment, Research & Evaluation*, 7(19), http://pareonline.net/getvn.asp?v=7&n=19 (retrieved 19 August 2010).

Sturgeon, K. and Winter, S. (1999) International marketing on the World Wide Web. New opportunities for research: what works, what does not and what is next. In *Proceedings of the ESOMAR Worldwide Internet Conference Net Effects*, London, pp. 191–200.

Taylor, H. (1999) The global Internet research revolution: A status report. *Quirk's Marketing Research Review*, 534, 1-3.

Taylor, H. (2000) Does Internet research work? Comparing online survey results with telephone survey. *International Journal of Market Research*, 42(1), 51–63.

Tourangeau, R. and Smith, T. W. (1996) Asking sensitive questions: The impact of data collection mode, question format, and question context. *Public Opinion Quarterly*, 60, 275–304.

Turner, C.F., Ku, L., Rogers, S.M., Lindberg, L.D., Pleck, J.H. and Sonenstein, F.L. (1998) Adolescent sexual behavior, drug use, and violence: Increased reporting with computer survey technology. *Science*, 280, 867–873.

US Department of Commerce (2000) Fall through the Net: Toward digital inclusion; a report on Americans' access to technology tools. Technical report.

White, E. (2000) Market research on the Internet has its drawbacks. *Wall Street Journal*, 2 March, B4.

7

The concept and assessment of customer satisfaction

Irena Ograjenšek and Iddo Gal

In all sectors of the economy, organizations need to know how well they are satisfying customer needs and expectations. To achieve this, they have to translate general conceptions of customer satisfaction into decisions about what to measure and what methods to use to collect data. A full understanding of the customer satisfaction assessment thus requires a look beyond technical (analytical methods and models) into substantive and methodological issues. Both conceptualization and operationalization of customer satisfaction, as well as the logic of measuring it, have to be examined. This chapter offers a roadmap for such examination. Following the introductory section, Section 7.2 addresses the interrelated concepts of customer satisfaction, perceived service quality and customer loyalty. Section 7.3 deals with methodological issues relevant in survey collection of customer satisfaction data. Section 7.4 evaluates the ABC ACSS questionnaire from the conceptual and methodological points of view discussed in Sections 7.2 and 7.3.

7.1 Introduction

How can organizations improve their assessment of customer satisfaction? We suggest that they have to properly translate general conceptions of customer satisfaction into decisions about what to measure and what methods to use to collect data. A full understanding of the issues involved in assessing customer satisfaction thus requires a look beyond the technical (analytical methods and models), focusing on the substantive and methodological. Both the conceptualization and operationalization of customer satisfaction, as well as the logic of measuring it, have to be examined.

The framework for this examination can be either the business-to-customer (B2C) or business-to-business (B2B) setting. The discussion in this chapter pertains to interactions

Modern Analysis of Customer Surveys: with applications using R, First Edition. Edited by Ron S. Kenett and Silvia Salini.
© 2012 John Wiley & Sons, Ltd. Published 2012 by John Wiley & Sons, Ltd.

within both of these. Even though some recent literature on customer satisfaction assessment focuses on issues in a B2B context, the larger body of literature has evolved in the context of B2C services and can offer both conceptual models and valuable methodological insights. In addition, B2B and B2C are not clearly delineated functional environments. Some organizations combine B2B and B2C features by serving businesses as well as end-users or individual customers simultaneously (e.g., Nespresso). Finally, customer satisfaction and related concepts of perceived service quality and customer loyalty are multifaceted constructs regardless of the framework used for their examination. Even in a B2B context (and consequently in the context of an establishment survey) they have an individual component. If we discuss loyalty or satisfaction of an organization (i.e., the organization is the customer), we should also consider loyalty or satisfaction of *an individual employee* (e.g., a purchasing agent) who, on behalf of the organization, interacts with the service provider and with its frontline service personnel. Thus, from the conceptual standpoint, a thorough understanding of customer satisfaction and related concepts in the B2B context must by default also involve ideas from the B2C context, because the B2B environment encompasses both organizational and individual-level sources of customer satisfaction and dissatisfaction.

The structure of this chapter is based on the premise that customer satisfaction assessment has to take into account both substantive and methodological issues, and that a focus on technical issues *per se* is not sufficient. Therefore, following this introduction, we address the concept of customer satisfaction and establish its relationships with perceived service quality and customer loyalty in Section 7.2. In Section 7.3 we reflect on methodological challenges of customer satisfaction surveys. In Section 7.4 we bring together conceptual and methodological issues discussed earlier and show their relevance in the evaluation of the ABC ACSS questionnaire introduced in Chapter 2. Finally, we summarize the contents of this chapter in Section 7.5.

7.2   The quality–satisfaction–loyalty chain

7.2.1   Rationale

Informally, it would appear to be easy to discuss 'satisfaction' or 'dissatisfaction' from a product, service, or experience point of view. However, when we aim for an *assessment* of customer satisfaction, in particular an assessment that is the basis for an organization's discussions, decision-making and actions, we need to be clear about the meaning of the concepts being used. Without conceptual clarity, it is difficult to consider the meaning and interpretation of *measurements* (i.e., numerical representations) of customer satisfaction levels, which are the focus of this book. Yet, as with many other concepts, multiple perspectives on the meaning of customer satisfaction have evolved over the years. In this section we briefly outline and evaluate some of these, and discuss the links between perceived service quality, customer satisfaction, and customer loyalty as addressed by Heskett *et al.* (1997).

7.2.2   Definitions of customer satisfaction

The scholarly and applied literature on service management refers to several concepts related to customers' reactions to a product or service. These are primarily 'customer satisfaction' and 'perceived service quality', but increasingly also 'customer experience' and 'customer engagement'.

At the most general level, perceived service quality and customer satisfaction are evaluation or appraisal variables that relate to customers' judgement about a product or service. Although research suggests that both customer satisfaction and perceived service quality are distinct constructs (Oliver, 1997; Taylor and Baker, 1994), and that there is a causal relationship between the two (Cronin and Taylor, 1992; Gotlieb *et al.*, 1994; Spreng and Mackoy, 1996; Salini and Kenett, 2009), they have in some cases been used interchangeably (Iacobuci *et al.*, 1994; Mittal *et al.*, 1998; Oliver, 1997; Parasuraman *et al.*, 1994; Taylor and Baker, 1994). However, some differences between the two concepts are acknowledged. Oliver (1997) argues as follows:

- Perceived service quality judgements are evaluations of specific cues or attributes, whereas satisfaction judgements are more global.

- Expectations of perceived service quality are based on perceptions of 'excellence', whereas satisfaction judgements include referents such as need and equity or fairness.

- Perceived service quality judgements are more cognitive, whereas satisfaction judgements are more affective and emotional reactions.

An additional aspect of conceptualizing customers' reactions to service is the time dimension. Lovelock and Wright (1999) define perceived service quality as 'customer's long-term, cognitive evaluations of a company's service delivery', and customer satisfaction as a 'short-term emotional reaction to a specific service performance'. They argue that satisfaction is by default experience-dependent, since customers evaluate their levels of satisfaction or dissatisfaction after each service encounter. In turn, this information is used to update customer perceptions on service quality. However, attitudes about service quality are not necessarily experience-dependent. They can also be based on the word of mouth or advertising. Thus, it may seem that satisfaction determines quality, and not vice versa.

Oliver's (1997) research on the direction of the causal relationship between perceived service quality and customer satisfaction adds a further layer of complexity by suggesting that the direction of influence of one variable over another depends on the level at which measurement is conducted. Oliver (1997) argues that at the single-transaction level there is a strong quality-affects-satisfaction relationship, while at the multiple-transaction level there is a strong satisfaction-affects-quality relationship, because overall satisfaction judgements influence customers' attitudes about perceived service quality.

The above discussion suggests that customer satisfaction is inherently a somewhat elusive construct, for several separate but related reasons: the construct has both cognitive and emotional components; its meaning has a relativistic aspect (i.e. people may have different psychological benchmarks in mind against which they compare their level of satisfaction or judgement of quality); and people's responses to satisfaction surveys depend on the time frame to which they, and not the survey, are referring. Further, it is important to note that while the research literature tries to distinguish between customer satisfaction and perceived service quality, in practice many service organizations do not: when designing customer satisfaction surveys, they often mix dimensions of these two constructs. For this reason, in the remainder of this chapter we often refer to customer satisfaction as a term encompassing all related constructs. For more on integrated models combining customer satisfaction with other dimensions, see Kenett (2004) and Chapter 5.

7.2.3   From general conceptions to a measurement model of customer satisfaction

In order to be able to measure customer satisfaction and perceived service quality, general conceptions have to be translated into decisions about *what to measure*. The conceptualization of these two concepts, and the resulting suggestions for measuring them, first emerged and matured in the scholarly and applied literature relating to the B2C setting. Seminal work in this regard has been done in the USA by Parasuraman *et al.* (1985, 1988, 1991a, 1994) and in Europe by Grönroos (1990, 2000). Two key outcomes of this work are outlined below.

Firstly, the research has advocated the use of a *gap* model. This means that customer satisfaction and perceived service quality are conceptualized as the gap between *expectations* (= what I want) and *perceptions* (= what I get) of relevant service attributes. Thus, if expectations exceed perceptions, perceived service quality is deemed as (relatively) poor or satisfaction is negative. If perceptions are higher than expectations (i.e., I got more than I expected), perceived service quality is deemed (relatively) high or satisfaction is positive; see, for example, Oliver (1997) or Zeithaml and Bitner (2002) for a more elaborate treatment of these ideas.

Secondly, the literature has pointed to several key service dimensions which customers care about and should therefore be measured in a survey. A useful starting point in this regard is the *SERVQUAL model* (Zeithaml *et al.*, 1990), a widely researched and used framework which specifies five key dimensions of perceived service quality:

- *Tangibles:* The conditions or appearance of physical facilities, equipment and personnel.

- *Reliability:* The ability to perform the promised service dependably and accurately.

- *Responsiveness:* Willingness to help customers and provide prompt service.

- *Assurance:* Knowledge and courtesy of employees as well as their ability to convey trust and confidence.

- *Empathy:* Individual care and a sense of attention to personal needs that a company provides its individual customers with.

Zeithaml *et al.* (1990) report that while all dimensions are important to customers, the reliability dimension appears to weight heavier than the others. It should also be noted that the first two dimensions relate to overall aspects of a service system, that is, they refer to an outcome of the integrated operation of organizational units, policies and technologies. The last three dimensions have an increasingly personal component as they reflect behaviours or attitudes of individual frontline employees.

Based on the basic ideas of a 'gap' model referring to five key dimensions, Parasuraman *et al.* (1985, 1988, 1991a, 1994) developed the SERVQUAL approach to assessing perceived service quality, whereby respondents are asked to answer two sets of questions covering the five dimensions listed above: one set regarding expectations from organizations in the given industry sector (e.g., what I expect with regard to cleanliness of the leading hotels in general), and one set of perceptions regarding a company in question (e.g., what I perceived or experienced with regard to cleanliness at the specific hotel).

Based on these ideas, and after extensive testing and refinements, Parasuraman et al. (1985, 1988, 1991a, 1994) developed a measurement instrument with 22 items (see Appendix). For each of these items (service attributes), a quality judgement is computed according to the following formula:

$$\text{Perception } (P_i) - \text{Expectation } (E_i) = \text{Quality } (Q_i).$$

The SERVQUAL score (perceived service quality) is then obtained from the following equation:

$$Q = \frac{1}{22} \sum_{i=1}^{22} (P_i - E_i).$$

To date, the SERVQUAL model has been used and examined in thousands of studies around the world, in different cultures and many different service industries in both B2B and B2C settings (see Ladhari, 2008). We briefly summarize some of the main concerns regarding the conceptual aspects of the SERVQUAL model, as they have important implications for the interpretation of results from customer satisfaction surveys and should be noted when organizations examine ways to improve their existing surveys. While the original focus of the SERVQUAL model has been on B2C issues, the same concerns can be raised with regard to its B2B applications:

- *Object of measurement*. Given the mix of systemic (organization-level) and individual (employee-level) issues, as well as cognitive and affective issues, to which the customer is asked to respond, it is not clear whether the scale measures perceived service quality or customer satisfaction.

- *Wording of the questionnaire*. Kasper *et al.* (1999) point out that it might be better not to use negatively worded questions since they are interpreted with more difficulty than the positively worded ones when it comes to perceptions and expectations.

- *Use of $P_i - E_i$ difference scores*. The value and purpose of using two separate data sets in the same questionnaire (perceptions and expectations) has been questioned (e.g., by Brown *et al.*, 1993). It has been argued that increasing $P_i - E_i$ scores may not always correspond to increasing levels of perceived service quality, and that this affects the construct validity of the assessment (Teas, 1993a, 1993b, 1994). Some researchers suggest that it might be better not to use difference scores since the factor structure of the answers given to the questions about expectations and perceptions, and the resulting difference scores, are not always identical. Additionally, research shows that performance perception scores alone give a good quality indication. Therefore, a modified scale using only performance perceptions to measure perceived service quality (SERVPERF) has been proposed (Cronin and Taylor, 1992, 1994).

- *Generalization of perceived service quality dimensions*. An empirically identified five-factor structure cannot be found in all service industries. Only the tangibles dimension has been confirmed in most studies. Otherwise the number of distinct perceived service quality dimensions varies from one to nine (Buttle, 1995). It seems that the dimensionality of perceived service quality might be determined by the type of service a researcher or an organization is dealing with. Sometimes, items belonging to seemingly separate dimensions appear to group together as one factor, suggesting that customers' reactions do not correspond with the assumed instrument design.

- *The static nature of the model.* There exist a number of long-term service processes (such as education or consulting) where both perceptions and expectations (and consequently quality evaluations) may change in time as customers change in terms of their expectations, qualifications, and understanding of their needs or qualifications (Haller, 1998). For these service processes, static or separate assessments cannot provide a good analytical framework. A dynamic model may be needed.

It is important to note that while the research literature tries to adequately address these concerns, in practice many service organizations do not. When designing customer satisfaction surveys, they often uncritically model them after the existing industrial 'templates', with all their strengths and weaknesses.

7.2.4   Going beyond SERVQUAL: Other dimensions of relevance to the B2B context

The five SERVQUAL dimensions discussed in the previous section are a useful starting point from which to examine the conceptualization and measurement of customer satisfaction in both the B2C and B2B contexts. However, they do not account for the full spectrum of issues of interest that pertain to the B2B context. The marketing and service sciences literature suggests a range of additional factors, which we sketch below:

- *Combination of individual and organizational levels.* A preliminary set of considerations pertains to an idea that was briefly noted in the introduction to this chapter – that satisfaction in a B2B context involves both individual and organizational components. As Habryn *et al.* (2010) explain, a B2B relationship is conducted across multiple organizational levels. Depending on how a service provider is organized, the relationship may involve individual employees (e.g., sales personnel, technical support or service specialists, supply-chain managers), as well as whole service teams, different business units, and at times senior managers representing the whole organization.

- *Greater focus on relationship constructs.* Several groups of factors with aspects specific to the B2B context stem from the realization that the B2B context often involves the creation of long-range relationships between business partners or clients and suppliers. These relationships are partly based on underlying constructs such as *trust, commitment,* or *shared values* (Theron *et al.*, 2010). While trust is somewhat related to the dimension of assurance included in the SERVQUAL B2C model, it arguably has a deeper meaning in the B2B context, given the underlying co-dependencies of service providers and their business clients. Such constructs, or related derivatives, should be included in an assessment of customer satisfaction to strengthen the coverage of affective or attitudinal issues that characterize customer satisfaction more than perceived service quality, and may underlie B2B customer loyalty. It is also important to examine customer engagement and customer experience, concepts that are rapidly gaining in importance with more and more B2B interactions taking place in a virtual environment.

- *Complaint satisfaction.* The complaint satisfaction construct was originally developed in the context of B2C but is relevant to B2B as well. According to Stauss and Seidel (2004), when a problem emerges in a service setting, it is essential to examine to what extent the client (the organization or its employee) is (dis)satisfied with the manner in which the complaint was handled and resolved. Stauss and Seidel (2004) outline several dimensions underlying complaint satisfaction: accessibility of the service provider; quality of interaction (e.g., politeness); speed and efficiency of complaint handling; and fairness of the outcome. Different dimensions of complaint satisfaction can contribute independently to overall satisfaction with the service. For example, a customer may perceive that a complaint was handled promptly, yet feel that the resolution was unjust, or communicated impolitely without acknowledging the anguish or damages that the client has suffered. These feelings can easily cause further irritation ('Not only did you deliver a defective product, but you blamed us instead of taking responsibility and were only willing to provide a refurbished part instead of the new one that we were supposed to receive.').

- *Usability of e-service technology and websites.* A separate group of factors is associated with the role of technology in the service process (Bitner *et al.*, 2010), and its contribution to the customer experience, specifically in the context of e-service provision or e-commerce. Increasingly, B2B as well as some B2C services are provided through electronic channels. The features of the web-based facilities through which customers interact with a website or a mobile application have been the subject of considerable research efforts. Santos (2003) identifies two groups of influential factors. 'Incubative' factors are of a more technical nature. Service providers can design and improve them via usability research well before customers ever use the service (e.g., ease of use, appearance, linkage, structure and layout, contents). On the other hand, 'active' factors, although they might be pre-designed, are eventually always evaluated by customers. Consequently, they are more closely associated with traditional constructs in customer satisfaction research (e.g., reliability, efficiency, support, communication, security, incentives). Further factors pertaining to the usability of technology-based services have been identified in ergonomics research and human factor studies (Hornbaek, 2006). For usability studies based on traffic web logs, see Harel *et al.* (2008).

The range and number of factors that may impact perceived service quality and customer satisfaction in a B2B setting (the list above is by no means exhaustive) suggest that the conceptual model underlying a modern assessment of customer satisfaction has to be carefully considered. Assessments have to encompass all levels of customer–service provider interactions, and be tailored to the specifics of each service organization, taking into account its unique mix of human-based and technology-based interaction channels.

7.2.5 From customer satisfaction to customer loyalty

Why do organizations measure customer satisfaction? Usually because they see it as a precursor to the achievement of relevant marketing goals. One of these is customer loyalty.

Customer loyalty theory is part of a larger framework called *relationship marketing* focused on the continuous long-term relationship between the customer and the company as opposed to individual (short-term) transactions. Gutek *et al.* (2000) are even more specific. They distinguish among *relationships* (if customers always return to the same company and to the same service provider within the company), *pseudorelationships* (if customers return to the same company, but on an *ad hoc* basis, that is, choose from different service providers within the company every time they need a specific service) and *encounters* (if customers turn to a different company and different service provider every time they need a specific service). In the B2B setting we are interested in relationships.

Kasper *et al.* (1999) differentiate between two types of commitment necessary to establish and continue a long-term relationship (similar reasoning can also be found in Bendapudi and Berry, 1997):

- *affective commitment*, reflecting the customer's affective motivation and positive will-ingness to continue a relationship – in other words, the customer's *desire* to maintain the relationship with the most reliable service provider;

- *calculative commitment*, reflecting the customer's calculative motivation based on the incurred losses and costs of switching to another service provider – in other words, the customer's *need* to maintain the relationship, for example, in the event of contractual prohibitions on replacing one insurance provider with another.

Emphasized as features of a relationship are interaction, dependence, reciprocity, lack of opportunism, all kinds of bonds, trust, commitment, [service] quality, etc. (Parvatiyar and Sheth, 2000). In this framework the importance of bonds should not be overlooked. Berry and Parasuraman (1991) differentiate among financial, social and structural bonds. *Financial bonds* are created with price incentives. These are used to encourage customers to make repurchases as often as possible. *Social bonds* are built in the processes of communication ('importance of staying in touch') and product customization. *Structural bonds* are created by provision of products that are valuable to customers and not readily available from other sources. All three types of bonds can be created in B2B relationships. The basic idea behind the concept is that long-term relationships are beneficial to both parties because the cost of transactions between them can be minimized. In other words, both parties *become loyal* to one another.

According to Kasper *et al.* (1999), customer loyalty can be defined as the degree to which the service-rendering activities of the company depend on one particular person or a group of interchangeable persons. This dependence affects the way in which other employees may replace a colleague in the service encounter. Moreover, the degree of customer loyalty to a par-ticular member of the contact personnel is also relevant, especially in a B2B setting. It may vary from very high to very low. A high degree of customer loyalty to a certain employee of the com-pany can be its main strength against competitors. However, it may also become dangerous if that person decides to leave the company as he/she might do so accompanied by his/her clients.

More generally, Dick and Basu (1994), Sheth *et al.* (1999) and Patterson and Ward (2000) define customer loyalty as the strength of the relationship between the individual's *relative attitude* (the way he/she perceives his/her bond with the company), and *repeat patronage* (the actual volume of purchase). Various levels of customer loyalty can be defined, ranging from an initial state (which is relatively shallow) to a deep and sustained level of commitment.

Management of customer loyalty normally happens in the framework of a customer service programme. Customer service programmes consist of all systematic company attempts to establish a close relationship with the customer and create a win–win situation (lower costs and higher satisfaction) for both sides. They can be either very simple (e.g., a 24-hour toll-free hotline) or very complex (e.g., fully developed loyalty programmes). Between the two poles are guest books in various forms (paper, electronic), warranty programmes, discount clubs, etc.

The most complex form of customer service programme (a loyalty programme) can (and usually does) include all other (simpler) forms. In its framework we are dealing with the multiple-transaction level, where overall satisfaction judgements influence customers' attitudes

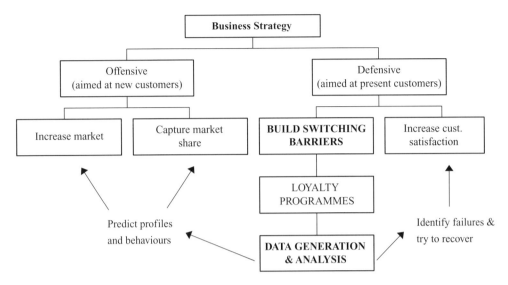

Figure 7.1   The role of customer service programmes in the process of business strategy implementation

Source: Ograjenšek and Žabkar (2010, p. 135).

about perceived service quality. Both quality and satisfaction seem to have a direct effect on customer retention and business performance (Zahorik and Rust, 1992; Zeithaml *et al.*, 1996). Consequently, loyalty programmes are an important strategic tool in the framework of both defensive and offensive business strategies, as shown in Figure 7.1.

Apart from demographics, data generated in the framework of a loyalty programme are hard facts (point-of-sale) as well as soft perceptions and motivations (survey) data. While it is important to acknowledge the fact that their combined analysis can offer valuable insights into customers' decision-making processes and actions, any further discussion goes beyond the scope of this chapter.

7.3   Customer satisfaction assessment: Some methodological considerations

7.3.1   Rationale

Over the last three decades, numerous authors have minutely examined various aspects of the customer satisfaction assessment (e.g., Crosby, 1993; Dutka, 1994; Vavra, 1997, 2002; Hill *et al.*, 2004; Hayes, 2008). Given the number of readily available sources, we limit ourselves to a few brief comments on selected methodological issues to bridge the gap between the conceptual model and the actual assessment.

7.3.2   Think big: An assessment programme

When it comes to assessment, it is imperative to scale up the discussion from the logic underlying a specific questionnaire to the logic underlying a company's overall customer satisfaction assessment programme (Hayes, 2008). This follows from the fact that companies usually derive little information value from a single survey, no matter how well planned and conceptually sound. At a minimum, a company should be able to monitor customer satisfaction levels and detect trends over time in an accurate and credible fashion. This requires that a series of comparable surveys are administered at pre-specified points in time using appropriate sampling methodology. 'Comparable' means that samples are sufficiently representative and similar to one another in their characteristics to allow credible inferences on statistically significant increases or declines in customer satisfaction levels (for additional aspects and analysis techniques see Chapters 3 and 20). Furthermore, companies often need not only to assess the satisfaction levels of their own customers. They might also want to know where they stand with regard to their competition. Consequently, they may need to carry out on their own, or participate in, an annual comparative (benchmarking) survey administered to a national sample that includes customers of several competing service providers. The latter is the underlying logic of the *customer satisfaction indices* such as the American Customer Satisfaction Index (Anderson and Fornell, 2000).

In sum, a company needs to plan a programme involving a series of surveys on a different temporal basis (e.g., monthly, quarterly, or annually). It needs to choose from a wide range of methods and technical issues associated with survey data collection. Data can be collected via different modes (e.g., paper-and-pencil surveys, phone surveys, or web/online surveys, discussed in Chapters 3 and 8). The survey length can vary considerably, from brief surveys involving 3–5 questions (e.g., after a customer's visit to a local branch or a website) to longer surveys with 20–40 items covering the full range of SERVQUAL and other dimensions. The conduct of such surveys has to be coordinated and planned in advance, to cover different types of customers (e.g., intermittent and frequent) who interact with the company through different channels, and whose loyalty or continued patronage may have different types of financial or other relationship value for the company. It goes without saying that the design of an assessment programme has to balance the cost of information gathering against the derived information value to the company.

Beyond these points, the notion of an assessment programme also implies that a company must acknowledge that customer satisfaction surveys, however well designed and carried out, cannot cover the full range of existing qualitative and quantitative information on customer satisfaction. Additional sources include the so-called 'listening posts' (Weiser, 1995) in the form of individual customer interviews, focus groups, and analysis of customer complaints. New developments in information and telecommunication technology support automated analysis of website logfiles and analysis of speech based on call centre recordings. In addition to monitoring the customer satisfaction levels, information from these sources also furthers understanding of the causes behind observed customer satisfaction levels and identifies possible directions for necessary work process changes and/or managerial actions. For more on an integrated view of customer satisfaction, see Chapter 5.

7.3.3   Back to basics: Questionnaire design

As shown in the previous section, a customer satisfaction assessment programme should include multiple tools and data collection points. But first and foremost, at its core there should exist a conceptual model that describes the key dimensions of the target construct. This is the basis for the creation of questionnaire items.

Given a range of possible customer satisfaction dimensions discussed in Section 7.2, it is clear that the actual content of an assessment instrument has to be carefully planned to capture those dimensions that a specific organization finds most relevant to its own situation and unique characteristics. Dimensions need to be ranked in terms of their relative importance as it may not always be possible to include all of them in a specific survey due to various constraints (e.g., respondent burden, time available for phone interview or web survey, cost).

Once the conceptual model is in place, the process of questionnaire design can begin. This process involves numerous principles, steps and decisions. The most important ones are briefly summarized below; for more details refer to Vavra (2002) and Ograjenšek (2007a, 2007b):

General questionnaire design principles pertain to individual questionnaire items regardless of the form of questionnaire and mode of administration. According to these principles, each questionnaire item should:

- be explicitly related to the goals and objectives of the survey;

- be understood by all respondents in the same way;

- ask for information that can be either immediately recalled by respondents or retrieved from their records;

- be constructed in such a way as to prevent introduction of a respondent and interviewer bias;

- be formatted in a way that minimizes data entry errors.

Specific questionnaire design principles should be followed for different forms of questionnaire and modes of administration. A self-administered questionnaire should provide the respondent with clear explanations and instructions for every questionnaire item or group of items. An interviewer-administered questionnaire should cater to the needs of both the interviewer and the respondent. To ensure that all interviewers ask the same question in the same manner, standards for interviewer behaviour, question presentation and probing should be established in advance, and discussed as part of interviewer training which should take place regardless of the questionnaire length and complexity (see also Chapter 3).

Design of actual survey questions involves decisions about their scope. A question may be both about specific attributes (cleanliness of toilets, of floors, of equipment) and overall holistic judgements about a general class (e.g., cleanliness). A questionnaire should contain a mix of closed (forced-choice) and open (constructed-response) items. By now, the use of Likert-type rating scales ranging from, say, 1 to 5 or 1 to 7, with proper labelling of end-points, is well established for closed questions. While responses to such questions can easily be statistically analysed, it is often difficult to interpret the meaning of the resulting statistics, especially for holistic ratings, without access to qualitative data that helps to explain why respondents provided particular ratings. In such cases the content analysis of answers to open questions can help to decide which steps need to be taken to deal with problems contributing to a lower than desired statistic.

Questionnaire testing and piloting are needed to evaluate the extent to which the questions yield the desired information as opposed to causing confusion, and to time the approximate length of questionnaire administration. A wide range of approaches can be used. We briefly indicate the logic of an expert panel or a peer evaluation, a pilot study, a split-panel design, and statistical analysis.

An expert panel or peer evaluation is applicable in the early stages of questionnaire design, and can provide preliminary feedback concerning the draft version of a questionnaire, as well as suggesting ways to improve construct coverage or item phrasing.

A pilot study with respondents from the target population should be used prior to actual data collection. It may involve detailed one-to-one cognitive interviews to probe the respondents' understanding of individual questions, and to examine the respondents' thinking processes as they provide their answers, thus pointing towards ways to eliminate potential problems associated with question wording. A split-panel design using different survey forms can help to evaluate the impact of question order, question wording, and data collection method on survey results.

Finally, statistical analysis is often useful or required to detect highly correlated items or questions that yield ceiling or floor effects (very high or very low average scores with little dispersion around them), to establish internal consistency or scale reliability, and to assess measurement error stemming from questionnaire design.

7.3.4   Impact of questionnaire design on interpretation

A vital concern in measurement instrument design is its impact on validity (Anastasi, 1997), a construct that has multiple meanings. For our purposes it will be linked to the type of permissible interpretations that can be made on the basis of collected information. In this context two issues should be considered by those who use results of customer satisfaction assessments:

- *Use of Likert-type scales.* The nature of the rating scales used in surveys, and their impact on respondent behaviour and on interpretations of results, are long-standing open issues in social science research (Krosnick and Fabrigar, 1997). No general agreement on the number of rating points and the inclusion of a neutral response option in the measurement instrument exists. Another problem is the equality of distances between points on the Likert scale as perceived by the respondent - one person's 'complete satisfaction' might be less than another's 'partial satisfaction'. It could be argued that none of these problems matter as long as the answers are normally distributed. In practice, however, the majority of customer satisfaction surveys tend to result in highly skewed customer responses. Consequently, the use of the arithmetic mean is likely to be a poor measure of central tendency, and may not be the best indicator of customer satisfaction levels.

- *Timing of questionnaire administration.* A seemingly technical issue, but one that has important implications for the interpretation of assessment results, is when to assess service expectations. Should expectations be solicited before the service experience, or away from the actual point of service delivery and unrelated to an encounter? Some researchers (e.g., Johns and Tyas, 1996) prefer to collect data after the service experience at the actual point of service delivery. Consequently, they fear this might load their results towards performance, while those of other researchers might have been loaded towards expectations.

Assumptions about what is being measured (expectations or perceptions), what is affecting customer satisfaction, and what is being affected by it have to take into account the time dimension and the customers' frame of reference. It is important to note that the latter can be affected either by the respondent choice, timing and mode of questionnaire administration, or (lack of) instructions – and *not* necessarily by the phrasing of actual items, which may be the only information analysed when item-by-item statistics are examined or reported.

7.3.5   Additional concerns in the B2B setting

Probably the most important concern in the B2B (establishment) setting is the *information retrieval process*. According to Forsyth *et al.* (1999) it involves:

- proper source identification (own memory, other employees' memories, organizational record systems);

- choice of strategy for information retrieval from identified sources;

- assessment of the match between retrieved information and information needed to select a response.

In addition, *motivational issues* as well as *efficiency of establishment survey completion* process have to be considered (see also Chapter 1).

Overall, the issues discussed in this section indicate that users of customer satisfaction assessment results should be aware of a host of apparently technical decisions which can affect reliability, validity, credibility, comparability, and interpretability of survey data, thus determining their value in decision-making processes.

7.4   The ABC ACSS questionnaire: An evaluation

7.4.1   Rationale

The improvement of customer satisfaction assessment requires that we take a critical look at existing assessments in order to identify areas where companies and assessors face dilemmas and where their decisions about the content and organization of a questionnaire affect the quality and usefulness of the resulting information. In this brief section we aim to show the relevance of conceptual and methodological issues discussed earlier in this chapter to the ABC ACSS questionnaire introduced in Chapter 2.

7.4.2   Conceptual issues

When assessing the ABC ACSS questionnaire from the conceptual point of view we limit ourselves to relational issues, web-based interactions, and complaint satisfaction:

Relational issues. Typically for a B2B provider, the majority of items in the ABC ACSS questionnaire relate to technical issues. The coverage of relational issues is relatively limited. Given that these can underlie the willingness of customers or buyers to engage and stay in a long-term relationship, in the future the questionnaire could also include dimensions such as *assurance* or *empathy*. However, this extension should be evaluated against the potential risk of higher non-response rate due to expanded questionnaire length.

Web-based interactions. The ABC ACSS questionnaire provides limited coverage of web-based interactions with the following five items:

---

45. The website resources are helpful.

46. The online ordering process is effective.

47. ABC reports provide valuable information for your operations.

48. The special application has helped you select consumables.

49. Overall satisfaction level with the customer website.

Given that supplier websites increasingly enable customers to access technical information and price lists, download training manuals, conduct commercial transactions, and place consultation requests, future versions of the ABC ACSS questionnaire will probably reflect these developments by including additional items in this category.

Complaint satisfaction. The ABC ACSS questionnaire is a typical annually administered survey that provides a concise coverage of complaint satisfaction in form of item 54 ('Complaints are handled promptly'). Weekly or monthly reports focused on complaint management should also include dimensions such as *accessibility*, *quality of interaction*, *speed*, *efficiency*, and *fairness*. Such questionnaires are typically distributed to customers who recently filed a complaint (see Chapter 3 for 'continuous' or 'event-driven' surveys).

7.4.3   Methodological issues

Among the issues which influence respondent's willingness to participate, and consequently the response rate, these are the most important ones:

Importance ratings. The ABC ACSS questionnaire asks respondents to intuitively rate the importance of each item, which some of them might be uncomfortable doing. This declared importance is important, however, because it can be compared to a generated importance derived from the multiple regression of responses on the concise coverage items, such as overall satisfaction or repurchasing intentions (see the analysis of decision drivers in Section 2.5 of Chapter 2).

The combined use of closed (forced-choice) and open (constructed-response) items. Rating items or Likert-type scales as used in the ABC ACSS questionnaire are followed by open-ended items asking for an explanation. Such comments were collected in the ABC ACSS but not considered in this book. In general, an open comment can be solicited in the following form:

Would you recommend ABC to other companies?

1 Very unlikely

. . .

5 Very likely

Why?

In some cases the request for specific comments depends on the given rating. Negative ratings usually trigger a request for customers to suggest possible improvements. Positive ratings call for an explanation which, if provided, may become a standard part of the internal best-practice manual.

Inclusion of double-barrelled items. While double-barrelled items are not tolerated in theory, organizations tend to use them in practice if the possibility of confounding is deemed inconsequential and they help to greatly reduce survey length, thereby increasing response rates. These are the examples of the double-barrelled items from the ABC ACSS questionnaire:

---

24. The technical staff is **courteous and helpful**.

28. The initial operation training was **productive and met your needs**.

35. Orders placed are **delivered when promised and are delivered complete**.

58. **Equipment and service contract terms** are clear.

Effect of item clustering on analytical logic. In the ABC ACSS questionnaire items related to SERVQUAL dimensions can be found under several functional areas (sales support, training, etc.). This form of questionnaire organization makes sense when each functional area is handled by a separate unit within the organization. In that case all items pertaining to a particular unit have to appear together and eventually yield a single summary score. However, the fact that items of the same underlying dimension (e.g., *reliability*) are spread all over the questionnaire should not stand in the way of analysing them as a group, thus providing an additional summary measure for the company to evaluate.

7.4.4   Overall ABC ACSS questionnaire asssessment

The ABC ACSS questionnaire is an example of a typical questionnaire with a format that, in one way or another, is often used by service organizations, and has several key strengths:

- It covers multiple aspects of the supplier's (i.e., ABC's) operations, using both specific questions andoverall holistic ratings.

- It allows for benchmarking, i.e., comparison of the satisfaction levels with the ABC company services to that of other suppliers.

- By its systematic mapping of key areas of concern to both supplier and customer it facilitates story-telling about the client–provider relationship.

The data collected with this measurement tool is analysed in subsequent chapters of this book.

7.5   Summary

In order to conduct effective and credible assessments of customer satisfaction, organizations need to have a solid conceptual model of customer satisfaction and a deep understanding of the key dimensions that affect satisfaction and loyalty of their customers. Only with a proper conceptual model in place (either based exclusively on the established SERVQUAL scale or enriched with additional dimensions such as complaint handling, accessibility, usability, human–technology interactions, engagement, experience, etc.) can organizations plan and implement an assessment programme. Such a programme should provide suitable (quantitative and qualitative) data on customer satisfaction levels. In addition, it should shed light on needs, expectations and other factors shaping customer reactions to products they purchase and services they receive.

With the goal of helping organizations establish their unique assessment programmes, our chapter reviews several issues associated with customer satisfaction measurement. We first discuss the quality–satisfaction–loyalty chain and then present some of the most important methodological considerations pertaining to customer satisfaction assessment. We illustrate the relevance of these conceptual and methodological issues by showcasing the ABC ACSS questionnaire, to which most chapters in this book refer.

What happens after an organization establishes its unique assessment programme? Whether and how it acts upon the results is determined by a wider range of systemic factors underlying its service processes. Key among them are *managerial and human resources practices* (Stauss and Seidel, 2004; Homburg and Fürst, 2005; Luria *et al.*, 2009). These not only shape the general service climate in a given organization. Regardless of the communication channel chosen, they also determine the *ability* of contact personnel or frontline workers to effectively handle customer requests, concerns and complaints (through service recovery guidelines and empowerment practices); and their *willingness* to invest time and effort in resolving service-related problems (through individual job characteristics and compensation levels). Hence, if service organizations want not only to *assess* customer satisfaction but also to *understand* which factors have to be addressed in order to improve customer satisfaction trends over time, they should systematically collect and evaluate information not only about their employees' performance, but also on their employees' motivational, behavioural and affective responses to different managerial practices (see Chapter 5 on integrated models).

References

Anastasi, A. (1997) *Psychological Testing*, 7th edn. Upper Saddle River, NJ: Prentice Hall.

Anderson, E.W. and C. Fornell (2000) Foundations of the American Customer Satisfaction Index. *Total Quality Management*, 7, 869–882.

Bendapudi, N. and Berry, L.L. (1997) Customers' motivations for maintaining relationships with service providers. *Journal of Retailing*, 1, 15–37.

Berry, L.L. and Parasuraman, A. (1991) *Marketing Services. Competing through Quality*. New York: Free Press.

Bitner, M.J., Zeithaml, V. and Gremler, D. (2010) Technology's impact on the gaps model of service quality. In P.P. Maglio et al. (eds), *Handbook of Service Science*, pp. 198–218. New York: Springer.

Brown, T.J., Churchill, G.A. and Peter, J.P. (1993) Improving the measurement of service quality. Research note. *Journal of Retailing*, 1, 127–139.

Buttle, F.A. (1995). What future for SERVQUAL? In M. Bergadaà (ed.), Proceedings of the 24th European Marketing Academy Conference. *ESSEC, Cergy-Pontoise*, pp. 211–230.

Cronin, J.J. and Taylor, S.A. (1992) Measuring service quality: a reexamination and extension. *Journal of Marketing*, 56(3), 55–68.

Cronin, J.J. and Taylor, S.A. (1994) SERVPERF versus SERVQUAL: Reconciling performance-based and perceptions-minus-expectations measurement of service quality. *Journal of Marketing*, 1, 125–131.

Crosby, L.A. (1993) Measuring customer satisfaction. In E.E. Scheuing and W.F. Christofer (eds), *The Service Quality Handbook*. New York: AMACOM.

Dick, A.S. and Basu, K. (1994) Customer loyalty: Toward an integrated conceptual framework. *Journal of the Academy of Marketing Science*, 2, 99–113.

Dutka, A.F. (1994) *AMA Handbook for Customer Satisfaction: Research, Planning, and Implementation*. Lincolnwood, IL: NTC Bisomess Books.

Forsyth, B.H., Levin, K. and Fisher, S.K. (1999) *Test of an Appraisal Method for Establishment Survey Questionnaires*, http://www.amstat.org/sections/srms/proceedings/papers/1999_021.pdf

Gotlieb, J.B., Grewal, D. and Brown, S.W. (1994) Consumer satisfaction and perceived quality: Complementary or divergent constructs? *Journal of Applied Psychology*, 79(6), 875–885.

Grönroos, C. (1990) *Service Management and Marketing. Managing the Moments of Truth in Service Competition*. Lexington, MA: Lexington Books.

Grönroos, C. (2000) Relationship marketing: The Nordic school perspective. In J.N. Sheth and A. Parvatiyar, (eds), *Handbook of Relationship Marketing*, pp. 95–117. Thousand Oaks, CA: Sage.

Gutek, B.A., Bennett, C., Bhappu, A.D. Schneider, S. and Woolf, L. (2000) Features of service relationships and encounters. *Work and Occupations*, 3, 319–352.

Harel, A., Kenett, R.S. and Ruggeri, F., (2008) modeling web usability diagnostics on the basis of usage statistics. In W. Jank and G. Shmueli (eds), *Statistical Methods in E-commerce Research*, pp. 131–172. Hoboken, NJ: John Wiley & Sons, Inc.

Habryn, F., Blau, B., Satzger, G. and Kölmel, B. (2010) Towards a model for measuring customer intimacy in B2B services. *Exploring Services Science*, 53, 1–14.

Haller, S. (1998) *Beurteilung von Dienstleistungsqualität. Dynamische Betrachtung des Qualitätsurteils im Weiterbildungsbereich*. Wiesbaden: Deutscher Universitäts-Verlag and Gabler Verlag.

Hayes, B.E. (2008) *Measuring Customer Satisfaction and Loyalty: Survey Design, Use, and Statistical Analysis Methods*. Milwaukee, WI: ASQ Quality Press.

Heskett, J.L., Sasser, W.E., and Schlesinger, L.A. (1997) *The Service Profit Chain: How Leading Companies Link Profit and Growth to Loyalty, Satisfaction, and Value*. New York: Free Press.

Hill, N., Brierley, J. and MacDougall, R. (2004) *How to Measure Customer Satisfaction*. Farnham: Gower.

Homburg, C. and Fürst, A. (2005) How organizational complaint handling drives customer loyalty: An analysis of the mechanistic and the organic approach. *Journal of Marketing*, 69(3), 95–114.

Hornbaek, K. (2006) Current practice in measuring usability: Challenges to usability studies and research. *International Journal of Human–Computer Studies*, 64(2), 79–102.

Iacobuci, D., Grayson, K.A. and Ostrom, A.L. (1994) The calculus of service quality and cutomer satisfaction: Theoretical and empirical differentiation and integration. In A. Swartz, D.E. Bowen and S.W. Brown (eds), *Advances in Services Marketing and Management: Research and Practice*. Stamford, CT: JAI Press.

Johns, N. and Tyas, P. (1996) Use of service quality gap theory to differentiate between foodservice outlets. *Service Industries Journal*, 3, 321–346.

Kasper, H., van Helsdingen, P. and de Vries, W. (1999) *Services Marketing Management. An International Perspective*. Chichester: John Wiley & Sons, Ltd.

Kenett, R.S. (2004). The integrated model, customer satisfaction surveys and Six Sigma. In *Proceedings of the First International Six Sigma Conference,* Center for Advanced Manufacturing Technologies, Wrocław University of Technology, Wrocław, Poland.

Krosnick, J.A. and Fabrigar, L.R. (1997) Designing rating scales for effective measurement in surveys. In L. Lyberg, P. Biemer, M. Collins, E. de Leeuw, C. Dippo, N. Schwarz and D. Trewin (eds), *Survey Measurement and Process Quality*, pp. 141–164. New York: John Wiley & Sons, Inc.

Ladhari, R. (2008) Alternative measures of service quality: A review. *Managing Service Quality*, 18(1), 65–86.

Lovelock, C. and Wright, L. (1999) *Principles of Service Marketing and Management*. Englewood Cliffs, NJ: Prentice Hall.

Luria, G., Gal, I. and Yagil, D. (2009) Employees' willingness to report service complaints. *Journal of Service Research*, 12(2), 156–174.

Mittal, V., Ross, W.T. and Baldasare, P.M. (1998) The asymmetric impact of negative and positive attribute-level performance on overall satisfaction and repurchase intentions. *Journal of Marketing*, 62(1), 33–47.

Ograjenšek, I. (2007a) Design and testing of questionnaires. In F. Ruggeri, R.S. Kenett and F.W. Faltin (eds), *Encyclopedia of Statistics in Quality and Reliability*, pp. 524–528. Chichester: John Wiley & Sons, Ltd.

Ograjenšek, I. (2007b) Selection and validation of response scales. In F. Ruggeri, R.S. Kenett and F.W. Faltin (eds), *Encyclopedia of Statistics in Quality and Reliability*, pp. 1778–1783. Chichester: John Wiley & Sons, Ltd.

Ograjenšek, I. and Žabkar, V. (2010) Enhancing the value of survey data on consumer satisfaction in the frame of a consumer loyalty programme: Case of a Slovenian retailer. *Quality Technology* and *Quantitative Management*, 7(2), 133–147.

Oliver, R.L. (1997) *Satisfaction: A Behavioral Perspective on the Consumer.* New York: McGraw-Hill.

Parasuraman, A., Zeithaml, V. and Berry, L. L. (1985) A conceptual model of service quality and its implications for future research. *Journal of Marketing*, Fall, 41–50.

Parasuraman, A., Zeithaml, V. and Berry, L. L. (1988) SERVQUAL: A multiple-item scale for measuring consumer perceptions of service quality. *Journal of Retailing*, 1, 12–40.

Parasuraman, A., Berry, L.L. and Zeithaml, V. (1991a) Understanding, measuring and improving service quality: Findings from a multiphase research program. In S.W. Brown, E. Gummesson, B. Edvardsson and B. Gustavsson (eds), *Service Quality. Multidisciplinary and Multinational Perspectives*, pp. 253–268. Lexington, MA: Lexington Books.

Parasuraman, A., Berry, L.L. and Zeithaml, V. (1991b) Refinement and reassessment of the SERVQUAL scale. *Journal of Retailing*, 67(4), 420–450.

Parasuraman, A., Zeithaml, V. and Berry, L.L. (1994) Reassessment of expectations as a comparison standard in measuring service quality: Implications for further research. *Journal of Marketing*, 58(1), 111–124.

Parvatiyar, A. and Sheth, J. N. (2000) The domain and conceptual foundations of relationship marketing. In J.N. Sheth and A. Parvatiyar (eds), *Handbook of Relationship Marketing*, pp. 3–38. Thousand Oaks, CA: Sage.

Patterson, P.G. and Ward, T. (2000) Relationship marketing and management. In T.A. Swartz and D. Iacobucci (eds), *Handbook of Services Marketing & Management*, pp. 317–342. Thousand Oaks, CA: Sage.

Salini, S. and Kenett, R. S. (2009) Bayesian networks of customer satisfaction survey data. *Journal of Applied Statistics*, 36(11), 1177–1189.

Santos, J. (2003) E-service quality: A model of virtual service quality dimensions. *Managing Service Quality*, 13(3), 233–246.

Sheth, J.N., Mittal, B. and Newman, B.I. (1999) *Customer Behavior*. Forth Worth, TX: Dryden Press.

Spreng, R. A. and Mackoy, R. D. (1996) An empirical examination of a model of perceived service quality and satisfaction. *Journal of Retailing*, 72(2), 201–214.

Stauss, B. and Seidel, W. (2004) *Complaint management: The heart of CRM*. New York: Thomson/South Western.

Taylor, S.A. and Baker, T. L. (1994) An assessment of the relationship between service quality and customer satisfaction in the formation of consumer's purchase intentions. *Journal of Retailing*, 70(2), 163–178.

Teas, R.K. (1993a) Consumer expectations and the measurement of perceived service quality. *Journal of Professional Services Marketing*, 2, 33–54.

Teas, R.K. (1993b) Expectations, performance evaluation, and consumers' perceptions of quality. *Journal of Marketing*, 57(4), 18–34.

Teas, R.K. (1994) Expectations as a comparison standard in measuring service quality: An assessment of a reassessment. *Journal of Marketing*, 58(1), 132–139.

Theron, E., Terblanche, N.S. and Boshoff, C. (2010) Trust, commitment and satisfaction: New perspectives from business-to-business (B2B) financial services relationships in South Africa. In P. Ballantine and J. Finsterwalder (eds), *Proceedings of the 2010 Australian and New Zealand Marketing Academy Conference*, http://anzmac2010.org/proceedings/pdf/anzmac10Final00097.pdf

Vavra, T.G. (1997) *Improving Your Measurement of Customer Satisfaction: A Guide to Creating, Conducting, Analyzing, and Reporting Customer Satisfaction Measurement Programs*. Milwaukee, WI: ASQ Quality Press.

Vavra, T.G. (2002) *Customer Satisfaction Simplified: A Step-by-step Guide for ISO 9001: 2000 Certification*. Milwaukee, WI: ASQ Quality Press.

Weiser, C.R. (1995) Championing the customer. *Harvard Business Review*, 73(6), 113–116.

Zahorik, A.J. and Rust, R.T. (1992) Modeling the impact of service quality on profitability. In A. Swartz, D.E. Bowen and S.W. Brown (eds), *Advances in Services Marketing and Management: Research and Practice*, pp. 247–276. London: JAI Press.

Zeithaml, V.A., and Bitner, M.J. (2002) *Services Marketing*. New York: McGraw Hill Higher Education.

Zeithaml, V.A., Berry, L.L. and Parasuraman, A. (1996) The behavioral consequences of service quality. *Journal of Marketing*, 60(2), 31–46.

Zeithaml, V.A., Parasuraman, A. and Berry, L.L. (1990) *Delivering Service Quality*. London: Free Press.

Appendix
SERVQUAL dimensions and items

Quality Dimension	Expectations (E_i)	Perceptions (P_i)
Tangibles	Excellent companies will have modern-looking equipment	XYZ has modern-looking equipment
	The physical facilities at excellent companies will be visually appealing	XYZ's physical facilities are visually appealing
	Employees of excellent companies will be neat in appearance	XYZ's employees are neat in appearance
	Materials associated with the service (such as pamphlets or statements) will be visually appealing in an excellent company	Materials associated with the service (such as pamphlets or statements) are visually appealing at XYZ
Reliability	When excellent companies promise to do something by a certain time, they will do so	When XYZ promises to do something by a certain time, it does so
	When customers have a problem, excellent companies will show a sincere interest in solving it	When you have a problem, XYZ shows a sincere interest in solving it
	Excellent companies will perform the service right the first time	XYZ performs its service right the first time
	Excellent companies will provide their services at the time they promise to do so	XYZ provides its services at the time it promises to do so
	Excellent companies will insist on error-free records	XYZ insists on error-free records
Responsiveness	Employees of excellent companies will tell customers exactly when services will be performed	Employees of XYZ tell you exactly when the service will be performed
	Employees of excellent companies will give prompt service to customers	Employees of XYZ give you prompt service
	Employees of excellent companies will always be willing to help customers	Employees of XYZ are always willing to help you

(Continued)

Quality Dimension	Expectations (E_i)	Perceptions (P_i)
	Employees of excellent companies will never be too busy to respond to customer requests	Employees of XYZ are never too busy to respond to your requests
Assurance	The behaviour of employees of excellent companies will instil confidence in customers	The behaviour of XYZ's employees instils confidence in you
	Customers of excellent companies will feel safe in their transactions	You feel safe in your transactions with XYZ
	Employees of excellent companies will be consistently courteous with customers	Employees of XYZ are consistently courteous with you
	Employees of excellent companies will have the knowledge to answer customer questions	Employees of XYZ have the knowledge to answer your questions
Empathy	Excellent companies will give customers individual attention	XYZ gives you individual attention
	Excellent companies will have operating hours convenient to all their customers	XYZ has operating hours convenient to you
	Excellent companies will have employees who give customers personal attention	XYZ has employees who give you personal attention
	Excellent companies will have their customers' best interests at heart	XYZ has your best interests at heart
	The employees of excellent companies will understand the specific needs of their customers	Employees of XYZ understand your specific needs

Adapted from Parasuraman *et al*. (1991b, pp. 446–449).

8

Missing data and imputation methods

Alessandra Mattei, Fabrizia Mealli and Donald B. Rubin

Missing data are a pervasive problem in many data sets and seem especially widespread in social and economic studies, such as customer satisfaction surveys. Imputation is an intuitive and flexible way to handle the incomplete data sets that result. We discuss imputation, multiple imputation (MI), and other strategies to handle missing data, together with their theoretical background. Our focus is on MI, which is a statistically valid strategy for handling missing data, although we also review other valid approaches, such as direct maximum likelihood and Bayesian methods for estimating parameters, as well as less sound methods. The creation of multiply-imputed data sets is more challenging than their analysis, but still relatively straightforward relative to other valid methods, and we discuss available software for MI. Some examples and advice on computation are provided using the ABC 2010 annual customer satisfaction survey. Ad hoc methods, including using singly-imputed data sets, almost always lead to invalid inferences and should be eschewed.

8.1 Introduction

Missing values are a common problem in many data sets and seem especially widespread in social and economic studies, including customer satisfaction surveys, where customers may fail to express their satisfaction level concerning their experience with a specific business because of lack of interest, unwillingness to criticize their sales representative, or other reasons. Unit nonresponse occurs when a selected unit (e.g., customer) does not provide any of the information being sought. Item nonresponse occurs when a unit responds to some items but not to others. Discussion of many issues related to missing data is included in the three volumes produced by the Panel on Incomplete Data of the Committee on National Statistics in 1983

Modern Analysis of Customer Surveys: with applications using R, First Edition. Edited by Ron S. Kenett and Silvia Salini.
© 2012 John Wiley & Sons, Ltd. Published 2012 by John Wiley & Sons, Ltd.

(Madaw and Olkin, 1983; Madaw *et al.*, 1983a,b), as well as in the volume that resulted from the 1999 International Conference on Survey Nonresponse (Groves *et al.*, 2002).

Methods for analyzing incomplete data can be usefully grouped into four main categories, which are not mutually exclusive (Little and Rubin, 1987, 2002). The first group comprises procedures based on subsets of the data set without missing data, either complete-case analysis (also known as 'listwise deletion'), which discards incompletely recorded units and analyzes only the units with complete data, or available-case analysis, which discards units with incomplete data on the variables needed to calculate certain statistics. These simple methods are generally easy to use and may be satisfactory with small amounts of missing data; however, they can often lead to inefficient and biased estimates.

The second group of methods comprises weighting procedures, which deal with unit nonresponse by increasing the survey weights for responding units in the attempt to adjust for nonresponse as if it was part of the sample design. Weighting is a relatively simple alternative for reducing bias from complete-case analysis. Because these methods drop the incomplete cases, they are most useful when sampling variance is not an issue.

The third group comprises imputation-based procedures, which fill in missing values with plausible values, where the resultant completed data are then analyzed by standard methods as if there never were any missing values. In order to measure and incorporate uncertainty due to the fact that imputed values are not actual values, alternative methods have been proposed, including resampling methods and multiple imputation (MI), as proposed by Rubin (1978a, 1987, 1996). MI is a technique that replaces each set of missing values with multiple sets of plausible values representing a distribution of possibilities. Each set of imputations is used to create a complete data set, which is analyzed by complete-data methods; the results are then combined to produce estimates and confidence intervals that incorporate missing-data uncertainty. Imputation methods were originally viewed as being most appropriate in complex surveys that are used to create public-use data sets to be shared by many users, although over the years they have been successfully applied in other settings as well.

The final group of methods comprises direct analyses using model-based procedures; models are specified for the observed data, and inferences are based on likelihood or Bayesian analysis. Using these methods is typically more complex than using previous methods.

A missing-data method is required to yield statistically valid answers for scientific estimands. By a scientific estimand we mean a quantity of scientific interest that can be calculated in the population and does not change its values depending on the data collection design used to measure it. Scientific estimands include population means, variances, correlation coefficients, and regression coefficients. Inferences for a scientific estimand are defined to be statistically valid if they satisfy the following three criteria (e.g., Rubin 1996; Rässler *et al*, 2008): (a) point estimation must be approximatively unbiased for the scientific estimand; (b) interval estimation must reach at least the nominal coverage: actual interval coverage $\geq$ nominal interval coverage, so that, for example, 95% intervals for a population mean should cover the true population mean at least 95% of the time; and (c) tests of hypotheses should reject at their nominal level or less frequently when the null hypothesis is true, so that, for example, a 5% test of a zero population correlation should reject at most 5% of the time when the population correlation is zero.

In general, only MI and model-based procedures can lead to valid inferences. Resampling methods, such as the bootstrap and jackknife, can satisfy criteria (b) and (c) asymptotically, while giving no guidance on how to satisfy criterion (a) in the presence of missing data, but rather implicitly assuming that approximately unbiased estimates for scientific estimands have already been obtained; see Efron (1994) and the discussion by Rubin (1994). Such methods do not represent a complete approach to the problem of missing data, and therefore we do not discuss them further here.

This chapter is organized as follows. First, we start with a basic discussion of missing-data patterns, which describe which values are observed in the data matrix and which are missing, and missing-data mechanisms, which concern the relationship between missingness and the values of variables in the data matrix. Second, we review the four classes of approaches to handling missing data briefly introduced above, with a focus on MI, which we believe is the most generally useful approach for survey data, including customer satisfaction data. Third, a simple MI analysis is conducted for the ABC ACSS data, and results are compared to those from alternative missing-data methods. We conclude the chapter with some discussion.

8.2   Missing-data patterns and missing-data mechanisms

In order to conduct statistical analyses appropriately in the presence of missing data, it is crucial to distinguish between the missing-data pattern and the missing-data mechanism. The missing-data pattern describes which values are missing and which are observed in the data matrix. The missing-data mechanism describes to what extent missingness depends on the observed and/or unobserved data values.

8.2.1   Missing-data patterns

Let $Y = [y_{ij}]$ denote the $N \times P$ rectangular matrix of complete data, with ith row $y_i = [y_{i1}, \ldots, y_{iP}]$, where y_{ij} is the value of variable Y_j for subject i. Define $R = [R_{ij}]$, the $N \times P$ observed-data indicator matrix, with $R_{ij} = 1$ if y_{ij} is observed and $R_{ij} = 0$ if y_{ij} is missing.

Figure 8.1 shows some examples of common missing-data patterns. A simple pattern is univariate missing data, where missingness is confined to a single variable (see Figure 8.1(a)). For instance, suppose we are interested in estimating the relationship between a dependent variable Y_P, such as the overall satisfaction level with ABC, and a set of covariates (independent variables) $Y_1, \ldots, Y_{P-1}$, such as the company's continent (European versus non-European) and country, company's segmentation, and age of ABC's equipment, all of which are intended to be fully observed. Although the covariates may be fully observed, the outcome Y_P for some customers may be missing, leading to univariate missingness.

Another common pattern is obtained when the single incomplete variable Y_P in Figure 8.1(a) is replaced by a set of variables $Y_{J+1}, \ldots, Y_P$, all observed or missing on the same set of units (see Figure 8.1(b)). An example of this pattern is unit nonresponse, which occurs, for instance, if a subset of ABC customers do not complete the questionnaire about their satisfaction with ABC.

Patterns (a) and (b) are special cases of (c), monotone missing data, which is a pattern of particular interest, because methods for handling it can be easier than methods for general patterns (d). Missingness in Y is monotone if the variables can be arranged so that all $Y_{J+1}, \ldots, Y_P$ are missing when Y_J is missing, for all $J = 1, \ldots, P - 1$. In other words, the first variable in Y is at least as observed as the second variable, which is at least as observed as the third variable, and so on. Such a pattern of missingness, or a close approximation to it, is not uncommon in practice. Monotone patterns often arise in repeated-measures or longitudinal data sets, because if a unit drops out of the study in one time period, then the data will typically be missing in all subsequent time periods. Sometimes a nonmonotone missing-data pattern can be made monotone, or nearly so, by reordering the variables according to their missingness rates.

Figure 8.1   *Examples of missing-data patterns. Rows correspond to units, columns to variables (adapted from Little and Rubin 2002, p. 5)*

Sorting rows and columns of the data matrix according to the missing data to see if a simple pattern emerges may be useful. Some methods of analysis are intended for particular patterns of missing data and use only standard complete-data analyses. Other methods may be applied to more general missing-data patterns, but usually require more computational effort than methods designed for special patterns.

8.2.2   Missing-data mechanisms and ignorability

A key component in a statistical analysis with missing data is the mechanism that leads to missing data: the process that determines which values are observed, and which are missing. Missing-data mechanisms are crucial because the properties of missing-data methods strongly depend on the nature of the dependencies in these mechanisms. The missing-data mechanism is characterized by the conditional probability of the indicator matrix, R, given Y and the unknown parameters governing this process, ξ: $p(R|Y, \xi)$.

Following Rubin (1976), who formalized the key concepts about missing-data mechanisms, the statistical literature (e.g., Little and Rubin 2002, p. 12) classifies missing-data mechanisms into three groups: missing completely at random (MCAR), missing at random (MAR), and missing not at random (MNAR). This language was chosen to be consistent with much older terminology in classical experimental design for completely randomized, randomized, and nonrandomized studies.

Missing data are said to be MCAR if missingness does not depend on the values of the data Y, missing or observed, that is, if the distribution of R is free of Y: $p(R|Y, \xi) = p(R|\xi)$ for all Y and ξ. In other words, missingness is MCAR if the probability that units provide data on a particular variable does not depend on the value of the variable or the value of any other variable. The MCAR assumption can be unrealistically restrictive, because it has testable implications and can be rejected by the observed data. For instance, in consumer satisfaction surveys, the MCAR assumption is contradicted by the data when companies working in different observed business areas are observed to have different rates of missing data on satisfaction.

It is generally more realistic to assume that missingness depends on observed values. For instance, the probability of missingness for satisfaction variables may depend only on completely observed variables, such as company's country, business area, age, legal status, or size, but not on any missing values. In such a case, the missing data are MAR, but not necessarily MCAR. MAR refers to missing data for which missingness can be explained by the observed values in the data set. Formally, let Y_{obs} denote the observed components of Y, and Y_{mis} the missing components. Missing data are MAR if the distribution of R depends only on Y_{obs}, and not on Y_{mis}: $p(R|Y, \xi) = p(R|Y_{obs}, \xi)$ for all Y_{mis} and ξ.

The missing-data mechanism is MNAR if the conditional distribution of R depends on the missing values in the data matrix Y, even given Y_{obs}. This could be the case with ABC if customers with lower satisfaction levels tend to be less likely to provide their satisfaction level than customers with higher overall satisfaction levels, even though they have exactly the same observed values of background covariates. The richer the data set in terms of observed variables, the more plausible the MAR assumption is.

In addition to formally defining the concepts underlying MCAR, MAR and MNAR, Rubin (1976) defined the concept of ignorability. Suppose that parametric models have been specified for both the distribution of the complete data, $p(Y|\psi)$, and the missing-data mechanism, $p(R|Y, \xi)$. The missing-data mechanism is ignorable for likelihood or Bayesian inference if: (i) the missing data are MAR; and (ii) the parameters of the data distribution, ψ, and the missing-data mechanism, ξ, are distinct, that is, the joint parameter space of (ψ, ξ) factorizes into the product of the parameter space of ψ and the parameter space of ξ, and when prior distributions are specified for ψ and ξ in a Bayesian setting, these are independent.

Ignorable missing-data mechanisms are desirable, because they allow us to obtain valid inferences about the estimands of interest, ignoring the process that causes missing data. Formally, the distribution of the observed data is obtained by integrating Y_{mis} out of the joint density of $Y = (Y_{obs}, Y_{mis})$ and R; and the full likelihood function of ψ and ξ is proportional to it:

$$\mathcal{L}_{full}(\psi, \xi|Y_{obs}, R) \propto p(Y_{obs}, R|\psi, \xi) = \int p(Y_{obs}, Y_{mis}|\psi) \, p(R|Y_{obs}, Y_{mis}, \xi) \, \mathrm{d}Y_{mis}. \quad (8.1)$$

Rubin (1976) showed that if the missing-data mechanism is ignorable (MAR and distinctness of ψ and ξ), then the full likelihood function (8.1) is proportional to a simpler likelihood function,

$$\mathcal{L}_{ign}(\psi|Y_{obs}) \propto \int p(Y_{obs}, Y_{mis}|\psi) \, \mathrm{d}Y_{mis}, \quad (8.2)$$

which does not depend on the missing-data mechanism. Therefore valid inference about the distribution of the data can be obtained using this simpler likelihood function, ignoring the missing-data mechanism.

MAR is typically regarded as the more important condition in considerations of ignorability, because if the missing data are MAR but distinctness does not hold, inferences based on the likelihood ignoring the missing-data mechanism are still potentially valid in the sense of satisfying criteria (a)–(c) of Section 8.1, but may not be fully efficient; see (Little and Rubin 2002, Section 6.2) and Rubin (1978b) for further discussion of these ideas. Also, in many cases, it is reasonable to assume that the parameters of the data distribution and the missing-data mechanism are distinct, so that the practical question of whether the missing-data mechanism is ignorable often reduces to a question about whether the MAR assumption is plausible. This argument requires some care, however, with random effects models, where there is a subtle interplay between the assumptions of MAR and distinctness, depending on the definition of the hypothetical complete-data (see Shih 1992).

In many missing-data contexts, it is not known whether or not the ignorability condition is correct; however, assuming it can be advantageous for a variety of reasons. First, ignorability can simplify analyses greatly. Second, the MAR assumption is often reasonable, especially when there are fully observed covariates available in the data set to 'explain' the reasons for the missingness. Unfortunately, data can never provide any direct evidence against MAR, so that MAR is not testable without auxiliary information, such as distributional assumptions; see the literature on selection models, for example, Heckman (1976) and Little (1985). Third, even if the missing data are MNAR, an analysis based on the MAR assumption can be helpful in reducing bias by effectively imputing missing data using relationships that are observed. Finally, if the missing data are MNAR, it is usually not at all easy to specify a plausible nonignorable missing-data model, because there is no direct evidence concerning the relationship of missingness to the missing values, since the missing values are, by definition, not observed (e.g., Rubin *et al.*, 1995). Moreover, ignorable models can lead to superior inferences than misspecified nonignorable models. In addition, even if the missing-data mechanism is correctly specified, information for estimating ψ and ξ jointly may be very limited, resting strongly on the untestable assumptions made about the distribution of Y.

8.3   Simple approaches to the missing-data problem

8.3.1   Complete-case analysis

A very simple approach to missing data is to exclude incomplete units, and to use only units with all variables observed. This means that all units (cases) with any missing variables are discarded, and complete-case analysis is carried out. Complete-case analysis (sometimes called listwise deletion) is the simplest approach to missing data, because standard complete-data statistical analysis can be directly applied without modification. In addition, it guarantees comparability of univariate statistics, since they are all calculated on a common sample of cases. However, complete-case analysis may have serious pitfalls, stemming from the potential loss of information in discarding incomplete cases. Specifically, the complete-case approach leads to a loss of precision and is generally biased when the missing-data mechanism is not MCAR. The degree of bias and loss of precision depends on (i) the amount and pattern of missing data; (ii) the degree to which the MCAR assumption is violated; and (iii) the estimand and the analysis being implemented.

Complete-case analysis is potentially wasteful for univariate analysis, because values of a particular variable are discarded when they belong to cases that are missing other variables. As a result, even when complete-case analysis is unbiased, it can be highly inefficient, especially with highly multivariate data sets. For example, consider a data set with 10 variables, each of which has probability of being missing of 0.1, and suppose that missingness on each variable is independent of missingness on the other variables. Then, the expected proportion of complete cases is $(1 - 0.1)^{10} = 0.35$, that is, the complete-case analysis would be expected to include only 35% of the units.

8.3.2   Available-case analysis

Another simple approach to missing data is available-case analysis, which uses only units with complete data on the variables that are needed for the analysis being considered. This approach can be regarded as complete-case analysis restricted to the variables of interest. Available-case analysis also arises when any variable with missing values is excluded from the analysis (sometimes called 'complete-variables analysis'). Available-case analysis retains at least as many of the data values as does complete-case analysis. A drawback of this approach is that the sample base generally changes from analysis to analysis. This variability in the sample base may be problematic, because different analyses will be based on different subsets of the data and thus will not necessarily be consistent with each other. For instance, when tables are computed for various conceptual sample bases (e.g., all customers, European customers, customers working in a specific business area), the changes in the sample bases in available-case analysis prevent associating a fixed sample size to each base. These changes in the sample bases also yield problems of comparability across analyses if the missing-data mechanism is not MCAR, and may lead to misleading results when estimates of quantities concerning different variables are combined. For instance, if summaries of different variables are to be compared, the set of units for which each variable is summarized can differ across variables, and the summaries can be incomparable if the assumption of MCAR is violated. As an extreme example in the context of combining estimates, consider the estimation of the covariance of two variables and their standard deviations using available-case analysis independently for each of the three statistics; when these estimates are combined to estimate the correlation between the two variables, the resulting estimated correlation can lie outside the range $[-1, 1]$.

Complete-case analysis and available-case analysis (and combinations thereof) are extremely common approaches to handling missing data, and either was often the default strategy for handling incomplete data in older software packages. Although they are simple to implement, which is undeniably seductive, they can have serious deficiencies, which can be avoided using more modern and appropriate methods.

8.3.3   Weighting adjustment for unit nonresponse

A relatively simple device for removing or reducing the bias from complete-case analysis when the missing-data mechanism is not MCAR is to assign a nonresponse weight to each complete case (i.e., each respondent). In probability sampling, sampled units are often weighted by the inverse of their probabilities of selection in order to adjust estimates of population quantities for differential selection probabilities. The basic idea underlying weighting adjustments is to treat the complete cases as an extra layer of selection, and then to weight each complete case by the product of the sampling weight and the inverse of the conditional probability of being a complete case given selection into the sample. Although sampling weights are determined by the sample design and hence are known, nonresponse weights are based on unknown nonresponse probabilities, which need to be estimated from the data.

Nonresponse weights are generally based on background information that is available for all of the units in the survey. For instance, when a nonrespondent matches a respondent with respect to background variables measured for both, the nonrespondent's weight is simply added to the matching respondent's weight, and the nonrespondent is discarded. Because the match is defined by observed variables, such adjustment implicitly assumes MAR: if the MAR assumption is satisfied, weighting, in principle, removes nonresponse bias. In order to increase the plausibility of the MAR assumption, it is important to attempt to record background characteristics of respondents and nonrespondents that are predictive of nonresponse and use these variables to define nonresponse weights. Background characteristics should also be predictive of survey outcomes to limit sampling variance of resulting estimates.

Weighting methods can be useful for removing or reducing the bias in complete-case analysis. However, weighting methods do have some serious pitfalls. First, weighted estimates can have unacceptably high sampling variance due to the possibility of large weights. Second, the computation of appropriate standard errors for weighted complete-case estimators is often problematic. Explicit formulas are available for simple estimators, but methods are not well developed for more complex situations. Finally, the use of such weighting adjustment when dealing with item nonresponse deletes all incomplete cases and so discards additional observed data, which are not used in creating the weighting adjustment. For further discussion of weighting procedures for nonresponse in general, see Bethlehem (2002), Gelman and Carlin (2002), Little and Schenker (1995), and Little and Rubin (2002, Section 3.3).

8.4   Single imputation

Both complete-case and available-case analysis generally discard units with some observed data. An attractive alternative approach for handling incomplete data is to impute (fill in) the values of the items that are missing. A variety of imputation approaches can be used that range from extremely simple to rather complex. These methods can be applied to impute one value for each missing item (single imputation) or, in some situations, to impute more than one value, to allow appropriate assessment of imputation uncertainty (multiple imputation). Imputations are typically created assuming that the missing-data mechanism is ignorable, and for simplicity, here we focus our discussion on the ignorable situation.

Good imputations are draws from the predictive distribution of the missing values. A method for creating a predictive distribution for the imputations based on the observed data is required. This distribution can be generated by using either an explicit or an implicit modeling approach or a combination of the two approaches. The first approach requires the specification of a formal statistical model, on which the predictive distribution is based with explicitly stated assumptions. In the implicit modeling approach, the focus is on an algorithm, which implies an underlying model. Although assumptions are now implicit, they still need to be carefully assessed to ensure that they are reasonable.

Explicit modeling methods include: *mean imputation*, where missing values are replaced by means from the responding units in the sample; *regression imputation*, which replaces missing values by predicted values from a regression of the missing item on items observed for the unit, usually calculated from units with both sets of variables present; and *stochastic regression imputation*, which replaces missing values by a value predicted by regression imputation plus a residual drawn to reflect uncertainty in the predicted value.

In survey practice a common implicit modeling method is *hot-deck imputation*. Hot-deck imputation replaces each missing value with a random draw from a 'donor pool' consisting of values of that variable observed on responding units similar to the unit with the missing value. A donor pool is selected, for instance, by choosing units with complete data who have 'similar' observed values to the unit with missing data, for example, by exact matching on their observed values, or using a distance measure (metric) on observed variables to define 'similar'.

Singly imputed data sets are straightforward to analyze using standard complete-data methods; however, creating decent imputations may require substantial effort. Little and Rubin (2002) suggest some guidelines for creating imputations. Specifically, imputations should be: (1) conditional on observed variables, to reduce bias due to nonresponse, improve precision, and reflect associations between missing and observed variables; (2) multivariate, to preserve associations between missing variables; and (3) randomly drawn from predictive distributions, rather than set equal to means, to account properly for variability.

Unconditional mean imputation, which replaces each missing value with the mean of the observed values of that variable, meets none of the three guidelines. Conditional mean imputation, which replaces missing values of each variable with the mean of that variable calculated within cells defined by observed categorical variables, and regression imputation can satisfy the first two guidelines. Only stochastic regression imputation and hot-deck imputation, when done properly, can meet all three guidelines for single imputation.

Singly imputed data sets, created following the three guidelines suggested by Little and Rubin (2002), can be analyzed using standard complete-data techniques. The resulting inferences can satisfy criterion (a) of Section 8.1, leading to approximately unbiased estimates under ignorability. However, such inferences nearly always fail to satisfy criteria (b) and (c), providing too small estimated standard errors, too narrow confidence intervals, and too significant p-values for hypothesis tests, regardless of how the imputations were created. The reason is that the automatic application of standard complete-data methods to singly-imputed data sets treats imputed values as if they were known, although they are actually not known. In other words, single imputation followed by a complete-data analysis that does not distinguish between real and imputed values is almost always statistically invalid, because inferences about estimands based on the filled-in data do not account for imputation uncertainty.

Special methods for sampling variance estimation following single imputation have been developed for specific imputation procedures and estimation problems (e.g., Schafer and Schenker 2000; and Lee *et al.*, 2002). However, such techniques need to be customized to the imputation method used and to the analysis method at hand, and they often require the user to have information about the imputation model that is not typically available in shared data sets.

As an alternative, multiple imputation can be carried out. Multiple imputation is less computationally intensive than the replication approach (e.g., Efron 1994; Shao 2002), and generally can lead to valid inferences in the sense of satisfying criteria (a)–(c) of Section 8.1. Multiple imputation accounts for missing data by not only restoring the natural variability in the missing data, but also incorporating the uncertainty created by predicting missing data. Maintaining the original variability of the missing data is done by creating imputed values that are based on variables correlated with the missing data and reasons for the missingness.

Uncertainty is accounted for by creating different versions of the missing data and using the variability between imputed data sets.

8.5   Multiple imputation

Multiple imputation was first proposed in Rubin (1977, 1978a) and discussed in detail in Rubin (1987, 1996, 2004a,b). MI is a simulation technique that replaces each missing value in Y_{mis} with a vector of $m > 1$ plausible imputed values. These m values are ordered in the sense that m completed data sets can be created by the set of vectors of imputations; replacing each missing value by the first component in its vector of imputation creates the first complete data set, replacing each missing value by the second component in its vector of imputation creates the second complete data set, and so on. Thus, m completed data sets are created: $Y^{(1)}, \ldots, Y^{(\ell)}, \ldots, Y^{(m)}$, where $Y^{\ell} = \left(Y_{\text{obs}}, Y_{\text{mis}}^{(\ell)}\right)$. Typically m is fairly small: $m = 5$ is a standard number of imputations to use. Rubin (1987) showed that the relative efficiency of an estimate based on m imputations to one based on an infinite number of them is approximately $(1 + \gamma_0/m)^{-1/2}$ in units of standard errors, where γ_0 is the population fraction of missing information.[1] Therefore, unless rates of missing information are unusually high, there is often limited practical benefit to using more than five to ten imputations, except when conducting multi-component tests.

MI retains the advantages of single imputation while allowing the uncertainty due to the process of imputation to be directly assessed and included to create statistically valid inferences. Specifically, each of the m completed data sets is analyzed using standard complete-data procedures. When the m sets of imputations are repeated random draws from the predictive distribution of the missing values under a particular missing-data mechanism, the m complete-data inferences can be easily combined to form one inference that appropriately reflects both sampling variability and missing-data uncertainty. When the imputations are from two or more models for nonresponse, the combined inferences under the models can be contrasted across models to display the sensitivity of inference to alternative missing-data mechanisms.

Most of the techniques presently available for creating MIs assume that the missing-data mechanism is ignorable, but it is important to note that the MI paradigm does not require ignorable nonresponse. MIs may, in principle, be created under any kind of model for the missing-data mechanism, and the resulting inferences will be valid under that mechanism (see Rubin 1987, Chapter 6). Schafer (1997) is an excellent source of computational guidance for creating multiple imputations under a variety of models for the data.

8.5.1   Multiple-imputation inference for a scalar estimand

The analysis of a multiply imputed data set is quite direct. First, each data set completed by imputation is analyzed using the standard complete-data method that would be used in the absence of nonresponse. Let θ be the scalar estimand of interest (e.g., the mean of a variable, or the proportion of customers who are highly satisfied with a service). Let $\widehat{\theta}$ and $\widehat{V}$ be the complete-data estimators of θ and the sampling variance of $\widehat{\theta}$, respectively. Also, let $\widehat{\theta}_\ell$ and $\widehat{V}_\ell$, $\ell = 1, \ldots, m$,

---

[1] In the simple case of univariate missingness and no covariates, γ_0 is equal to the expected fraction of missing units (see Section 8.5.1). When there are many variables in a survey, however, γ_0 is typically smaller than this fraction because of the dependence between variables and the resulting ability to improve prediction of missing values from observed ones.

be m complete-data estimates of θ and their associated sampling variances, calculated from m repeated imputations under one missing-data model. The m sets of statistics are combined to produce the final point estimate: $\widehat{\theta}_{\mathrm{MI}} = m^{-1} \sum_{\ell=1}^{m} \widehat{\theta}_\ell$. The variability associated with this estimate has two components: the average within-imputation variance $W_m = m^{-1} \sum_{\ell=1}^{m} \widehat{V}_\ell$, and the between-imputation variance $B_m = (m-1)^{-1} \sum_{\ell=1}^{m} \left(\widehat{\theta}_\ell - \widehat{\theta}_{\mathrm{MI}}\right)^2$. The total variability associated with $\widehat{\theta}_{\mathrm{MI}}$ is $T_m = W_m + \left(1 + m^{-1}\right) B_m$, where the factor $1 + m^{-1}$ reflects the fact that only a finite number of completed-data estimates $\widehat{\theta}_\ell, \ell = 1, \ldots, m$, are averaged together to obtain the final point estimate (Rubin 1987, pp. 87–94).

The reference distribution for interval estimates and significance tests for θ is a Student t distribution: $\left(\theta - \widehat{\theta}_{\mathrm{MI}}\right) T_m^{-1/2} \sim t$. Under the assumption that, with complete data, a normal reference distribution would be appropriate, the degrees of freedom of the t distribution can be approximated by the value $\nu = (m-1)(1 + r_m^{-1})^2$, where $r_m = \left(1 + m^{-1}\right) B_m / W_m$ is the relative increase in variance due to nonresponse (Rubin 1987; Rubin and Schenker 1986). Barnard and Rubin (1999) relaxed the normality assumption to allow Student t reference distributions with complete data, and proposed the small-sample adjusted value $\nu_{\mathrm{BR}} = \left(\nu^{-1} + \nu_{\mathrm{obs}}^{-1}\right)^{-1}$ for the degrees of freedom of the t distribution in the MI analysis, where $\nu_{\mathrm{obs}} = (1 + r_m)^{-1} \nu_{\mathrm{com}} (\nu_{\mathrm{com}} + 1)(\nu_{\mathrm{com}} + 3)^{-1}$, and ν_{com} is the complete-data degrees of freedom. Another useful statistic about the nonresponse is the fraction of missing information due to nonresponse: $\gamma_m = (1 - (\nu+1)/(\nu+3)W_m/t_m)$, or the generalization proposed by Barnard and Rubin (1999) $\gamma_m = (1 - \nu_{\mathrm{BR}} + 1)/(\nu_{\mathrm{BR}} + 3)W_m/T_m)(\nu_{\mathrm{com}} + 3)/(\nu_{\mathrm{com}} + 1)$.

Inferential questions that cannot be cast in terms of a one-dimensional estimand can be handled through multivariate generalizations of this rule. See, for instance, Li *et al.* (1991a, 1991b), Rubin and Schenker (1991), Meng and Rubin (1992), and Little and Rubin (2002, Section 10.2) for additional methods for combining vector-valued estimates, significance levels, and likelihood ratio statistics.

8.5.2   Proper multiple imputation

The great virtues of MI are its simplicity and its generality. The user may analyze the data by virtually any technique that would be appropriate if the data were complete. The validity of the method, however, hinges on how the imputations $Y_{\mathrm{mis}}^{(1)}, \ldots, Y_{\mathrm{mis}}^{(m)}$ are generated. Clearly it is not possible to obtain valid inferences in general if imputations are created arbitrarily. The imputations should, on average, give reasonable predictions for the missing data, and the variability among them must reflect an appropriate degree of uncertainty. Rubin (1987) provides technical conditions under which repeated-imputation methods lead to statistically valid answers. An imputation method that satisfies these conditions is said to be 'proper'. The term 'proper' basically means that the summary statistics $\widehat{\theta}_{\mathrm{MI}}$, W_m and B_m, previously defined, yield approximately valid inference for the complete-data statistics, $\widehat{\theta}$ and $\widehat{V}$, over repeated realizations of the missing-data mechanism. Specifically, a multiple-imputation procedure is proper for the complete-data statistics, $\widehat{\theta}$ and $\widehat{V}$, if the following three conditions are satisfied: (1) as $m \to \infty$, $\left(\widehat{\theta}_{\mathrm{MI}} - \widehat{\theta}\right)/\sqrt{B_m}$ converges in distribution to a $\mathcal{N}(0, 1)$ random variable over the distribution of the response indicators R with Y held fixed; (2) W_m is a consistent estimate of $\widehat{V}$ as $m \to \infty$, with R regarded as random and Y regarded as fixed; (3) treating Y as fixed, the variability of the variance of $\widehat{\theta}_{\mathrm{MI}}$ over an infinite number of multiple imputations is of lower order than that of $\widehat{\theta}$.

These conditions are useful for evaluating the properties of an imputation method but provide little guidance for one seeking to create such a method in practice. For this reason, it is recommended that imputations be created through Bayesian arguments. For notational simplicity, assume that the missing-data mechanism is ignorable. Proper imputations are often most easily obtained as independent random draws from the posterior predictive distribution of the missing data given the observed data. Given a parametric model for the complete data, $p(Y_{obs}, Y_{mis} | \psi)$, and a prior distribution for the unknown model parameters, $p(\psi)$, the posterior predictive distribution of Y_{mis} given Y_{obs} can be formally written as $p(Y_{mis} | Y_{obs}) = \int p(Y_{mis}, \psi | Y_{obs}) d\psi = \int p(Y_{mis} | Y_{obs}, \psi) p(\psi | Y_{obs}) d\psi$. The distribution $p(Y_{mis} | Y_{obs})$ is a 'posterior' distribution because it is conditional on the observed data, Y_{obs}, and it is a 'predictive' distribution because it predicts the missing data, Y_{mis}. Imputations crated as independent realizations from $p(Y_{mis} | Y_{obs})$ can be proper because they reflect uncertainty about Y_{mis} given the parameters of the complete-data model, as well as uncertainty about the unknown model parameters, by taking draws of ψ from its posterior distribution, $p(\psi | Y_{obs})$, before using ψ to impute the missing data, Y_{mis}, from $p(Y_{mis} | Y_{obs}, \psi)$.

Imputations methods that do not account for all sources of variability are defined to be improper by Rubin (1987, Chapter 4). Thus, for instance, fixing ψ at a point estimate $\hat{\psi}$ and then drawing m imputations for Y_{mis} independently from $p(Y_{mis} | Y_{obs}, \hat{\psi})$ would constitute an improper MI procedure.

8.5.3   Appropriately drawing imputations with monotone missing-data patterns

When there are many variables to be imputed, drawing random samples from the posterior predictive distribution, $p(Y_{mis} | Y_{obs})$, may be difficult and require high-level expertise in both statistical computing methodology and software development. In a principled modeling approach, filling in the entire set of missing data, Y_{mis}, requires postulating a joint model for all variables with any missingness given the other variables, which has to be flexible enough to reflect the structure of complex data, which may include continuous, semicontinuous, ordinal, binary, and categorical variables. However, when the missing-data pattern is monotone, creating multiple imputations is relatively straightforward because the joint distribution of all variables can be specified sequentially as the product of conditional distributions.

Specifically, suppose that $(Y_{obs}, Y_{mis}) = (Y_{obs}^*, Y_{mis,1}, Y_{mis,2}, \dots, Y_{mis,k})$ follows a monotone pattern of missingness, where Y_{obs}^* represents the fully observed variables, and $Y_{mis,1}$ is the incompletely observed variable with the fewest missing values, $Y_{mis,2}$ the variable with the second fewest missing values, and so on, $Y_{mis,k}$ being the incompletely observed variable with the most missing values. Proper imputation with a monotone missing-data pattern begins by fitting an appropriate model to predict $Y_{mis,1}$ from Y_{obs}^* and then using this model to impute the missing values in $Y_{mis,1}$. For example, a regression model of $Y_{mis,1}$ on Y_{obs}^* can be fitted using units with $Y_{mis,1}$ observed, then the regression parameters of this model are drawn from their posterior distribution, and the missing values of $Y_{mis,1}$ are drawn from the posterior distribution of $Y_{mis,1}$ given these drawn parameters and the observed values of Y_{obs}^*. Next, the missing values for $Y_{mis,2}$ are imputed using Y_{obs}^* and the observed and imputed values of $Y_{mis,1}$; for example, if $Y_{mis,2}$ is a binary variable, a logistic regression model for $Y_{mis,2}$ given $(Y_{obs}^*, Y_{mis,1})$ could be used. Continue to impute the next most complete variable until all missing values have been imputed. In the case of monotone missing-data patterns, the product of the univariate prediction models defines the implied full imputation model, $p(Y_{mis} | Y_{obs})$, and the collection of imputed values is a proper imputation of the missing data, Y_{mis}, under this model.

8.5.4   Appropriately drawing imputations with nonmonotone missing-data patterns

When missingness is not monotone, creating imputations generally involves applying iterative simulation techniques, because directly drawing from $p\left(Y_{\mathrm{mis}}|Y_{\mathrm{obs}}\right)$ is generally intractable. In this case, Markov chain Monte Carlo (MCMC) provides a flexible set of tools for creating MIs from parametric models. Schafer (1997) describes MCMC methods to multiply-impute rectangular data sets with arbitrary patterns of missing values when the missing-data mechanism is ignorable, and also provides data examples and practical advice. These methods are applicable when the rows of the complete-data matrix can be modeled as independent and identically distributed observations from the following multivariate models: multivariate normal for continuous data, multinomial (including log-linear models) for categorical data, and the general location model for mixed multivariate data.

One MCMC method well suited to missing-data problems is the data augmentation (DA) algorithm of Tanner and Wong (1987). Briefly, letting t index iterations, DA involves iterating between (i) randomly sampling missing data from their conditional posterior predictive distributions, $Y_{\mathrm{mis}}^{(t)} \sim p\left(Y_{\mathrm{mis}}|Y_{\mathrm{obs}}, \psi^{(t-1)}\right)$, where $\psi^{(t-1)}$ is the current draw of unknown parameters; and (ii) randomly sampling unknown parameters from a simulated current complete-data posterior distribution, $\psi^{(t)} \sim p\left(\psi|Y_{\mathrm{obs}}, Y_{\mathrm{mis}}^{(t)}\right)$. Given an initial value for ψ, say $\psi^{(0)}$, this algorithm defines a Markov chain $\left\{\left(Y_{\mathrm{mis}}^{(t)}, \psi^{(t)}\right), t = 1, 2, \ldots\right\}$, which, under quite general conditions, converges to the stationary distribution of interest, $p\left(Y_{\mathrm{mis}}, \psi|Y_{\mathrm{obs}}\right)$.

Executing these steps until the Markov chain has reached effective convergence produces a draw of ψ from its observed data posterior distribution, $p\left(\psi|Y_{\mathrm{obs}}\right)$, and a draw of Y_{mis} from $p\left(Y_{\mathrm{mis}}|Y_{\mathrm{obs}}\right)$, the distribution from which MIs are to be generated. In many cases, the second step of the algorithm, $\psi^{(t)} \sim p\left(\psi|Y_{\mathrm{obs}}, Y_{\mathrm{mis}}^{(t)}\right)$, is straightforward. In more complicated situations, this step is intractable and may be replaced by one or more cycles of another MCMC algorithm that converges to $p\left(\psi|Y_{\mathrm{obs}}, Y_{\mathrm{mis}}^{(t)}\right)$. Much software presently available for creating multiple imputations uses DA (or variants of DA) to fill in missing values. Other algorithms that use MCMC methods for imputing missing values include the Gibbs sampler (Geman and Geman 1984) and the Methopolis–Hastings algorithm (Hastings 1970; Metropolis et al., 1953; Metropolis and Ulam 1949). See also Gelman et al. (2003) and Gilks et al. (1996) for more details on these algorithms, and Schafer (1997) for a complete exposition of MCMC methods in the imputation setting.

An alternative and popular approach to the creation of imputations in nonmonotone incomplete multivariate data uses the Gibbs sampler with fully conditionally specified models, where the distribution of each variable given all the other variables is the starting point. For each variable, a draw of parameters estimated using units with that variable observed is made, and then the missing data are imputed for that variable, and the procedure cycles through each variable with missing values, replacing missing values that are being conditioned on in a regression by the previously imputed values. Practical implementations of this idea include Kennickell (1991), Van Buuren and Oudshoorn K. (1999), Van Buuren and Oudshoorn C.G.M. (2000), Raghunathan et al. (2001), Münnich and Rässler (2005), and Van Buuren et al. (2006). The theoretical weakness of this approach is that the specified conditional densities may be incompatible, in the sense that they cannot be derived from a single joint distribution, and therefore the stationary distribution to which the Gibbs sampler attempts to converge may not exist.

In order to minimize or eliminate such incompatibility, Baccini *et al.* (2010) apply the 'multiple imputation by ordered monotone blocks' (IMB) strategy to the Anthrax Vaccine Adsorbed Trial data. This approach extends the theory for monotone patterns to arbitrary missing patterns, by breaking the problem into a collection of smaller problems where missing data do form a monotone pattern. The proposal of monotone blocks is a natural extension of using a single major monotone block (Rubin 2003). The IMB algorithm can be briefly described as follows. The variables and units in the data set are first rearranged such that the missing values not forming part of a monotone block are identified as minimal. The part that is monotone is labelled the 'first' monotone block. For those missing values that do not belong to the first monotone block, the process is repeated, identifying a rearrangement forming a monotone block, with the rest of the missing values being minimal. The process continues until all missing values have been identified with a monotone block. After the monotone blocks are obtained, the missing data within each block are multiply imputed. MI proceeds as follows: the missing data of all but the first monotone block are filled in with preliminary values; Bayesian sequential models are used to simulate the missing values for the first monotone block; the data imputed for the first monotone block are treated as observed and the missing values for the second monotone block are imputed, again using Bayesian sequential models. This process is performed across all the monotone blocks, and iterated until apparent convergence.

8.5.5 Multiple imputation in practice

A key feature of either single or multiple imputation is that the imputation phase is operationally distinct from subsequent analysis phases. As a result, imputations may be created by one person or organization and the ultimate analyses carried out by another, and the implicit or explicit model used for creating imputations may differ from the implicit or explicit model used in subsequent analyses of the completed data. In many cases, imputations are created just once by an expert in missing-data techniques (e.g., the data collector), who may have detailed knowledge or even additional confidential data that cannot be made available to the ultimate analysts but which may be relevant to the prediction of missing values. The ultimate user of multiply-imputed data could apply a variety of potentially complicated complete-data analyses, and then use the combining rules and combined results even though the multiple imputations were created under different models.

This feature gives MI great inherent flexibility and it is especially attractive in the context of public-use data sets that are shared by many ultimate users, but raises the possibility that the statistical model or assumptions used to create the imputed data sets may be incompatible with those used to analyze them. Meng (1994) defines the imputer's and analyst's models as 'congenial' if the resulting inference is fully valid. When the imputer makes fewer assumptions than the analyst, then MI generally leads to valid inferences with perhaps some loss of efficiency, because the additional generality of the imputation model may increase variability among the imputed data sets. When the imputer makes more assumptions than the analyst – and the extra assumptions are true – then imputations may turn out 'superefficient' from the perspective of the data analyst (Rubin 1996), in the sense that MI inferences may be more precise than any inference derived from the observed data and the analyst's model alone, because they reflect the imputer's better knowledge about the process that creates nonresponse. The only serious negative effect of inconsistency arises when the imputer makes more assumptions than the analyst and these additional assumptions are false, because the multiple imputations created under an incorrect model can lead to erroneous conclusions. For further discussions on the validity of multiple-imputation inference when the imputer's and analyst's models differ, see Fay (1992), Meng (1994), and Rubin (1996).

Clearly, congeniality is more easily satisfied when the imputer and the analyst are the same entity or communicate with each other. In the context of shared data sets, however, in order to warrant near-congeniality of the imputer's and user's models, the imputation model should include a set of variables as rich as possible (e.g, Rubin 1996). In practice, this means that an imputation model should reasonably preserve those distributional features (e.g., associations) that will be the subject of future analyses. It is especially important to include design variables, such as variables used to derive sampling weights, or the sampling weights themselves, and domain indicators when domain estimates are to be obtained by subsequent users. When such critical variables are excluded from the imputation model, point estimates, as well as sampling variance estimates based on this model, will generally be biased (e.g., Kim *et al.*, 2006).

The performance of a MI procedure depends on several factors, including the posited missing-data mechanism, the (implicit or explicit) imputation model specified for the data, and the complete-data analyses the ultimate user performs. Therefore, in order to obtain valid inference from multiply-imputed data sets, the performance of the imputation strategy should be carefully assessed (e.g., Baccini *et al.*, 2009; Schafer *et al.*, 1996; Tang *et al.*, 2005).

8.5.6   Software for multiple imputation

Many statistical software packages have built-in or add-on functions for creating multiply-imputed data sets, managing the results from each imputed data set, and combining the inferences using the method described in Section 8.5.1 and its multivariate generalizations. Joseph Schafer has produced the S-Plus libraries NORM (which is also available as stand-alone Windows package), CAT, MIX and PAN for multiply imputing normal, categorical, mixed and panel data, respectively. These libraries are freely available (see http://www.stat.psu.edu/~jls/misoftwa.html).

Multiple Imputation by Chained Equations (MICE) is another freely-available library distributed for S-Plus, which may be downloaded from the www.multiple-imputation.com Web site. Procedures to impute missing data using MICE are also implemented in other software packages, including IVEware, the R environment, STATA, and SPSS. For instance, the *mi* (Gelman *et al.*, 2011; Su *et al.*, forthcoming) and *mice* (Van Buuren and Groothuis-Oudshoorn, forthcoming) packages implement the MICE procedure within R, and STATA provides the *ice* command to impute missing data based on the chained equation approach (Royston 2004, 2005). MICE-MI is a stand-alone version of mice, and it is downloadable from http://web.inter.nl.net/users/S.van.Buuren/mi/hmtl/mice.htm. The S-Plus missing data library extends S-Plus to support model-based missing data models, by using the EM (Dempster *et al.*, 1977) and DA algorithms (Tanner and Wong, 1987).

IVEware (http://www.isr.umich.edu/src/smp/ive) by Raghunathan *et al.* (2001) is very flexible and freely available software for MI; it is an SAS version 9 callable routine built using the SAS macro language or a stand-alone executable. In addition to supporting chained equations, IVEware extends multiple imputation to support complex survey sample designs.

In SAS/STAT, multiple imputation is implemented by two procedures. The imputation step is carried out by PROC MI. Then, complete-data methods are employed using any of the SAS procedures for complete-data analysis. Finally, the results are combined using PROC MIANALYZE.

SOLAS is commercially available software designed specifically for creating and analyzing multiply-imputed data sets (http://www.statsol.ie/solas/solas.htm). SOLAS is most appropriate for data sets with a monotone or nearly monotone pattern of missing data.

Other packages that provide some support for MI are currently available. We refer to the www.multiple-imputation.com website for more information, or to Horton and Lipsitz (2001) and Horton and Kleinman (2007) for some historical perspective.

8.6   Model-based approaches to the analysis of missing data

We now describe model-based missing-data methods where an explicit model for the complete data is specified and inferences are based on the likelihood or posterior distribution under that model. In full generality, statistical models are developed by specifying $p(Y, R|\psi, \xi)$, the joint distribution of Y and R (Rubin 1976). Two classes of models have been proposed, based on alternative factorizations for this distribution. Selection models (e.g., Heckman, 1976; Little and Rubin, 2002) specify the joint distribution of Y and R as

$$p(Y, R|\psi, \xi) = p(Y|\psi) p(R|Y, \xi), \tag{8.3}$$

where $p(Y|\psi)$ represents the complete-data model for Y, $p(R|Y, \xi)$ represents the model for the missing-data mechanism, and ψ and ξ are unknown distinct parameters. Pattern-mixture models (e.g., Rubin, 1977, 1978a; Glynn *et al.*, 1986, 1993; Little, 1993) specify

$$p(Y, R|\phi, \pi) = p(Y|R, \phi) p(R|\pi) \tag{8.4}$$

where $p(Y|R, \phi)$ represents the conditional distribution of Y given the missing-data pattern R, $p(R|\pi)$ models the missing-data indicator, and ϕ and π are unknown distinct parameters. Pattern-mixture models partition the data with respect to the missingness of the variables, and the resulting marginal distribution of Y is a mixture of distributions. If R is independent of Y, that is, missingness is MCAR, then these two model forms are easily seen to be equivalent with $\psi = \phi$ and $\xi = \pi$. When the missing data are not MCAR, the two specifications generally yield different models because of the different distinctness of parameters.

Little and Rubin (2002, Chapter 15) discuss the use of selection and pattern-mixture approaches in the context of nonignorable missing-data mechanisms for different types of data. As discussed previously, a crucial point about the use of nonignorable models is that they are difficult to specify correctly, because there is no direct evidence in the data about the relationship between the missing-data mechanism and the missing values themselves. For this reason, selection models and pattern-mixture models for nonignorable missing data generally depend strongly on assumptions about specific distributions, which are not directly testable. Consequently, sensitivity to model specification is a serious scientific problem for both selection and pattern-mixture models. Thus, whenever possible, it is advisable to consider several nonignorable models, rather than to rely exclusively on one model, and to explore the sensitivity of answers to the choice of the model, using a baseline analysis under ignorability as a primary point of comparison.

As discussed in Section 8.2.2, when the missing-data mechanism is ignorable, statistically valid inferences for the parameters of the data distribution, ψ, can be based on the likelihood function for ψ ignoring the missing-data mechanism: $\mathcal{L}_{ign}(\psi|Y_{obs})$ (see equation (8.2)) (Rubin 1976). Little and Rubin (2002, Chapters 11–14) provide a complete exposition of ignorable likelihood and Bayesian methods, also describing their application to solve different analytic problems, as well as reviewing several examples where analyses of incomplete data are carried out under the assumption of an ignorable missing-data mechanism.

Once maximum likelihood (ML) estimates of parameters have been obtained, inferences can be derived by applying standard methods. In many incomplete-data problems, however, the likelihood function (8.2) is a complicated function, and explicit expressions for the ML estimates of ψ are difficult to derive. Standard numerical ML algorithms, such as the Newton–Raphson algorithm, can be applied, but other iterative procedures, exploiting the missing-data aspect of the problem, may have advantages. The best known of these algorithms is the expectation–maximization (EM) algorithm (Dempster *et al.*, 1977), which takes advantage of the facts that: (1) if ψ were known, estimation of many functions of Y_{mis} would be relatively easy; and (2) if the data were complete, computation of ML estimates would be relatively simple. Several extensions of EM have been also proposed, including ECM (Meng and Rubin 1993), ECME (Liu and Rubin 1994), AECM (Meng and van Dyk 1997), and PX-EM (Liu *et al.*, 1998). For detailed discussions of the theoretical properties of the EM algorithm, examples of its use, methods for obtaining standard errors based on the algorithm, and its extensions, see Dempster *et al.* (1977), McLachlan and Krishnan (1997), Schafer (1997), and Little and Rubin (2002, Chapters 8, 9 and 11–15).

ML techniques are most useful when sample sizes are large, because then the log-likelihood is approximately quadratic in the neighborhood of the ML estimates, and can be summarized well using the ML estimates and their asymptotic covariance matrix. In small samples, ML methods may have unsatisfactory properties because the assumption of asymptotic normality of the likelihood may be unreasonable. Thus, alternatives to ML inference may be preferable. When sample sizes are small or ML techniques are intractable, simulation methods can be used, which are often easier to implement than analytic methods. From the Bayesian perspective, the focus is on iterative simulation methods for approximating the posterior distribution of the parameters of interest, ψ.

Under the assumption of an ignorable missing-data mechanism, Bayesian inferences for ψ are based on the observed-data posterior distribution with density $p(\psi|Y_{obs}) \propto p(\psi) p(Y_{obs}|\psi)$, where $p(\psi)$ is the prior density for ψ. As with ML estimation, working explicitly with this observed-data posterior distribution can be difficult. The DA algorithm (Tanner and Wong 1987), introduced in Section 8.5.4, may facilitate drawing ψ from $p(\psi|Y_{obs})$. For discussions of the theoretical properties of the DA algorithm, its extensions, and examples of the use of Bayesian iterative simulation methods, see Tanner and Wong (1987), Gelfand and Smith (1990), Schafer (1997), and Little and Rubin (2002, Chapters 10–14). Although we focus on ignorable nonresponse, the EM and DA algorithms can be also applied in the context of the nonignorable missing-data mechanism (e.g., Little and Rubin 2002, Chapter 15).

If the sample size is large, likelihood-based analyses and Bayesian analyses under diffuse prior distributions are expected to provide similar results because the likelihood dominates the prior distribution and is nearly multivariate normal. If the sample size is small, a Bayesian analysis may be preferable, because it allows us to avoid the usual assumption of asymptotic normality of the likelihood.

8.7 Addressing missing data in the ABC annual customer satisfaction survey: An example

Since 2001, ABC has conducted an annual customer satisfaction survey (ACSS) to gather information on its touch points and interactions with customers through a questionnaire consisting of 81 questions. The ABC ACSS suffers from both unit nonresponse and item nonresponse. Over the years, and across geographical areas, respondents to the ACSS questionnaire range from 10% to 80%, with a typical response rate of 45%. Here, we focus on item nonresponse, using data from the 2010 ACSS, which provides information on satisfaction levels of 266 ABC customers (see Chapter 2 for a detailed description of the ACSS and preliminary analyses of the 2010 ACSS). For simplicity, we select a subset of 12 questions, including overall satisfaction with ABC, willingness to recommend ABC to other companies, repurchasing intentions, and overall satisfaction with each of the following topics: equipment, sales support, technical support, training, supplies, software solutions, customer website, purchasing support, and contracts and pricing. Each satisfaction item is an ordinal variable with five categories ($1 =$ very low satisfaction level, $\ldots$, $5 =$ very high satisfaction level). Six customer background variables are also included: country, segmentation, age of ABC's equipment, company's profitability, customer's position in the company and customer seniority. Country and customer's position are completely observed, age of ABC's equipment and customer seniority are missing for one customer, company's profitability is missing for 25 customers, and segmentation is missing for 43 customers.

The first column of Table 8.1 presents the proportion of missing values for each satisfaction variable, which shows that missingness rates are somewhat high for some variables. The missing-data pattern is not monotone and only 67 (25%) customers provide complete data. Thus, a complete-case analysis may lead to a substantial loss of information, implying a loss of precision and potential bias. An available-case analysis may be a simple alternative using more information, although, as previously stated (see Section 8.3.2), this approach has many potential problems.

Rather than removing variables or units with missing data, we can impute missing values. We handle the problem of missing data in the ABC ACSS using both single and multiple imputation. We initially use the naive unconditional mean imputation approach to impute one value for each missing item: missing values in each satisfaction variable are filled in by the median of the recorded values of that variable. The median, rather than the mean, is used because each satisfaction item is an ordinal categorical variable. Unfortunately, this strategy can severely distort the distribution for variables, leading to complications with summary measures including, most notably, underestimation of variances. Moreover, this method will distort associations between variables.

As an alternative to unconditional median imputation, we use the R environment and functions from the `mice` library to create $m = 5$ multiply-imputed data sets for each of three multiple-imputation methods. The first imputes the missing values in each variable using a simple random sample from the observed values of that variable. This method is useful if the data are assumed to be MCAR, but it does not account for the associations between variables. The second method fills in missing values using Bayesian polytomous logistic regression for nominal and ordinal categorical variables, and Bayesian linear regression for quantitative variables. The third method is like the second, but ordinal categorical variables (such as the satisfaction items) are modeled using Bayesian linear regression, and the imputed values are then rounded to the nearest level observed in the data.

In our simple application, we focus on two estimands: the proportion of satisfied or highly satisfied (satisfaction level ≥ 4) customers for each satisfaction item, and the association between the overall satisfaction variables (overall satisfaction with ABC, willingness to recommend ABC to other companies, and repurchasing intentions) and satisfaction with sales support, technical support, and software solutions. Association is measured using

Table 8.1 Univariate analyses of ABC ACSS data using complete-case analysis (CCA), available-case analysis (ACA), single imputation (SI) and multiple imputation (MI): proportion of highly or very highly satisfied customers (standard errors in parentheses)

Variable	Missing-data Proportion	CCA	ACA	SI[†]	MI[‡] (a)	(b)	(c)
Overall satisfaction with ABC							
Overall satisfaction	0.015	0.507	0.595	0.602	0.592	0.590	0.592
		(0.061)	(0.030)	(0.030)	(0.031)	(0.030)	(0.031)
Willingness to recommend ABC	0.019	0.537	0.632	0.639	0.632	0.629	0.630
		(0.061)	(0.030)	(0.029)	(0.030)	(0.030)	(0.030)
Repurchasing intentions	0.015	0.552	0.649	0.654	0.647	0.644	0.647
		(0.061)	(0.029)	(0.029)	(0.029)	(0.029)	(0.030)
Overall satisfaction with . . .							
Equipment	0.034	0.657	0.623	0.635	0.623	0.616	0.617
		(0.058)	(0.030)	(0.030)	(0.030)	(0.030)	(0.030)
Sales support	0.068	0.493	0.472	0.440	0.464	0.471	0.471
		(0.061)	(0.032)	(0.030)	(0.033)	(0.031)	(0.033)
Technical support	0.008	0.716	0.686	0.688	0.683	0.685	0.683
		(0.055)	(0.029)	(0.028)	(0.029)	(0.029)	(0.029)
Training	0.338	0.657	0.733	0.823	0.737	0.709	0.714
		(0.058)	(0.033)	(0.023)	(0.034)	(0.035)	(0.035)
Supplies	0.053	0.388	0.472	0.447	0.471	0.467	0.471
		(0.060)	(0.031)	(0.030)	(0.031)	(0.032)	(0.032)
Software solutions	0.312	0.388	0.393	0.271	0.401	0.375	0.371
		(0.060)	(0.036)	(0.027)	(0.032)	(0.036)	(0.030)
Customer website	0.214	0.433	0.455	0.357	0.450	0.432	0.459
		(0.061)	(0.034)	(0.029)	(0.033)	(0.036)	(0.038)
Purchasing support	0.086	0.537	0.519	0.560	0.531	0.517	0.511
		(0.061)	(0.032)	(0.030)	(0.032)	(0.032)	(0.032)
Contracts and pricing	0.049	0.224	0.332	0.316	0.334	0.325	0.328
		(0.051)	(0.030)	(0.029)	(0.031)	(0.029)	(0.029)

[†]Unconditional median imputation was used.
[‡]Three imputation methods were used: (a) imputations are randomly drawn from the observed values; (b) a separate univariate regression model is specified for each variable, taking into account the measurement scale of that variable (Bayesian polytomous logistic regression for categorical variables and Bayesian linear regression for quantitative variables); (c) as with (b), but ordinal categorical variables are modeled using Bayesian linear regression, and the imputed values are rounded to the nearest level observed in the data.

the Goodman–Kruskal gamma statistic. Table 8.1 (last six columns) and Table 8.2 present estimates and their standard errors for these univariate and bivariate statistics, using the alternative missing-data methods mentioned above.

As can be seen in these tables, the standard error estimates for the complete-case estimators are twice as large as those of the available-case analysis and the single and multiple imputation-based analyses, showing the large loss of precision due to discarding incomplete cases. Moreover, the differences between the complete-case and available-case estimates

Table 8.2   Bivariate analyses of ABC ACSS data using complete-case analysis (CCA), available-case analysis (ACA), single imputation (SI) and multiple imputation (MI): the Goodmann–Kruskal gamma statistic (standard errors in parentheses)

				MI[‡]		
Variable	CCA	ACA	SI[†]	(a)	(b)	(c)
Target variable: overall satisfaction with ABC						
Willingness to recommend ABC	0.889	0.890	0.893	0.883	0.894	0.889
	(0.047)	(0.025)	(0.025)	(0.026)	(0.024)	(0.025)
Repurchasing intentions	0.676	0.764	0.766	0.749	0.766	0.763
	(0.095)	(0.039)	(0.039)	(0.045)	(0.039)	(0.039)
Overall satisfaction with . . .						
Sales support	0.541	0.418	0.409	0.381	0.390	0.382
	(0.124)	(0.068)	(0.068)	(0.069)	(0.069)	(0.068)
Technical support	0.595	0.646	0.636	0.631	0.650	0.641
	(0.115)	(0.051)	(0.052)	(0.057)	(0.051)	(0.053)
Software Solutions	0.372	0.487	0.460	0.353	0.401	0.413
	(0.137)	(0.078)	(0.073)	(0.079)	(0.081)	(0.077)
Target variable: willingness to recommend ABC to other companies						
Overall satisfaction	0.889	0.890	0.893	0.883	0.894	0.889
	(0.047)	(0.025)	(0.025)	(0.026)	(0.024)	(0.025)
Repurchasing intentions	0.767	0.889	0.890	0.876	0.889	0.888
	(0.081)	(0.028)	(0.028)	(0.030)	(0.028)	(0.028)
Overall satisfaction with . . .						
Sales support	0.583	0.457	0.438	0.410	0.434	0.434
	(0.105)	(0.061)	(0.061)	(0.063)	(0.063)	(0.061)
Technical support	0.409	0.481	0.474	0.473	0.483	0.473
	(0.126)	(0.058)	(0.057)	(0.059)	(0.057)	(0.060)
Software solutions	0.389	0.496	0.480	0.349	0.433	0.478
	(0.127)	(0.076)	(0.072)	(0.084)	(0.067)	(0.063)
Target variable: repurchasing intentions						
Overall satisfaction	0.676	0.764	0.766	0.749	0.766	0.763
	(0.095)	(0.039)	(0.039)	(0.045)	(0.039)	(0.039)
Willingness to recommend ABC	0.767	0.889	0.890	0.876	0.889	0.888
	(0.081)	(0.028)	(0.028)	(0.030)	(0.028)	(0.028)
Overall satisfaction with . . .						
Sales support	0.675	0.477	0.463	0.430	0.459	0.457
	(0.082)	(0.059)	(0.059)	(0.063)	(0.061)	(0.059)
Technical support	0.222	0.381	0.378	0.380	0.390	0.376
	(0.140)	(0.065)	(0.065)	(0.066)	(0.065)	(0.067)
Software solutions	0.336	0.416	0.412	0.301	0.353	0.398
	(0.134)	(0.082)	(0.077)	(0.079)	(0.071)	(0.070)

[†]Unconditional median imputation was used.

[‡]Three imputation methods were used: (a) imputations are randomly drawn from the observed values; (b) a separate univariate regression model is specified for each variable, taking into account the measurement scale of that variable (Bayesian polytomous logistic regression for categorical variables and Bayesian linear regression for quantitative variables); (c) as with (b), but ordinal categorical variables are modeled using Bayesian linear regression, and the imputed values are rounded to the nearest level observed in the data.

suggest that the MCAR assumption is not plausible for the 2010 ACSS data, and so complete-case estimates may be biased. As we might expect, multiple-imputation methods lead to standard error estimates slightly larger than those of single imputation, because they incorporate all uncertainty due to predicting missing data.

Consider now the univariate statistics in Table 8.1. The differences among the point estimates provided by the alternative missing-data methods are not striking irrespective of the proportion of missing values in each item. However, complete-case and available-case analyses, as well as unconditional median imputation, cannot generally be recommended, given their theoretical pitfalls.

Larger differences between inferences provided by the alternative missing-data methods can be observed for the bivariate analyses. Specifically, the two multiple-imputation methods, which incorporate information on the relationships between variables (last two columns of Table 8.2), lead to quite different results than the marginal multiple-imputation method (column named MI (a)) and the other more naive approaches. However, these results do not appear to be sensitive to the models used for imputation represented in the last two columns.

As a closing note, we offer two reminders. First, as previously noted, missing data in the ABC ACSS are not monotone, and therefore the MICE strategy we applied might theoretically use incompatible fully conditional distributions for imputing missing values. Alternative imputation methods, such as IMB (e.g., Baccini *et al.*, 2010), could be used and the results compared with those from MICE. Second, although not a focus of this review, sensitivity analyses are an important component of modeling when relatively many data are missing, and they should be routinely conducted (see also Chapter 9 of this book on dealing with problems of outlier detection). Such additional analyses require effort, but allow insight into the impact of missing data assumptions (e.g., Baccini *et al.*, 2009).

8.8 Summary

Missing data are a prevalent problem in many social and economic studies, including customer satisfaction surveys. Here, we have reviewed some important general concepts regarding missing-data patterns and mechanisms that lead to missing data, and then discussed some common, but naive, techniques for dealing with missing data, as well as less naive and more principled methods. Simple approaches, such as complete-case analysis and available-case analysis, provide inefficient, though valid, results when missing data are MCAR, but generally biased results when missing data are MAR. Other frequently used methods for handling missing data, such as unconditional mean imputation, provide generally biased results even under the MCAR assumption.

Multiple imputation is a principled method and a useful tool for handling missing data because of its flexibility allowing the imputation and subsequent analysis to be conducted separately. This key feature of MI is especially attractive in the context of public-use data sets to be shared by many users. However, the validity of the analysis can depend on the cability of the imputation model to capture the missingness mechanism correctly. Also, the separation of imputation model and analysis model raises the issue of compatibility between these two models. If an appropriate imputation model is employed, MI generally leads to statistically valid inferences. MI is also becoming popular because of the availability of easy-to-use statistical software. Such software helps users to apply MI in a broad range of missing-data settings.

MI is not the only principled method for handling missing values, nor is it necessarily the best for a specific problem. In some cases, good estimates can be obtained through a weighted estimation procedure. In fully parametric models, ML or Bayesian inferences can often be conducted directly from the incomplete data by specialized numerical methods, such as the EM or DA algorithms, or their extensions. MI has the advantages of flexibility over direct analyses, by allowing the imputer's and analyst's models to differ.

A crucial issue arising in the analysis of incomplete data concerns the uncertainty about the reasons for nonresponse. Therefore it is useful to conduct sensitivity analyses under different modeling assumptions. Multiple imputation allows the straightforward study of the sensitivity of inferences to various missing-data mechanisms simply by creating repeated randomly drawn imputations under more than one model and using complete-data methods repeatedly (e.g., Rässler, 2002; Rubin, 1977, 1986; Baccini *et al.*, 2009, 2010).

Many of the approaches discussed here may be applied under either an ignorable or nonignorable model for the missing-data mechanism. The observed data can never provide any direct evidence against ignorability, and procedures based on ignorable missing-data models typically lead to at least partial corrections for the bias due to nonresponse.

Acknowledgements

This chapter borrows in some places from previous reviews written by other combinations of co-authors. These reviews are presented in publications by Cook and Rubin (2005), Rässler *et al.* (2007) and Rässler *et al.* (2008). We thank the authors of these papers for their generosity.

References

Baccini, M., Cook, S., Frangakis, C.E., Li, F., Mealli, F., Rubin, D.B. and Zell, E.R. (2009) Evaluating multiple imputation procedures using simulations in a Bayesian prospective. Paper presented at the 2009 Joint Statistical Meetings, Washington, DC.

Baccini, M., Cook, S., Frangakis, C.E., Li, F., Mealli, F., Rubin, D.B. and Zell, E.R. (2010) Multiple imputation in the anthrax vaccine research program. *Chance*, 23, 16–23.

Barnard, J. and Rubin, D.B. (1999) Small-sample degrees of freedom with multiple imputation, *Biometrika*, 86, 948–955.

Bethlehem, J.G. (2002) Weighting nonresponse adjustments based on auxiliary information. In R.M. Groves, D.A. Dillman, J.L. Eltinge and R.L.A. Little (eds), *Survey Nonresponse*, pp. 275–287. New York: John Wiley & Sons, Inc.

Cook, S. and Rubin, D.B. (2005) Multiple imputation in designing medical device trials. In K.M. Becker and J.J. Whyte (eds), *Clinical Evaluation of Medical Devices*, pp. 241–251. Washington, DC: Humana Press.

Dempster, A.P., Laird, N.M. and Rubin, D.B. (1977) Maximum likelihood from incomplete data via the EM algorithm. *Journal of the Royal Statistical Society, Series B*, 39, 1–22.

Efron, B. (1994) Missing data, imputation, and the bootstrap, *Journal of the American Statistical Association*, 89, 463–475.

Fay, R.E. (1992) When are inferences from multiple imputation valid? *Proceedings of the Survey Research Methods Section of the American Statistical Association*, 227–232.

Gelfand, A.E. and Smith, A.F.M. (1990) Sampling-based approaches to calculating marginal densities, *Journal of the American Statistical Association*, 85, 398–409.

Gelman, A. and Carlin, J.B. (2002) Poststratification and weighting adjustment. In R.M. Groves, D.A. Dillman, J.L. Eltinge and R.L.A. Little (eds), *Survey Nonresponse*, pp. 289–302. New York: John Wiley & Sons, Inc.

Gelman, A., Carlin J.B., Stern, H.S. and Rubin, D.B. (2003) *Bayesian Data Analysis*, 2nd edn. Boca Raton, FL: Chapman & Hall.

Gelman, A., Hill, J., Yajima, M., Su, Y. and Pittau, M. (2011) mi: Missing data imputation and model checking. Package for the R statistical software. http://lib.stat.cmu.edu/R/CRAN/.

Geman, S. and Geman, D. (1984) Stochastic relaxation, Gibbs distributions, and the Bayesian restoration of images, *IEEE Transactions on Pattern Analysis and Machine Intelligence*, 6, 721–741.

Gilks, W.R., Richardson, S., and Spiegelhalter, D.J.E. (1996) *Markov Chain Monte Carlo in Practice*. New York: Chapman & Hall.

Glynn, R.J., Laird, N.M. and Rubin, D.B. (1986) Selection modeling versus mixture modeling with noningnorable nonresponse. In H. Wainer (ed.), *Drawing Inferences from Self-Selected Samples*, pp. 115–142. New York, Springer.

Glynn, R.J., Laird, N.M. and Rubin, D.B. (1993) Multiple imputation in mixture models for nonignorable nonresponse with follow-ups. *Journal of the American Statistical Association*, 88, 984–993.

Groves, R.M., Dillman, D.A., Eltinge, J.L. and Little, R.J.A. (eds) (2002) *Survey Nonresponse*. New York: John Wiley & Sons, Inc.

Hastings, W.K. (1970) Monte Carlo sampling methods using Markov chains and their applications. *Biometrika*, 57, 97–109.

Heckman, J.J. (1976) The common structure of statistical models of truncation, sample selection and limited dependent variables and a simple estimator for such models. *Annals of Economic and Social Measurement*, 5, 475–492.

Horton, N.J. and Kleinman, K.P. (2007) Much ado about nothing: A comparison of missing data methods and software to fit incomplete data regression models. *American Statistician*, 61(1), 79–90.

Horton, N.J. and Lipsitz, S.R. (2001) Multiple imputation in practice: Comparison of software packages for regression models with missing variables. *American Statistician*, 55, 244–254.

Kennickell, A.B. (1991) Imputation of the 1989 survey of consumer finances: Stochastic relaxation and multiple imputation. *American Statistical Association Proceedings of the Section on Survey Research Methods*, 1–10.

Kim, J.K., Brick, J.M., Fuller, W.A. and Kalton, G. (2006) On the bias of the multiple-imputation variance estimator in survey sampling. *Journal of The Royal Statistical Society, Series B*, 68(3), 509–521.

Lee, H., Rancourt, E. and Särndal, C.E. (2002) Variance estimation for survey data under single imputation. In R.M. Groves, D.A. Dillman, J.L. Eltinge and R.L.A. Little (eds), *Survey Nonresponse*, pp. 315–328. New York: John Wiley & Sons, Inc.

Li, K.H., Meng, X.-L., Raghunathan, T.E. and Rubin, D.B. (1991a) Significance levels from repeated *p*-values with multiply-imputed data. *Statistica Sinica*, 1, 65–92.

Li, K.H., Raghunathan, T.E. and Rubin, D.B. (1991b) Large-sample significance levels from multiply-imputed data using moment-based statistics and an F reference distribution. *Journal of the American Statistical Association*, 86, 1065–1073.

Little, R.J.A. (1985) A note about models for selectivity bias. *Econometrica*, 53, 1469–1474.

Little, R.J.A. (1993) Pattern-mixture models for multivariate incomplete data. *Journal of the American Statistical Association*, 88, 125–134.

Little, R.J.A. and Rubin, D.B. (1987) *Statistical Analysis with Missing Data*. New York: John Wiley & Sons, Inc.

Little, R.J.A. and Rubin, D.B. (2002) *Statistical Analysis with Missing Data*, 2nd edn. Hoboken, NJ: John Wiley & Sons, Inc.

Little, R.J.A. and Schenker, N. (1995) Missing data. In G. Arminger, C.C. Clogg and M.E. Sobel (eds), *Handbook of Statistical Modeling for the Social and Behavioral Sciences*, pp. 39–75. New York: Plenum Press.

Liu, C. and Rubin, D.B. (1994) The ECME algorithm: A simple extension of EM and ECM with faster monotone convergence. *Biometrika*, 81, 533–648.

Liu, C., Rubin, D.B. and Wu, Y.N. (1998) Parameter expansion to accelerate EM: The PX-EM algorithm. *Biometrika*, 85, 755–770.

Madaw, W.G. and Olkin, I. (1983) *Incomplete Data in Sample Surveys. Volume 3: Proceedings of the Symposium*. New York: Academic Press.

Madaw, W.G., Nisselson, H. and Olkin, I. (1983a) *Incomplete Data in Sample Surveys. Volume 1: Report and Case Studies*. New York: Academic Press.

Madaw, W.G., Olkin, I. and Rubin, D.B. (1983b) *Incomplete Data in Sample Surveys. Volume 2: Theory and Bibliographies and Case Studies*. New York: Academic Press.

McLachlan, G.J. and Krishnan, T. (1997) *The EM Algorithm and Extensions*. New York: John Wiley & Sons, Inc.

Meng, X.-L. (1994) Multiple-imputation inferences with uncongenial sources of input (with discussion). *Statistical Science*, 10, 538–573.

Meng, X.-L. and Rubin, D.B. (1992) Performing likelihood ratio tests with multiply imputed data sets. *Biometrika*, 79, 103–111.

Meng, X.-L. and Rubin, D.B. (1993) Maximum likelihood estimation via the ECM algorithm: A general framework. *Biometrika*, 80, 267–278.

Meng, X.-L. and van Dyk, D. (1997) The EM algorithm: An old folk-song sung to a fast new tune. *Journal of the Royal Statistical Society, Series B*, 59, 511–567.

Metropolis, N., Rosenbluth, A.W., Rosenbluth, M.N., Teller, A. and Teller, E. (1953) Equations of state calculations by fast computing machines. *Journal of Chemical Physics*, 21, 1087–1091.

Metropolis, N. and Ulam, S. (1949) The Monte Carlo method. *Journal of the American Statistical Association*, 49, 335–341.

Münnich, R. and Rässler, S. (2005) PRIMA: A new multiple imputation procedure for binary variables. *Journal of Official Statistics*, 21, 325–341.

Raghunathan, T.E., Lepkowski, J.M., Van Hoewyk, J. and Solenberger, P. (2001) A multivariate technique for multiply imputing missing values using a sequence of regression models. *Survey Methodology*, 27, 85–95.

Rässler, S. (2002) *Statistical Matching: A Frequentist Theory, Practical Applications, and Alternative Bayesian Approaches*, Lecture Notes in Statistics 168. New York: Springer.

Rässler, S., Rubin, D.B. and Schenker, N. (2007) Incomplete data: Diagnosis, imputations, and estimation. In E. de Leeuw, J. Hox and D. Dillman (eds), *The International Handbook of Survey Research Methodology*. Thousand Oaks, CA: Sage.

Rässler, S., Rubin, D.B. and Zell, E.R. (2008) Incomplete data in epidemiology and medical statistics. In C.R. Rao, J.P. Miller and D.C. Rao (eds), *Handbook of Statistics*, 27, pp. 569-601. Amsterdam: Elsevier.

Royston, P. (2004) Multiple imputation of missing values. *Stata Journal*, 4(3), 227–241.

Royston, P. (2005) Multiple imputation of missing values: update. *Stata Journal*, 5(2), 188–201.

Rubin, D.B. (1976) Inference and missing data. *Biometrika*, 63(3), 581–592.

Rubin, D.B. (1977) Formalizing subjective notions about the effect of nonrespondents in sample surveys. *Journal of the American Statistical Association*, 72, 538–543.

Rubin, D.B. (1978a) Multiple imputations in sample surveys: A phenomenological Bayesian approach to nonresponse. *Proceedings of the Survey Research Methods Section of the American Statistical Association*, 20–34.

Rubin, D.B. (1978b) A note on Bayesian, likelihood, and sampling distribution inferences. *Journal of Educational Statistics*, 3, 189–201.

Rubin, D.B. (1986) Statistical matching using file concatenation with adjusted weights and multiple imputations. *Journal of Business and Economic Statistics*, 4, 87–95.

Rubin, D.B. (1987) *Multiple Imputation for Nonresponse in Surveys*. New York: John Wiley & Sons, Inc.

Rubin, D.B. (1994) Comment on 'Missing data, imputation, and the bootstrap' by B. Efron, *Journal of the American Statistical Association*, 89, 475–478.

Rubin, D.B. (1996) Multiple imputation after 18+ years (with discussion). *Journal of the American Statistical Association*, 91, 473–489.

Rubin, D.B. (2003) Nested multiple imputation of NMES via partially incompatible MCMC. *Statistica Neerlandica* 57(1), 3–18.

Rubin, D.B. (2004a) *Multiple Imputation for Nonresponse in Surveys*, 2nd edn. Hoboken, NJ: John Wiley & Sons, Inc.

Rubin, D.B. (2004b) The design of a general and flexible system for handling nonresponse in sample surveys. *American Statistician*, 58, 298–302.

Rubin, D.B. and Schenker, N. (1986) Multiple imputation for interval estimation from simple random sample with ignorable nonresponse. *Journal of the American Statistical Association*, 81, 366–374.

Rubin, D.B. and Schenker, N. (1991) Multiple imputation in health-care databases: An overview and some applications. *Statistics in Medicine*, 10, 585–598.

Rubin, D.B., Stern, H. and Vehovar, V. (1995) Handling 'don't know' survey responses: The case of Slovenian plebiscite. *Journal of the American Statistical Association*, 90, 822–828.

Schafer, J.L. (1997) *Analysis of Incomplete Multivariate Data*. New York: Chapman & Hall.

Schafer, J.L. and Schenker, N. (2000) Inference with imputed conditional means, *Journal of the American Statistical Association*, 95, 144–154.

Schafer, J.L., Ezzati-Rice, T.M., Johnson, W., Khare, M., Little, R.J.A. and Rubin, D.B. (1996) The NHANES III Multiple Imputation Project. *Proceedings of the Survey Research Methods Section, American Statistical Association*.

Shao, J. (2002) Replication methods for variance estimation in complex surveys with imputed data. In R.M. Groves, D.A. Dillman, J.L. Eltinge and R.L.A. Little (eds), *Survey Nonresponse*, pp. 303–314. New York: John Wiley & Sons, Inc.

Shih, W.J. (1992) On informative and random dropouts in longitudinal studies. *Biometrics*, 48, 970–972.

Su, Y.-S., Gelman, A., Hill, J., Yajima, M. (forthcoming) Multiple imputation with diagnostics (mi) in R: Opening windows into the black box. *Journal of Statistical Software*. http://www.stat.columbia.edu/gelman/research/published/mipaper.rev04.pdf.

Tang, L., Song, J., Belin, T.R. and Unuetzer, J. (2005) A comparison of imputation methods in a longitudinal randomized clinical trial. *Statistics in Medicine*, 24, 2111–2128.

Tanner, M.A. and Wong, W.H. (1987) The calculation of posterior distributions by data augmentation (with discussion). *Journal of the American Statistical Association*, 82, 528–550.

Van Buuren, S. and Groothuis-Oudshoorn K. (forthcoming) MICE: Multivariate imputation by chained equations in R. *Journal of Statistical Software*. http://lib.stat.cmu.edu/R/CRAN/web/packages/mice/index.htm.

Van Buuren, S. and Oudshoorn, C.G.M. (2000) *Multivariate Imputation by Chained Equations: MICE v1.0 User's Manual*, Report PG/VGZ/00.038. Leiden: TNO Preventie en Gezondheid.

Van Buuren, S. and Oudshoorn, K. (1999) *Flexible Multivariate Imputation by MICE*, TNO/VGZ/PG 99.054. Leiden: TNO Preventie en Gezondheid.

Van Buuren, S., Brand, J.P.L., Groothuis-Oudshoorn, C.G.M. and Rubin, D.B. (2006) Fully conditional specification in multivariate imputation. *Journal of Statistical Computation and Simulation*, 76, 1049–1064.

9

Outliers and robustness for ordinal data

Marco Riani, Francesca Torti and Sergio Zani

This chapter tackles the topics of robustness and multivariate outlier detection for ordinal data. We initially review outlier detection methods in regression for continuous data and give an example which shows that graphical tools of data analysis or traditional diagnostic measures based on all the observations are not sufficient to detect multivariate atypical observations. Then we focus on ordinal data and illustrate how to detect atypical measurements in customer satisfaction surveys. Next, we review the generalized linear model of ordinal regression and apply it to the ABC survey. The chapter concludes with an analysis of a set of diagnostics to check the goodness of the suggested model and the presence of anomalous observations.

9.1 An overview of outlier detection methods

There are several definitions of outliers in the statistical literature (see Barnett and Lewis, 1994; Atkinson *et al.*, 2004; Hadi *et al.*, 2009). A commonly used definition is that outliers are a minority of observations in a data set that is represented by a common pattern which can be captured by some statistical model. The assumption here is that there is a core of at least 50% of observations that is homogeneous and a set of remaining observations (hopefully few) which has patterns that are inconsistent with this common pattern. Awareness of outliers in some form or another has existed for at least 2000 years. Thucydides, in his third book about the Peloponnesian War (III 20, 3–4), describes how in 428 BC the Plataeans used concepts of robust statistics in order to estimate the height of the ladder which was needed to overcome the fortifications built by the Peloponnesians and the Boeotians who were besieging their city.

> The same winter the Plataeans, who were still being besieged by the Pelopon-
> nesians and Boeotians, distressed by the failure of their provisions, and seeing no

Modern Analysis of Customer Surveys: with applications using R, First Edition. Edited by Ron S. Kenett and Silvia Salini.
© 2012 John Wiley & Sons, Ltd. Published 2012 by John Wiley & Sons, Ltd.

hope of relief from Athens, nor any other means of safety, formed a scheme with the Athenians besieged with them for escaping, if possible, by forcing their way over the enemy's walls; the attempt having been suggested by Theaenetus, son of Tolmides, a soothsayer, and Eupompides, son of Daimachus, one of their generals. At first all were to join: afterwards, half hung back, thinking the risk great; about two hundred and twenty, however, voluntarily persevered in the attempt, which was carried out in the following way. Ladders were made to match the height of the enemy's wall, which they measured by the layers of bricks, the side turned towards them not being thoroughly whitewashed. These were counted by many persons at once; and though some might miss the right calculation, most would hit upon it, particularly as they counted over and over again, and were no great way from the wall, but could see it easily enough for their purpose. The length required for the ladders was thus obtained, being calculated from the breadth of the brick.

Rejection of outliers prior to performing classical statistical analysis has thus been regarded as an essential preprocessing step for almost as long as the methods have existed. For example, a century and a half ago Edgeworth stated that: 'The method of Least Squares is seen to be our best course when we have thrown overboard a certain portion of our data – a sort of sacrifice which often has to be made by those who sail upon the stormy seas of Probability'.

For univariate data, Grubbs (1969) defined an outlier as an observation that 'appears to deviate markedly from other members of the sample in which it occurs'. For multivariate data, however, the examination of each dimension by itself or in pairs does not work, because it is possible for some data points, as we will see in the next section, to be outliers in multivariate space, but not in any of the original univariate dimensions. When the variables under study are not measured on a quantitative scale, as often happens in customer satisfaction surveys, the need to jointly consider several variables becomes of paramount importance in detecting atypical observations. For example, in the 11 September attacks on World Trade Center in New York, five out of the eighty passengers on one of the flights displayed unusual characteristics with respect to a set of qualitative variables. These five passengers were not US citizens, but had lived in the USA for some period of time, were citizens of a particular foreign country, had all purchased one-way tickets, had purchased these tickets at the gate with cash rather than credit cards, and did not have any checked luggage. One or two of these characteristics might not be very unusual, but, taken together, they could be seen as markedly different from the majority of airline passengers.

Identification of outlying data points is often by itself the primary goal, without any intention to fit a statistical model. The outliers themselves sometimes can be points of primary interest, drawing attention to unknown aspects of the data, or especially, if unexpected, leading to new discoveries. For example, one of the purposes of the Institute for the Protection and Security of the Citizen, one of the institutes of the European Commission's Joint Research Centre, is to find patterns of outliers in data sets including millions of trade flows grouped in a large number of small to moderate size samples. The statistically relevant cases, which can be due to potential cases of tax evasion or activities linked to money laundering, are presented for evaluation and feedback to subject-matter experts of the anti-fraud office of the European Commission and its partner services in the member states. In the context of international trade, the unexpected presence of a set of atypical transactions from or towards a particular country can be an indication of trade carousels (repetitive trade of the same good among the same persons), potential illegal activities or stockpiling.

The examples above clearly demonstrate the need for outlier identification both on small and very large data sets, whether the available variables are quantitative or qualitative. See Su and Tsai (2011) for a recent overview of the different approaches to outlier detection.

9.2   An example of masking

The purpose of this section is to show that when multivariate (multiple) outliers are present in the data, traditional tools of visual inspection or traditional data analysis based on least squares (LS) in regression may lead us to incorrect conclusions. Atkinson and Riani (2000, pp. 5–9) give an example of a regression data set with 60 observations on three explanatory variables (X1, X2 and X3) where there are six masked outliers that cannot be detected using standard analyses. The scatter plot of the response y against the three explanatory variables (see Figure 9.1) simply shows that y is increasing with each of X1, X2 and X3, but does not reveal particular observations far from the bulk of the data. The traditional plot of residuals against fitted values (see left panel of Figure 9.2) shows no obvious pattern and the largest residual is for a unit (no. 43) which lies well within the envelopes of the quantile–quantile plot of studentized residuals (see the right-hand panel of Figure 9.2).

In the regression context, the best-known robust methods are based on the use of least trimmed squares (LTS), least median of squares (LMS) or on forward search (FS) – see Morgenthaler (2007) and Rousseeuw and Hubert (2011) for recent reviews of robust methods in regression. The LTS regression method searches for an estimate of the vector of regression coefficients which minimizes the sum of the h smallest squared residuals, where h must be at least half the number of observations. On the other hand, the LMS estimator minimizes the median of the squares of the residuals. LTS and LMS are very robust methods in the sense that the estimated regression fit is not unduly influenced by outliers in the data, even if there are

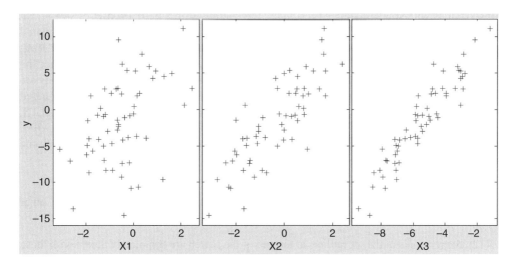

Figure 9.1   Plot of y *against each of the explanatory variables.*

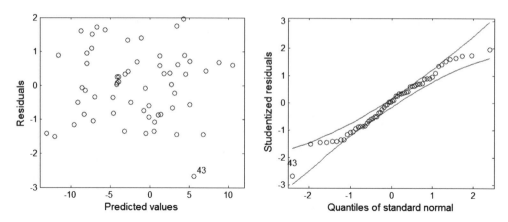

Figure 9.2   Traditional uninformative diagnostic plots. Left: least squares residuals against fitted values. Right: quantile–quantile plot of studentized residuals.

several atypical observations. Due to this robustness, we can detect outliers by their large LTS (LMS) residuals. Given that explicit solutions for LMS and LTS do not exist, approximate solutions are sought using elemental subsets, that is, subsets of p observations, where p is the number of explanatory variables including the intercept (Rousseeuw, 1984; Rousseeuw and Van Driessen, 2006).

Because of the way in which models are fitted, whether with LS, LTS or LMS, we lose information about the effect of individual observations on inferences about the form and parameters of the model. In order to understand the effect that each unit, outlier or not, exerts on the fitted model, it is necessary to start with a subset of data and monitor the required diagnostics. In the example above, if we start with a least squares fit to four observations, robustly chosen, we can calculate the residuals for all 60 observations and next fit to the five observations with smallest squared residuals. In general, given a fit to a subset of size m, we can order the residuals and take, as the next subset, the $m + 1$ cases with smallest squared residuals. This gives a forward search through the data (Atkinson and Riani 2000; Riani *et al.*, 2009), ordered by closeness to the model. We expect that the last observations to enter the search will be those that are furthest from the model and so may cause changes once they are included in the subset used for fitting. Figure 9.3 shows the monitoring of the scaled squared residuals for the 60 units of the data set. In this case we have initialized the search with LTS, investigating all possible $\binom{60}{4}$ subsets and taking the one with the smallest sum of the 50% smallest residuals, although this is not necessary. This fascinating plot reveals not only the presence of six masked outliers but also that:

(1) the six outliers form a cluster, because their trajectories are very similar – in other words, they respond in a similar way to the introduction of units into the subset;

(2) the residuals of the six outliers at the end of the search are completely mixed with those of the other units, therefore traditional methods based on single deletion diagnostics cannot detect them;

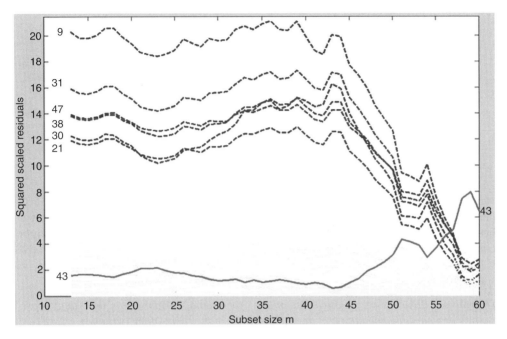

Figure 9.3   Monitoring of squared scaled residuals. The outliers have been drawn with dotted lines, while the trajectory of the case which in the final step shows the largest residual has been drawn with a solid line. All the other unimportant trajectories have been shown in faint grey.

(3) the entry of the six outliers causes a big increase in the trajectory of the residual for unit 43. Indeed this is the unit which in the final step has the largest residual and may be wrongly considered as an outlier from the traditional plot of residuals against fitted values (see Figure 9.2).

9.3   Detection of outliers in ordinal variables

The problem of defining and detecting outliers for ordinal variables has received scant attention in the literature. For example, the recent book of Agresti (2010) on the analysis of ordinal data does not mention this topic. Only a few papers deal with ordinal outliers in multivariate statistical methods (Zijlstra *et al.*, 2007; Pardo, 2010; Dong, 2010; Liu *et al.*, 2010). One of the reasons may be due to the difficulty of defining outliers in ordinal variables. Clearly, if we define as univariate outliers the observations that are different from the majority of the observations in a data set, for an ordinal variable corresponding to a ranking, no unit can be considered as an outlier, because the observations take on values (ranks) from 1 to n. In an ordered categorical variable with k levels, a unit may have each of k, *a priori*, defined categories and therefore no outlier could be detected. However, in a few special cases the frequency distribution of a variable may show univariate outliers. Consider, for example, the fictitious distribution of the responses of 200 customers on overall satisfaction given in Table 9.1. The two 'very unsatisfied' customers may be considered as univariate outliers, because this category of the

Table 9.1   Overall satisfaction of 200 customers.

Levels	Frequencies
Very unsatisfied	2
Unsatisfied	0
Fair	30
Satisfied	120
Very satisfied	48
	200

variable has a very low frequency and is separated from the other categories (the 'unsatisfied' label has zero frequency). If we assume that there is an underlying quantitative variable which has produced the observed ordinal categories (continuum hypothesis), the first category may be considered as coming from a different distribution. However, distributions like the one in Table 9.1 are very unusual in real situations. For example, in the ABC survey, no variable presents such a distribution. Therefore, for ordinal variables we suggest the following definition:

Definition   *Bivariate and multivariate outliers for ordinal variables are those units representing an unusual combination of the categories or of the ranks of the variables.*

The previous definition of multivariate outliers can be applied both to rankings and to ordered categorical variables or to a set of nominal and ordinal variables. Notice that, similarly to what we saw in the previous section, for categorical variables, a multivariate outlier may not necessarily be a univariate outlier with respect to single ordinal categorical variables.

9.4   Detection of bivariate ordinal outliers

The joint distribution of two ordinal categorical variables, with k_1 and k_2 categories, can be presented in a $k_1 \times k_2$ contingency table. Consider, for example, the data in Table 9.2, showing the bivariate distribution of the 255 (of 266) not-missing respondents of the ABC survey with respect to the variables 'Overall satisfaction level with ABC' (question 1) and 'Overall

Table 9.2   Contingency table of the 255 non-missing respondents of questions 'Overall satisfaction level with ABC' (q1) and 'Overall satisfaction level with the equipment' (q11) in the ABC survey.

q1\q11	Very unsatisfied	Unsatisfied	Fair	Satisfied	Very satisfied	Total
Very unsatisfied	4	3	3	1	0	11
Unsatisfied	1	8	6	8	⇒2	25
Fair	0	5	40	21	1	67
Satisfied	⇒1	2	20	86	7	116
Very satisfied	0	0	3	26	7	36
Total	6	18	72	142	17	255

satisfaction level with the equipment' (question 11). The marginal distributions of the two variables show no outlier, but the pairs marked with the arrow symbol, (satisfied, very unsatisfied) with frequency 1 and (unsatisfied, very satisfied) with frequency 2, may be considered as unusual (non-coherent) combinations of the categories, that is, bivariate ordinal outliers.

Kendall's tau-b rank correlation index computed on all 255 units is equal to 0.480, with standard error 0.049. If we delete the three potential outliers, the tau-b index becomes 0.518 with a standard error of 0.044. The rank correlation without the outliers is higher and may be more appropriate to describe the association between the two variables for the majority of units.

9.5   Detection of multivariate outliers in ordinal regression

In this section, we apply a generalized linear model for ordinal variables using 'Overall satisfaction level with ABC' (q1) as response variable (see Bradlow and Zaslavsky, 1999, for an approach based on a different model for ordinal data). In Section 9.5.1, we briefly review the building blocks of ordinal regression for better understanding of our results. In Section 9.5.2 we define our model, we analyse the discrepancies between the observed and predicted values for detecting the eventual presence of atypical observations, and finally, we propose a detailed analysis of some anomalous units, or subgroups of units, for better understanding of the key features of the data set.

9.5.1   Theory

In most cases it is implausible to assume normality and homogeneity of variance for an ordered categorical outcome when (as in the case of our application) the ordinal outcome contains only a small number of discrete categories (Chu and Ghahramani, 2005). Thus, the ordinal regression model becomes the preferred modelling tool, because it does not assume normality and constant variance. The model is based on the assumption that there is a latent continuous outcome variable and that the observed ordinal outcome (which in our case is the global satisfaction index) arises from discretizing the underlying continuum into k ordered groups.

To be more precise, the ordinal regression model (McCullagh, 1980), which is nothing more than a generalized linear model (McCullagh and Nelder, 1989), may be written in the following form:

$$\text{link}(\gamma_{ij}) = \alpha_j - [b_1 x_{i1} + b_2 x_{i2} + \ldots + b_p x_{ip}], \quad j = 1, \ldots, k-1 \text{ and } i = 1, \ldots, n.$$

$$(9.1)$$

The parameter α_j represents the threshold value of the jth category of the underlying continuous variable. The α_j terms often are not of much interest in themselves, because their values do not depend on the values of the independent variables for a particular case. They are like the intercept in a linear regression, except that each level j of the response has its own value. In our application the number of categories $k-1$ is equal to 4 (1 = very unsatisfied, 2 = unsatisfied, 3 = fair, 4 = satisfied). As usual, one of the classes of the response (in this case the category 'very satisfied') must be omitted from the model, because it is redundant.

The parameter γ_{ij} is the cumulative distribution function for the jth category of the ith case. For example, in our case γ_{i2} is the cumulative probability that the ith subject is 'very unsatisfied' or 'unsatisfied'.

In (9.1), $\mathrm{link}(\gamma_{ij})$ is the so-called 'link function' which is typical of generalized linear models. In the case of ordinal regression, it is a transformation of the cumulative probabilities of the ordered dependent variable that allows for estimation of the model. The most commonly used links are logit, probit, negative log-log, complementary log-log and Cauchit. While the negative log-log link function is recommended when the probability of the lowest category is high, the complementary log-log is particularly suitable when the probability of the highest category is high. Finally, the logit, probit and Cauchit links assume that the underlying dependent variable respectively has a logistic, normal or Cauchy distribution. In general the Cauchit link is mainly used when extreme values are likely to be present in the data.

The number of regression coefficients is p: $b_1, \ldots, b_p$ are a set of regression coefficients and $x_{i1}, x_{i2}, \ldots, x_{ip}$ are a set of p explanatory variables for the ith subject.

With regard to the $-[b_1 x_{i1} + b_2 x_{i2} + \cdots + b_p x_{ip}]$ term in the model, it is interesting to note two aspects. First, the negative sign before the square brackets ensures that larger coefficients indicate an association with larger scores. In our case, for example, a positive coefficient of an explanatory variable means that people who are in that class have a greater probability of showing a global level of satisfaction.

Second, the expression inside the square brackets does not depend on j. In other words, if the link is logit, for example, this implies that the effect of the independent variables is the same for all the logits. Thus the results are a set of parallel lines or hyperplanes, one for each category of the outcome variable. It is possible to test this assumption using the so-called 'test of parallel lines' against the alternative that the relationships between the independent variables and logits are not the same for all logits. If the output of the testing procedure leads us to reject the hypothesis of parallel lines, it is necessary to introduce into the model a scale component and to modify the link function as follows:

$$\mathrm{link}(\gamma_{ij}) = \frac{\alpha_j - [b_1 x_{i1} + b_2 x_{i2} + \ldots + b_p x_{ip}]}{\exp(\tau_1 z_{i1} + \ldots + \tau_k z_{ik})}. \tag{9.2}$$

The numerator of equation (9.2) is known in the literature as the *location* component of the model, while the denominator specifies the *scale*. The $\tau_1, \tau_2, \ldots, \tau_k$ are coefficients for the scale component and the $z_{i1}, z_{i2}, \ldots, z_{ik}$ are predictor variables for the ith subject for the scale component (chosen from the same set of variables as the xs). The scale component accounts for differences in variability for different values of the predictor variables. For example, if certain groups have more variability than others in their ratings, using a scale component to account for this may improve the model.

In ordinal regression, in order to evaluate the goodness of fit of the model, the indices which are typically used are Cox and Snell's R^2 (R^2_{CS}) and Nagelkerke's pseudo-R^2(R^2_N) (Nagelkerke, 1991). The second index is a modification of the first which adjusts the scale of the statistic to cover the full range from 0 to 1. These two indexes are expressed as follows:

$$R^2_{CS} = 1 - \left(\frac{l(0)}{l(\hat{\beta})}\right)^{2/n}, \quad R^2_N = \frac{R^2_{CS}}{\max(R^2_{CS})}, \tag{9.3}$$

where $l(\hat{\beta})$ is the log-likelihood of the current model, $l(0)$ is the log-likelihood of the initial model which does not contain explanatory variables (null model), and $\max(R^2_{CS}) = 1 - \{l(0)\}^{2/n}$.

9.5.2   Results from the application

We initially considered all the 133 variables present in the ABC 2010 ACSS data set as explanatory variables. We then repeated the analysis considering only the 21 overall satisfaction level variables (questions 11, 17, 25, 31, 38, 42, 43, 49, 57, 65, 67, 70 and 73–81). Here we present only the results of this last model, since, for the purposes of our exposition, the difference between the two sets of results is not appreciable.

The data set has a large number of missing values. Thus we decided to keep in the analysis only the six variables with less than 10% missing data (questions 11, 17, 25, 38, 57, 65, i.e. overall satisfaction level with the equipment, sales support, technical support, ABC's supplies and orders, purchasing support and contracts and pricing) and 216 observations obtained with the listwise criteria (see Chapter 10 of this volume for a systematic treatment of missing values).

Figure 9.4, which shows the bar-plot of the response variable 'Overall satisfaction level' (q1), highlights that its empirical distribution is slightly negatively skewed. Therefore, as mentioned in Section 9.5.1, suitable link functions, *a priori*, are the logit and the complementary log-log. After fitting the model with these two links, we noticed that all coefficients have the expected sign, with the exception of q57. This variable was always highly non-significant, even after suppressing the observations with the most anomalous combinations with the response. Hence, we decided to remove q57 from all subsequent analyses.

With regard to the goodness of fit, the R^2 indices in equation (9.3) indicated that the model with the complementary log-log link was more satisfactory. Using this link we both tested model (9.1) with just the location component and model (9.2) with the scale component. The introduction of the scale component (in this case only q25 was significant in the scale) made Nagelkerke's pseudo-R^2 much lower than the original one, (0.499 against 0.818). We therefore decided to adopt the specification without the scale. We also investigated all the pairwise interactions among the explanatory variables; we did not find statistical evidence for their inclusion.

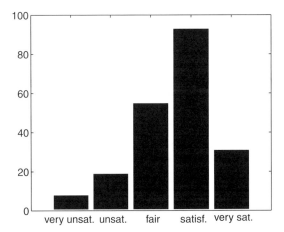

Figure 9.4   Bar-plot of the 'Overall satisfaction level' (q1): the 'satisfied' (4) category is the most frequent one.

Table 9.3   Results of the ordinal generalized linear model: list of thresholds, list of the variables in the location component, estimated coefficients and relative *p*-values.

	Estimated coefficients	Significance (*p*-value)
Thresholds		
$y = 1$ (Very unsatisfied)	1.931	0.000
$y = 2$ (Unsatisfied)	3.335	0.000
$y = 3$ (Fair)	4.979	0.000
$y = 4$ (Satisfied)	6.885	0.000
Position		
q11	0.533	0.000
q17	0.058	0.488
q25	0.528	0.000
q38	0.258	0.044
q65	0.266	0.018

The estimated coefficients of our final model together with their significance values are given in Table 9.3. All coefficients, except that for q17, are significant at the 5% level. However, from the contingency table between variables q1 and q17 (Table 9.4), we can see that there are 12 observations, marked with arrows, showing a global level of satisfaction that is different from that collected in q17 by three or more categories. If we suppress these, the *p*-value of q17 decreases as far as 0.001 and Nagelkerke's pseudo-R^2 becomes equal to 0.971. Therefore, q17 must be kept in the regression model.

The value of the pseudo-R^2 (0.818) in this model is very high even if there is still about 20% of variance that is left unexplained. Now we wish to evaluate the stability of the model in the presence of some units with an unusual combination of the categories, as we did before for q1 and q17, taking into account that, as happens in ordinary regression models, an observation with a large residual does not necessarily mean that it is influential with respect to the fitted equation, and vice versa. To start this check, in Table 9.5 we report the contingency table

Table 9.4   Contingency table of the 216 respondents of questions 1, 'Overall satisfaction level with ABC', and 17, 'Overall satisfaction level with the sales support' in the ABC survey.

		q17					
		Very unsatisfied (1)	Unsatisfied (2)	Fair (3)	Satisfied (4)	Very satisfied (5)	Total
q1	Very unsatisfied (1)	7	2	0	1⇐	0	10
	Unsatisfied (2)	1	9	7	0	4⇐	21
	Fair (3)	6	9	19	19	4	57
	Satisfied (4)	3⇐	13	27	37	15	95
	Very satisfied (5)	1⇐	3⇐	7	12	10	33
	Total	18	36	60	69	33	216

Table 9.5   Contingency table of 'Overall satisfaction level' (q1)' and 'Predicted class of satisfaction level' (PRE).

| | Predicted class for q1 (PRE) | | | | | |
	Very unsatisfied (1)	Unsatisfied (2)	Fair (3)	Satisfied (4)	Very satisfied (5)	Total
q1 Very unsatisfied (1)	2	2	5←	1⇐	0	10
Unsatisfied (2)	1	4	10	6←	0	21
Fair (3)	1←	1	21	34	0	57
Satisfied (4)	0	0	9	77	9	95
Very satisfied (5)	0	0	1←	15	17	33
Total	4	7	46	133	26	216

of the 'Overall satisfaction level' (q1)' and the 'Predicted class of satisfaction level' by the model (PRE). This table clearly shows that the predicted values are in general very close to the observed levels of satisfaction. Most of the frequencies lie around the main diagonal of the table.

In Table 9.5, the cells where the absolute difference between predicted and observed class of satisfaction is greater than 1 are indicated with arrows. For example, in the cell with the fat open arrow, the model overestimates q1 by three points: it predicts category 4 instead of category 1. We will label this case with its position in our data subset, 163 (201 in the original data set). The reason why the model largely misclassifies this unit is partially explained in Table 9.6: the ranking assigned by the respondent to the dependent variable q1 is lower than the rankings mainly given by the same respondent to the other questions; q1 is even lower than the minimum category taken by the other variables. If we apply our ordinal regression model to the data set without this unit, Nagelkerke's pseudo-R^2 increases from 0.818 to 0.849 and the p-values of the regression parameters remain approximately invariant.

In Table 9.5, four cells are also indicated by thin arrows, where q1 differs from its predicted values by two categories: there are six cases where the model predicts 4 instead of 2, five cases where the model predicts 3 instead of 1, one case where the model predicts 3 instead of 5, and one case where the model predicts 1 instead of 3. Concerning the first six units for which q1 = 2 and PRE = 4 (units 33, 96, 101, 128, 186, 207 in the data subset, corresponding to units 39, 121, 127, 157, 229, 256 in the original data set), Table 9.7 shows that also in this case, the respondents assign to q1 a category almost always lower than or equal to the minimum assigned for all the other variables. If we now apply our regression model to the data set without unit 163 and without these six units, Nagelkerke's pseudo-R^2 increases from 0.849 to 0.918.

Table 9.6   Unit 163 (q1 = 1, PRE = 4): the 'Overall satisfaction level' (q1) is not in agreement with the other partial satisfaction levels.

	q1	q11	q17	q25	q38	q65
Unit 163	1	2	4	4	3	4

Table 9.7   Units 33, 96, 101, 128, 186, 207 (q1 = 2, PRE = 4). In all cases the 'Overall satisfaction level' (q1) is not in agreement with the other satisfaction levels.

	q1	q11	q17	q25	q38	q65
Unit 33	2	4	5	3	3	3
Unit 96	2	4	5	4	3	3
Unit 101	2	5	2	1	5	4
Unit 128	2	5	5	3	2	5
Unit 186	2	3	3	5	3	2
Unit 207	2	4	2	3	3	3

The same considerations about the inconsistency of the overall level of satisfaction with the partial levels are valid for the other three groups of units for which q1 = 1 and PRE = 3, q1 = 5 and PRE = 3, and q1 = 3 and PRE = 1, as is clear from Tables 9.8–9.10. If we apply our regression model to the data set without unit 163 and without respectively the units shown in Tables 9.8, 9.9 and 9.10, Nagelkerke's pseudo-R^2 increases from 0.849 to 0.883, 0.918 and 0.888. If, instead, we drop all 14 cases mentioned from the data set, Nagelkerke's pseudo-R^2 increases as far as 0.967.

The units discussed so far can be considered influential (anomalous) with respect to our ordinal regression model. For a better interpretation of these results, we can also investigate the observations with anomalous combination of the categories, leaving aside the defined model. Taking all the variables on the same level, we could, for example, compare for each observation i the 'Overall satisfaction level' q1 with the median of the other five variables, as follows:

$$d(i) = q1(i) - \text{median}[q11(i), q17(i), q25(i), q38(i), q65(i)] \tag{9.4}$$

The result of this proposal on the ABC data subset is given in Figure 9.5. The x-axis is the observation number while the y-axis plots quantity (9.4). The observations of interest are labelled with their data subset row number. As expected, for unit 163, which was the most anomalous in Table 9.5, $d(i)$ takes a very low value. Among the six units of Table 9.7, for which q1 = 2 and PRE = 4, units 128, 96 and 101 deserve a mention for the low values of $d(i)$. Of the other five units reported in Table 9.8 for which q1 = 1 and PRE = 3, units 22, 49 and

Table 9.8   Units 14, 22, 49, 70, 77 of our data subset (units 14, 25, 63, 90, 98 in the original data set): q1 = 1 and PRE = 3. In most of the cases the 'Overall satisfaction level' (q1) is not in agreement with the other satisfaction levels.

	q1	q11	q17	q25	q38	q65
Unit 14	1	3	1	2	3	2
Unit 22	1	4	1	2	3	3
Unit 49	1	3	1	4	3	1
Unit 70	1	1	1	5	3	1
Unit 77	1	1	2	4	3	3

Table 9.9   Unit 124 of our data subset (unit 152 in the original data set): q1 = 5 and PRE = 3. The 'Overall satisfaction level' (q1) is higher than the maximum of the partial satisfaction levels.

	q1	q11	q17	q25	q38	q65
Unit 124	5	3	2	4	3	2

Table 9.10   Unit 60 of our data subset (unit 79 in the original data set): q1 = 3 and PRE = 1. The 'Overall satisfaction level' (q1) is higher than the maximum of the partial satisfaction levels.

	q1	q11	q17	q25	q38	q65
Unit 60	3	2	2	1	1	1

77 take low values. Finally, the values of $d(i)$ for observations 124 and 60, for which q1 = 5 and PRE = 3 and q1 = 3 and PRE = 1, are very large.

One of the highest values reported in Figure 9.5 corresponds to unit 135 (unit 164 in the original data set), which was not identified as anomalous in Table 9.5 as it belongs to the combination q1 = 5 and PRE = 4. As clearly shown by Table 9.11, the value assigned by the respondent to the 'Overall satisfaction level' (q1) is greater than the rankings mainly given by the same respondent to the other questions. However, in our regression model, the largest coefficients are those relative to variables q11 and q25. For them unit 135 takes high categories

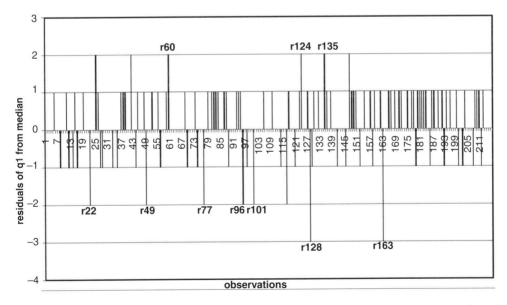

Figure 9.5   Difference between q1 and the median of the other five variables for each observation of the data subset.

Table 9.11   Unit 135 (q1 = 5, PRE = 4). Only variables q11 and q25 are in line with q1.

	q1	q11	q17	q25	q38	q65
Unit 135	5	4	1	5	2	3

(4 and 5, respectively). Therefore, while an exploratory analysis would suggest that unit 135 is anomalous, with our regression analysis such a unit appears to be coherent with the model. The same considerations could be entertained for all the other high values that are plotted, but not labelled.

Remark 1   In Section 9.3, using an exploratory data analysis approach, we defined as outliers for ordinal variables the units which presented unusual combinations of the categories. Here, using a modelling approach, we can consider as outliers the units which deviate markedly from the fitted model, that is, those with the largest residuals.

Remark 2   In this chapter we did not consider the approach of outlier detection based on robust cluster analysis (see Riani *et al.*, 2010).

9.6   Summary

In this chapter we have investigated the topics of robustness and multivariate outlier detection for ordinal data. In the first part of the chapter we presented an overview of methods for outlier detection for continuous data in regression. In this context we illustrated an example of a data set with masking effects and showed the difficulties presented by these kinds of data. In the second part of the chapter, we suggested a definition of outliers in ordinal data and proposed methods for identifying them. We also recalled the ordinal regression model and applied it to the ABC 2010 ACSS data, previously cleansed of missing values. We highlighted how the presence of anomalous observations can lead to erroneous conclusions about the choice of the best model. Having chosen the final model, we analysed its stability and highlighted a series of potential anomalous/influential observations.

References

Agresti, A. (2010) *Analysis of Ordinal Categorical Data*, 2nd edition. Hoboken, NJ: John Wiley & Sons, Inc.

Atkinson, A.C. and Riani, M. (2000) *Robust Diagnostic Regression Analysis*. New York: Springer.

Atkinson, A.C., Riani, M. and Cerioli A. (2004) *Exploring Multivariate Data with the Forward Search*. New York: Springer.

Barnett, V. and Lewis, T. (1994) *Outliers in Statistical Data*. Chichester: John Wiley & Sons, Ltd.

Bradlow, E.T. and Zaslavsky, A.M. (1999) A hierarchical latent variable model for ordinal data from a customer satisfaction survey with 'no answer' responses. *Journal of the American Statistical Association*, 94, 43–52.

Chu, W. and Ghahramani, Z. (2005) Gaussian processes for ordinal regression. *Journal of Machine Learning Research*, 6, 1019–1041.

Dong, F. (2010) Bayesian method to detect outliers for ordinal data. *Communications in Statistics – Simulation and Computation*, 39, 1470–1484.

Grubbs, F.E. (1969) Procedures for detecting outlying observations in samples. *Technometrics* 11, 1–21.

Hadi, A.S., Imon, A. and Werner, M. (2009) Detection of outliers. *Wiley Interdisciplinary Reviews: Computational Statistics*, 1, 37–70.

Liu, Y., Wu, A.D. and Zumbo, B.D. (2010) The impact of outliers on Cronbach's coefficient alpha estimate of reliability: Ordinal/rating scale item responses, *Educational and Psychological Measurement*, 70, 5–21.

McCullagh, P. (1980) Regression models for ordinal data. *Journal of the Royal Statistical Society, Series B*, 42, 109–142.

McCullagh, P. and Nelder, J. (1989) *Generalized Linear Models*, 2nd edn. London: Chapman & Hall.

Morgenthaler, S. (2007) A survey of robust statistics. *Statistical Methods and Applications*, 15, 271–293; *Statistical Methods and Applications*, 16, 171–172 (erratum).

Nagelkerke, N. J. D. (1991) A note on a general definition of the coefficient of determination. *Biometrika*, 78, 691–692.

Pardo, M.C. (2010) High leverage points and outliers in generalized linear models for ordinal data. In C.H. Skiadas (ed.), *Advances in Data Analysis*, pp. 67–80. Boston: Birkhäuser.

Riani, M., Atkinson, A.C. and Cerioli, A. (2009) Finding an unknown number of multivariate outliers. *Journal of the Royal Statistical Society, Series B*, 71, 447–466.

Riani, M., Cerioli, A. and Rousseeuw, P.J. (eds) (2010) Special issue on robust methods for classification and data analysis. *Advances in Data Analysis and Classification*, 4.

Rousseeuw, P.J. (1984) Least median of squares regression. *Journal of the American Statistical Association*, 79, 871–880.

Rousseeuw, P.J. and Hubert, M. (2011) Robust statistics for outlier detection, *Wiley Interdisciplinary Reviews: Data Mining and Knowledge Discovery*, 1, 73–79.

Rousseeuw, P.J. and Van Driessen, K. (2006) Computing LTS regression for large data sets. *Data Mining and Knowledge Discovery*, 12, 29–45.

Schlatter, R. (ed.) (1975) *Hobbes's Thucydides*, translated by Thomas Hobbes. New Brunswick, NJ: Rutgers University Press.

Su, X. and Tsai, C.-L. (2011) Outlier detection. *Wiley Interdisciplinary Reviews: Data Mining and Knowledge Discovery*, 1 (3), 261–268.

Zijlstra, W.P., van der Ark, L.A. and Sijtsma, K. (2007) Outlier detection in test and questionnaire data. *Multivariate Behavioral Research*, 42, 531–555.

Part II

MODERN TECHNIQUES IN CUSTOMER SATISFACTION SURVEY DATA ANALYSIS

10

Statistical inference for causal effects

Fabrizia Mealli, Barbara Pacini and Donald B. Rubin

Research questions motivating many scientific studies are causal in nature. Causal questions arise in medicine (e.g., how effective is a drug treatment?), economics (e.g., what are the effects of job training programs?), sociology (e.g., is there discrimination in labor markets?), customer satisfaction (e.g., what are the effects of different ways of providing a service?), and many other fields. Causal inference is used to measure effects from experimental and observational data. Here, we provide an overview of the approach to the estimation of such causal effects based on the concept of potential outcomes, which stems from the work on randomized experiments by Fisher and Neyman in the 1920s and was then extended by Rubin in the 1970s to non-randomized studies and different modes of inference. Attractions of this approach include the allowance for general heterogeneous effects and the ability to clarify the assumptions underlying the estimation of causal effects in complex situations.

10.1 Introduction to the potential outcome approach to causal inference

A large number of questions arising in applied fields are causal questions. For example, in customer satisfaction analysis, one may be concerned about the effect on customer behavior of providing a new product or service. Typical statistical tools that are used to answer such questions are regression models. However, if one wants to give some estimates a causal interpretation, a wider framework must be adopted; in this chapter, we present an approach and the related statistical methods for properly answering causal questions. The approach is commonly referred to as *Rubin's causal model* (Holland, 1986; Imbens and Rubin, 2008), after a series of papers by Rubin (1974, 1975, 1976, 1977, 1978, 1979, 1980). In this framework, causal

---

Modern Analysis of Customer Surveys: with applications using R, First Edition. Edited by Ron S. Kenett and Silvia Salini.
© 2012 John Wiley & Sons, Ltd. Published 2012 by John Wiley & Sons, Ltd.

questions are linked to explicit manipulations, that is, they are tied to specific interventions or treatments, having at least two different levels.

The potential outcome approach has received increasing attention in recent years, but it has its roots in the work on randomized experiments by Fisher (1925), who introduced randomization as the basis for *reasoned inference*, and a 1923 paper by Neyman (see Neyman, 1990), who formally introduced the potential outcome notation in randomized experiments.

Despite the almost immediate acceptance in the late 1920s of Fisher's proposal for randomized experiments and Neyman's notation for potential outcomes in randomized experiments, the framework was apparently not used in nonrandomized settings until the 1970s. Indeed, the potential outcome approach was first extended by Rubin (1974, 1976, 1977, 1978, 1990), and subsequently exploited by others, to apply to nonrandomized studies and forms of inference other than those that are randomization-based. It took another quarter of a century before the language and reasoning of potential outcomes became broadly accepted as the way to define and assess causal effects.

The potential outcome approach seems to have been basically accepted and almost generally adopted by most researchers by the end of the twentieth century in many fields (epidemiology, medical statistics, as well as in the behavioral sciences), and it has also contributed to a substantial convergence of methods from statistics and econometrics (Angrist et al, 1996; Dehejia and Wahba, 1999; Heckman, 2000; Imbens and Wooldridge, 2009).

The framework has two key parts. First, causal effects are viewed as comparisons of potential outcomes, each corresponding to a level of the treatment and each observable, had the treatment taken on the corresponding level, but at most one outcome actually observed, the one corresponding to the treatment level realized. Second, the assignment mechanism is explicitly defined as a probability model for how units receive the different treatment levels. In this perspective, a causal inference problem is thus viewed as a problem of missing data, where the assignment mechanism is explicitly modeled as a process for revealing the observed data. The assumptions on the assignment mechanism are crucial for identifying and deriving methods to estimate causal effects.

Other views of causality have been proposed in the literature. For example, Pearl (1995, 2000) uses a different approach, which combines aspects of structural equation models and path diagrams. The approach uses graphs to represent the assumptions underlying causal statements and to see which causal effects can or cannot be inferred from the data. We find the approach interesting, although it does not seem to enhance the drawing of causal inference, as discussed for instance in Rubin (2004, 2005). Another view is the Granger–Sims causality, which considers causality essentially as a prediction problem and is mainly used by economists in the analysis of time series and will not be discussed in this chapter. For a review of other approaches to causal inference, see also Cox and Wermuth (2004).

The potential outcome approach has several advantages over other approaches to causal inference, which are mainly based on the analysis of realized outcomes. It allows the definition of causal effects before, and independently of, the assignment mechanism, that is, the way we get to observe the potential outcomes, without making functional form assumptions or considering distributional properties of the outcomes or assignments. This also implies that the modeling of potential outcomes is separated from the modeling of the assignment mechanism, unlike the modeling of realized outcomes, which combines the two. Also, critical assumptions for inferring causal effects are formulated in terms of potentially observable variables, rather than in terms of unobserved components (i.e., error terms); this has clear advantages because assumptions are linked to specific individual behaviors, which are easier to discuss and assess.

This chapter is organized as follows. After describing the framework in more detail in the rest of this section, in Section 10.2 we formally define the assignment mechanism and assumptions about it, which characterize randomized and observational studies. Various modes of inference in classical randomized experiments are presented in Section 10.3, and extended to observational studies with *regular* assignment mechanisms in Sections 10.4.1–10.4.3. Sections 10.4.4–10.4.6 discuss further extensions and complications with the analysis of *irregular* designs, an area creating superb issues for research. The perspective we present in this chapter is fully developed in the forthcoming text by Imbens and Rubin (2012).

10.1.1   Causal inference primitives: Units, treatments, and potential outcomes

In real life, causal language is widely used in an informal way. Many common statements appear as an attempt to answer questions that have a causal nature; sentences such as 'my headache went away because I took an aspirin' and 'he got a good job because he participated in a training program' are typically informed by observations on past exposures (of headache outcomes after taking aspirin or not, of labor market performance of people with or without training). These statements are usually the result of informal statistical analyses, drawing conclusions from associations between measurements of some characteristics (variables) of different individuals. Until recently, when dealing with observational studies, statisticians would only talk about associations between potential causes and outcomes, leaving causation to a researcher's subjective judgment.

The need to answer questions that essentially have a causal nature arises in many fields, and the same is true for some issues posed in customer satisfaction settings. Satisfying the customers is crucial to staying in business in this modern world of global competition. Companies must satisfy and even delight their customers with the value of their products and services to gain their loyalty and repeat purchases. Customer satisfaction is therefore a primary goal of process improvement programs. A company may want to assess the impact of different ways to provide a service, or what aspects of the service really affect customer attitude or customer behavior – either positively and negatively. Relevant outcomes could be customers' subjective evaluations (attitudes) or different measures of customers' behavior, such as repeat-purchase behavior, loyalty behavior, or recommending the company to others. As a motivating example, suppose a software company wishes to evaluate the effect of different modes of technical assistance (e.g., on-line or on-site) on customer satisfaction. This is an example of a causal question.

Making causal inference requires thinking about an explicit action (manipulation, intervention) applied to a *unit* under which an outcome variable can be observed (the *potential outcomes*). There are several primitive concepts that are basic to defining causal effects in the potential outcome approach to causal inference. A *unit* is a physical object (an individual person, a family, a firm, a specific market, or a customer in our context) at a particular place and time. The same unit can be subject or exposed to a specific *treatment* (a particular action or intervention) or to alternative treatments, which could be different active treatments, or no treatment at all. Here we consider only settings with two treatments, say an active treatment and a control treatment, although the extension to more than two treatments is conceptually simple. Let W_i denote which treatment unit i received: $W_i = 1$ if the unit is exposed to the active treatment, $W_i = 0$ if it is exposed to the control treatment. In our example, the provision of on-site technical assistance is the active treatment, as opposed to the control (standard) treatment (on-line support).

For each unit–treatment pair we define a *potential outcome*: for each unit there are thus two associated potential outcomes at a future point in time after treatment. They are the value of some outcome measurement Y (e.g., the overall satisfaction level) if the unit was exposed to the active treatment, $Y_i(1)$, and the value of Y at the same future point in time if the unit was exposed to the control treatment, $Y_i(0)$. The causal effect of the treatment on each unit is defined to be a comparison of the treatment and control potential outcomes, $Y_i(1)$ and $Y_i(0)$, typically their difference, log-difference, or ratio.

The fundamental problem of causal inference (Holland, 1986; Rubin, 1978) is that we can only observe at most one of the potential outcomes for each unit, either $Y_i(0)$ or $Y_i(1)$, depending on the treatment actually received. In our example, we can observe the satisfaction level of some customers after providing on-site technical support, whereas for some others the outcome variable is observable after providing on-line support. However, we can imagine that, although the customer was exposed at a particular point in time to a particular treatment, at the same point in time the same customer could have been exposed to the alternative treatment.

The idea of the potential outcomes forces us to think about an intervention as an explicit manipulation, which must occur temporally before the observation of any associated potential outcomes. A correct definition of causal effect depends on the specification of potential outcomes, and this definition is distinct from the definition of the treatment assignment mechanism that determines which potential outcomes are realized and possibly observed.

This definition of causal effects as comparisons of potential outcomes is not only implicit in the informal statements exemplified at the beginning of this section, but is also used to formalize *but-for* concepts in legal settings. Suppose someone committed an action that is illegal, and, as a result of this action, someone else suffered damage. The causal effect of the illegal action (i.e., the amount of damage) is defined by the comparison between the potential outcome in the presence of the illegal action and the counterfactual-world potential outcome had the illegal action not occurred.

10.1.2   Learning about causal effects: Multiple units and the stable unit treatment value assumption

Although a causal effect is defined on a single unit, multiple units are required to make causal inference: for each unit we can observe at most a single potential outcome, so that only by relying on multiple units can we have some units exposed to the active treatment and other units exposed to the control treatment. Informal causal statements are commonly based on repeated past observations, concerning the same physical object at different times or different objects at the same time. For example, a company might argue that adopting a price reduction policy will avoid the decrease in sales during the summer period, based on its previous experience, that is, considering both the level of sales at times when the policy was adopted and the level of sales in the absence of such a policy.

The comparison of multiple units provides a basis for learning about causal effects, even though the presence of multiple units does not solve the problem of causal inference. Suppose we consider two different units – for example, two different customers exposed to on-site or on-line technical support at the same time. We now have four potential outcomes for each unit: the outcome for unit 1 if both units receive on-site support (active treatment), $Y_1(1, 1)$, the outcome for unit 1 if both units receive on-line support (control treatment), $Y_1(0, 0)$; the outcome for unit 1 if unit 1 receives on-site support and unit 2 receives on-line support, $Y_1(1, 0)$; the outcome for unit 1 if unit 1 receives on-line support and unit 2 receives on-site support, $Y_1(0, 1)$. Clearly, for each unit we can only observe at most one of these four potential outcomes, namely the one realized depending on which kind of technical support the two customers actually receive. The number of potential outcomes grows with the number of units considered. In addition, the number of potential outcomes increases if we contemplate different *doses* of the treatments (e.g., different abilities of technicians in providing on-site support, leading to potentially different levels of customer satisfaction). Unless we can restrict the number of potential outcomes to be compared, the availability of multiple units does not help to infer causal effects. A common assumption, which simplifies the analysis considerably, is the *stable unit treatment value assumption* (SUTVA; Rubin, 1980, 1990a), under which the potential outcomes for the ith unit depend only on the treatment the ith unit received. SUTVA assures that units do not interfere with one another, and that there are no hidden variations of the treatments.

In many situations it may be reasonable to assume that treatments applied to one unit do not affect the outcome for another unit: the type of technical support received by a customer should not affect the level of satisfaction of a different customer. Nevertheless, it may also be plausible to think that the overall satisfaction level of a customer might be influenced by learning about the experience of other customers with the same company. This situation could occur if, for example, customers belong to a network or simply have the occasion to communicate with each other. In such cases a more restrictive form of SUTVA can be invoked by defining the unit of interest to be a larger set within which individuals interact, for example investigating causal effects on specific target markets instead of on single customers. Such aggregation is likely to make the assumption more plausible, although it may lead to reduced apparent precision because of the reduced number of units.

SUTVA also implies that for each unit there is only a single version of each treatment level, meaning that, for example, on-site support for one customer is provided by technicians with the same ability. This does not imply that all the technicians have the same ability across all customers, but only that for customer i exposed to on-site support, a well-defined (stable) potential outcome, $Y_i(1)$, is specified.

Under SUTVA, the presence of multiple units can be exploited to estimate causal effects. Note that we cannot retrieve any information about the validity of SUTVA from observed data. We can only rely on subject-matter knowledge. SUTVA is a substantive assumption, which is usually maintained, even though it is not always appropriate. Researchers can attempt to make the assumption reasonable when designing randomized experiments, while in observational studies its plausibility should be assessed in the context of the empirical setting.

10.1.3   Defining causal estimands

Causal effects are defined at the unit level; due to the fundamental problem of causal inference, they can never be observed and must be inferred. Estimands of interest may be simple differences, $Y_i(1) - Y_i(0)$, or ratios, $Y_i(1)/Y_i(0)$, but in general comparisons can take different forms. Consider a finite population of N units. We often wish to summarize the N individual causal effects for the entire population of N units or for subpopulations; more generally, population or subpopulation causal effects are comparisons of the potential outcomes on a common set of units.

Commonly defined causal estimands are average treatment effects, either for the entire population or for some subpopulation. Some causal estimands correspond to other features of the joint distribution of potential outcomes. For example, the finite population average treatment effect is

$$\tau_{\text{ate}}^{\text{FP}} = \frac{1}{N} \sum_{i=1}^{N} [Y_i(1) - Y_i(0)].$$

Often the N units are considered as a random sample from a large superpopulation, so that a causal estimand of interest is the population average treatment effect,

$$\tau_{\text{ate}}^{\text{P}} = E[Y_i(1) - Y_i(0)],$$

that is, the expectation of the unit-level causal effect on the whole population. This estimand is particularly relevant when all the units in the population are potentially involved in the intervention. A different estimand is the population average treatment effect on the treated, defined by averaging over the subpopulation of treated units,

$$\tau_{\text{att}}^{\text{P}} = E[Y_i(1) - Y_i(0)|W_i = 1].$$

In many observational studies $\tau_{\text{att}}^{\text{P}}$ is a more interesting estimand than $\tau_{\text{ate}}^{\text{P}}$. As an example, consider the case of evaluating the effect of smoking on health: In this case it makes sense to assess the effect only on the subpopulation of those who choose to smoke.

In a customer satisfaction context, consider a company that is designing a marketing intervention intended to improve brand loyalty for targeted customers, for example, offering a 30% discount to customers whose orders exceed some threshold. The causal effect of the intervention on all customers may not be of interest to the company. In general, $\tau_{\text{att}}^{\text{P}}$ is appropriate when in the overall population there are units who are very unlikely to receive the same treatment in the future.

Sometimes average treatment effects are defined as averages over subpopulations defined in terms of pretreatment variables, called covariates. Covariates are specific background attributes of units that are determined before assignment to treatment and take the same value whatever treatment each unit is exposed to. In a customer satisfaction survey involving a sample of individuals, such variables could include specific permanent or pretreatment characteristics (e.g., age, sex, education, income level, family and socio-economic status, past consumption behavior). Covariates can also include pretreatment measures of satisfaction level or of past loyalty behavior, thus differing from the potential outcomes only by the time of measurement.

For example, we may be interested in the average effect of a marketing campaign for the subgroup of European customers and the subgroup of US customers, separately. The software company may want to evaluate the effects of providing on-site versus on-line technical support separately for customers with different types of commercial contract (e.g., private or business), or located in different areas. Indeed, some subpopulations of customers may be of more interest to the company, so that estimating causal effects on these specific subgroups of units may be useful to target future interventions better. We define the conditional average treatment effect,

$$\tau_{\text{cate}}^{x} = E[Y_i(1) - Y_i(0)|X_i = x],$$

and, analogously, the conditional average causal effect on the treated,

$$\tau_{\text{catt}}^{x} = E[Y_i(1) - Y_i(0)|X_i = x, W_i = 1].$$

If the treatment effect is constant ($Y_i(1) - Y_i(0) = \tau$, for some constant τ), all these estimands are identical. However, in the presence of heterogeneous effects of the treatment, they generally differ.

Although average causal effects are especially easy to estimate without bias using standard statistical tools in randomized experiments, there is no reason to focus solely on these quantities. As an alternative to average treatment effects, we can focus on more general functions of potential outcomes. For example, we may be interested in the median of $Y_i(1)$ versus the median of $Y_i(0)$ across all N units, or in the median of the difference $Y_i(1) - Y_i(0)$ (in general, different from the difference in medians), over the entire population or over a subpopulation. Another possible class of estimands concerns quantile treatment effects: These are usually defined as differences between quantiles of the two marginal potential outcome distributions ($\tau_q = F_{Y(1)}^{-1}(q) - F_{Y(0)}^{-1}(q)$). Although introduced in the statistics literature in the 1970s (Doksum, 1974; Lehman, 1974), these estimands have only recently received attention in the applied literature. More generally, comparisons involving the entire distributions of potential outcomes under treatment and under control for some subpopulation may be of interest (Imbens and Rubin, 1997b) and may be formally defined in Rubin's causal model.

10.2   Assignment mechanisms

Statistical inference for causal effects requires the specification of an assignment mechanism, that is, the process that determines which units receive which treatments, and so which potential outcomes are realized and which are missing. The assignment mechanism is a probabilistic model for the conditional probability of receiving the treatment as a function of potential outcomes and covariates. We first illustrate it in an artificial example and then present formal notation for it and define different classes of assignment mechanisms.

10.2.1   The criticality of the assignment mechanism

Suppose again that a software company is considering one of two treatments to apply to each of its eight customers: on-line support or on-site support. This company knows its customers very well, so it chooses the treatment that better satisfies each customer. We refer to such a company as the *perfect company*. Table 10.1 gives the hypothetical potential outcomes in satisfaction scores (on a scale from 0 to 15) under each treatment for these customers, so that their individual causal effects can be calculated. The column labeled W shows which treatment each customer received, $W_i = 0$ (on-line support) or $W_i = 1$ (on-site support) for the ith customer, and the final columns show the observed potential outcomes. Notice that the averages of the $Y_i(0)$ and $Y_i(1)$ potential outcomes indicate that the typical customer will be more satisfied with the on-line support: the average causal effect is two points in favor of the on-line support. But the company, which is providing support for the benefit of its customers, reaches the opposite conclusion from the examination of the observed data: the customers assigned the on-site support are, on average, twice as satisfied as the customers assigned the on-line support, with absolutely no overlap in their distributions! Moreover, if the company now applies on-site support to all customers in a population of customers who are just like the eight in the study, it will be disappointed: the average score will be closer to 5 under the on-site support rather than the 11 seen in this study.

Table 10.1 Perfect company example

	Potential outcomes		Observed data		
	$Y(0)$	$Y(1)$	W	$Y(0)$	$Y(1)$
	13	14	1	?	14
	6	0	0	6	?
	4	1	0	4	?
	5	2	0	5	?
	6	3	0	6	?
	6	1	0	6	?
	8	10	1	?	10
	8	9	1	?	9
True averages	7	5	Observed averages	5.4	11

What is wrong? The simple comparison of observed results assumes that the treatments were randomly assigned, rather than, as they were, to provide maximal satisfaction to the customers. We will say more about randomized experiments, but the point here is that the assignment mechanism is crucial to valid inference about causal effects, and the company used a *nonignorable* assignment mechanism, formally defined in subsequent sections. With a posited assignment mechanism it is possible to draw causal inferences; without one it is impossible. In this sense, when drawing inferences, a model for the assignment mechanism is more fundamental than a *scientific* model for the potential outcomes.

Notice that the company, by comparing observed means, is using the three observed values of $Y_i(1)$ to represent the five missing values of $Y_i(1)$, effectively imputing the mean of observed $Y_i(1)$, $\bar{y}_1$, for the five $Y_i(1)$ question marks, and analogously effectively filling in the mean of observed $Y_i(0)$, $\bar{y}_0$, for the three $Y_i(0)$ question marks. This process makes sense for point estimation if the three observed values of $Y_i(1)$ were randomly chosen from the eight values of $Y_i(1)$ and the five observed values of $Y_i(0)$ were randomly chosen from the eight values of $Y_i(0)$. But under the actual assignment mechanism, it does not. It would obviously make much more sense, under the actual assignment mechanism, to impute the missing potential outcomes for each customer to be less than or equal to that customer's observed outcome.

10.2.2   Unconfounded and strongly ignorable assignment mechanisms

The assignment mechanism gives the conditional probability of each vector of assignments given the covariates and potential outcomes,

$$\Pr(W|X, Y(0), Y(1)),$$

where W, $Y(1)$ and $Y(0)$ are N-component column vectors and X is a matrix with N rows. A specific example of an assignment mechanism is a completely randomized experiment with N units, where $n < N$ are assigned to the active treatment, and $N-n$ to the control treatment:

$$\Pr(W|X, Y(0), Y(1)) = \begin{cases} 1/\binom{N}{n} & \text{if } \sum_{i=1}^{N} W_i = n \\ 0 & \text{otherwise.} \end{cases}$$

An unconfounded assignment mechanism (Rubin, 1990b) is free of dependence on either $Y(0)$ or $Y(1)$:

$$\Pr(W|X, Y(0), Y(1)) = \Pr(W|X).$$

The assignment mechanism is *probabilistic* if each unit has a positive probability of receiving either treatment:

$$0 < \Pr(W_i = 1|X, Y(0), Y(1)) < 1.$$

If the assignment mechanism is unconfounded and probabilistic, it is called *strongly ignorable* (Rosenbaum and Rubin, 1983b), a stronger version of ignorable, defined in Section 10.2.3.

10.2.3   Confounded and ignorable assignment mechanisms

A confounded assignment mechanism is one that depends on the potential outcomes:

$$\Pr(W|X, Y(0), Y(1)) \neq \Pr(W|X).$$

A special class of possibly confounded assignment mechanisms is ignorable assignment mechanisms (Rubin, 1978). Ignorable assignment mechanisms are defined as being free of dependence on any missing potential outcomes:

$$\Pr(W|X, Y(0), Y(1)) = \Pr(W|X, Y_{obs}),$$

where $Y_{obs} = \{Y_{obs,i}\}$ and $Y_{obs,i} = W_i Y_i(1) + (1 - W_i)Y_i(0)$. Ignorable confounded assignment mechanisms arise in practice especially in sequential experiments, where, for example, the next unit's probability of being exposed to the active treatment depends on the observed outcomes of those previously exposed to the active versus the observed outcomes of those previously exposed to the control treatment.

All unconfounded assignment mechanisms are ignorable, but not all ignorable assignment mechanisms are unconfounded. The critical role played by ignorable and strongly ignorable assignment mechanisms in causal inference is easily seen in the trivial example of the *perfect company* (see Table 10.1), which involved a nonignorable assignment mechanism.

10.2.4   Randomized and observational studies

Having introduced some possible restrictions on the assignment mechanism, let us now use them to characterize randomized and nonrandomized studies.

A randomized study is a probabilistic assignment mechanism that is under the researcher's control and is a known function of its arguments. The software company in our example might design a randomized study by selecting a sample of its customers and by tossing a fair coin to randomly assign a subset of them to receive the additional on-site support.

An observational study, in contrast, is an assignment mechanism whose functional form is unknown. This would be the case if the software company left the choice between on-site and on-line support to the customers. To draw a statistical causal inference, an assignment mechanism must be posited; even if the assignment mechanism is unknown, we need to posit one, and this defines the template into which we can map the data from an observational study. That is, we need to posit a particular form for the assignment mechanism. The usual template that we try to use in observational studies is the class of classical randomized experiments. In order to characterize this class of experiments, we need to define regular designs.

Suppose that the assignment mechanism is unconfounded:

$$\Pr(W|X, Y(0), Y(1)) = \Pr(W|X)$$

(e.g., senior European customers have probability 0.8 of being assigned the new service, on-site support; new European customers 0.5; senior non-European customers 0.7; new non-European customers 0.3). Because of the random indexing of units, by appealing to de Finetti's (1963) theorem, we can write $\Pr(W|X)$ as

$$\Pr(W|X) \propto \int \prod_{i=1}^{N} e(X_i|\phi)\pi(\phi)d\phi, \quad \text{for } W \in \mathcal{W}, \tag{10.1}$$

where the function $e(X_i|\phi)$ gives the probability that a unit with value X_i of the covariates has $W_i = 1$ as a function of the parameter ϕ with prior (or marginal) probability density function $\pi(\phi)$. Assignment mechanisms for which the representation in expression (10.1) is true and that have $0 < e(X_i|\phi) < 1$ (for all X_i, ϕ) are called *regular*, whether or not the functions $e(.|\phi)$ and $\pi(\phi)$ are known, and are strongly ignorable. The unit-level assignment probabilities, $e(X_i|\phi)$, are called propensity scores (Rosenbaum and Rubin, 1983a).

Regular mechanisms with known functional form for the propensity scores are the class of classical randomized experiments, which often allow particularly straightforward estimation of causal effects from all inferential perspectives, as we shall shortly see. Therefore, these assignment mechanisms form the basis for inference for causal effects in more complicated situations, such as when assignment probabilities depend on covariates in unknown ways. Strongly ignorable assignment mechanisms, which essentially are collections of separate completely randomized experiments at each value of X_i that has a distinct probability of treatment assignment, form the basis for the analysis of observational studies. Designs that are known to be regular but have unknown propensity scores are the most critical template for inference of causal effects from observational data. That is, we attempt to assemble data with enough covariates that it becomes plausible (or initially arguable) that the unknown assignment mechanism is unconfounded given these covariates. Then the observational study can be analyzed using techniques for a regular design with unknown propensity scores. Without unconfoundedness there is no general approach for estimating causal effects, and so this restriction has great practical relevance. In Section 10.4.4 we will see how inference may proceed in observational studies when relaxing unconfoundedness with or without replacing it by additional assumptions.

10.3   Inference in classical randomized experiments

Fundamentally, there are three formal statistical modes of causal inference; two are based only on the assignment mechanism and treat the potential outcomes as fixed unknown quantities. One is due to Neyman (1990) and the other to Fisher (1925). Both will first be described in the absence of covariates. The assignment-based modes as developed by Fisher and Neyman, and generalized in our presentation, are randomization-based modes of inference because they both assume randomized experiments. The third mode is Bayesian, and treats the potential outcomes as random variables; it will be discussed in Section 10.4, together with the use of regression models.

10.3.1 Fisher's approach and extensions

Fisher (1925) was interested in deriving exact p-values regarding treatment effects for a finite population of size N. The first element in Fisher's approach is the null hypothesis regarding treatment effects, which is usually that of no effect of the treatment for any unit in this population, that is, $Y_i(1) = Y_i(0)$, for all i. Such null hypotheses are called *sharp* because under them all potential outcomes are known from the observed values of the potential outcomes: $Y_i(1) = Y_i(0) = Y_{\text{obs},i}$. As a result, under this type of null hypotheses, the value of any statistic S, that is, any function of the observed potential outcomes, such as the difference of the observed averages for units exposed to treatment and units exposed to control, $\bar{y}_1 - \bar{y}_0$, is known not only for the observed assignment, but for all possible assignments. Thus, the distribution of any statistic can be deduced, and p-values can be calculated as the probability (under the assignment mechanism and under the null hypothesis) that we would observe a value of S as *unusual* as, or more unusual than, the observed value of S, S_{obs}. 'Unusual' is defined *a priori*, typically by how discrepant S_{obs} is from zero. The p-value represents the plausibility of the observed value of the statistic S, had the null hypothesis been true. Fisher's approach is thus closely related to the mathematical idea of proof by contradiction. It is basically a *stochastic proof by contradiction*, where the p-value provides the probability of a result as rare as, or rarer than, the actual observed result occurring if the null hypothesis were true, where the probability is over the distribution induced by the assignment mechanism. Fisher's approach is elegant and provides a powerful alternative to typical analyses of experimental data, which are usually limited to simple mean differences by treatment status.

Fisher's approach can be extended to other sharp null hypotheses, for example, an additive null, which asserts that for each unit, $Y_i(1) = Y_i(0)$ is a specified constant. The collection of such null hypotheses that do not lead to an extreme p-value can be used to create an interval estimate of the causal effect assuming additivity. Although Fisher never discussed extensions of his approach, it can indeed be extended immediately to other statistics and a variety of fully specified assignment mechanisms, including unconfounded and even nonignorable ones, because all potential outcomes are known and thus the probability of any assignment is known. Fisher's perspective, however, has some limitations, for example it has no ability to generalize beyond the units in the experiment. Some of these limitations are not present in Neyman's approach, which is also approximate.

10.3.2 Neyman's approach to randomization-based inference

Neyman (1990, 1934) was interested in the long-run operating characteristics of a statistical procedure under repeated (sampling and) randomized assignment of treatments. His focus was on the average effect across a population of units, which may be equal to zero even when some or all unit-level effects are different from zero. He attempted to find point estimators that were unbiased and interval estimators that had the specified nominal coverage over the distribution induced by the assignment mechanism. Specifically, an unbiased estimator of the causal estimand is first created; second, an unbiased (or upwardly biased) estimator of the variance of that unbiased estimator is found. Then, an appeal to the central limit theorem is made for the normality of the estimator over its randomization distribution, so that a confidence interval for the causal estimand can be obtained.

To be more explicit, the causal estimand is typically the average causal effect $\tau_{\text{ate}}^{\text{FP}}$, where the averages are over all units in the population being studied, and the traditional statistic for estimating this effect is the difference in observed sample averages for the two groups, $\bar{y}_1 - \bar{y}_0$, which can be shown to be unbiased for $\tau_{\text{ate}}^{\text{FP}}$ in a completely randomized experiment. A common choice for estimating the sampling variance of $\bar{y}_1 - \bar{y}_0$ over its randomization distribution, in completely randomized experiments with $N = n_1 + n_0$ units, is $se^2 = s_1^2/n_1 + s_0^2/n_0$ where s_1^2, s_0^2, n_1, and n_0 are the observed sample variances and sample sizes in the two treatment groups. Neyman (1990) showed that se^2 overestimates the actual variance of $\bar{y}_1 - \bar{y}_0$ unless additivity holds, that is, unless all individual causal effects are constant, in which case se^2 is unbiased for the variance of $\bar{y}_1 - \bar{y}_0$. The standard 95% confidence interval for $\tau_{\text{ate}}^{\text{FP}}$ is $\bar{y}_1 - \bar{y}_0 \pm 1.96se$, which, in large enough samples, includes $\tau_{\text{ate}}^{\text{FP}}$ in at least 95% of the possible random assignments.

Although less direct than Fisher's, Neyman's method of inference forms the theoretical foundation for much of what is done in important areas of application, including medical and industrial experiments. However, Neyman's approach is not prescriptive in the sense of telling us what to do to create an inferential procedure, but rather, it tells us how to evaluate a proposed procedure for drawing causal inferences. Thus, it really is not well suited to dealing with complicated problems except in the sense of telling us how to evaluate proposed answers that are obtained by any method. Fisher's approach also suffers from the lack of prescription.

10.3.3   Covariates, regression models, and Bayesian model-based inference

In classical randomized experiments, if a covariate is used in the assignment mechanism, that covariate must be reflected in the analysis because it affects the randomization distribution induced by the assignment mechanism. For instance, in a randomized block design one covariate (or more) is used as a blocking variable, that is, strata are created and a completely randomized experiment is conducted in each stratum. Also, covariates can be used to increase efficiency of estimation, even when not used in the assignment mechanism. This point is shown by simple examples in Rubin (2008a, para. 3.3). In the Fisherian perspective, this reduced variance translates into more significant p-values when the null hypothesis is false; in the Neymanian perspective, the reduced variance translates into smaller estimated variance and therefore shorter confidence intervals.

A common and easy way of incorporating covariates in the analysis is with regression methods. Regression methods are a typical way of analyzing both experimental and observational data in many fields. Linear regression models are typically specified with a set of explanatory variables, which are the treatment indicator and covariates. Parameters are estimated by standard ordinary least squares, and the coefficient of primary interest is that of the treatment indicator. The framework within which regression methods are used is different from the Fisherian and Neymanian ones presented above. The data are viewed as a sample from an infinite population, and the properties of the estimators of models' parameters are assessed over the distribution induced by repeated sampling from that population. As a consequence, outcomes in the sample are random; in settings of randomized experiments, the ordinary least squares estimate of the coefficient of the treatment indicator can be given a causal interpretation as an estimate of an average causal effect, although these results are usually valid only in large samples.

Regressions are useful because they can be easily extended to the analysis of observational data; they have, however, some disadvantages relative to the Bayesian model-based methods that we will briefly describe now, because the restrictions to linear regression limit the set of causal estimands and thus limit the substantive questions that can be addressed. The disadvantages of regression methods relative to other methods are particularly relevant for the analysis of observational data, as we shall discuss in Section 10.4.1.

As with regression methods, Bayesian inference for causal effects views the potential outcomes as random variables, and any function of them, including causal estimands of interest, are also random variables. Bayesian inference requires a model for the underlying data, $\Pr(X, Y(0), Y(1))$, and this model is where the science enters, and is the third but optional part of Rubin's causal model (Rubin, 1978).

Bayesian inference for causal effects directly and explicitly confronts the missing potential outcomes, $Y_{\mathrm{mis}} = \{Y_{\mathrm{mis},i}\}$, where $Y_{\mathrm{mis},i} = W_i Y_i(0) + (1 - W_i) Y_i(1)$. This perspective takes the specification for the assignment mechanism and the specification for the underlying data, and derives the posterior predictive distribution of Y_{mis}, that is, the distribution of Y_{mis} given all observed values:

$$\Pr(Y_{\mathrm{mis}} | X, Y_{\mathrm{obs}}, W). \qquad (10.2)$$

This distribution is posterior because it is conditional on all observed values (X, Y_{obs}, W) and it is predictive because it predicts the missing potential outcomes. The posterior distribution of any causal estimand $\tau = \tau(Y(0), Y(1), X, W)$ can, in principle, be calculated from (a) the distribution in expression (10.2) , (b) the observed values of the potential outcomes, Y_{obs}, (c) the observed assignments, W, and (d) the observed covariates, X. This conclusion is immediate if we view the posterior predictive distribution in (10.2) as specifying how to take a random draw of Y_{mis}. Once a value of Y_{mis} is drawn, any causal effect can be directly calculated from the drawn value of Y_{mis} and the observed values of X and Y_{obs}. Repeatedly drawing values of Y_{mis} and calculating the causal effect for each draw generates the posterior distribution of the causal effect of interest. Thus, from this perspective causal inference is entirely viewed as a missing-data problem, where we multiply impute (see also Chapter 8 in this volume) the missing potential outcomes to generate a posterior distribution for the causal effects. A complete description of how to generate these imputations is given in Rubin (2008a).

A great advantage of this general approach is that we can model the data on one scale (e.g., log(income) as normal), impute on that scale, but transform the imputations before drawing inferences on another scale (e.g., raw euros). Unlike the usual regression approach, the estimand of interest is not in general a particular parameter of the statistical model. A critical issue in the model-based approach is the choice of the model for imputing the missing potential outcomes. Although in an experimental setting inferences for the estimands of interest are usually robust to the parametric model chosen, as long as the specification is flexible, in observational settings with many covariates, model specification is more difficult and results are generally more sensitive to such specifications. This issue also raises the need for appropriate design of observational studies, and we shall return to it in more detail in subsequent sections.

10.4   Inference in observational studies

The gold standard for the estimation of causal effects is to conduct randomized experiments; an alternative is to design and carefully execute an observational study. In a typical business setting interventions cannot be fully randomized to units: almost all interventions are specifically targeted or even voluntarily selected by customers. To draw statistical inferences in observational studies, a model for the assignment mechanism is needed, and this defines the template for analyzing the data. That is, we need to posit a particular form for the assignment mechanism, and a major template that we try to use is the class of regular designs, as defined in Section 10.2.4. Given a large enough set of pretreatment variables, unconfoundedness is viewed as a reasonable approximation to the actual assignment mechanism. The assumption of unconfoundedness is the most critical requirement of a regular assignment mechanism; most observational studies fundamentally rely on unconfoundedness, often implicitly, and often in combination with other assumptions. Inference in irregular designs is more challenging and some examples will be briefly reviewed in Section 10.4.6.

10.4.1   Inference in regular designs

Under the assumptions for a regular assignment mechanism (see Section 10.2.4), we can give a causal interpretation to the comparison of observed potential outcomes for treated and control units, within subpopulations defined by values of the observed pretreatment variables. The assignment mechanism can be interpreted as if, within subpopulations of units with the same value for the covariates, a completely randomized experiment was carried out. For example, suppose that the access to on-line versus on-site support depends solely on customer seniority. Now, consider a subpopulation of all customers with the same seniority. The observed difference in average satisfaction levels for treated and control units, within this subgroup, is unbiased for the average effect of the treatment in this subpopulation. The overall average effect can be estimated by a weighted average of conditional estimated average effects, following the same strategy as in randomized block experiments.

Problems arise if a covariate takes on many values or if there is more than one key covariate to condition on, because we have an increasing number of groups to be considered, and it becomes more difficult to find both treated and control units to make comparisons in each group. It may be necessary to compare outcomes for treated and control units with similar, but not identical, values of the pretreatment variables. Several methods to adjust for observed covariates may be found in the literature, which lead to different inferential procedures with different credibility and/or robustness, depending on the plausibility of the assumptions maintained.

Statistical methods adjusting for differences in covariates may be invalidated by substantial differences in the distributions of covariates between treated and control units. Even if unconfoundedness holds, it may be that there are regions of the space of covariates with relatively few treated or control units; as a result, inferences may rely largely on extrapolation and are less credible than in regions with substantial overlap in treated and control covariate distributions. Methods for constructing subsamples with good balance in covariate distributions between treatment groups may be implemented, where 'balance' means that the covariate distributions of treated and control units are approximately the same. In completely randomized experiments, covariate distributions are exactly balanced, at least in expectation, while in observational studies, there are often substantial differences between covariate distributions in the two treatment groups.

10.4.2   Designing observational studies: The role of the propensity score

Although the validity of causal conclusions from randomized experiments is generally accepted, inference from observational studies typically faces problems that compromise the validity of the resulting causal conclusions. To deal with such difficulties, one should attempt to design observational studies to reproduce, as well as possible, a randomized experiment. Here, by *design* we mean 'all contemplating, collecting, organizing, and analyzing of data that takes place prior to seeing any outcome data' (Rubin, 2008b). Note that the design phase does not involve the outcome data, which need not be available (or collected) at this stage. This is an advantage of the approach, which allows us to avoid any conflict of interest, which may be a relevant problem with traditional approaches to causal inference involving fitting regression models where the estimated causal effect is given by the estimated coefficient of an indicator variable for exposure to an intervention and the estimated answers are constantly being seen and modified as models are fitted and refitted. This process may lead to possibly different answers from which the analyst can choose depending on what the company would expect to see (Rubin and Waterman, 2006).

Consider designing a study to evaluate the effects of a monthly e-mail newsletter service offered by a company on overall customer loyalty, where only some customers have opted in to receive e-mails. The company aims to make investment decisions to produce the newsletter based of the comparison of the expected consumer behavior (buying or recommending) when making the investment and when not making the investment. Customers who decide to use the new service may systematically differ from other customers and a rich set of covariates should be collected to control for these differences. First, we suggest assessing the degree of balance in the covariate distributions by comparing the distributions of covariates in the treated and control subsamples. Scaled differences in average covariate values by treatment status can be used; alternatively, one can focus on the distribution of the propensity score. In fact, the propensity score, formally defined (see Section 10.2.4) as the probability of a unit receiving the treatment condition rather than the control condition, as a function of observed covariates, is the most basic ingredient of an unconfounded assignment mechanism. It plays a relevant role both for designing observational studies and for estimating and assessing causal effects under unconfoundedness. Rosenbaum and Rubin (1983a) showed that the propensity score is a *balancing* score, that is, $\Pr(X|W, e(X|\phi)) = \Pr(X|e(X|\phi))$, so that the distribution of the covariates is the same among treated and controls units with the same value of the propensity score. In addition, if unconfoundedness holds, then it is also true that $\Pr(W|e(X|\phi), Y(0), Y(1)) = \Pr(W|e(X|\phi))$. This implies that all biases due to observable covariates can be removed by conditioning solely on the propensity score.

The propensity score is rarely known in an observational study, and therefore must be estimated, typically using a model (such as logistic regression). The goal is to obtain an estimated propensity score that balances the covariates between treated and control subpopulations, rather than one that estimates the true propensity score as accurately as possible.

The estimated propensity score is a powerful tool for assessing balance and overlap in the design phase. If covariate distributions are identical in the two treatment groups, the propensity score must be constant, and vice versa. If there is a substantial imbalance in the original sample, the estimated propensity score can be used to construct a better subsample, for example by matching each treated unit to the closest control unit in terms of the estimated propensity score, and then discarding unmatched units (e.g., Rosenbaum and Rubin, 1985).

A different way to improve the quality of the original sample is to focus on estimating causal effects in the subpopulation of units with propensity score values bounded away from zero and one, by eliminating units with covariate values such that the propensity score is close to zero or one. Sometimes we may even decide that the data set at hand is inadequate to make the study worthwhile, because of lack of data on key covariates, or because of lack of overlap in the distributions of key covariates between treatment groups (Rubin, 2008b).

If substantial overlap in the multivariate distribution of covariates in the treatment and control group is achieved, several methods can be used for the estimation of causal effects. A good design phase makes the inference more robust to the choice of the adjusting method, as well as less sensitive to possible model specification.

10.4.3   Estimation methods

Many of the available procedures for estimating and assessing causal effects use the estimated propensity score. Here we introduce some broad classes of estimation methods: subclassification on the propensity score, matching, and model-based imputation methods. Estimators that combine aspects of some of these strategies can also be implemented and are strongly recommended in some cases to reduce bias or improve robustness (Imbens and Wooldridge, 2009). We briefly sketch how different procedures work, and we refer to Imbens (2004) for a discussion on what we can expect in terms of achievable precision.

Given knowledge of the propensity score, one can directly use some of the strategies developed for randomized experiments, such as blocking or subclassification: the sample is divided into subclasses, based on the value of the propensity score, and within the subclasses the data are analyzed as if they came from a completely randomized experiment. Obviously, a preliminary step in this approach is to estimate the propensity score.

Matching methods do not necessarily require the estimation of the propensity score. Matching methods used for estimation proceed to impute missing potential outcomes by finding, for a treated unit, a control unit as similar as possible in terms of covariates. For example, suppose we want to evaluate the effect of on-site versus on-line support on customer satisfaction level for a senior European customer with a business contract who received on-site support. In matching, we would look for a similar customer (senior, European, with a business contract) who instead received the standard on-line support. However, exact matching on all covariates is usually infeasible, in particular when there are many covariates compared to the number of units, so that some decisions have to be taken (selecting relevant covariates for exact matching, choosing a distance measure, etc.). The choice of a specific metric characterizes each matching algorithm.

As a third broad class of estimation methods, we rely on the specification of a conditional model for the joint distribution of $(Y_i(0), Y_i(1))$ given X_i and a vector of unknown parameters. Missing outcomes can be imputed using this model as described in Section 10.3.3. Standard linear, or more flexible, regression analysis is frequently used. As stated already, the model-based analysis can be easily conducted within the Bayesian framework.

10.4.4   Inference in irregular designs

Irregular assignment mechanisms allow the assignment to depend not only on covariates but also on observed and unobserved potential outcomes. Irregular designs include also assignment probabilities that are equal to zero or one, that is, nonprobabilistic assignments. An important special case of the latter are regression discontinuity settings (Thistlethwaite and Campbell, 1960; Cook, 2008), where the treatment status depends, either completely or partly, on the value of an observed covariate being above or below a specific threshold. These settings can be analyzed essentially as if a randomized experiment took place around the threshold and are reviewed in Lee and Lemieux (2010).

Irregular assignment mechanisms also include *broken* randomized experiments, that is, randomized experiments suffering from post-assignment complications such as noncompliance and missing outcome data, which are not under experimental control and break randomization in unknown ways. These assignment mechanisms are particularly challenging, and without additional assumptions, there is limited scope for inferring causal effects. We now briefly discuss some specific settings and related strategies to address these complications, providing some useful references for interested readers.

10.4.5   Sensitivity and bounds

Because observational studies are rarely known to be unconfounded, we are usually concerned with the sensitivity of responses to unobserved covariates. Typically sensitivity analyses utilize the idea of a fully missing covariate, U, such that treatment assignment is ignorable given U and X, but not given only the observed covariates X. The relationships between U and W, and between $U, Y(0)$ and $Y(1)$, all given X are then varied, to see how estimates of causal estimands would change under specific deviations from ignorability. See, for example, Rosenbaum and Rubin (1983b) and Ichino *et al.* (2008). Extreme versions of sensitivity analyses examine bounds (Manski *et al.*, 1992; Manski, 2003); bounds on point estimates show the range of values that would be obtained over extreme distributions of the unobserved covariate U. Although often very broad, bounds can play an important role in informing us about how different assumptions help sharpen inference.

10.4.6   Broken randomized experiments as templates for the analysis of some irregular designs

The advantages of randomization are partially lost when post-assignment complications arise. For example, units may not fully comply with their treatment assignment. Noncompliance is especially common in social experiments, which are typically open-label and often designed as randomized encouragement experiments. For instance, the software company could encourage a randomly chosen subgroup of customers to use the new service by sending emails explaining the characteristics of the service, but only some of these encouraged customers might decide to accept the offer. Another complication is *truncation by death* (Rubin, 2006), which arises when the outcome is not defined for all randomized units. This happens, for example, if a customer is a firm, and the firm ceases its activity. The satisfaction of a no longer existing firm is not well defined. This complication is different from unintended missing outcomes, which occur if a firm is lost in the follow-up interview.

Each of these complications, or combinations of them, can be viewed as special cases of *principal stratification* (Frangakis and Rubin, 2002) where the principal strata are defined by partially unobserved intermediate potential outcomes, namely compliance behavior under both treatment assignments, survival under both treatment assignments, nonresponse under both treatment assignments. These complications can be handled much more flexibly with the likelihood or Bayesian approach, than with assignment-based methods (e.g., Imbens and Rubin, 1997a; Zhang *et al.*, 2009; Mattei and Mealli, 2007; Rubin and Zell, 2010). Sensitivity of inference to prior assumptions can be severe, and the Bayesian approach is well suited not only to revealing this sensitivity, but also to formulating reasonable prior restrictions.

The analysis of broken randomized experiments can be used as a template to analyze some observational studies. For example, randomized experiments with noncompliance can be used as a template to analyze observational studies with so-called *instrumental variables*, as shown in Angrist *et al.* (1996), where the instrumental variable plays the same role as random assignment.

References

Angrist, J., Imbens, G.W. and Rubin, D.B. (1996) Identification of causal effects using instrumental variables. *Journal of the American Statistical Association*, 91, 444–472.

Cook, T. (2008) Waiting for life to arrive: A history of the regression-discontinuity design in psychology, statistics and economics. *Journal of Econometrics*, 142, 636–654.

Cox, D.R. and Wermuth N. (2004) Causality: a statistical view. *International Statistical Review*, 72, 285–305.

De Finetti B. (1963) Foresight: Its logical laws, its subjective sources. In H.E. Kyburg and H.E. Smokler (eds), *Studies in Subjective Probability*. New York: John Wiley & Sons, Inc.

Dehejia, R.H. and Wahba, S. (1999) Causal effects in nonexperimantal studies: Reevaluating the evaluation of training program. *Journal of the American Statistical Association*, 95, 1053–1062.

Doksum, K. (1974) Empirical probability plots and statistical inference for nonlinear models in the two-sample case. *Annals of Statistics*, 2, 267–277.

Fisher, R.A. (1925) *Statistical Methods for Research Workers*. Edinburgh: Oliver and Boyd.

Frangakis, C. and Rubin, D.B. (2002) Principal stratification in causal inference. *Biometrics*, 58, 21–29.

Heckman J.J. (2000) Causal parameters and policy analysis in economics: A twentieth century retrospective. *Quarterly Journal of Economics*, 115, 45–97.

Holland, P.W. (1986) Statistics and causal inference (with discussion). *Journal of the American Statistical Association*, 81, 945–970.

Ichino, A., Mealli, F. and Nannicini, T. (2008) From temporary help jobs to permanent employment: What can we learn from matching estimators and their sensitivity? *Journal of Applied Econometrics*, 23, 305–327.

Imbens, G.W. (2004) Nonparametric estimation of average treatment effects under exogeneity: A review. *Review of Economics and Statistics*, Vol 86, 4–30.

Imbens, G.W. and Rubin, D.B. (1997a) Bayesian inference for causal effects in randomized experiments with noncompliance. *Annals of Statistics*, 25, 305–327.

Imbens, G.W. and Rubin, D.B. (1997b) Estimating outcome distributions for compliers in instrumental variables models. *Review of Economic Studies*, 64, 555–574.

Imbens, G.W. and Rubin, D.B. (2008) Rubin causal model. In S.N. Durlauf and L.E. Blume (eds), *The New Palgrave Dictionary of Economics*, 2nd edn. Basingstoke: Palgrave Macmillan.

Imbens, G.W. and Rubin, D.B. (2012) *Causal Inference in Statistics and Social Sciences*. Cambridge: Cambridge University Press, forthcoming.

Imbens, G.W. and Wooldridge, J.M. (2009) Recent developments in the econometrics of program evaluation. *Journal of Economic Literature*, 47, 5–86.

Lee, D.S. and Lemieux, T. (2010) Regression discontinuity designs in economics. *Journal of Economic Literature*, 48, 281–355.

Lehman, E. (1974) *Nonparametrics: Statistical Methods Based on Ranks*. San Francisco: Holden-Day.

Manski C. (2003) *Partial Identification of Probability Distributions*. New York: Springer.

Manski, C., Sandefur, G., McLanahan, S. and Powers, D. (1992) Alternative estimates of the effect of family structure during adolescence on high school. *Journal of the American Statistical Association*, 87, 25–37.

Mattei, A. and Mealli, F. (2007) Application of the principal stratification approach to the Faenza randomized experiment on breast self-examination. *Biometrics*, 63(2), 437–446.

Neyman, J. (1934) On the two different aspects of the representative method: The method of stratified sampling and the method of purposive selection. *Journal of the Royal Statistical Society, Series A*, 97, 558–606.

Neyman, J. (1990) On the application of probability theory to agricultural experiments. Essay on principals. Section 9 (with discussion). *Statistical Science*, 5(4), 465–480.

Pearl, J. (1995) Causal diagrams for empirical research. *Biometrika*, 82(4), 669–710.

Pearl, J. (2000) *Causality: Models, Reasoning and Inference*, Cambridge: Cambridge University Press.

Rosembaum, P.R. and Rubin, D.B. (1983a) The central role of the propensity score in observational studies for causal effects. *Biometrika*, 70, 41–55.

Rosembaum, P.R. and Rubin, D.B. (1983b) Assessing sensitivity to an unobserved binary covariate in an observational study with binary outcome. *Journal of the Royal Statistical Society, Series B*, 45, 212–218.

Rosenbaum, P. and Rubin, D.B. (1985) Constructing a control group using multivariate matched sampling incorporating the propensity score. *American Statistician*, 39, 33–38.

Rubin, D.B. (1974) Estimating causal effects of treatments in randomized and nonrandomized studies. *Journal of Educational Psychology*, 66, 668–701.

Rubin, D.B. (1975) Estimating causal effects of treatments in randomized and nonrandomized studies. In *Proceedings of the Social Statistics Section*, pp. 233–239. Alexandria, VA: American Statistical Association.

Rubin D.B. (1976) Inference and missing data (with discussion). *Biometrika*, 63, 581–592.

Rubin D.B. (1977) Assignment to treatment group on the basis of a covariate. *Journal of Educational Statistics*, 2, 1–26.

Rubin, D.B. (1978) Bayesian inference for causal effects: The role of randomization. *Annals of Statistics*, 6, 34–58.

Rubin, D.B. (1979) Discussion of 'Conditional independence in statistical theory' by A.P. Dawid. *Journal of the Royal Statistical Society, Series B*, 41, 27–28.

Rubin, D.B. (1980) Comment on 'Randomization analysis of experimental data: The Fisher randomization test' by D. Basu. *Journal of the American Statistical Association*, 75, 591–593.

Rubin, D.B. (1990a) Comment: Neyman (1923) and causal inference in experiments and observational studies. *Statistical Science*, 5, 472–480.

Rubin, D.B. (1990b) Formal modes of statistical inference for causal effects. *Journal of Statistical Planning and Inference*, 25, 279–292.

Rubin, D.B. (2004) Direct and indirect causal effects via potential outcomes. *Scandinavian Journal of Statistics*, 31, 161–170.

Rubin, D.B. Causal inference using potential outcomes: Design, modeling, decisions, *Journal of the American Statistical Association*, 100, 322–331 (2005).

Rubin, D.B. Causal inference through potential outcomes and Principal Stratification: application to studies with censoring due to death. *Statistical Science*, 21, 299–321 (2006).

Rubin, D.B. (2008a) Statistical inference for causal effects, with emphasis on application in epidemiology and medical statistics. In C.R. Rao, J.P. Miller and D.C. Rao (eds), *Handbook of Statistics: Epidemiology and Medical Statistics*, Vol. 27, pp. 28–62. Amsterdam: Elsevier.

Rubin, D.B. (2008b) For objective causal inference, design trump analysis. *Annals of Applied Statistics*, 2, 808–840.

Rubin, D.B. and Waterman, R. (2006) Estimating the causal effects of marketing interventions using propensity score methodology. *Statistical Science*, 21, 206–222.

Rubin, D.B. and Zell E.R. (2010) Dealing with noncompliance and missing outcomes in a randomized trial using Bayesian technology: Prevention of perinatal sepsis clinical trial, Soweto, South Africa. *Statistical Methodology*, 7, 338–350.

Thistlethwaite, D. and Campbell, D. (1960) Regression-discontinuity analysis: An alternative to the ex-post facto experiment. *Journal of Educational Psychology*, 51, 309–317.

Zhang, J.L., Rubin, D.B. and Mealli, F. (2009) Likelihood-based analysis of the causal effects of job-training programs using principal stratification. *Journal of the American Statistical Association*, 104, 166–176.

11

Bayesian networks applied to customer surveys

Ron S. Kenett, Giovanni Perruca and Silvia Salini

Bayesian networks are gaining popularity in a wide range of application areas such as risk management, web data analysis, and management science. Availability of software for analysing Bayesian networks is further expanding their role in decision analysis and decision support systems. This chapter introduces Bayesian networks and their application to customer satisfaction surveys. The chapter begins with a theoretical introduction to Bayesian networks. Then Bayesian networks are applied to the ABC survey data set and to the Eurobarometer transportation survey. A summary section concludes the chapter. The chapter proceeds with an overview of publicly available software programs that can be used to implement Bayesian networks and a section comparing the goals of predicting and explaining with Bayesian Networks.

11.1 Introduction to Bayesian networks

A survey with n questions produces responses that can be considered random variables, $X_1, \ldots, X_n$. Some of these variables, q of them, are responses to general questions such as overall satisfaction, recommendation or repurchasing intention. These are considered target variables. Responses to the other $n - q$ questions can be analysed under the hypothesis that they are positively dependent with target variables. The combinations (X_i, X_j), $X_i, \in X_1, \ldots, X_{n-q}$, $X_j \in X_{n-q+1}, \ldots, X_n$, are either positive dependent or independent, for each pair of variable (X_i, X_j), $i \leq n - q$, $n - q < j \leq n$. In general, dependency patterns can be extracted from data by using data mining techniques and statistical models (Hand *et al.*, 2001).

Different sources of knowledge, such as subjective information from expert opinions, statistics from the literature and customer survey data, can be integrated with Bayesian networks (BNs). For comprehensive examples of applications of Bayesian networks to survey

Modern Analysis of Customer Surveys: with applications using R, First Edition. Edited by Ron S. Kenett and Silvia Salini.
© 2012 John Wiley & Sons, Ltd. Published 2012 by John Wiley & Sons, Ltd.

data analysis, see Kenett and Salini (2009) and Salini and Kenett (2009). See also Chapter 5 for general integrated models.

BNs belong to the family of probabilistic *graphical models* used to represent knowledge about an uncertainty domain. Specifically, each node in the graph represents a random variable, and the edges between the nodes represent probabilistic dependencies among the corresponding random variables. These conditional dependencies can be estimated using known statistical and computational methods. Hence, BNs combine principles from graph theory, probability theory, computer science and statistics.

BNs implement a graphical model structure known as a *directed acyclic graph* (DAG) that is popular in statistics, machine learning and artificial intelligence. They are both mathematically rigorous and intuitively understandable, and enable an effective representation and computation of the joint probability distribution over a set of random variables (Pearl, 2000).

The structure of a DAG is defined by two sets: the set of nodes and the set of directed edges. The nodes represent random variables and are drawn as circles labelled by the variable names. The edges represent direct dependences among the variables and are represented by arrows between nodes. In particular, an edge from node X_i to node X_j represents statistical dependence between the corresponding variables. Thus, an arrow indicates that a value taken by variable X_j depends on the value taken by variable X_i, or, roughly speaking, that variable X_i 'influences' X_j. Node X_i is then referred to as a 'parent' of X_j and, similarly, X_j is referred to as the 'child' of X_i. An extension of these genealogical terms is often used to define the sets of 'descendants', the set of nodes from which the node can be reached on a direct path. The structure of the acyclic graph guarantees that there is no node that can be its own ancestor or its own descendent. Such a condition is of vital importance to the factorization of the joint probability of a collection of nodes, as shown below. Note that although the arrows represent direct causal connection between the variables, a *reasoning process* can operate on a BN by propagating information in any direction.

A BN can effectively reflect a simple conditional independence statement, namely that each variable is independent of its non-descendents in the graph, given the state of its parents. This property is used to reduce, sometimes significantly, the number of parameters that are required to characterize the joint probability distribution (JPD) of the variables. Such a reduction provides an efficient way to compute the posterior probabilities, given the evidence present in the data (Lauritzen and Spiegelhalter, 1988; Pearl, 2000; Jensen, 2001).

In addition to the DAG structure, which is often considered as the 'qualitative' part of the model, one needs to specify the 'quantitative' parameters of the model. These parameters are obtained by applying the Markov property, where the conditional probability distribution (CPD) at each node depends only on its parents. For discrete random variables, this conditional probability is often represented by a table, listing the local probability that a child node takes on each of the feasible values, for each combination of values of its parents. The joint distribution of a collection of variables can be determined uniquely by these local conditional probability tables.

Formally, a Bayesian network, B, is an annotated acyclic graph that represents a joint probability distribution over a set of random variables $\mathbf{V}$. The network is defined by a pair $B = \langle G, \Theta \rangle$, where G is the directed acyclic graph whose nodes $X_1, X_2, \ldots, X_n$ represents random variables, and whose edges represent the direct dependencies between these variables. The graph G encodes independence assumptions, by which each variable X_i is independent of its non-descendants given its parents in G, denoted generically as π_i. The second component, Θ, denotes the set of parameters of the network. This set contains the parameter $\theta_{x_i|\pi_i} = P_B(x_i|\pi_i)$

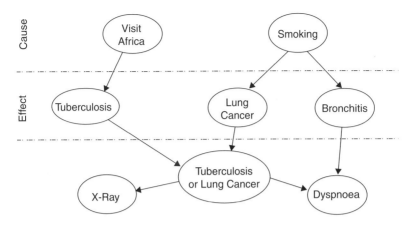

Figure 11.1   An example of a Bayesian network (adapted from Lauritzen and Spiegelhalter, 1988).

for each realization x_i of X_i conditioned on π_i, the set of parents of X_i in G. Accordingly, B defines a unique joint probability distribution over **V**, namely,

$$P_B(X_1, X_2, \ldots, X_n) = \prod_{i=1}^{n} P_B (X_i \,|\pi_i) = \prod_{i=1}^{n} \theta_{X_i|\pi_i} \,.$$

For simplicity of representation, we omit the subscript B. If X_i has no parents, its local probability distribution is said to be *unconditional*, otherwise it is *conditional*. If the variable represented by a node is *observed*, then the node is said to be an evidence node, otherwise the node is said to be *hidden* or *latent*.

The complexity of a domain may be reduced by models and algorithms that describe an approximated reality. When variable interactions are too intricate to apply in an analytic model, we can still represent current knowledge about the problem, such as a cause generating at least one effect where the final effect is the target of the analysis (Pearl, 2000). For example, in Figure 11.1 adapted from Lauritzen and Spiegelhalter (1988), a network topology of cause and effect is built by choosing a set of variables, 'Visit Africa' and 'Smoking', that describe a domain where a patient presents some problems and the physician wants to identify his/her disease and the correct therapy.

The domain knowledge allows experts to draw an arc to a variable from each of its direct causes (i.e. 'visiting Africa' may cause 'tuberculosis'). Given a BN that specifies the JPD in a factored form, one can evaluate all possible inference queries by marginalization, that is, summing over 'irrelevant' variables. Two types of inference support are often considered: *predictive support* for node X_i, based on evidence nodes connected to X_i through its parent nodes (also called *top-down reasoning*), and *diagnostic support* for node X_i, based on evidence nodes connected to X_i through its child nodes (also called *bottom-up reasoning*). In general, the full summation (or integration) over discrete (continuous) variables is called *exact inference* and known to be an NP-hard problem. Some efficient algorithms exist to solve the exact inference problem in restricted classes of networks.

In many practical settings, the BN DAG is unknown and one needs to learn it from the data. This problem is known as the BN *learning problem*, which can be stated informally as follows:

given training data and prior information (e.g., expert knowledge on causal relationships), estimate the graph topology (network structure) and the parameters of the JPD in the BN. The BN model selection procedure can therefore combine data-driven learning with expert opinion elicitation.

Learning the BN structure is considered a harder problem than learning the BN parameters. Moreover, another obstacle arises in situations of *partial observability* when nodes are hidden or when data is missing. In the simplest case of known BN structure and full observability, the goal of learning is to find the values of the BN parameters (in each CPD) that maximize the (log-)likelihood of the training data set. This data set contains m cases that are often assumed to be independent. Given training data set $\Sigma = \{\mathbf{x}_1, \ldots, \mathbf{x}_m\}$, where $\mathbf{x}_l = (x_{l1}, \ldots, x_{ln})^T$, and the parameter set $\Theta = (\theta_1, \ldots, \theta_n)$, where θ_i is the vector of parameters for the conditional distribution of variable X_i (represented by one node in the graph), the log-likelihood of the training data set is a sum of terms, one for each node:

$$\log L(\Theta \,|\, \Sigma) = \sum_m \sum_n \log P(x_{li} \,|\, \pi_i, \theta_i).$$

The log-likelihood scoring function is decomposed according to the graph structure, hence one can maximize the contribution to the log-likelihood of each node independently. Another alternative is to assign a prior probability density function to each parameter vector, and use the training data to compute the posterior parameter distribution and the Bayes estimates.

During quantitative learning of a BN, conditional and unconditional prior probability distributions represent the belief on θ. The distribution functions are expressed for each node of the DAG G. By updating priors with likelihood functions (i.e. data collection) the corresponding posterior distributions are generated. Available data updates prior belief and prior distributions so that posterior distributions converge to a likelihood shape with increasing available of data.

The choice of prior distributions gives a mathematical shape to *a priori* belief on θ (Bernardo and Smith, 1994). This can be achieved by different methods such as exponential power distributions, vague priors, and empirical Bayes distributions. Typically, only one method is employed for prior selection, which limits the subjective contribution of a model developer. Under this hypothesis, priors have to be developed without any assumption on shape and parameter.

Among various possible approaches, the best solution to elicit the prior appears to be the *maximum entropy method* proposed by Jaynes (2003). With this method, the shape distribution is derived from available information and no assumptions are made. Alternatively, one can apply *conjugate priors* where prior and posterior distributions belong to the same family of distribution (Bernardo and Smith, 1994).

The *maximum entropy prior* function is what maximizes the Shannon–Jaynes information entropy defined as:

$$S = -\sum_{i=1}^{n} p_i \ln p_i, \qquad \sum_{i=1}^{n} p_i = 1,$$

where S is the knowledge over the distribution and S is equal to 0 for a degenerate distribution, that is, full knowledge over the value of interest. The larger S, the greater is the ignorance about the uncertain value of interest.

In the case of no constraint other than normalization, S is maximized by the uniform distribution, $p_i = 1/n$. See Jaynes (2003), Fomby and Carter Hill (1997), and Grandy and Schick (1999) for the proof with Lagrange multipliers.

For continuous variables, and when the only information about the distribution is either the expected value or both the expected value and the variance, two maximum-entropy priors are usually applied that are based on the moments of the distribution M_r. For $r = 1$, M_r is equal to the expected value. Furthermore, the first and second moment together provide the variance. The knowledge about the expected value is the extension to continuous variables of the discrete case.

The *conjugate prior family* is chosen on the basis of mathematical convenience. The posterior distribution belongs to the same family as the prior distribution – the family is a closed class under transformation from prior to posterior (Raiffa, 1997).

In the next section we focus on practical applications of Bayesian networks, including a review of software programs and case studies.

11.2   The Bayesian network model in practice

Several software programs implement algorithms and models for constructing BNs. In this chapter we mostly refer to the `bnlearn` library in R (Scutari, 2010). Compared with other available BN software programs, the advantage of `bnlearn` is its ability to perform both constraint-based and score-based methods. This free R package implements five constraint-based learning algorithms (grow-shrink, incremental association, fast incremental association, interleaved incremental association, max-min parents and children), two score-based learning algorithms (hill climbing, tabu) and two hybrid algorithms (MMHC, phase restricted maximization).

The main disadvantage of `bnlearn` is that it does not allow the mixing of continuous and categorical variables. Some existing libraries handle networks with mixed variables, however their learning procedure is still experimental and hardly applicable to complex models and large data sets (`deal`: Bottcher and Dethlefsen, 2007).

11.2.1   Bayesian network analysis of the ABC 2010 ACSS

In this section we apply a BN to the ABC 2010 annual customer satisfaction survey (ACSS). The data consists of responses to a questionnaire comprising of 81 questions. Chapter 2 presents background on the ABC company and a basic analysis of the data. The application of BN to the analysis of customer satisfaction survey data was suggested and implemented in Kenett and Salini (2009) and Salini and Kenett (2009). Other related references include Gasparini (2000) and Renzi *et al.* (2009).

We first consider the three target variables *overall satisfaction, recommendation,* and *repurchasing intention*, represented in bivariate form by a 'yes' or 'no' response. This provides a simple example of a BN with two parents and shows how the probabilities and log-likelihoods are worked out. In this example we have three nodes, each with two possible values, 'yes' or 'no'. Table 11.1 shows the learning table with the sufficient statistics of the maximum-likelihood estimators.

Figure 11.2 shows a BN produced by the hill climbing algorithms with Akaike information criterion (AIC) as score function, and Figure 11.3 shows the conditional probabilities for the

Table 11.1   Learning table with maximum likelihood estimators

		Satisfaction			
		No		Yes	
		Repurchase		Repurchase	
		No	Yes	No	Yes
Recommendation	No	7	3	4	1
	Yes	0	4	5	76

recommendation child node. A satisfied customer with positive intention of repurchasing the product/service has a probability of 0.98 of recommending the product/service to others. A graphical representation facilitates the visual understanding of the network conditional probability.

Bayesian network for two parent example

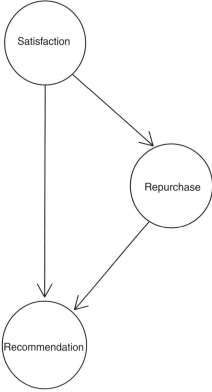

Figure 11.2   A Bayesian network with three nodes: satisfaction, recommendation and repurchasing intention

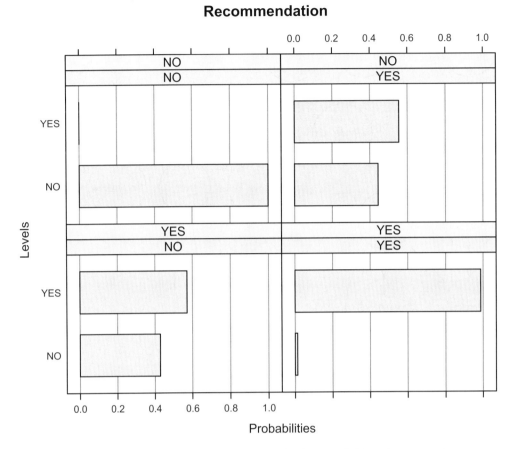

Figure 11.3   Conditional probabilities of a three-node Bayesian network

We now consider the six overall satisfaction questions for specific topics – (1) equipment and system, (2) sales support, (3) technical support, (4) supplies and orders, (5) purchasing support and (6) contracts and pricing – and the output (target) variables overall satisfaction, recommendation and repurchasing intentions. We now also include the variable *country*. There are 266 records. The objective is to understand what dimensions have a direct influence on *Overall Satisfaction, Recommendation* and *Repurchasing Intentions*. The data is analysed with a constraint-based algorithm and with a score-based algorithm (AIC criterion). The two networks are shown in Figure 11.4.

The variables that directly influence satisfaction, recommendation, and repurchasing intention are suppliers, equipment, and technical support. Moreover, satisfaction, recommendation, and repurchasing intentions are influenced by the country of the customer. Purchasing support, contract and prices and sales support do not influence the three output variables. In the BN constructed by a score-based algorithm, these three variables are related to each other and influenced by equipment and technical support. This information has practical relevance. An intervention to improve satisfaction levels from technical support or equipment can increase repurchasing intentions, recommendation, and overall satisfaction.

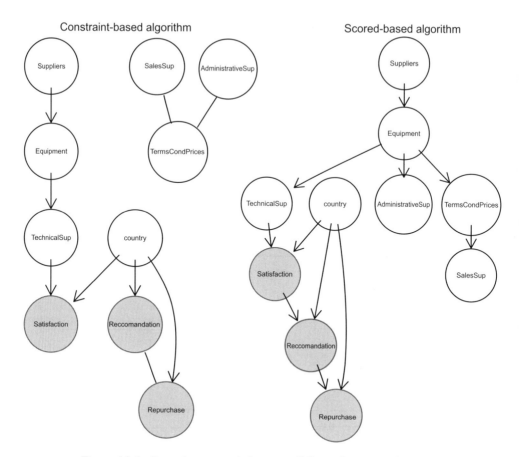

Figure 11.4   Bayesian network from two different learning algorithms

On the basis of the conditional probability of the estimated network, we can perform various simulations and assess various scenarios. As an example, we can change the satisfaction profile of technical support and show how the overall satisfaction level changes (see Tables 11.2 and 11.3 and Figure 11.5). As expected, there is a relationship between satisfaction with technical support and overall satisfaction level. This relationship is, however, not linear and different

Table 11.2   Original distribution of overall satisfaction with technical support

Category	Percentage
1	5.09
2	12.04
3	14.35
4	40.74
5	27.78

Table 11.3   Technical support simulated distribution

Category	Percentage
1	19.91
2	18.52
3	21.30
4	20.37
5	19.91

for each country. A classical linear model, using linear coefficients and dummy variables for representing the countries, would therefore not be able to detect this relationship.

Using the BN, it is possible to simulate different scenarios. In this sense, a BN is similar to a cause-and-effect diagram and is easy to interpret (Kenett, 2007; Renzi *et al.* 2009). In addition, a BN summarizes subject-matter knowledge and data-derived information. These can be integrated by both learning the multidimensional structure of the data and assigning conditional distributions to the variables. Finally, a BN can be constructed with computationally efficient algorithms for evidence propagation. This means that various possible improvement scenarios can be easily simulated and evaluated. A BN can therefore be considered an innovative approach to support strategic decisions.

11.2.2   Transport data analysis

Our second example is an analysis of customer satisfaction with transportation systems in Europe. In particular, we focus on consumer satisfaction with railway transport in 14 European countries. Our data is composed of three Eurobarometer (EB) surveys conducted in 2000, 2002 and 2004. These surveys were sponsored by the European Commission, and designed for monitoring and planning purposes. They record levels of satisfaction with services of general interest (SGI), combined with information about respondents' individual characteristics.

In recent years, interest in consumers' experience with SGI has significantly increased. Several surveys have been conducted in order to assess whether and how consumer satisfaction differs across European countries. At the same time, a broad literature has started investigating the factors affecting individual perceptions and evaluations.

Intuitively, if a sample of train passengers are asked to score their satisfaction with the service on a scale of 1 to 10, they are unlikely to report the same score, even after sharing an identical travel experience. This is due to the fact that their judgements are based on a mixture of expectations, attitudes and other features which vary across individuals. Perceptions are affected by individual-specific characteristics, such as income, gender, education, but also by group-specific features, social norms and the like (Bertrand and Mullainathan, 2001). For example, a recurrent finding in the literature links old people with higher patterns of satisfaction compared with young respondents, whilst women apparently tend, *ceteris paribus*, to complain more than men (Fiorio and Florio, 2008). For more on these issues see Chapters 3, 7 and 14.

While some of the literature interprets customer satisfaction surveys by exploiting econometric models for categorical dependent variables, just a few studies have exploited BNs (Anderson *et al.*, 2004). The goal of our analysis here is to investigate the determinants

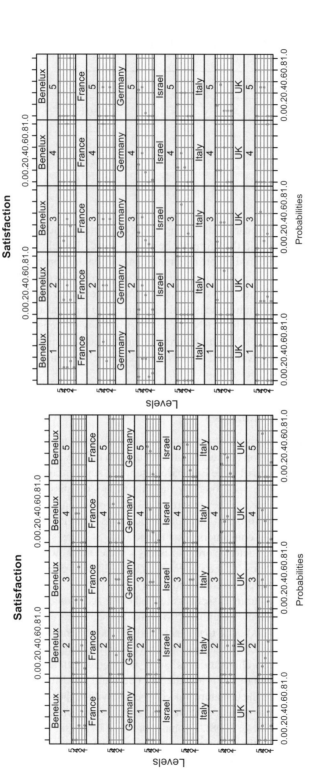

Figure 11.5   Conditional probability of satisfaction per technical support and country (the two-parent node) in the case of the Table 11.2 and 11.3.

of consumer satisfaction with railway transport through BN methods. The EB data set is composed of three main groups of variables: individual characteristics, socioeconomic and railways indicators.

EB surveys. Since 1973, the European Commission has been monitoring the evolution of public opinion in European Union member states, helping in the preparation of reports and the evaluation of different policies. EB surveys collect a large amount of information about individuals' actual attitudes, such as vote intention or media use, their satisfaction with life in general and with some specific issues, such as government policies. Satisfaction with SGI has also been widely investigated in special reports: EB special issues 53 in 2000, 58 in 2002 and 62 in 2004 analyse the quality perceived by consumers in public services such as gas, electricity, telecommunications and transportation. Here we focus on individuals' satisfaction with rail transport between cities, which has been analysed in all the EB reports mentioned above.

EB special issues 53, 58 and 62 present one question concerning both price and quality satisfaction. In particular, respondents are asked to express their opinion about prices, by choosing a category on a three-point scale: '3' if they find the service pricing to be 'fair', '2' if the price is judged as 'unfair' and '1' if it is 'excessive'. Similarly, the quality of the rail transportation service can be defined according to the following scale: '1' if it is perceived as 'very good', '2' if it is just 'fairly good', '3' if it is 'fairly bad' and '4' for 'very bad'. Figure 11.6

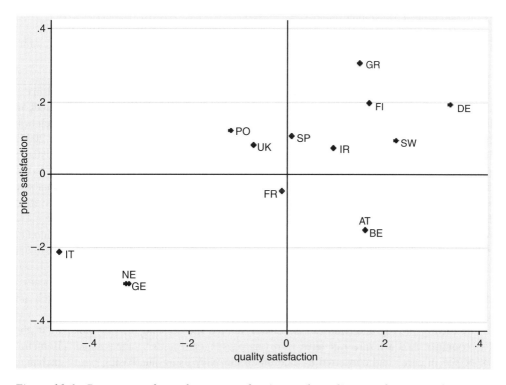

Figure 11.6 Departures from the mean of price and quality satisfaction with railway transport in EU countries

Source: Eurobarometer special issues 53 (2000), 58 (2002) and 62 (2004).

presents the country average values for satisfaction with both price and quality, obtained by pooling the three samples. Countries in the lower left-hand corner are characterized by lower than average levels, while the opposite holds for countries in the top right-hand corner.

At the same time, respondents are usually asked about many aspects of their lives. In our model we include some individual characteristics in order to analyse their links with satisfaction levels. Specifically, the variables used in the learning procedure of our BN are as follows:

- age (*age*), gender (*gender*) and nationality (*country*);

- education (*educ*), coded into four groups – low (for those who stopped education before age 14), medium (14–17), high (more than 18), with a fourth group consisting of those who were still studying at the time of the interview;

- occupation (*occupation*), classified in seven categories – homemaker, student, retired, unemployed, self-employed, employed, manual workers;

- marital status (*marital*) – married, unmarried, widowed, and separated or divorced;

- political views (*political*), ranked in three groups (left, centre and right voters);

- attitudes toward railway transport – in particular, the ease of access (*access*) to the service and the transparency of the information (*info*) offered by the service provider about fares, timetables, etc.;

- official complaint submitted (*complaint*) in the last 12 months either to any complaint-handling body or to the service provider.

Socioeconomic indicators. The inclusion of some macroeconomic indicators is based on the assumption that the economic and social environment in which individuals live has an effect on their perception of satisfaction. For example, it seems reasonable to guess a negative relation between high unemployment rates and satisfaction. Other researchers (Fiorio *et al.*, 2007) found significant results concerning GDP, GDP growth rate and employment rates. Usually macroeconomic indicators are country-specific. Since EB surveys classify respondents based on their region of residence (at NUTS II level) we use the Eurostat Regional Statistics database as a source for macroeconomic indicators. The NUTS classification is a hierarchical system for dividing up the economic territory of the EU. The NUTS II level corresponds to the basic regions. The socioeconomic indicators include net disposable income of households (*income*) expressed in parity purchasing power terms, population density (*popdens*) and unemployment rates (*unempl*).

Railway indicators. These add to individual characteristics and macroeconomic indicators three variables representative of some features of the railway transport service, which are supposed to influence individuals' preferences.

Following Fitzroy and Smith (1995), who studied the demand for rail transport in the EU countries, we measure rail fares (*fare*) as passenger revenue per passenger-kilometre, expressed in euros and converted to a common currency using purchasing parity exchange rates. Again, according to Fitzroy and Smith (1995), the demand for rail transport (*demand*) is defined as passenger-kilometres per capita. These two variables are defined at country level, since regional data are not available. The source of these two indicators is the World Bank's Railway Database.

The variable *railways* is defined as the ratio between the length of railways and the total area of each NUTS II region, and can be interpreted as a *proxy* for the supply of railway transport.

Finally, the variable *year* discriminates between the three EB volumes (2000, 2002 and 2004).

As in the analysis of the ABC 2010 ACSS, we use the R `bnlearn` library.

We first analyse the link between the variables listed above and satisfaction with both price and quality of the railway service. Since we cannot mix continuous and categorical variables, our first step is to adapt our data set to this restriction. In our case the only individual characteristic coded as a continuous variable is the age of the respondents, while the problem arises for all the socioeconomic and railway indicators. Hence, we recode into decile groups the variable *age* and all the indicators defined at regional level. *Fare* and *demand*, which are country-specific, are classified into quartile groups.

At this stage our data set is ready to be analyzed using BN techniques. Before starting the learning procedure, we have to carefully consider our data with reference to the structure of our hypothetical network. Intuitively, some connections between variables are *a priori* reasonable, while others are not. For instance, the level of satisfaction cannot have an influence on respondents' age or gender, but the opposite may hold. Hence, we can integrate the learning procedure with prior information by building a blacklist, specifying a list of arcs which must be excluded from the network. In other words, the learning procedure takes this set of constraints into account and will not, for example, set an arc from satisfaction to age.

Once we have a blacklist, we are ready to start the learning procedure. As already pointed out, compared with other software, an advantage of `bnlearn` is that it relies on the number of learning algorithms which are implemented. This feature makes it possible to iterate the learning procedure and to test the robustness of our results.

We estimate a BN via all eight available algorithms. Moreover, for some of them (hill climbing, tabu and MMHC) we specify two different score functions: the Akaike (AIC) and Bayesian (BIC) information criteria.

The results are summarized in Table 11.4. In the first column all the nodes which have at least one child across the repeated learning procedures are listed. For each of them the second column reports the children, while in the other columns the occurrence of a directed arc between the two nodes is highlighted with a 1. A value of 0.5 indicated that the two nodes are linked by an undirected arc.

As we can see, the results are not consistent across the different learning procedures. Some algorithms tend to over-specify the network (hill climbing and tabu) while for others the opposite is the case (e.g., hybrid algorithms). This evidence conveys two messages. The first concerns the usefulness of a learning procedure which implements more than one algorithm. The second involves the weakness of the causal relationships among some of the variables included in our model. In particular, the connection between individual characteristics and macro indicators is rarely verified.

The last column in Table 11.4 reports the number of times each arc is verified across the learning procedures, from a minimum of 1 to a maximum of 11. According to these values, the BN which seems to score best is the one based on the phase restricted maximization algorithm (PHM). Concerning the structure of the networks, we can observe how the strongest relationships link individual characteristics to each other, whilst socioeconomic and railway indicators are poorly connected to other variables.

Table 11.4 BN learning procedures for transportation data

Parent nodes	Child nodes	Score-based algorithms				Constraint-based algorithms				Hybrid algorithms			TOT.
		HC (BIC)	HC (AIC)	TABU (BIC)	TABU (AIC)	GS	IAMB	FAST IAMB	INTER IAMB	MMHC (BIC)	MMHC (AIC)	PHM	
Country	Fare	1	1	1	1	1	1		1	1	1	1	10
	Educ	1	1	1	1	1	1			1	1	1	9
	Demand	1	1	1	1		1		1	1	1		8
	Pricesat	1	1	1	1			1		1	1		7
	Railways	1	1	1	1				1				5
	Income	1	1	1	1								4
	Unempl	1	1	1	1								4
	Popdens	1	1	1	1								4
	Political	1	1	1	1								4
	Qualsat		1		1			1			1		4
	Info	1	1	1	1								4
	Complaint		1		1								2
	Access		1		1								2
Year	Fare	1	1	1	1	1	1		1			1	8
	Unempl	1	1	1	1					1	1		6
	Occupation		1		1		1	1	1		1		6
	Railways		1		1		1	1	1		1		6
	Access	1	1	1	1								4
	Demand	1	1	1	1								4
	Qualsat		1		1	1							3
	Info		1		1								2
	Income		1		1								2
	Popdens		1		1								2
Age	Marital	1	1	1	1	1	1	1	1	1	1	1	11
	Occupation	1	1	1	1	1	1		1	1	1	1	10
	Educ	1		1		1				1			4

Gender	Marital	1	1	1	1	1	1	1	1		11
	Occupation	1	1	1		1	1		1		7
Educ	Occupation	1	1	1	1	1	1	1	1		11
	Political	1						1			2
	Complaint				1	1					1
Occupation	Marital	1	1	0.5		0.5	0.5	1	1		5
	Complaint				1	1	1				1
Marital	Political		0.5	0.5	0.5	0.5	0.5	1			3.5
	Occupation			0.5	0.5	0.5	0.5				1.5
Political	Pricesat		1	1				1			1
	Marital		0.5	0.5	0.5	0.5	0.5				1
Access	Qualsat	1	1	1	1	1	1	1	1		11
	Info	1	1	1	1	1	1	1	1		11
	Pricesat			1	1	1	1	1	1		4
Pricesat	Qualsat	1	0.5	0.5	0.5	0.5	0.5	1	1		5
	Pricesat		0.5	0.5	1	0.5	0.5				6
Qualsat	Complaint	1	1	1	1				1		4
Info	Pricesat	1	1	1	1	1	1	1	1		11
	Qualsat	1	1	1	1	1	1	1	1		11
	Complaint	1	1	1	1	1	1	1	1		9
Demand	Qualsat		1	1	1	1	1	1	1		2
	Pricesat		1	1	1	1	1	1	1		2
	Railways								0.5	0.5	0.5
	Access	1								1	2

(Continued)

Table 11.4 (Continued)

Parent nodes	Child nodes	Score-based algorithms				Constraint-based algorithms				Hybrid algorithms			TOT.
		HC (BIC)	HC (AIC)	TABU (BIC)	TABU (AIC)	GS	IAMB	FAST IAMB	INTER IAMB	MMHC (BIC)	MMHC (AIC)	PHM	
Fare	Complaint					1	1		1	1	1	1	6
	Demand		1		1			1		1	1		5
	Qualsat					1	1		1			1	4
	Pricesat					1							1
Railways	Access		1					1			1		3
	Demand					0.5		1				1	2.5
	Fare									1	1		2
Income	Popdens	1	1	1						1		1	5
	Railways			1		1	1			1		1	5
	Unempl							1	1				2
	Political							1					1
Popdens	Income	1	1	1		1	1	1					6
	Railways			1	1					1		1	4
	Access						1		1				2
	Unempl					1		1					2
	Political										1		1
Unempl	Income	1	1	1	1	1	1		1	1		1	9
	Railways			1			1		1	1	1	1	6
	Complaint						1		1				2
	Political						1	1					2
	Popdens											1	1
	Info										1		1
Info	Pricesat	1	1	1	1	1	1	1	1	1	1	1	11
	Qualsat	1	1	1	1	1	1	1	1	1	1	1	11
	Complaint	1		1		1	1	1	1	1	1	1	9

This result was somehow predictable, as it underlines one of the greatest disadvantages of BN analysis compared with other methods. The impossibility of mixing categorical and continuous variables forced us to recode our macro indicators using decile and quartile groups. Since, in our model, the macro variables represent an approximation of an unobserved latent variable (e.g., per capita GDP is a proxy for individual income), their translation into categorical terms constitutes a further approximation which weakens the connection with individual characteristics.

The main findings of our analysis are more evident in Figure 11.7. Here we constructed a BN by including all the arcs that obtained a score higher than 7 in Table 11.4. Two graphical representations for the same network are plotted. Figure 11.7a was obtained using Rgraphviz, a library that interfaces R with Graphviz, a package for drawing graphs, while Figure 11.7b was drawn using Graphviz. Both programs are open source tools.

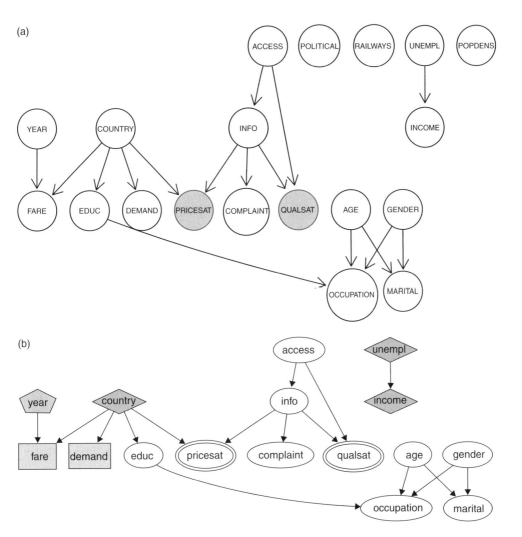

Figure 11.7   Network representation of the most robust results reported in Table 11.4.

`Graphviz` can be downloaded at www.graphviz.org, while `Rgraphviz` can be found at www.bioconductor.org/packages/release/bioc/html/Rgraphviz.html. Both websites provide useful manuals and documentation.

Among all individual characteristics, only political views are not included in the network. Considering the socioeconomic and railway indicators, both population density and length of railways are excluded due to the weakness of their connections with other variables (Table 11.4). Moreover, Figure 11.7 points out how the macro indicators stay outside the 'core' of the network. The unemployment rate has an influence on the income, which is a reasonable result, but none of them is linked to respondents' individual characteristics. The same applies for the railway indicators: in this case the indirect connection to individual features is represented by the variable *country*. Again, this finding makes sense as *country* is a hybrid variable, since it can be interpreted both as an individual characteristic (respondents' nationality) and as a macro indicator (*fare* and *demand* are defined at country level).

The rest of our BN can be analysed as if it were composed of two sub-networks. The first includes those variables concerning individuals' attitudes toward railway services. The *access* to the service influences the judgements about the transparency of the *information* offered by the provider and about the *quality* of the service. *Info* has an effect on the official *complaints* submitted by travellers and on both *price* and *quality satisfaction*.

The second sub-network includes the individual characteristics not connected to railway transport. Respondents' occupation depends on their *education*, *age* and *gender*, while marital status is influenced by *age* and *gender*. Intuitively, these findings are sensible and consistent with previous literature. The link between these two sub-networks is represented, again, by respondents' nationality, which is linked to both *education* and *price satisfaction*.

Having constructed the networks shown in Figure 11.7 using the evidence reported in Table 11.4, we note that the `bnlearn` package provides a set of useful tools for checking the robustness of networks. These tools are helpful since they help the researcher to better understand the proper specification of a BN and to provide information on the effects of a change in one or more variables on the network's structure.

We next summarize some of these commands, with a practical example of their use provided on the website for this chapter.

By using the command *arc.strength* we can attach a numerical value to each arc of the network, as a representation of the strength of the linkage between couples of nodes. The command *strength.plot* provides a graphical representation of the network where the robustness of the connections is highlighted through arcs having different thicknesses.

Another method for checking our results involves associating a score with our BN (`bnlearn` command *score*). The `bnlearn` library implements different kinds of score. For a detailed description of the scoring options see www.bnlearn.com/documentation/man/score.html.

Finally, another option to verify the robustness of a network component involves the cross-validation procedure (`bnlearn` command `bn.cv`). This method is particularly useful when we wish to compare different networks, since it provides a standard way to obtain unbiased estimates of a model's goodness of fit.

11.2.3   R packages and other software programs used for studying BNs

In the previous sections we used R packages for our applications. This choice reflects the quality and variety of R packages available for BN estimation and their worldwide popularity. Other software packages are also used to study BNs in scientific papers and empirical projects:

- GeNIe (Graphical Network Interface) is the graphical interface to SMILE (Structural Modeling, Inference, and Learning Engine), a fully portable Bayesian inference engine developed by the Decision Systems Laboratory and thoroughly field tested since 1998. GeNIe can be freely downloaded (http://genie.sis.pitt.edu/) with a user guide and related documentation.

- Hugin (http://www.hugin.com/index.php) is commercial software that provides a variety of products for both research and non-academic use. The Hugin graphical user interface allows the user to build BNs, learning diagrams, etc.

- IBM PASW SPSS Modeler (http://www.spss.com/) includes several tools that enable the user to deal with a list of features and statistical methods such as BN. Again, IBM SPSS is not free software.

Compared with R, these three software packages incorporate intuitive graphical interfaces that reduce the required programming effort. For a detailed review, the reader is referred to the web page of the University of British Columbia: www.cs.ubc.ca/~murphyk/Bayes/bnsoft.html.

11.3   Prediction and explanation

Bayesian networks provide models for determining cause and effect and prediction under alternative scenarios. As mentioned in Section 11.1, two types of inference support are often considered: predictive support for node X_i, based on evidence nodes connected to X_i through its parent nodes (top-down reasoning); and diagnostic support for node X_i, based on evidence nodes connected to X_i through its child nodes (bottom-up reasoning). Exploratory power providing explanations on causality relationships is, however, different from predictive power (Shmueli, 2010). Providing explanatory power is derived from unbiased models. The explanatory view is retrospective and focused on understanding the complex relationships between various aspects covered by a survey. These dual goals of BNs are related to the concepts of enumerative and analytic studies introduced by Deming (1953). Enumerative studies lead to actions on the frame being studied. In the context of survey data, estimating differences between overall satisfaction levels of customers from different countries is enumerative. Analytic studies, on the other hand, focus on actions to be taken on the process or cause system that is at the basis of the frame under study. For example, setting improvement goals for technical support satisfaction levels can be done by using BN conditioned on very satisfied customers. Such conditioning indicates the maximum level we should realistically strive for. This exploits the BN and our knowledge to determine the level of satisfaction with technical support that characterizes very satisfied customers.

Survey data is naturally quantified on nominal or ordinal scales. BNs, as presented above, are naturally applied to such data since they build on JPDs derived from contingency tables linking nominal or ordinal data. When the data is continuous a discretization procedure needs to be implemented. We leave out the details of such discretization procedure and refer the user to the various alternatives provided by the software products mentioned in the previous section.

In this section we compare the exploratory approach with the predictive approach using a simple continuous data example originally discussed in Shmueli (2010). As mentioned, applying BNs to such data requires discretization. The example below presents the difference between modelling for predictive purposes and for explanatory purposes, prior to such a discretization.

In general, predictive power is achieved by minimizing mean square error (i.e., model bias plus sampling variance). This view is prospective – we would like to make correct predictions. A basic method of evaluating predictive power is to use data that has not been used for constructing the prediction model.

Denote the observed output by Y, the true model by $f(X)$, the incorrect model specification by $f^*(X)$, and the estimated model, based on the incorrect model specification, by $\hat{f}^*(X)$. We have

$$
\begin{aligned}
MSE = E(Y - \hat{Y}|x)^2 &= E(Y - \hat{f}^*(x))^2 \\
&= E(Y - f(x) + f(x) - f^*(x) + f^*(x) - \hat{f}^*(x))^2 \\
&= E(Y - f(x))^2 - (f^*(x) - f(x))^2 + E(\hat{f}^*(x) - f^*(x))^2 \\
&= \mathrm{Var}(Y) + \mathrm{bias}^2
\end{aligned}
$$

Assume that the true model is

$$
f(x) = \beta_1 x_1 + \beta_2 x_2.
$$

The estimated model is then

$$
\hat{f}(x) = \hat{\beta}_1 x_1 + \hat{\beta}_2 x_2.
$$

However, for the underspecified model,

$$
f^*(x) = \gamma_1 x_1,
$$

the estimated model is

$$
\hat{f}^*(x) = \hat{\gamma}_1 x_1.
$$

Let

$$
MSE_1 = E(Y - f(x))^2 = \sigma^2 + 0 + \mathrm{Var}\left(x_1 \hat{\beta}_1 + x_2 \hat{\beta}_2\right),
$$

$$
MSE_2 = E\left(Y - f^*(x)\right)^2 = \sigma^2 + x_2^2 \beta_2^2 + \mathrm{Var}(x_1 \hat{\gamma}_1).
$$

Then $MSE_2 < MSE_1$ when σ^2 is large, $|\beta^2|$ is small, and $\mathrm{corr}(X_1, X_2)$ is high on a limited range of Xs. A misspecification of the model can therefore bias our prediction error (Shmueli, 2010).

A BN sensitivity analysis was studied in Cornalba *et al.* (2007) using statistically designed experimental runs defined by conditioning various demographic variables, and generating corresponding conditioned BNs. The model sensitivity was determined by studying the goodness of fit between data generated from the conditioned networks and the original data used to construct the network.

The simulated data represents a different context from the one used to build the specific BN. By conducting such designed experiments, and evaluating a goodness-of-fit statistic, we can assess the dependence of network-based predictions on various variables, such as demographic variables. In a sense, this evaluates the predictive power of the BN. The BN explanatory power is a direct consequence of the robustness of the generated BN. In Section 11.2.1 we conducted such a robustness study which was used to determine how well the network explains specific cause and effect relations.

Other models can be used to map relationships between variables measured by a survey questionnaire. Log-linear models (Chapter 12) and decision trees (Chapter 15) are both predictive and explanatory. Structural equation and partial least squares models (Chapter 16) rely on existing theoretical models of causality relationships. Multilevel models (Chapter 19) and fuzzy methods (Chapter 21) offer yet other approaches.

11.4 Summary

Customer satisfaction surveys are used to identify the drivers of customer satisfaction, set up improvement goals and initiate focused actions. This chapter presents a powerful tool for analysing customer satisfaction surveys, Bayesian networks. These can be used effectively for attaining these goals. A Bayesian network provides a visual cause-and-effect map of the survey variables and shows clearly which variables affect customer satisfaction. In providing such a map, or directed acyclic graph, we produce knowledge that provides insights for managers and specialists. For a comprehensive study of cause and effect models and the role of Bayesian networks, see Frosini (2006). Godfrey and Kenett (2007) provide an overall review of the contributions of Juran to management science including operational methods that build on cause-and-effect relationships. Both papers give a historical review of causality methodologies, including milestone contributions such as those of Bacon and Hume.

In studying the robustness properties of the Bayesian network we are evaluating potential bias in the model and focus on the explanatory power of the analysis (see Section 11.2.1).

Another benefit of a Bayesian network is that it provides an opportunity to conduct 'what if' sensitivity scenarios. Such scenarios allow us to set up goals and improvement targets. We can determine the expected improvement in overall customer satisfaction if we improve technical service (see Section 11.3). This type of study is based on a predictive analysis. Our objective here is to assess prediction error of the cause-and-effect model represented by the Bayesian network. For more on Bayesian network sensitivity analysis, see Cornalba *et al.* (2007).

Finally, another advantage of the Bayesian network is that it ensures anonymity of respondents. This is critical in employee surveys and is often promised in customer surveys in order to increase response rates.

The chapter includes two case studies. The first is an application of Bayesian networks to the ABC 2010 ACSS data. The second is an application to Eurobarometer data on railway transportation systems throughout Europe. The data analysis examples are conducted using the R system. Section 11.2.3 reviews other available software packages. Section 11.3 expands on the dual role of Bayesian networks in terms of providing cause-and-effect relationships and predictions on alternative scenarios. The comprehensive list of references provides links to additional resources for the interested reader.

References

Anderson, R.D., Mackoy, R.D., Thompson, V.B. and Harrell G. (2004) A Bayesian-network estimation of the service-profit chain for transport service satisfaction. *Decision Sciences*, 35(4), 665–689.

Bernardo, J.M. and Smith, A.F.M. (1994) *Bayesian Theory*. New York: John Wiley and Sons, Inc.

Bertrand, M. and Mullainathan, S. (2001) Do people mean what they say? Implications for subjective survey data. *American Economic Review*, 91(2), 67–72.

Boettcher, S.G. and Dethlefsen, C. (2003) deal: A package for learning Bayesian networks. *Journal of Statistical Software*, 8(20). http://www.jstatsoft.org/v08/i20/.

Cornalba, C., Kenett, R.S. and Giudici, P. (2007) Sensitivity analysis of Bayesian networks with stochastic emulators. Paper presented to Joint ENBIS-DEINDE 2007 Conference on 'Computer Experiments versus Physical Experiments',University of Turin, 11 April.

Deming, W.E. (1953) On the distinction between enumerative and analytic surveys. *Journal of the American Statistical Association*, 48, 244–255.

Fiorio, C.V. and Florio, M. (2008) *Do You Pay a Fair Price for Electricity? Consumers' Satisfaction and Utility Reform in the EU*. DEAS Working Paper 2008-12, Department of Economics, University of Milan. http://www.economia.unimi.it/uploads/wp/Deas2008_12wp.pdf

Fiorio, C.V., Florio, M., Salini, S. and Ferrari, P.A. (2007) *Consumers' Attitudes on Services of General Interest in the EU: Accessibility, Price and Quality 2000–2004*. FEEM Working Paper No. 2.2007. Available at SSRN: http://ssrn.com/abstract=958939.

Fitzroy, F. and Smith, I. (1995) The demand for rail transport in European countries. *Transport Policy*, 2(3), 153–158.

Fomby, T. B. and Carter Hill, R. (1997) *Advances in Econometrics, Vol. 12: Applying Maximum Entropy to Econometric Problems*. Greenwich, CT: JAI Press.

Frosini, B.V. (2006) Causality and causal models: A conceptual perspective. *International Statistical Review*, 74(3), 305–334.

Gasparini, M. (2000) Controllo della qualità di lotti in uscita in presenza di dati ordinali che esprimano il grado di soddisfazione del consumatore. In *Valutazione della qualità e customer satisfaction: il ruolo della statistica*, pp. 337–347. Milano: Vita e Pensiero.

Godfrey, A.B. and Kenett, R.S. (2007) Joseph M. Juran, a perspective on past contributions and future impact. *Quality and Reliability Engineering International*, 23, 653–663.

Grandy, W. T. and L. H. Schick, *Maximum Entropy and Bayesian Methods*. Kluwer, Academic Publishers: Dordrecht, 1999.

Hand, D., Mannila H. and Smyth, P. (2001) *Principles of Data Mining (Adaptive Computation and Machine Learning)*. Cambridge, MA: MIT Press.

Jaynes, E.T. (2003) *Probability Theory: The Logic of Science*. Cambridge: Cambridge University Press.

Jensen, F.V. (2001) *Bayesian Networks and Decision Graphs*. New York: Springer.

Kenett, R.S. (2007), Cause and effect diagrams. In F. Ruggeri, R.S. Kenett and F. Faltin (eds), *Encyclopaedia of Statistics in Quality and Reliability*. Chichester: John Wiley & Sons, Ltd.

Kenett, R.S. and Salini S. (2009) New frontiers: Bayesian networks give insight into survey-data analysis. *Quality Progress*, August, 31–36.

Lauritzen, S.L. and Spiegelhalter, D.J. (1988) Local computations with probabilities on graphical structures and their application to expert systems. *Journal of the Royal Statistical Society, Series B*, 50(2), 157–224.

Pearl, J. (2000) *Causality: Models, Reasoning, and Inference*. Cambridge: Cambridge University Press.

Raiffa, H. (1997) *Decision Analysis: Introductory Readings on Choices Under Uncertainty*. New York: McGraw-Hill.

Renzi, M.F., Vicard, P., Guglielmetti, R. and Musella, F. (2009) Probabilistic expert systems for managing information to improve services. *TQM Journal*, 21(4), 429–442.

Salini, S. and Kenett, R.S. (2009) Bayesian networks of customer satisfaction survey data. *Journal of Applied Statistics*, 36(11), 1177–1189.

Scutari, M. (2010) Learning Bayesian networks with the bnlearn R package. *Journal of Statistical Software*, 35(3).

Shmueli, G. (2010) To explain or to predict? *Statistical Science*, 25(3), 289–310.

12

Log-linear model methods

Stephen E. Fienberg and Daniel Manrique-Vallier

The chapter introduces the structuring of categorical data in the form of contingency tables, and then turns to a brief introduction to log-linear models and methods for their analysis, followed by their application in the context of customer satisfaction surveys. The focus is on the adaption of methods designed primarily for nominal data to the type of ordinal data gathered in the ABC annual customer satisfaction survey (ACSS). The chapter outlines some basic methodology based on maximum likelihood methods and related model search strategies, and then puts these methodological tools to work in the context of data extracted from the ACSS.

12.1   Introduction

Categorical data are ubiquitous in virtually all branches of science, but especially in the social sciences and in marketing. A contingency table consists of counts of units of observation cross-classified according to the values of several categorical (nominal or ordinal) variables. Thus standard survey data, which are largely categorical in nature, are best thought of as forming a very large contingency table. If we have collected data from n individuals or respondents and there are p questions in the survey questionnaire, then the table is of dimension p and the counts total n.

Suppose there are $p = 16$ questions with binary response categories on a questionnaire. Then the corresponding contingency table would contain $2^{16} = 65\,536$ cells. With a sample of $n = 1000$ respondents, a value not atypical for a large customer satisfaction survey, we can expect to see few large counts since the average cell count would be 0.01! Thus methods for the analysis of categorical survey data need to be able to cope with such sparseness, especially since there are likely to be closer to 100 questions on the survey questionnaire, rather than 16. One strategy for coping involves latent variable structures with relatively few parameters, such at the Rasch model and its generalizations (see Chapter 14, as well as some alternatives

Modern Analysis of Customer Surveys: with applications using R, First Edition. Edited by Ron S. Kenett and Silvia Salini.
© 2012 John Wiley & Sons, Ltd. Published 2012 by John Wiley & Sons, Ltd.

described briefly at the end of this chapter). Another strategy is to take small groups of variables and look more closely at their interrelationships. That is the strategy on which we focus here using log-linear models.

Hierarchical log-linear models provide an important and powerful approach to examining the dependence structure among categorical random variables. There are now many books describing these methods and computer programs in standard packages such as R and SAS for implementing them. See, for example, the books by Agresti (2002, 2007, 2010), Christensen (1997), Bishop *et al.* (1975), Fienberg (1980), Goodman (1978), and Haberman (1974).

In the next section, we describe log-linear models and their interpretation, along with some basic methodology that allows for the analysis of survey data. Then we put the models and tools to work on an excerpt of the ABC annual customer satisfaction survey (ACSS) to illustrate their use and interpretation. We describe some related latent class models in a final section and explain their potential uses.

12.2   Overview of log-linear models and methods

12.2.1   Two-way tables

Virtually all readers will be familiar with the usual chi-squared test for the independence of the row and column variables in a two-way contingency table, which was first proposed by Pearson (1900) and appears in most introductory statistics textbooks:

$$X^2 = \sum_{i,j} \frac{(\text{Observed}(i, j) - \text{Expected}(i, j))^2}{\text{Expected}(i, j)}. \tag{12.1}$$

Suppose that the row variable is gender, male or female, and the column variable is product approval, yes or no. Then the X^2 test focuses on whether approval is independent of gender, where

$$\text{Expected Count}(i, j) = \frac{(\text{Row Total } i) \times (\text{Column Total } j)}{\text{Grand Total}} \tag{12.2}$$

is the expected value of the (i, j)th cell. We usually refer the value of X^2 to the tabulated upper-tail values of the chi-squared (χ^2) distribution on 1 degree of freedom and compute a p-value – for example, a value of X^2 of 3.64 corresponds to a p-value of 5%.

Log-linear models can help extend analyses such as this (see Bishop *et al.*, 1975; Fienberg, 1980). If we let x_{ij} be the count in the (i, j)th cell of the table, indexed by i for gender and j for product approval, we display the 2×2 tables of cell counts and probabilities $p_{ij} = P[\text{Gender} = i, \text{Approval} = j]$ as follows:

		Approval					Approval		
		Yes	No	Totals			Yes	No	Totals
Gender	Male	x_{11}	x_{12}	x_{1+}	Gender	Male	p_{11}	p_{12}	p_{1+}
	Female	x_{21}	x_{22}	x_{2+}		Female	p_{21}	p_{22}	p_{2+}
	Totals	x_{+1}	x_{+2}	n		Totals	p_{+1}	p_{+2}	1

The same notation works for more categories, that is, for an $I \times J$ table. We can now write the log-linear model for the expected cell values, $\{m_{ij} = E[x_{ij}]\}$, in such tables as

$$\log m_{ij} = \log(np_{ij}) = u + u_{1(i)} + u_{2(j)} + u_{12(ij)} \tag{12.3}$$

where $n = \sum_i \sum_j x_{ij}$, with constraints such as

$$\sum_{i=1}^{2} u_{1(i)} = \sum_{j=1}^{4} u_{2(j)} \sum_{i=1}^{2} u_{12(ij)} = \sum_{j=1}^{4} u_{12(ij)} = 0. \tag{12.4}$$

The version of the model in equation (12.3) has has the form of an analysis of variance model. When $u_{12(ij)} = 0$ for all i and j, we can rewrite equation (12.3) as $m_{ij} = m_{i+}m_{+j}/n$ and the maximum likelihood estimate of m_{ij} is simply the result of setting $m_{i+} = x_{i+}$ and $m_{+j} = x_{+j}$. In this notation, the X^2 statistic of expression (12.1),

$$X^2 = \sum_{i,j} \frac{(x_{ij} - x_{i+}x_{+j}/n)^2}{x_{i+}x_{+j}/n}, \tag{12.5}$$

is equivalent to testing whether the interaction terms $u_{12(ij)}$ are all zero in the log-linear model. Alternatively, would could use the likelihood ratio statistic comparing the log-additive model $(u_{12(ij)} = 0)$

$$G^2 = 2 \sum_{i,j} x_{ij} \log \left[\frac{x_{ij}}{x_{i+}x_{+j}/n} \right], \tag{12.6}$$

due originally to Wilks (1935). The X^2 and G^2 statistics are usually close in value, differing mainly in the presence of small cell counts. Cressie and Read (1988) provide a detailed account of a class of goodness-of-fit measures that include both the X^2 and G^2 statistics as special cases and have the same asymptotic χ^2 distribution under the null hypothesis that the model of independence holds.

For a 2×2 table, a common measure of dependence between the row and column variables is the odds ratio introduced by Yule (1900):

$$OR = \frac{\text{odds}(X_2 = 2|X_1 = 2)}{\text{odds}(X_2 = 2|X_1 = 1)} = \frac{P[X_2 = 2|X_1 = 2]/(1 - P[X_2 = 2|X_1 = 2])}{P[X_2 = 2|X_1 = 1]/(1 - P[X_2 = 2|X_1 = 1])} = \frac{p_{11}p_{22}}{p_{12}p_{21}}.$$

We can estimate OR by

$$\widehat{OR} = \frac{x_{11}x_{22}}{x_{12}x_{21}}. \tag{12.7}$$

From (12.3) and (12.4), it is easy to show that

$$u_{12(11)} = \frac{1}{4} \log(OR). \tag{12.8}$$

Hence we can use the log-linear model in equation (12.3) to estimate the odds ratio. When $OR > 1$ $(u_{12(11)} > 0)$, the variables in the table are positively associated; when $OR < 1$ $(u_{12(11)} < 0)$ they are negatively associated. $OR = 1$ $(u_{12(11)} = 0)$ corresponds to statistical independence.

Because for the 2×2 table there is only a single parameter associated with independence, either $u_{12(11)}$ or OR, we say that there is 1 degree of freedom (d.f.) associated with this model. Under the model of independence, both the X^2 and G^2 statistics have an asymptotic χ^2 distribution on 1 d.f. For the more general $I \times J$ contingency table, there are $(I - 1)(J - 1)$ such odds ratios of the form (12.7), and the test statistics have an asymptotic distribution that follows the χ^2 distribution on $(I - 1)(J - 1)$ d.f. under independence.

12.2.2   Hierarchical log-linear models

All the ideas and notation in the preceding subsection carry over into the representation and analysis of higher-dimensional tables. Now we add doubly subscripted u-terms to capture higher-order interactions among sets of variables. Thus for a three-way $I \times J \times K$ table, we write the general log-linear model as

$$\log m_{ijk} = \log E[x_{ijk}] = u + u_{1(i)} + u_{2(j)} + u_{3(k)} \qquad (12.9)$$
$$+u_{12(ij)} + u_{13(ik)} + u_{23(jk)} + u_{123(ijk)},$$

with side-constraints that all doubly-subscripted u-terms add to zero across any index. The model where we set all three-factor terms $u_{123(ijk)}$ equal to zero is the 'no-second-order interaction model' studied first by Bartlett (1935) and later by Birch (1963) and Goodman (1969).

Many authors use a form of shorthand notation to specify interpretable log-linear models. Here we use the notation [1][2] to refer to the additive model $u + u_{1(i)} + u_{2(j)}$, and the notation [12] to refer to the model with $u_{12(jk)}$ and all lower-order terms, $u + u_{1(i)} + u_{2(j)} + u_{12(ij)}$, exactly as we wrote in equation (12.3). We say that these models are hierarchical because the inclusion of interaction terms like [12] implies inclusion of all lower-order terms, [1] and [2]; this is known as the *hierarchy principle*. In general the quantities in square brackets in this notation then refer to the highest-order u-terms in the model.

Thus for the no-second-order interaction model for the three-way $I \times J \times K$ table we describe the model as [12][13][23]. The three components, or highest-order u-terms, correspond to the two-way marginal totals ($\{m_{ij+}\}, \{m_{i+k}\}, \{m_{+jk}\}$), and the estimated expected are the solutions of the equations that set these equal to their expectations:

$$m_{ij+} = x_{ij+}, \quad \text{for } i = 1, 2, \ldots, I; j = 1, 2, \ldots, J,$$
$$m_{i+k} = x_{i+k}, \quad \text{for } i = 1, 2, \ldots, I; k = 1, 2, \ldots, K,$$
$$m_{+jk} = x_{+jk}, \quad \text{for } j = 1, 2, \ldots, J; k = 1, 2, \ldots, K.$$

We solve these equations to get maximum likelihood estimates, $\{\hat{m}_{ijk}\}$, using some form of iterative method such as iterative proportional fitting (see Bishop *et al.*, 1975), or compute them using a generalized linear model program.

All of these ideas and notation generalize to four-way and higher-dimensional tables. For example, for a four-way table, the model [134][234] includes the two three-way interaction terms $u_{134(ikl)}$ and $u_{234(jkl)}$, as well as all two-way interactions except for $u_{12(ij)}$ and all 'main' effects:

$$\log m_{ijkl} = \log E[x_{ijkl}] = u + u_{1(i)} + u_{2(j)} + u_{3(k)} + u_{4(l)}$$
$$+u_{13(ik)} + u_{14(il)} + u_{34(kl)} + u_{23(jk)} + u_{24(jl)}$$
$$+u_{134(ikl)} + u_{234(jkl)}.$$

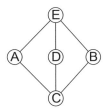

Figure 12.1   Graphical representation of the final model, $[AE][BE][BC][DE][AC][CD]$, *for the equipment data.*

If we hold variables 3 and 4 at fixed levels k_0 and l_0 in this model, most terms in the model are constant and the only terms that vary with variables 1 and 2 are additive:

$$\log m_{ijk_0l_0} = \log E[x_{ijk_0l_0}] = \{u + u_{3(k_0)} + u_{4(l_0)} + u_{34(k_0l_0)}\}$$
$$+\{u_{1(i)} + u_{13(ik_0)} + u_{14(il_0)} + u_{134(ik_0l_0)}\}$$
$$+\{u_{2(j)} + u_{23(jk_0)} + u_{24(jl_0)} + u_{234(jk_0l_0)}\}.$$

This representation thus implies that variables 1 and 2 are conditionally independent given variables 3 and 4. Models where we set exactly one first-order interaction term equal to zero are often referred to as *partial association* models.

The computation of the degrees of freedom of general hierarchical log-linear models is analogous to the case of the $I \times J$ table, although more involved as we consider more complex models. Fortunately, most standard computer programs implementing log-linear methods provide the degrees of freedom of the model as part as their standard output, so we seldom have to compute them directly. When tables are sparse, however, we may need to adjust them 'by hand', as we shall see below.

We often represent conditional independence models graphically, as we do in Figures 12.1 and 12.2, where each node is a variable in the model, and absence of edges represents conditional independence according to the graphical Markov property: if A, B and C are (possibly empty) subsets of nodes (variables), we say that A and B are conditionally independent given C, if and only if deleting the nodes in C from the graph (and any edges incident to them) results in disjoint subgraphs, one containing the nodes in A and another the nodes in B. A hierarchical log-linear model whose conditional independence relations are precisely those that we can read off its graph in this way is known as a *graphical model*. For further details, see Whitaker (1990) and Edwards (2000). Lauritzen (1996) provides a detailed theoretical presentation on graphical models more generally.

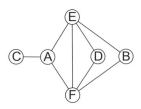

Figure 12.2   Graphical representation of the final model, $[DEF][BEF][AEF][AC]$, *for the overall satisfaction data.*

12.2.3   Model search and selection

We use hierarchical log-linear models to assess dependence among three or more variables in a contingency table. Special cases of these models, the log-linear graphical and decomposable models, allow us to search large contingency tables efficiently for independence, conditional independence, and dependence relationships that are suggestive of the causal structure of the underlying data generating process (Edwards, 2000).

Extending the ideas from the two-way table case, we define the general G^2 statistic or *deviance* for a general hierarchical loglinear model, (a), as

$$G^2(a) = -2 \sum_i x_i \log \left[\frac{\hat{m}_i^{(a)}}{x_i} \right],$$
(12.10)

where $\hat{m}_i^{(a)}$ is the maximum likelihood estimate of the expected value of the ith cell, computed under model (a). The G^2 statistic is asymptotically χ^2 distributed, with the appropriate degrees of freedom for model (a) and can be used to test the goodness of fit of model (a).

We say that two hierarchical log-linear models, (1) and (2), are nested when model (2) is a special case of model (1), obtained by setting some of the u-terms of (1) to zero, respecting the hierarchy principle – for example, [1][2][3] is nested within [13][23], which is in turn nested within [12][23][13]. When this is the case, we can compute the G^2 of model (2), conditional on model (1) being correct, as

$$\Delta G^2 = G^2(2|1) = -2 \sum_i \hat{m}_i^{(1)} \log \left[\frac{\hat{m}_i^{(2)}}{\hat{m}_i^{(1)}} \right]$$
$$= G^2(2) - G^2(1),$$
(12.11)

which we can use to formally test the difference between models (1) and (2). If models (1) and (2) have respectively ν_1 and ν_2 degrees of freedom, then $G^2(2|1)$ is asymptotically distributed as χ^2 with $\nu_2 - \nu_1$ degrees of freedom. Although conditional tests based on G^2 statistics are a powerful tool when comparing competing models, we should always keep in mind that the theoretical results are only valid for nested models, thus they cannot guide us when comparing models that do not belong to the same hierarchy, such as [1][23] and [2][13].

As a general principle, more complex models will always fit the data better than simpler models (less bias), but at the price of less precise estimates (more variance), while simpler models allow for more precise estimates at the price of higher bias. A popular technique to balance fit and parsimony is to penalize the G^2 with some measure that expresses complexity. The Akaike information criterion (AIC) and the Bayesian information criterion (BIC), defined as

$$AIC = G^2 - 2 \cdot df,$$
(12.12)
$$BIC = G^2 - \log(N) \cdot df,$$
(12.13)

where N is the sample size of the contingency table and df the degrees of freedom implied by the model, are two common choices and can be used to compare nonnested models. AIC tends to favor more complex models than BIC.

12.2.4   Sparseness in contingency tables and its implications

As we consider larger number of variables, the number of elementary cells in the corresponding contingency tables increases exponentially, and with it the probability of having cells with zero counts. One powerful feature of log-linear models is that parsimonious models can yield positive expected values for elementary cells, even if the observed counts are zero. However, if some of those observed zeros result in a pattern that produces a zero in a margin corresponding to a minimal configuration required by a model, the corresponding parameter cannot be estimated. For example, if $x_{+01} = 0$, then $u_{23(01)}$ cannot be estimated from the data.

We can still use log-linear models in these cases, provided that we take the precaution of excluding the problematic zero elementary cells from analysis, by setting their expected values to zero *a priori*, or equivalently, conditioning on the fact that the marginal counts are zero and thus that the cell values that add to them must be zero as well. Corresponding to these zero cell values are nonestimable parameters in the log-linear model and these essentially are excluded from the model or set equal to zero. The resulting test statistics, X^2 and G^2, will still be asymptotically χ^2, but we must adjust the degrees of freedom to account for the loss of data and parameters, using the formula

$$df' = df - z_e + z_p, \tag{12.14}$$

where df are the degrees of freedom implied by the original model, z_e the number of cells with zero estimates and z_p the number of parameters that cannot be estimated; see Bishop *et al.* (1975) and Fienberg (1980) for details. Most of the times the adjustment will have to be done by hand, as most computer programs do not yet do this automatically.

Unfortunately, depending on the model, there exist patterns of zeros that do not result in marginal tables with zero entries but also lead to the nonexistence of the maximum likelihood estimates. For example, in a $2 \times 2 \times 2$ table with zeros in the $(1, 1, 1)$th and $(2, 2, 2)$th cells,

$$
\begin{array}{|c|c|c|c|c|c|}
\hline
\;0\; & \; & \; & \; & \; & \; \\
\hline
\; & \; & \; & \; & \;0\; & \; \\
\hline
\end{array}
$$

under the no-second-order interaction model [12][13][23], the two cells with zero entries are constrained to be zero. The problem associated with zeros depends only on the *location* of the zeros, and not on the magnitude of the nonzero cells. There are many more complex examples of these kinds of 'zero cell' problems, but no trivial characterization exists. Tools from algebraic geometry do, however, provide a vehicle for understanding and dealing with the problem; see, for example, Rinaldo (2005), Eriksson *et al.* (2006), and Dobra *et al.* (2008).

12.2.5   Computer programs for log-linear model analysis

Most standard statistical packages have one or more programs or sets of routines that are useful for implementing the methodology described above. In particular, SAS, STATA, and R have generalized linear model routines, and R also has multiple versions of iterative proportional fitting and separate programs for assessing goodness of fit.

MIM is a software system for graphical modeling written for the Windows operating system, available from http://www.hypergraph.dk/, and it is compatible with graphical search methods described in Edwards (2000). MIM is useful for generating the graphs that go with

Table 12.1 Variables in equipment section of ACSS

Label	Description
A	The equipment's features and capabilities meet your needs
B	Improvements and upgrades provide value
C	Output quality meets or exceeds expectations
D	Uptime is acceptable
E	Overall satisfaction with the equipment

the models it fits. A version of MIM is also available in a special R library, mimR, but again it runs only under the Windows operating system.

As we noted in the previous subsection, none of the existing log-linear model or generalized linear model programs deals especially well with the problems of sparsity, that is, the identification of problematic zero cells and the attendant corrections required for computing degrees of freedom. Fienberg and Rinaldo (2007) give further details of what happens in such circumstances.

12.3 Application to ABC survey data

To illustrate the log-linear methodology summarized in the preceding section, we turn to the second section of the ABC 2010 customer satisfaction survey. This section asks customers for their evaluation (on a scale from 1 to 5) of a series of characteristics related to their level of satisfaction with the equipment provided by the company (Table 12.1). The data form a 5^5 contingency table, with a total of 3125 cells but with only $n=220$ observations – a sparse table.

In order to reduce the sparsity of the resulting table we recoded the answers as binary variables by grouping levels 1–3 and 4–5 into single categories. Table 12.2 shows the resulting five-way contingency table. Most of the alternative choices for collapsing categories did not adequately address the sparsity issue.

We begin our search for a descriptive model by fitting each of the log-linear models that include all the uniform order terms (Table 12.3), as well as all of the partial association models (Table 12.4). We note that two of the four-way margins contain one zero ($[ACDE]$ and $[ABDE]$), thus we must be very careful with models that include terms that require those configurations.

In the absence of zeros, all the partial association models in Table 12.4 would have 8 degrees of freedom. However, six of them depend on either margin $[ACDE]$ or $[ABDE]$ (models 1, 2, 6, 7, 8 and 9) and one (model 5) on both. Thus we need to adjust the degrees of

Table 12.2 Data from equipment section of ACSS

E	A								
Low	No	27	10	6	2	4	8	1	3
	Yes	7	1	4	3	4	1	2	2
High	No	0	1	0	1	0	4	2	5
	Yes	3	10	4	21	8	11	7	65

Table 12.3   All k-way interactions models for the equipment data. (Note that model with all four-way interactions is saturated, with 0 degrees of freedom, due to constraints induced by zeros in the margins corresponding to the minimal configurations).

Model	G^2	d.f.	p-value
All 1-way	278.042	26	≈ 0
All 2-way	18.964	16	0.2705
All 3-way	4.419	6	0.6202
All 4-way	0	0	–

freedom to assess the significance of any goodness-of-fit test statistic. For the first group, the marginal zero prevents us from estimating one of the parameters, while forcing two elementary cells to have zero expected counts; thus we have to adjust the degrees of freedom by subtracting 1. In the case of model $[ACDE][ABDE]$, the marginal zeros prevents the estimation of two parameters, while producing three expected elementary counts with zeros; then we have to adjust the degrees of freedom again by subtracting 1, producing 7 degrees of freedom.

Inspecting the results reported in Table 12.3, we see that we need at least some first-order interactions to obtain a well-fitting model. Results in Table 12.4 suggest starting with a model that incorporates the two-way interactions u_{AE}, u_{BE}, u_{DE}, u_{AC}, and u_{CD}. The minimal model that includes those interactions, $[AE][BE][DE][AC][CD]$, does not fit the data very well by itself ($G^2 = 33.92$, d.f. $= 21$), but we use it as a starting point for a model search.

We used the computer program MIM and its bidirectional stepwise procedure based on conditional G^2 tests to search in the space of graphical models. We find a well-fitting model in just one step, by adding the term corresponding to the configuration $[BC]$. The final model, $[AE][BE][BC][DE][AC][CD]$, provides a good fit to the data, with $G^2 = 28.42$ and 20 degrees of freedom. The corresponding graph is shown in Figure 12.1.

We can repeat the model selection procedure using AIC as our criterion for assessing fit, and we get the sequence of models $[AE][BE][DE][AC][CD] \rightarrow [CDE][BE][ACE] \rightarrow [CDE][BDE][ACE]$. The selected model also fits the data very well ($G^2 = 17.63$ on 16 degrees of freedom), but it is much more complex, thus we choose to retain the simpler model.

Table 12.4   All partial association models for the equipment data.

	Model	Test	G^2	d.f.	p-value
1	$[BCDE][ACDE]$	$u_{AB} = 0$	6.914	7	0.4379
2	$[BCDE][ABDE]$	$u_{AC} = 0$	14.841	7	0.0381
3	$[BCDE][ABCE]$	$u_{AD} = 0$	11.582	8	0.1709
4	$[BCDE][ABCD]$	$u_{AE} = 0$	60.143	8	4.369×10^{-10}
5	$[ACDE][ABDE]$	$u_{BC} = 0$	4.829	7	0.6809
6	$[ACDE][ABCE]$	$u_{BD} = 0$	11.545	7	0.1165
7	$[ACDE][ABCD]$	$u_{BE} = 0$	15.698	7	0.0280
8	$[ABDE][ABCE]$	$u_{CD} = 0$	14.836	7	0.0382
9	$[ABDE][ABCD]$	$u_{CE} = 0$	8.978	7	0.2542
10	$[ABCE][ABCD]$	$u_{DE} = 0$	27.401	8	6.025×10^{-4}

Table 12.5   Overall satisfaction variables

Label	Description
A	Overall satisfaction level with equipment
B	Overall satisfaction level with sales support
C	Overall satisfaction level with technical support
D	Overall satisfaction level with training
E	Overall satisfaction level with supplies
F	Overall satisfaction level with contracts and pricing

A stepwise search based on the BIC criterion results in the same model selected by conditional G^2 tests.

Since our final model, $[AE][BE][BC][DE][AC][CD]$, is graphical we can read some of interesting structural features directly from the graph in Figure 12.1. First, we see that given variables C and E, all the rest are mutually independent. This means that for each combination of the overall satisfaction with the equipment and whether or not the output quality meets the client's expectations, the rest of the characteristics measured by the other questions are mutually independent. Second, given the answers to questions A, B and D, the overall level of satisfaction with the equipment turns out to be independent of whether or not the output quality meets the client's expectations.

As a further illustration, we analyze the overall satisfaction questions together. Table 12.5 shows the selected variables that result in a $5^6 = 15\,625$ contingency table. Again, because this is even sparser than in the previous illustrative extract, we dichotomize the variables by grouping the first three levels (1,2,3 = low) and last two levels (4 and 5 = high). Table 12.6 shows the resulting cross-classification.

Even the collapsed 2^6 table presents challenges for our analysis, with three of the 20 three-way margins containing zero counts (configurations $[ADE]$, $[BDF]$ and $[DEF]$), 10 of the 15 four-way margins with zeros, and all of the five-way margins containing zeros. As a

Table 12.6   Data from overall satisfaction questions

			D High				Low			
			E High		Low		High		Low	
			F High	Low	High	Low	High	Low	High	Low
A	B	C								
High	High	High	22	9	6	2	0	2	1	3
		Low	2	1	0	3	0	1	1	3
	Low	High	2	8	3	7	0	1	0	5
		Low	2	2	0	2	0	0	0	3
Low	High	High	2	5	0	7	0	0	1	1
		Low	0	1	0	2	0	0	0	1
	Low	High	0	1	0	8	0	0	0	6
		Low	2	4	1	10	0	0	0	12

Table 12.7 All k-way interactions models for Overall Satisfaction data, for $k = 1, 2$ and 3.

k	Deviance	d.f.	p-value
1	187.733	57	7.772×10^{-16}
2	42.259	42	0.4598
3	7.085	9*	0.6283

*Adjusted for sparsity

consequence, we have to be extra careful when fitting high-order models. For this same reason, this time we do not try to fit the partial association models to begin our model search.

By fitting all the models that include all the same order terms (Table 12.7), we can start our search for a good model by examining models that include two-way interactions.

Starting from the independence model ($[A][B][C][D][E][F]$), we used MIM to perform a bidirectional stepwise search using conditional likelihood ratio tests, and the AIC and BIC criteria. The final models are summarized in Table 12.8. Note that again we have had to adjust the degrees of freedom due to the presence of a zero in the $[DEF]$ margin. This time, stepwise search based on conditional tests of the form ΔG^2 and AIC select the same model, while BIC selects a different, more parsimonious one, but whose fit is not so good. Furthermore, these two models are nested and the conditional test between them gives conditional test value of $\Delta G^2 = 24.368$ with 13 degrees of freedom, suggesting that the inclusion of the extra terms has a significant effect. Thus we select model $[DEF][BEF][AEF][AC]$ and display its independence graph in Figure 12.2.

We can read some interesting structural features of the data directly from the corresponding independence graph. First, given satisfaction with the equipment (A), the overall satisfaction with technical support (C) is independent of all other items. Second, joint knowledge of the satisfaction level with supplies (E) and contracts and pricing (F) makes satisfaction with sales support (B) and satisfaction with training (D) independent of all other items. Furthermore, it also makes the pair A and C independent of the rest of the variables.

12.4   Summary

In this chapter we have summarized the log-linear model approach to the analysis of customer survey data and illustrated it in the context of excerpts from the ACSS data. As we noted in the introduction, while log-linear models are often useful in exploring relatively large sparse

Table 12.8 Final models selected by stepwise search using conditional likelihood ratio tests, as well as the AIC and BIC criteria for the overall satisfaction data.

Criterion	Final Model	G^2	d.f.
ΔG^2 and AIC	$[DEF][BEF][AEF][AC]$	54.910	39*
BIC	$[EF][DE][BF][AF][AC]$	79.278	52

*Adjusted for sparsity.

contingency tables, they typically can only cope with subsets of variables from lengthy survey questionnaires.

A powerful extension to these models involves the introduction of latent categorical variables, and then use the model of conditional independence of the observed (or manifest variables) given the latent variable. These *latent class models* often provide a more parsimonious interpretation for the relationships observed in the contingency table, and with relatively small numbers of parameters. Goodman (1974, 1979, 1984) is largely responsible for the basic methodology used with latent class models, but their likelihood functions involve complexities such as multiple modes and some identification problems (see Fienberg *et al.*, 2010).

Another related class of models introduced more recently utilizes a different form of latent structure through the notion of mixed membership of individuals in idealized classes, or grade of membership, and has been applied to relatively large contingency table problems by Erosheva *et al.* (2007) and Manrique-Vallier and Fienberg (2010).

Both the latent class and the mixed membership modeling approaches provide natural supplements to the methods described in this chapter as well as the to the Rasch models described in Chapter 14 of this book. Erosheva (2005) gives a low-dimensional geometric comparison of these models that may interest some readers.

References

Agresti, A. (2002) *Categorical Data Analysis*, 2nd edn. New York: John Wiley & Sons, Inc.

Agresti, A. (2007) *An Introduction to Categorical Data Analysis*, 2nd edn. Hoboken, NJ: John Wiley & Sons, Inc.

Agresti, A. (2010) *Analysis of Ordinal Categorical Data*, 2nd edn. Hoboken, NJ: John Wiley & Sons, Inc.

Bartlett, M.S. (1935) Contingency table interactions. *Supplement to the Journal of the Royal Statistical Society*, 2, 248–252.

Birch, M.W. (1963) Maximum likelihood in three-way contingency tables. *Journal of the Royal Statistical Society, Series B*, 25, 220–233.

Bishop, Y.M.M., Fienberg, S.E., and Holland, P.W., with contributions by Light, R.J. F. Mosteller, F. (1975) *Discrete Mulivariate Analysis: Theory and Practice*. Cambridge, MA: MIT Press. Reprinted (2007) by Springer.

Christensen, R. (1997) *Log-linear Models and Logistic Regression*, 2nd edn. New York: Springer.

Cressie, N.A.C., and Read, T.R.C. (1988). *Goodness-of-Fit Statistics for Discrete Multivariate Data*. New York: Springer.

Dobra, A., Fienberg, S.E., Rinaldo, A., Slavkovic, A.B. and Zhou, Y. (2008) Algebraic statistics and contingency table problems: Log-linear models, likelihood estimation, and disclosure limitation. In M. Putinar and S. Sullivant (eds), *Emerging Applications of Algebraic Geometry*, IMA Series in Applied Mathematics, pp. 63–88. New York: Springer.

Edwards, D. (2000) *Introduction to Graphical Modelling*, 2nd edn. New York: Springer.

Eriksson, N., Fienberg, S.E., Rinaldo, A. and Sullivant, S. (2006) Polyhedral conditions for the nonexistence of the MLE for hierarchical log-linear models. *Journal of Symbolic Computation*, 41, 222–233.

Erosheva, E.A. (2005) Comparing latent structures of the grade of membership, Rasch, and latent class models. *Psychometrika*, 70, 619–628.

Erosheva, E.A., Fienberg, S.E, and Joutard, C. (2007) Describing disability through individual-level mixture models for multivariate binary data. *Annals of Applied Statistics*, 1, 502–537.

Fienberg, S.E. (1980) *The Analysis of Cross-Classified Categorical Data*, 2nd edn. Cambridge, MA: MIT Press. Reprinted(2007) by Springer.

Fienberg, S.E., Hersh, P., Rinaldo, A. and Zhou, Y. (2010) Maximum likelihood estimation in latent class models for contingency table data. In P. Gibilisco, E. Riccomagno, M.P. Rogantin, and H.P. Wynn (eds), *Algebraic and Geometric Methods in Statistics*, pp. 31–66. New York: Cambridge University Press.

Fienberg, S.E. and Rinaldo, A. (2007) Three centuries of categorical data analysis: Log-linear models and maximum likelihood estimation. *Journal of Statistical Planning and Inference*, 137, 3430–3445.

Goodman, L.A. (1969) On partitioning χ^2 and detecting partial association in three-way contingency tables. *Journal of the Royal Statistical Society, Series B*, 31, 486–498.

Goodman, L.A. (1974) Exploratory latent structure analysis using both identifiable and unidentifiable models. *Biometrika*, 61, 215–231.

Goodman, L.A. (1979) On the estimation of parameters in latent structure analysis, *Psychometrika*, 44, 123–128.

Goodman, L.A. with contributions by Clogg, C.C. and a foreword by Duncan, O.D. (1984) *The Analysis of Cross-classified Data Having Ordered Categories*. Cambridge, MA: Harvard University Press.

Goodman, L.A. with contributions by Davis, J.A. and Magidson, J. (1978) *Analyzing Qualitative/Categorical Data: Log-linear Models and Latent-structure Analysis*. Cambridge, MA: Abt Books.

Haberman, S.J. (1974) *The Analysis of Frequency Data*. Chicago: University of Chicago Press.

Lauritzen, S.F. (1996) *Graphical Models*. New York: Oxford University Press.

Manrique-Vallier, D. and Fienberg, S.E. (2010) Longitudinal mixed-membership models for survey data on disability. *Longitudinal Surveys: from Design to Analysis, XXV International Methodology Symposium*, Statistics Canada.

Pearson, K. (1900) On the criterion that a given system of deviation from the probable in the case of a correlated system of variables is such that it can reasonably be supposed to have arisen from random sampling. *Philosophical Magazine*, 50, 59–73.

Rinaldo, A. (2005) Maximum likelihood estimates in large sparse contingency tables. PhD thesis, Department of Statistics, Carnegie Mellon University.

Whittaker, J. (1990) *Graphical Models in Applied Multivariate Statistics*. New York: John Wiley & Sons, Inc.

Wilks, S.S. (1935) The likelihood test of independence in contingency tables. *Annals of Mathematical Statistics*, 6, 190–196.

Yule, G.U. (1900) On the association of attributes in statistics: with illustration from the material of the childhood society, &c. *Philosophical Transaction of the Royal Society*, 194, 257–319.

13

CUB models: Statistical methods and empirical evidence

Maria Iannario and Domenico Piccolo

This chapter is devoted to a new class of statistical models, called CUB models, introduced for the purpose of interpreting and fitting ordinal responses. After a brief discussion of psychological foundations and statistical properties of such mixture random variables, CUB models are generalized by introducing subjects' and objects' covariates and also by taking account of contextual effects. Special emphasis is given to the graphical tools generated by such models which allow an immediate visualization of the effects of covariates with respect to space, time and circumstances. Some applications to ABC 2010 annual customer satisfaction survey data set and to students' satisfaction with a university orientation service confirmed their usefulness in real situations. An R program for CUB model inference is presented in an appendix.

13.1 Introduction

This chapter introduces a class of statistical models based on the psychological mechanism that induces customer to choose a definite item or to manifest an expressed preference towards some object or brand. We propose an integrated approach related to customer choice and analyse the unobservable constructs which generate expressed evaluations. Over the years, there have been several efforts to develop a measure of how products and services supplied by a company or public institution meet or exceed customer expectations or preferences. They derive from a qualitative perception and thus they require specific methods since classical statistical models are neither suitable nor effective for their study. In the real world, alternative discrete choices are finite and the statistical interest in this area is mainly devoted to non-parametric methods or probability structures adequate to interpret, fit and forecast human choices.

Customer satisfaction, in fact, is the result of several components related to the respondent and generated by internal and external forces which depend on a number of psychological

Modern Analysis of Customer Surveys: with applications using R, First Edition. Edited by Ron S. Kenett and Silvia Salini.
© 2012 John Wiley & Sons, Ltd. Published 2012 by John Wiley & Sons, Ltd.

and physical facts. It is an abstract concept analysed through economic indicators, such as the American Customer Satisfaction Index (ACSI). This measure includes several latent variables of endogenous and exogenous type (Anderson and Fornell, 2000; Fornell *et al.*, 1996) and its importance stems from the connections among customer expectations, perceived quality and perceived value. It is updated quarterly and is considered a significant indicator of economic performance of US individual firms and the wider economy, with accurate predictive power with respect to consumer spending and stock market growth.

In our context, we register responses concerning evaluation or satisfaction as expressed during surveys. Respondents are usually asked to rate a well-defined item or object on a point scale or to order a homogeneous list of these.

There are two different selection methods for evaluation studies: *ranking* and *rating*. In *ranking analysis*, each person gives an indirect and compared evaluation which is the result of a paired or sequential selection process. Indeed, the positioning of an object is necessarily conditioned by all the others, and a correct study of the responses should consider the joint distribution of responses. In fact, the analysis of the marginal distribution of a single object is worthy of interest in several marketing contexts, such as advertisement effects, brand loyalty, and efficacy of innovations. A different approach is concerned with the study of distances among ranking vectors by means of a naturally defined metric (Critchlow, 1985; Critchlow *et al.*, 1991; Feigin, 1993; Feigin and Cohen, 1978; Tubbs, 1993; van Blokland-Vogelsang, 1993).

In *rating analysis* we register a judgement involving preference, evaluation, assessment, perceived value, concern, pain, for example, expressed on a Likert scale by n subjects and related to the support $\{1, 2, \ldots, m\}$, for a given m. To be specific, we interpret 'rating' as the ordered measure of a subjective perception generated by an act of cognition. Quite often, people are requested to express preferences on several aspects related to the item or some definite product or service. In customer satisfaction studies, respondents are conditioned by their global experience of the topic and their expectation (by observing the gap between their expected and experienced performance of the product, for instance). People usually tend to evaluate (in a consistent manner) several aspects of the item although they concern different features. Thus, we observe a strong correlation among responses to similar topics.

The current literature discriminates between *stated* and *revealed preferences*, as discussed by Louviere *et al.* (2000) and Train (2003). However, since we are mainly interested in the study of data sets resulting from customer satisfaction surveys, our modelling approach is concerned with stated preferences. In the following, we will focus on rating analysis since this approach to customer motivation and satisfaction is more diffuse in marketing researches.

In any case, our main interest is in the study of the stochastic structure that leads to the final decision. This framework supports several empirical situations in many different area where survey data are collected as ordinal data. In this regard a vast and encouraging experience supports such an approach and thus, in this contribution, we confine ourselves to those aspects mostly related to customer satisfaction analysis.

In the next section we discuss logical and psychological foundations of the approach; then, in Section 13.3 we derive statistical consequences in terms of parsimonious modelling and point to the main characteristics of this class of models aimed at interpreting and visualizing data. In Section 13.4 we discuss inferential issues, and in Sections 13.5 and 13.6 we consider the role of subjects' covariates. Section 13.7 considers models that include items' and contextual covariates within the the the same inferential framework. The approach is applied with reference to ABC 2010 annual customer satisfaction survey data set and to students' satisfaction towards their university orientation service in Section 13.8. Sections 13.9 and 13.10 provide some further generalizations and concluding remarks. The documentation for an R program is given in the Appendix.

13.2   Logical foundations and psychological motivations

Logical and psychological considerations motivate the introduction of a class of models to explain survey responses when people are requested to make an ordered choice expressing their satisfaction. For the sake of conciseness, we briefly discuss some of these.

- When faced with discrete alternatives, people adhere to a definite choice by pairwise comparison of items or by sequential removals. Statistical models should take these strategies into account.

- Current models entertain both uncertainty in the choice and randomness of the experiment as a common source generating indecision. Indeed, the origin and interpretation of these components are quite different and deserve disjoint analyses.

- Expectation does not exhaust ratings data since distributions with completely different shapes maintain the same mean value; similar considerations apply also to high-order moments. As a consequence, it is convenient to move to models able to consider the whole distribution of values and not only a synthetic index (e.g. average, mode, and so on).

- Empirical evidence shows that expressed preferences range from symmetric to highly skewed, flat and almost bell-shaped distributions, with modes ranging everywhere on the support. Sometimes, *shelter* effects should be considered for effective models.

- Decisions depend on personal, objective and contextual factors and a statistical approach should be able to consider all these aspects in the perspective of a joint modelling approach.

- The freedom of human choice is considerable, so we have to accept a limited predictability of models aimed at measuring individual satisfaction. Looking for clusters, subgroups, and selected categories of customers should improve the understanding of homogeneity in consumer behaviour and the related prediction/expectation of preferences.

Such points legitimate the introduction of stochastic models for the study of sampling surveys where subjects express a definite opinion selected from an ordered list of categories.

13.3   A class of models for ordinal data

The discrete choice from a limited ordinal list of m alternatives is the result of complex human decisions which we synthesize by considering liking or attractiveness towards the item (opinion of the subject on the object) and fuzziness surrounding the final choice (external circumstances). We will define such unobservable components as *feeling* and *uncertainty*, respectively, although in a given specific context some more correct wording would be necessary.

Feeling is the result of several factors related to the life of a person, his/her gender and age, education and job, previous experience, quality of personal relationships, and so on. As a consequence, we may summarize them as the approximation of a sum of several random variables which converges to a unimodal continuous distribution. If we look for a discretization of this component, we refer to the support $\{1, 2, \ldots, m\}$ and introduce the shifted binomial random variable, characterized by a ξ parameter, whose probability mass is

$$b_r(\xi) = \binom{m-1}{r-1} \xi^{m-r}(1-\xi)^{r-1}, \qquad r = 1, 2, \ldots, m.$$

This random variable is very flexible since we are able to derive discrete distributions with different shapes by adequately varying cutting points of the latent continuous component.[1]

Uncertainty is a component of any human choice and derives from several circumstances: amount of time devoted to the response, degree of involvement with the problem, use of a limited set of information, nature of the chosen scale, willingness to joke and fake, tiredness and/or fatigue, partial understanding of the item, lack of self-confidence, laziness, apathy, boredom, and so on. The simplest solution for its parametrization is to assume a totally random choice that assigns constant probability to each category, that is, a discrete uniform random variable: $U_r(m) = 1/m, r = 1, 2, \ldots, m$. It is well known that this distribution maximizes entropy in the class of all discrete distributions with finite support.

In order to reduce some ambiguity, we find it useful to discern between two similar concepts related to the indecision about the selection process. We speak of:

- *uncertainty* with reference to the subjective respondents' indecision. This aspect is related to the very nature of human choices and in our modelling approach we will explicitly consider this structural component.

- *randomness* with reference to the way data are collected from a subset of a given population. This aspect is related to sampling selection, measurement errors and limited knowledge. In any modelling approach this issue is considered by using the random variable paradigm.

A fundamental issue in our proposed method is the assumption that we are modelling a single respondent's behaviour; thus, we are not necessarily conjecturing the existence of two subgroups of respondents (e.g., a thoughtful and a wavering population). In fact, *each respondent* includes a proportion (π) of *feeling* and a proportion $(1 - \pi)$ of *uncertainty* in his/her decision. As a consequence, the final discrete choice is a mixture:

$$Pr(R = r) = \pi \, b_r(\xi) + (1 - \pi) U_r(m), \qquad r = 1, 2, \ldots, m.$$

Such a distribution will be referred to as a CUB random variable[2] and it is well defined for parameters $\boldsymbol{\theta} = (\pi, \xi)'$ belonging to the parameter space

$$\Omega(\boldsymbol{\theta}) = \{(\pi, \xi) : \ 0 < \pi \leq 1, \ 0 \leq \xi \leq 1\}.$$

---

[1] More elaborate solutions are possible (e.g., by introducing a shifted beta-binomial random variable); however, any increased complexity should be checked against parsimony and identifiability.

[2] Originally, such a mixture was referred to as MUB, but this later became CUB. Indeed, we may interpret it as a Combination of Uniform and shifted Binomial random variables. Such models were introduced by Piccolo (2003) and have been further discussed by D'Elia and Piccolo (2005), Iannario (2008a), Piccolo and D'Elia (2008) and Iannario and Piccolo (2010b, 2011b).

Iannario (2010a) proved that such a model is identifiable for any $m > 3$.

It is correct to assume that both parameters are subject-dependent (see Section 13.5) but for the moment we suppose a unique random variable for the whole experiment. Indeed, experience confirms that there are objects, topics, items that receive high/intermediate/low feeling with extreme/moderate/low confidence. The selected distribution is able to synthesize and discriminate items with respect to such aspects.

For interpreting the CUB distribution parameters, as mentioned before, each respondent acts with a *propensity* to adhere to a thoughtful and to a completely uncertain choice (measured by π and $1 - \pi$, respectively). In addition, the quantity $1 - \pi$ may be interpreted as a *measure of uncertainty*, whereas $1 - \xi$ is a *measure of adherence* to the proposed choice.

In customer satisfaction surveys, the quantity $1 - \xi$ increases with agreement with the item, and this consideration may be formally assessed by considering that a positively (negatively) skewed distribution implies $\xi < \frac{1}{2}$ ($\xi > \frac{1}{2}$). We may deduce that ξ is related to the predominance of unfavourable responses (lower than the midrange). Briefly, the *feeling parameter* (ξ) may be interpreted as mostly related to location measures and strongly determined by the skewness of expressed ratings: it *increases when respondents choose low ratings*, and vice versa.

We observed that uncertainty of the choice increases with $1 - \pi$. In fact, *uncertainty* adds dispersion to the shifted binomial distribution and thus should be related to entropy concepts. Specifically, the frequency in each category increases with $1 - \pi$ and thus modifies the heterogeneity of the distribution.

Formally, we denote the normalized *Gini heterogeneity index* by

$$\mathcal{G} = \frac{m}{m-1} \left(1 - \sum_i p_i^2 \right),$$

where p_i, $i = 1, 2, \ldots, m$, is a discrete probability distribution, and by $\mathcal{G}_{CUB}$ and $\mathcal{G}_{SB}$ the Gini indices for the CUB model and the shifted binomial components, respectively. Then, the following relationship has been proved (Iannario, 2010b):

$$\mathcal{G}_{CUB} = 1 - \pi^2 \left(1 - \mathcal{G}_{SB} \right).$$

This identity shows that, for a given shifted binomial component, heterogeneity is inversely related to π and increases with uncertainty (expressed by $1 - \pi$).

CUB random variables are characterized by a large number of different shapes and variability (left panel of Figure 13.1). This happens with only two parameters and considerably improves the fitting of observed data. In addition, there is a one-to-one correspondence among CUB models and points in the space $\Omega(\boldsymbol{\theta})$, where coordinates are expressed by the *uncertainty* and *feeling* parameters, respectively (right panel of Figure 13.1).

In fact, on the basis of the previous interpretation, the more π is located to the right of the unit square, the more respondents are inclined to give definite responses (uncertainty is low). Moreover, since $1 - \xi$ measures the *strength of feeling* of the subjects for a positive evaluation of the object, the closer ξ is located to the border of the upper region of the unit square, the less the item is preferred. Such visualization of CUB models is an important aspect of this modelling approach. It adds value to experimental surveys because it offers an immediate tool for an effective interpretation of ordinal data.

For the analysis of the random variable R, we may compute standard indices, but some words of caution are necessary since we are modelling qualitative (ordinal) phenomena. Most

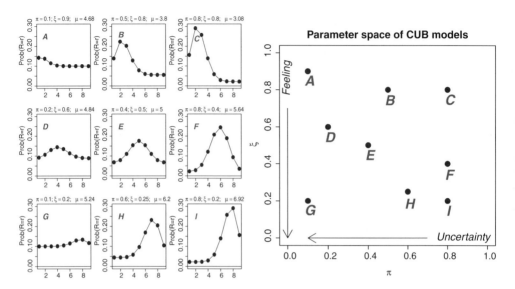

Figure 13.1   One-to-one correspondence between CUB models and parameter space

of the indices may be interpreted by assuming their numerical values related to continuous latent variables which generate the final qualitative assessment. These computations may be of interest not *per se* but when one compares ordinal responses given by subgroups of subjects to the same questionnaire.

The first four cumulants of a CUB random variable were obtained by Piccolo (2003). The expected value of the random variable R is

$$\mathbb{E}(R) = \frac{(m+1)}{2} + \pi\,(m-1)\left(\frac{1}{2} - \xi\right).$$

It moves towards the central value (midrange) of the support depending on the sign of $(\frac{1}{2} - \xi)$. Thus, we expect higher (smaller) mean values when $\xi \to 0$ ($\xi \to 1$) as confirmed by the skewness of the distribution, which is governed by $(\frac{1}{2} - \xi)$.

13.4   Main inferential issues

For inferential purposes, sample data in the form of individual ratings $\mathbf{r} = (r_1, r_2, \ldots, r_n)'$ are strictly equivalent to those contained in the vector $(n_1, n_2, \ldots, n_m)'$ of observed frequencies, where n_r is the frequency of $R = r$.

In fact, if we let $p_r(\pi, \xi) = Pr(R = r \mid \pi, \xi)$, $r = 1, 2, \ldots, m$, the *log-likelihood function* of a CUB model (without covariates) is:

$$\log L(\pi, \xi) = \sum_{r=1}^{m} n_r \log\left[\pi\binom{m-1}{r-1}(1-\xi)^{r-1}\xi^{m-r} + (1-\pi)\frac{1}{m}\right].$$

Then, for a given sample of size n, any sub-vector of $(n_1, n_2, \ldots, n_m)'$ of dimension $m - 1$ is a sufficient statistic for (π, ξ).

Maximum likelihood (ML) estimates of parameters can be obtained by means of the EM algorithm (McLachlan and Krishnan, 2008) as specified for CUB models by Piccolo (2006). Standard results for asymptotic inference from ML theory apply both for estimation and testing hypotheses on the basis of the observed information matrix. Specifically, such results may be usefully exploited for the construction of asymptotic confidence regions.[3]

A significant improvement for accelerating the EM algorithm convergence has been obtained by the introduction of starting values (preliminary estimators) of parameters based on the frequency distribution of the sample data, as in Iannario (2009b, 2010b). Given $m > 3$, an observed random sample $(r_1, r_2, \ldots, r_n)'$ and the relative frequency distribution $\{f_1, f_2, \ldots, f_m\}$, we get the sample mode M_n defined as the integer value of the support where the frequency is maximum, that is:

$$M_n = j \quad \Longleftrightarrow \quad f_j > f_k, \quad \forall k \neq j = 1, 2, \ldots, m.$$

Then we compute the joint parameter estimators:

$$\tilde{\xi} = 1 + \frac{0.5 - M_n}{m}; \qquad \tilde{\pi} = \min \left(\sqrt{\frac{\sum_{r=1}^m f_r^2 - \frac{1}{m}}{\sum_{r=1}^m [b_r(\tilde{\xi})]^2 - \frac{1}{m}}}, \quad 1 \right).$$

Observe that $\tilde{\xi}$ makes no reference to uncertainty parameter whereas $\tilde{\pi}$ is strictly conditioned by $\tilde{\xi}$. In addition, these estimators arise from the very *qualitative nature* of ordinal data and they may be derived from the sampled ordinal data without referring to any arbitrary quantitative rescaling.

For the goodness-of-fit indices of qualitative data, it is not convenient to rely on classical measures such as X^2 tests (invariably significant for moderate and large sample size even though there is a 'quite good fit'). Thus, we prefer the exploratory normalized measure

$$\mathcal{F}^2 = 1 - \frac{1}{2} \sum_{r=1}^m | f_r - p_r(\hat{\pi}, \hat{\xi}) |, \qquad 0 \leq \mathcal{F}^2 \leq 1,$$

where f_r and $p_r(\hat{\pi}, \hat{\xi})$ are the observed relative frequencies and the probabilities estimated by the CUB model, respectively. The index $\mathcal{F}^2$ is the complement of a normed dissimilarity and measures the proportion of subjects correctly predicted by the estimated model.

Measures of fit for more general CUB models (pseudo-R^2, AIC, BIC and other likelihood-based indices) have been discussed in Iannario (2009a). Moreover, some recent results (Di Iorio and Piccolo, 2009) obtained by defining generalized residuals (Gouriéroux *et al.*, 1987) for CUB models may be usefully exploited in the validation step in order to detect significant violations and/or atypical situations.

Finally, we observe that a strength of the CUB model approach derives from the inclusion of *uncertainty*, a component generally present in customer satisfaction studies. However, if $\pi \to 1$, the mixture distribution simplifies to a shifted binomial random variable (D'Elia, 1999, 2000a,b). In this case, from the likelihood function, by letting $\bar{R}_n = \sum_i r_i/n$, we get an explicit formula for the ML estimator of ξ,

$$\hat{\xi} = \frac{m - \bar{R}_n}{m - 1},$$

---

[3] Alternative approaches – permutation tests (Arboretti Giancristofaro *et al.*, 2011) and Bayesian inference (Deldossi and Paroli, 2011) – are currently under investigation.

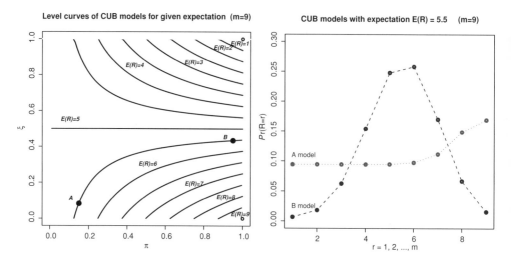

Figure 13.2   Level curves of expectations and CUB models with the same expectation

for which unbiasedness and consistency are immediate. Moreover, the asymptotic distribution of

$$Z_n = \sqrt{n} \, \frac{\hat{\xi} - \xi}{\sqrt{\hat{\xi}(1 - \hat{\xi})/(m-1)}} \xrightarrow{d} N(0, 1)$$

allows for effective testing and confidence intervals.

13.5   Specification of CUB models with subjects' covariates

The formula of the expected value in Section 13.3 shows that the mean value of CUB models is not a useful measure of feeling and uncertainty since infinitely many parameters generate the same expectation. This point is illustrated in Figure 13.2, where we let $m = 9$. Models A and B (left panel) have the same expectation but present a different probability structure (right panel) with respect to decision-making processes.

Therefore, we introduce CUB models with covariates by means of a straightforward relationship between parameters and covariates,[4] according to a more general paradigm (King *et al.*, 2000).

Subjective data are covariates which are functions of respondents. Such covariates may be quantitative (personal measurements such as age, income, height, health indicator) or qualitative (dichotomous or polytomous covariates such as gender, residence, profession). In the following, we will denote *subject's covariates* by $\mathbf{y}$ (for uncertainty) and $\mathbf{w}$ (for feeling).

Suppose that we have collected a sample of ordinal data $\mathbf{r} = (r_1, r_2, \dots, r_n)'$, and the matrices $\mathbf{Y} = \|1, y_{i1}, y_{i2}, \dots, y_{ip}\|_{i=1,\dots,n}$ and $\mathbf{W} = \|1, w_{i1}, w_{i2}, \dots, w_{iq}\|_{i=1,\dots,n}$ which include

---

[4] In generalized linear models (McCullagh, 1980; McCullagh and Nelder, 1989; Nelder and Wedderburn, 1972) the specification for ordinal data models relates expectations to cumulative probabilities of the outcome as a monotone increasing transformation of a linear predictor onto the unit interval, assuming logit, probit or similar link functions. Several variants are possible, as discussed by Agresti (2002) and Dobson and Barnett (2008).

information on respondents which can be used to characterize *uncertainty* and *feeling*, respectively. The extended sample data for explaining the rating r_i of the ith subject becomes

$$S_n = (r_i \mid \mathbf{y}_i \mid \mathbf{w}_i) = (r_i \mid 1, y_{i1}, y_{i2}, \dots, y_{ip} \mid 1, w_{i1}, w_{i2}, \dots, w_{iq}), \quad i = 1, 2, \dots, n.$$

Then, for a given $m > 3$, a general CUB model is defined by:

1. a *stochastic component*,

$$Pr(R_i = r \mid \mathbf{y}_i; \mathbf{w}_i) = \pi_i \binom{m-1}{r-1} \xi_i^{m-r} (1 - \xi_i)^{r-1} + (1 - \pi_i) \left(\frac{1}{m} \right),$$
$$r = 1, 2, \dots, m;$$

   for any $i = 1, 2, \dots, n$.

2. two *systematic components*,

$$\pi_i = \frac{1}{1 + e^{-\mathbf{y}_i \, \beta}}, \qquad \xi_i = \frac{1}{1 + e^{-\mathbf{w}_i \, \gamma}}, \qquad i = 1, 2, \dots, n,$$

   where $\mathbf{y}_i$ and $\mathbf{w}_i$ denote the covariates of the ith subject, selected to explain π_i and ξ_i, respectively.[5]

Data contained in $\mathbf{Y}$ and $\mathbf{W}$ are completely general (made up of a few or several dummy, discrete and continuous covariates, even with interactions) and they may partially or completely overlap. In addition, the presence of covariates with limited support implies that the admissible space for the parameters π_i and ξ_i is now the open unit square.

Since CUB models with covariates assume a deterministic link among the parameters (explaining *uncertainty* and *feeling*, respectively) and subjects' characteristics, this framework considers both $\mathbf{Y}$ and $\mathbf{W}$ as subjects' variables.

Explicitly, for $r_i \in \{1, 2, \dots, m\}$ and $i = 1, 2, \dots, n$, a CUB model with subjects' covariates is defined by the following probability distribution:

$$Pr(R_i = r \mid \mathbf{y}_i; \mathbf{w}_i) = \frac{1}{1 \mid e^{-y_i \beta}} \left[\binom{m-1}{r_i - 1} \frac{\left(e^{-\mathbf{w}_i \gamma}\right)^{r_i - 1}}{(1 + e^{-\mathbf{w}_i \gamma})^{m-1}} - \frac{1}{m} \right] + \frac{1}{m}.$$

A general EM algorithm has been derived (Piccolo, 2006) and implemented for the log-likelihood function:

$$\ell(\boldsymbol{\theta} \mid S_n) = \sum_{i=1}^{n} \log \left[Pr(R_i = r \mid \mathbf{y}_i; \mathbf{w}_i) \right].$$

It is possible to apply standard asymptotic ML theory (Pawitan, 2001) for any inferential purpose. Specifically, the significance of the effect of covariates in the models may be effectively tested. Given the estimated log-likelihoods $\ell_{\mathcal{M}_1}, \ell_{\mathcal{M}_0}$ for two nested models $\mathcal{M}_0 \subset \mathcal{M}_1$, with parameter $g_1 > g_0$, respectively, we compare differences in deviances obtained by different models as follows:

$$\text{Test}(\mathcal{M}_1 \text{ versus } \mathcal{M}_0) = 2 \left(\ell_{\mathcal{M}_1} - \ell_{\mathcal{M}_0} \right).$$

---

[5] We know from extensive experimentation that the logistic function is adequate; in some circumstances, log-log-complementary or probit distribution functions may be of interest as well as any one-to-one function mapping real numbers into the unit interval.

To assess its asymptotic significance, this statistic should be compared with percentiles of a χ^2 random variable with $g = g_1 - g_0$ degrees of freedom.

13.6   Interpreting the role of covariates

Since covariates are directly related to parameters, there are many ways to interpret their effects (when they are statistically significant) by making reference to unobserved components (as specified by uncertainty and feeling parameters) or by studying the probabilities of ordered choices.

For simplicity, we will discuss a CUB model with q covariates for *feeling*, so that

$$\xi_i = \frac{1}{1 + e^{-(\gamma_0 + \gamma_1 w_{1i} + \ldots + \gamma_q w_{1q})}}, \qquad i = 1, 2, \ldots, n.$$

A plot of $1 - \xi$ against the range of w_k may be considered in order to take the effect of the covariate w_k on the feeling into account. This function should be computed by fixing the other covariates at pertinent values (the averages, for instance). More specifically, for a positive increasing $w_k, i = 1, 2, \ldots, q$, we see that feeling decreases (increases) if $\gamma_k > 0$ ($\gamma_k < 0$).

A plot of probabilities $Pr(R = r \mid \mathbf{w}_i)$ for varying covariates may be of interest.[6] Sometimes researchers are interested in specific probabilities of selecting a given item (the best or worst, for instance).

All these considerations may be adapted straightforwardly to interpret the effect of covariates on the *uncertainty*.

The role of covariates in the CUB modelling framework may be usefully exploited in presence of clusters. Indeed, if subgroups behave in significantly different ways, observed distributions manifest different modal preferences (if *feeling* is different) or mixed heterogeneity (if *uncertainty* is different). In these circumstances, a CUB model with covariates induces different modes and heterogeneity. Such dichotomous or polytomous aspects are effectively estimated by data, without imposing further components on the mixture to improve the fitting. This result increases parsimony and interpretation since it allows to test statistical evidence for possible causes of selective behaviour among respondents.[7]

Finally, the introduction of covariates enables the prediction of profiles of consumers for given subjects' covariates. In this case, we get conditioned probability distributions so that the marketing researcher may assess standard indices (such as modal values and averages) or choose intervals of values for preassigned probabilities. Moreover, it is possible to investigate for which values of covariates some probabilities of interest are optimized in order to assess marketing strategies.

---

[6] Sometimes the concept of *quasi-elasticity* (as defined by Franses and Paap 2001, among others) is considered since it measures the percentage point change in the probability of a given category owing to a percentage increase in w_k. Given the explicit formulation of components via parameters, we prefer to rely on them for more immediate interpretations.

[7] It is possible to generalize this framework by introducing hierarchical/multilevel CUB models and interpreting the probability structure at different levels of the hierarchy (see Section 13.9).

13.7   A more general sampling framework

Assume that n subjects (judges, raters, consumers, customers, students, ...) express ratings r_{ij} pertaining to a collection of J items (objects, services, brands, questions, ...)

$$\mathcal{I}_1, \; \mathcal{I}_2, \ldots, \mathcal{I}_j \ldots, \mathcal{I}_J.$$

Such data are generally gathered together in the $n \times J$ matrix $\mathbf{R}$ of the given ratings r_{ij}, $i = 1, 2, \ldots, n$; $j = 1, 2, \ldots, J$, expressed by the n raters (the rows) with respect to the J objects (the columns):

$$\mathbf{R} = (\mathbf{r}_1, \mathbf{r}_2, \ldots \mathbf{r}_j \ldots \mathbf{r}_J) = \begin{pmatrix} r_{11} & r_{12} & \cdots & r_{1j} & \cdots & r_{1J} \\ r_{21} & r_{22} & \cdots & r_{2j} & \cdots & r_{2J} \\ \cdots & \cdots & \cdots & \cdots & \cdots & \cdots \\ r_{n1} & r_{n2} & \cdots & r_{nj} & \cdots & r_{nJ} \end{pmatrix}.$$

The matrix $\mathbf{R} = \|r_{ij}\|$ is an *observed sample* of size n, a realization of a *multivariate random variable* $(R_1, R_2, \ldots, R_J)$. The collection of ratings assigned by the n raters to a prefixed item $\mathcal{I}_j$, for some $j = 1, 2, \ldots, J$, is the *column* $\mathbf{r}_j = (r_{1j}, r_{2j}, \ldots, r_{nj})'$. Thus, when we study a single item, we *only* analyse the *marginal distribution* of this multivariate random variable.

The standard assumption of independence among subjects (*row independence*) may be maintained if we choose our sample according to randomized experimental designs. Sometimes, it is more correct to assume some relationships within clusters of respondents characterized by similar subjective and/or objective characteristics.

In contrast, the assumption of independence among ratings R_j (= *column independence*) is not realistic since surveys are generally organized with reference to related questions or in order to compare similar items. Thus, both logical arguments and empirical evidence should reject independence assumptions among rating variables.

For several items, we will extend the standard CUB models in order to take such real situations into account within the same inferential paradigm by implementing the EM algorithm for ML estimation of parameter vectors.

A starting point for this extension is the consideration that a rating is determined by both objective and subjective covariates which modify the location and variability of expressed ratings. In fact, in sensometric analysis customer satisfaction is analysed by considering sensory preferences with regard to several foods with comparable attributes. Sometimes, variables related to subjects and objects are defined as *choices' covariates* and *choosers' covariates*, respectively (Agresti, 2002).

Thus, a multi-object analysis is especially useful for studying how ratings are modified by both subjects' and items' covariates. The motivation for including this real situation in the multi-objects paradigm is that *the same covariates are measured on all subjects*. Finally, we will consider also *contextual covariates* which play an important role in many circumstances. The idea behind the formal definition of this kind of item parameter is related to space and time. The probability distribution and inferential attitudes are invariant with respect to the objects' covariate statistical model.

13.7.1   Objects' covariates

Objective information relates to covariates which are functions of items/objects and which specify both quantitative and qualitative aspects (related to content, appearance, structure, texture, ingredients, etc.). *Objects' covariates* are denoted by **x**. Specifically, for each object $j = 1, 2, \ldots, J$, we denote the H covariates characterizing the jth object as $\mathbf{x}_j = (x_{j1}, x_{j2}, \ldots, x_{jH})$. Thus, we are measuring H variables on J objects. Such information is summarized in the $J \times H$ matrix **X**, whose generic elements are

$$\|\mathbf{X}\| = \{x_{jh}\} \quad j = 1, 2, \ldots, J; \ h = 1, 2, \ldots, H.$$

We maintain the previous paradigm with regard to subjects' covariates (Piccolo and D'Elia, 2008) and introduce objects' covariates by means of a joint modelling approach that includes all the objects to be considered. The entire information set for explaining the rating r_{ij} of the ith subject on the jth item becomes

$$(r_{ij} \mid 1, y_{i1}, y_{i2}, \ldots, y_{ip} \mid 1, w_{i1}, w_{i2}, \ldots, w_{iq} \mid x_{j1}, x_{j2}, \ldots, x_{jH}).$$

and such information is related to model parameters by the *systematic* links

$$\begin{cases} \pi_{ij} & = (\pi \mid \mathbf{y}_i, \mathbf{x}_j) = \frac{1}{1+\exp(-\mathbf{y}_i \, \boldsymbol{\beta} - \mathbf{x}_j \, \boldsymbol{\nu})}, \\ \xi_{ij} & = (\xi \mid \mathbf{w}_i, \mathbf{x}_j) = \frac{1}{1+\exp(-\mathbf{w}_i \, \boldsymbol{\gamma} - \mathbf{x}_j \, \boldsymbol{\eta})}, \end{cases} \quad i = 1, 2, \ldots, n; \ j = 1, 2, \ldots, H,$$

where $\boldsymbol{\nu} = (\nu_1, \nu_2, \ldots, \nu_H)'$ and $\boldsymbol{\eta} = (\eta_1, \eta_2, \ldots, \eta_H)'$ are further parameters to be estimated.[8] Here, π_{ij} (ξ_{ij}) is related to *uncertainty* (*feeling*) expressed by the ith subject, whose relevant characteristics are specified by $\mathbf{y}_i$ ($\mathbf{w}_i$) when he/she is asked to rate the jth object, whose characteristics are specified in turn by $\mathbf{x}_j$.

In this extended framework, the probability of a rating r_{ij} as a function of the entire information set is expressed by

$$Pr\left(R_j = r_{ij} \mid \mathbf{y}_i, \mathbf{w}_i, \mathbf{x}_j\right) = \frac{1}{1 + e^{-\mathbf{y}_i \boldsymbol{\beta} - \mathbf{x}_j \boldsymbol{\nu}}} \left[\binom{m-1}{r_{ij}-1} \frac{\left(e^{-\mathbf{w}_i \boldsymbol{\gamma} - \mathbf{x}_j \boldsymbol{\eta}}\right)^{r_{ij}-1}}{\left(1 + e^{-\mathbf{w}_i \boldsymbol{\gamma} - \mathbf{x}_j \boldsymbol{\eta}}\right)^{m-1}} - \frac{1}{m} \right] + \frac{1}{m}.$$

If we write as $\boldsymbol{\theta} = (\boldsymbol{\beta}, \boldsymbol{\gamma}, \boldsymbol{\nu}, \boldsymbol{\eta})'$ the parameter vector to be estimated, given the sample information $\mathcal{S}_n = (\mathbf{r} \mid \tilde{\mathbf{Y}} \mid \tilde{\mathbf{W}} \mid \tilde{\mathbf{X}})$, the log-likelihood function is

$$\ell(\boldsymbol{\theta} \mid \mathcal{S}_n) = \sum_{j=1}^{H} \sum_{i=1}^{n} \log \left[Pr\left(R_j = r_{ij} \mid \mathbf{y}_i, \mathbf{w}_i, \mathbf{x}_j\right) \right].$$

This likelihood function cannot be factorized with respect to subjects' and objects' parameters. Thus, we are looking for a joint maximization over the parameter space, and the proposed framework implies a genuine multivariate approach to multi-objects modelling.

From a computational point of view, we organize information in the following format:

$$\begin{pmatrix} \mathbf{r}_1 \\ \mathbf{r}_2 \\ \cdots \cdots \\ \mathbf{r}_J \end{pmatrix} \middle| \begin{pmatrix} \mathbf{Y} \mid \mathbf{W} \\ \mathbf{Y} \mid \mathbf{W} \\ \cdots \cdots \\ \mathbf{Y} \mid \mathbf{W} \end{pmatrix} \middle| \begin{pmatrix} \mathbf{x}_1 \otimes \mathbf{1}_n \\ \mathbf{x}_2 \otimes \mathbf{1}_n \\ \cdots \cdots \\ \mathbf{x}_J \otimes \mathbf{1}_n \end{pmatrix}$$

where $\mathbf{1}_n$ is a unit vector of length n and $\otimes$ denotes the Kronecker product.

---

[8] For ease of notation, we assume that the set of objects' covariates $\mathbf{x}_j$ for explaining both parameters is the same.

Each row of this system shows the entire information for studying the rating r_{ij} expressed by the ith subject on the jth item as a joint function of: *subject's covariates* $\mathbf{y}_i = (1, y_{i1}, y_{i2}, \ldots, y_{ip})$ and $\mathbf{w}_i = (1, w_{i1}, w_{i2}, \ldots, w_{iq})$, for $i = 1, 2, \ldots, n$, and *object's covariates* $\mathbf{x}_j = (x_{j1}, x_{j2}, \ldots, x_{jH})$, for $j = 1, 2, \ldots, J$.

We apply the implemented EM algorithms for ML estimation of parameter vectors in this extended framework. However, from a computational point of view, the multi-objects approach is more demanding since the dimension of the response vector increases from n to nJ, and the size of the matrix of data connecting subjects' and objects' covariates increases from $n \times (p + q + 2)$ to $n J \times (p + q + 2 + H)$. Of course, we can exploit the plotting feature of estimated CUB models as points in the parameter space $\Omega(\boldsymbol{\theta})$ and study how these points are modified both on the basis of subjects' and objects' characteristics.

13.7.2 Contextual covariates

As already mentioned, *contextual* information is concerned with covariates which are related to external circumstances and which specify the conditions that accompany or influence the choice. These facts may be related to time, space, environment and any characteristic which surrounds the experiment. Typical instances of this framework are as follows:

- we collect evaluations on a specific service in different contexts (units, departments, schools, faculties, public and private institutions, . . .);

- all respondents are requested to fill in forms with the same set of covariates (gender, age, job, . . .);

- characteristics of contexts are well known to researchers (location, service hours, sizes, number of employers, period of submission of survey, . . .).

Contextual covariates are denoted by $\mathbf{z}$. Specifically, for each circumstance $t = 1, 2, \ldots, T$, we denote the K covariates related to the context by $\mathbf{z}_t = (z_{t1}, z_{t2}, \ldots, z_{tK})$. Finally, with regard to context,[9] we are measuring K variables on T circumstances. Such information is summarized in the $T \times K$ matrix $\mathbf{Z}$, whose generic elements are

$$\|\mathbf{Z}\| = \{Z_{tk}\} \quad t = 1, 2, \ldots, T; \, k = 1, 2, \ldots, K.$$

Formally, $n_1, n_2, \ldots, n_T$ subjects are sampled in relation to T contexts and these circumstances are characterized by K variables; in addition, the same p and q subjective covariates are measured/collected on the subjects. For convenience, let $n = n_1 + n_2 + \ldots + n_T$. We define:

- *ratings* $\mathbf{r}^{(t)} = (r_1^{(t)}, r_2^{(t)}, \ldots, r_{n_t}^{(t)})'$, for $t = 1, 2, \ldots, T$;

- *subjects' covariates* $\mathbf{y}_i^{(t)} = (1, y_{i1}^{(t)}, y_{i2}^{(t)}, \ldots, y_{ip}^{(t)})$ and $\mathbf{w}_i^{(t)} = (1, w_{i1}^{(t)}, w_{i2}^{(t)}, \ldots, w_{iq}^{(t)})$, for $i = 1, 2, \ldots, n_t$;

- *contextual covariates* $\mathbf{z}_t = (z_{t1}, z_{t2}, \ldots, z_{tK})$ for $t = 1, 2, \ldots, T$ – these rows are replicated n_t times.

---

[9] We assume that both *objective* and *contextual* covariates are the same for latent components (*uncertainty* and *feeling*) of the choice. However, they may be different if they concern subjects' characteristics. This simplification is not essential from a formal point of view, but we will maintain it throughout.

The sample information set is given by $S_n^{(T)} = (\mathbf{r} \mid \tilde{\mathbf{Y}} \mid \tilde{\mathbf{W}} \mid \tilde{\mathbf{Z}})$ or, more explicitly, by

$$
\begin{pmatrix} \mathbf{r}^{(1)} \\ \mathbf{r}^{(2)} \\ \cdots \\ \mathbf{r}^{(T)} \end{pmatrix} \left| \begin{array}{l} \begin{pmatrix} \mathbf{Y}^{(1)} \mid \mathbf{W}^{(1)} \\ \mathbf{Y}^{(2)} \mid \mathbf{W}^{(2)} \\ \cdots\cdots\cdots \\ \mathbf{Y}^{(T)} \mid \mathbf{W}^{(T)} \end{pmatrix} \end{array} \right| \begin{pmatrix} \mathbf{z}_1 \otimes \mathbf{1}_{n_1} \\ \mathbf{z}_2 \otimes \mathbf{1}_{n_2} \\ \cdots\cdots \\ \mathbf{z}_T \otimes \mathbf{1}_{n_T} \end{pmatrix} .
$$

Each row of the system is generated as

$$
S_n^{(T)} = (r_i^{(t)} \mid 1, y_{i1}^{(t)}, y_{i2}^{(t)}, \ldots, y_{ip}^{(t)} \mid 1, w_{i1}^{(t)}, w_{i2}^{(t)}, \ldots, w_{iq}^{(t)} \mid z_{t1}, z_{t2}, \ldots, z_{tK}),
$$

for $i = 1, 2, \ldots, n_t$ and $t = 1, 2, \ldots, T$.

In this framework, the probability of a rating $r_i^{(t)}$ as a function of the whole information set is expressed by

$$
Pr\left(R = r_i^{(t)} \mid \mathbf{y}_i^{(t)}, \mathbf{w}_i^{(t)}, \mathbf{z}_t\right)
$$

$$
= \frac{1}{1 + e^{-\mathbf{y}_i^{(t)}\boldsymbol{\beta} - \mathbf{z}_t\boldsymbol{\lambda}}} \left[\binom{m-1}{r_i^{(t)} - 1} \frac{\left(e^{-\mathbf{w}_i^{(t)}\boldsymbol{\gamma} - \mathbf{z}_t\boldsymbol{\delta}}\right)^{r_i^{(t)}-1}}{\left(1 + e^{-\mathbf{w}_i^{(t)}\boldsymbol{\gamma} - \mathbf{z}_t\boldsymbol{\delta}}\right)^{m-1}} - \frac{1}{m} \right] + \frac{1}{m}.
$$

Then, if we write as $\boldsymbol{\theta} = (\boldsymbol{\beta}, \boldsymbol{\gamma}, \boldsymbol{\lambda}, \boldsymbol{\delta})'$ the set of all parameters to be estimated, given the sample information $S_n^{(T)}$, the log-likelihood function is

$$
\ell(\boldsymbol{\theta} \mid S_n^{(T)}) = \sum_{t=1}^{T} \sum_{i=1}^{n_t} \log\left[Pr\left(R = r_i^{(t)} \mid \mathbf{y}_i^{(t)}, \mathbf{w}_i^{(t)}, \mathbf{z}_t\right) \right].
$$

In order to fit the whole information set in the framework of the standard CUB model, we organize it as follows:

- *ratings* in the $\mathbf{R}$ matrix should be vectorized by $\tilde{\rho} = \text{vec}(\mathbf{R})$;

- *subjective* covariates in the $\mathbf{Y}$ and $\mathbf{W}$ matrices should be expanded by $\tilde{\mathbf{Y}} = (\mathbf{1}_J \otimes \mathbf{Y})$ and $\tilde{\mathbf{W}} = (\mathbf{1}_J \otimes \mathbf{W})$;

- *objective* and *contextual* covariates in $\mathbf{X}$ and $\mathbf{Z}$ should be expanded in order to fit ratings dimensions, as $\tilde{\mathbf{X}} = (\mathbf{X} \otimes \mathbf{1}_n)$ and $\tilde{\mathbf{Z}} = (\mathbf{Z} \otimes \mathbf{1}_n)$.

We emphasize the difference between the sampling structure assumed in Sections 13.7.1 and 13.7.2, respectively. In fact, in the former subjects' data are constant (all respondents express satisfaction towards J objects) while in contextual analysis we have different groups of respondents who express satisfaction in different situations.

13.8   Applications of CUB models

In order to explain the ratings expressed by respondents in different contexts, we apply the previous approach to two case studies. Both are concerned with customer satisfaction services, the first relating to the private sector (the ABC annual customer satisfaction survey) and the second in a public institution (a university orientation service). In both cases, we first analyse

Table 13.1 CUB model estimates of customer satisfaction overall and with various specific items

Items	$\hat{\pi}$	$\hat{\xi}$	log-likelihood	Index $\mathcal{F}^2$
Overall	0.875 *(0.060)*	0.338 *(0.020)*	−314.18	0.925
Equipment	0.999 *(0.043)*	0.363 *(0.018)*	−283.55	0.820
Sales	0.640 *(0.091)*	0.389 *(0.028)*	−345.38	0.982
Technical	0.719 *(0.067)*	0.235 *(0.023)*	−321.12	0.919
Supplies	1.000 *(0.081)*	0.404 *(0.017)*	−272.38	0.832
Terms	1.000 *(0.082)*	0.490 *(0.017)*	−309.20	0.895

Standard errors of parameter estimates are in parentheses.

the model without covariates and then include subjective and contextual variables which improve the interpretation of the fitted models. Results show the determinants of customer satisfaction and how public and private sectors manifest similar behaviour by focusing on perceived service quality.

13.8.1 Models for the ABC annual customer satisfaction survey

CUB models for 'overall' satisfaction were introduced by Iannario and Piccolo (2010b) based on the 2010 survey, and the inclusion of the dichotomous covariate for question 3, 'Is ABC your best supplier?', demonstrated to be a relevant issue for the feeling of customers towards the company. In addition, preliminary results showed low uncertainty in the responses and the ability of these models to capture and measure the connection between customers' satisfaction and loyalty.

In this section, we refer to the ABC 2010 annual customer satisfaction survey whose objectives, questionnaire and exploratory results have been already analysed in Chapter 2. Specifically, we compare expressed *overall satisfaction* with the characteristics of $n = 227$ customers.[10]

In Table 13.1 we present the main results obtained by fitting univariate CUB models for customer satisfaction for the items 'overall', 'equipment', 'sales', 'technical', 'supplies' and 'terms'. Fitting measures were derived by using the previously defined index $\mathcal{F}^2$.

We observe that customers express a judgement with no uncertainty parameters with regard to 'equipment', 'supplies' and 'terms' and with limited uncertainty with regard to the other items.[11] As Figure 13.3 (left panel) confirms, satisfaction is higher for questions involving hard services (maximum feeling is for 'technical' and 'equipment' items) and this is reflected in a high 'overall' satisfaction. Thus, for this case study, customers are satisfied with products and mechanical components produced by ABC and to a lesser extent with services and related products.

Exploiting the flexibility of this framework, we might ask how much the global evaluation (expressed by *overall satisfaction*) is determined by single aspects by estimating a CUB model which includes partial satisfaction with items as covariates of the whole. Again, 'technical'

---

[10] Our analysis is limited to just six items, given that there were too many missing values observed in other items.

[11] We are aware of inferential issues related to parameters on or near the boundary of the parameter space, as stressed by Molenberghs and Verbeke (2007); however, the models we are discussing have been also estimated by shifted binomial models and results presented here have been confirmed.

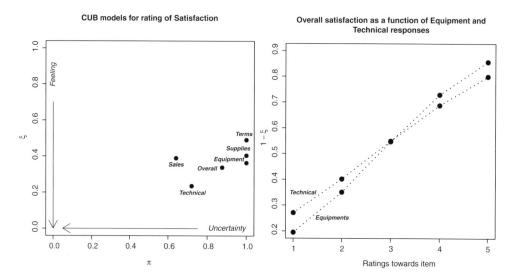

Figure 13.3   CUB models for single items and effects on the 'Overall' customer satisfaction

and 'equipment' items turn out to be the best explanations of *overall satisfaction*, and the estimated relationships between feeling/satisfaction and covariates are:

$$\begin{cases} \xi_i(\textit{Techn}) = \dfrac{1}{1 + \exp(-1.585 + 0.593\, \textit{Techn}_i)}, \\[2ex] \xi_i(\textit{Equip}) = \dfrac{1}{1 + \exp(-2.220 + 0.802\, \textit{Equip}_i)}, \end{cases} \qquad i = 1, 2, \ldots, n.$$

In Figure 13.3 (right panel) we show how estimated models interpret variation in feeling of *overall satisfaction* (measured by $1 - \xi$) as a function of satisfaction with specific items. The effects of both items on the response are quite similar, but 'equipment' seems more extreme in determining the score of *overall satisfaction* response.

Ratings of such items are strongly related but it is interesting to observe that their joint estimates significantly improve likelihood estimation. In fact, maximized log-likelihood functions for models with feeling covariates 'Technical', 'Equipment' and 'Technical and Equipment' are -275.76, -276.81, and -257.00, respectively. Thus, a better model (with no uncertainty, since $\hat{\pi} = 1$) implies that

$$\xi_i(\textit{Techn, Equip}) = \frac{1}{1 + \exp(-1.585 + 0.593\, \textit{Techn}_i + 0.802\, \textit{Equip}_i)}; \quad i = 1, 2, \ldots, n.$$

Figure 13.4 shows this joint effect on the *Overall satisfaction*.

13.8.2   Students' satisfaction with a university orientation service

An extensive survey of students' opinions has been carried out by University of Naples Federico II with reference to the orientation service which operates in the 13 faculties. Our study will focus on a data set gathered in 2008 by means of a questionnaire submitted to a sample of users. Each student was asked to score his/her satisfaction with various aspects of the

Overall satisfaction as a joint function of Equipment and Technical responses

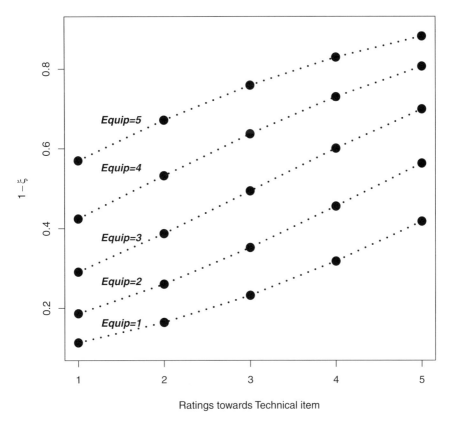

Figure 13.4   'Overall' satisfaction as a joint effect of 'Technical' and 'Equipment' responses

orientation service. Judgements were expressed using an ordinal scale with $m = 7$ ranging from 1 (= 'completely unsatisfied') to 7 (= 'completely satisfied'). The results refer to $n = 4042$ validated questionnaires, and previous investigations confirmed a substantial homogeneity of respondents satisfaction (more refined analyses would distinguish among some faculties). In this data set, the responses are characterized by uniformly low uncertainty.

We experienced that the satisfaction (*response*) expressed by several students (*subjects*) towards university orientation service may be related to the characteristics of different Faculties (*contexts*). In the following discussion, we confine ourselves to evaluations of *global satisfaction* with the service as a function of subjective (*Gender* and *Age*) and contextual covariates (proportion of long-term students,[12] = *LTstudents*).

Table 13.2 summarizes estimated models of increasing complexity obtained by adding subjects' covariates (*Gender* and *Age*) and then contextual covariates with respect to a standard CUB model. Parameters are stable in this stepwise procedure and regularly significant; moreover, likelihood ratio tests (to be checked against the χ^2 percentiles at 5%, listed in the

---

[12] We define as long-term students those enrolled for a number of years greater than the regular course.

Table 13.2   CUB model estimation with subjects' and contextual covariates

Covariates	$\hat{\pi}$	$\hat{\gamma}_0$ (or $\hat{\xi}$)	$\hat{\gamma}_1$	$\hat{\gamma}_2$	$\hat{\delta}$	$\ell(\hat{\boldsymbol{\theta}})$	$2\,(\ell(\hat{\boldsymbol{\theta}}) - \ell(\boldsymbol{\theta}_0))$
No covariates	0.971	0.226				−5845.7	—
Gender	0.972	−1.288	0.109			−5840.0	11.4
Age	0.970	−0.507		−0.033		−5821.5	48.4
Gender *Age*	0.971	−0.562	0.110	−0.033		−5815.8	59.8
LTstudents	0.972	0.112			−3.241	−5739.0	213.4
Gender *Age* *LTstudents*	0.972	0.567	0.087	−0.026	−3.075	−5720.3	250.8

$\chi^2_{(g=1)} = 3.841$; $\chi^2_{(g=2)} = 5.991$; $\chi^2_{(g=3)} = 7.815$

bottom of the table) confirm the importance of subjective and especially contextual covariates. Thus, satisfaction of students improves with age, although women significantly express more negative results than men; significantly, the proportion of *LTstudents* is an important variable for explaining satisfaction. The (convincing) conclusion is that long-term university students require more counselling support.

From a formal point of view, the estimated comprehensive CUB model expresses the probability of an ordinal score as a function of a constant *uncertainty* parameter $\hat{\pi} = 0.972$ and a feeling given by, for any $i = 1, 2, \ldots, n_t$,

$$\xi_i^{(t)} = \frac{1}{1 + e^{-0.0567 - 0.087\,Gender_i + 0.026\,Age_i + 3.075\,LTstudents^{(t)}}}, \qquad t = 1, 2, \ldots, 13.$$

It is immediate to ascertain that, *ceteris paribus*, the feeling increases with *Age* (ranging from 18 up to 65 years) and *LTstudents* (a proportion ranging from 0.20 up to 0.50) whereas it decreases for women.

In Figure 13.5, different representations are shown in order to provide different interpretations from the estimated models. Specifically, for a given *Age* = 30 and by varying the proportion of *LTstudents*, the satisfaction of students is compared with respect to *Gender* (upper plots). Then, bottom left plot shows the effect of this contextual proportion on the parameter space confirming that it acts on feeling (satisfaction increases when proportion of *LTstudents* increases). Finally, bottom right plot shows the variation of feeling parameter with respect to proportion of *LTstudents* for given *Gender* and some prefixed *Age*, as expressed by $\xi_i^{(t)}$.

13.9   Further generalizations

We list some lines of research leading to generalizations and useful variants of CUB models which are currently under investigation.

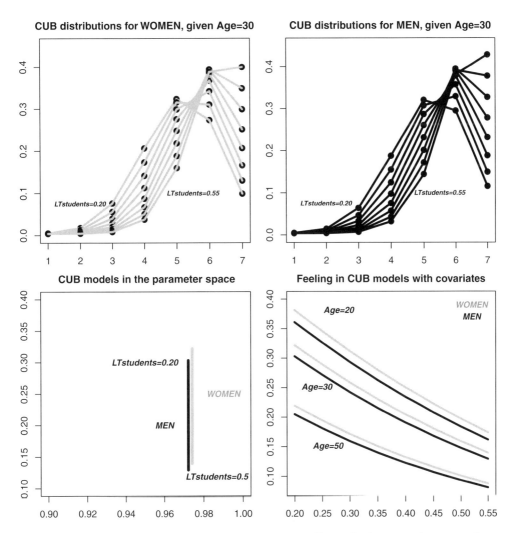

Figure 13.5   Representations for interpreting the joint effects of subjects and contextual co-variates

- *CUB models and shelter choice.* When for some reason (privacy, laziness, rounding effects, improper wording, and so on), customers select a specific category c with a frequency larger than that expected by a model, it is convenient to extend CUB models (by considering a dummy variable at $R = c$) to motivate such *shelter choice*. Thus, for a given $m > 4$ and known $c \in \{1, 2, \ldots, m\}$, this model is defined by

$$p_r(\boldsymbol{\theta}) = \pi_1 \binom{m-1}{r-1} \xi^{m-r}(1-\xi)^{r-1} + \pi_2 \frac{1}{m} + (1 - \pi_1 - \pi_2) D_r^{(c)},$$
$$r = 1, 2, \ldots, m,$$

where $\boldsymbol{\theta} = (\pi_1, \pi_2, \xi)'$ is the parameter vector which characterizes this distribution and $D_r^{(c)}$ is a degenerate random variable whose probability mass is concentrated at $r = c$.

Then the parameter space becomes

$$\Omega(\boldsymbol{\theta}) = \{(\pi_1, \pi_2, \xi) : \ \pi_1 > 0, \ \pi_2 \geq 0, \ \pi_1 + \pi_2 \leq 1, \ 0 \leq \xi \leq 1\}.$$

This extended CUB model collapses to the standard one if $\pi_1 + \pi_2 = 1$. In addition, it is able to account for the (rare) situation where most respondents' choices are concentrated at a single intermediate category (if $\pi_2 = 0$ and $\pi_1 \to 0$).

This extension significantly improves the fitting and helps in interpreting the weight of such choices, as discussed by Corduas *et al.* (2009); Iannario (2010c); Iannario and Piccolo (2010a). In addition, a CUB model with *shelter* solves a critical issue raised by the choice of a shifted binomial distribution which allows a collapsed mode only at $R = 1$ (when $\xi = 0$) or $R = 0$ (when $\xi = 1$).

- *Hierarchical CUB models.* In multilevel models, the behaviour of subgroups is more correctly defined by including random effects as intercept and/or slope. We define HCUB models where parameters to be estimated are considered as random variables. If we consider, for instance, the random effects for the feeling parameter with a single covariate w_i,

$$\xi_i = \frac{1}{1 + e^{-\Gamma_0 - \Gamma_1 w_i}} \qquad i = 1, 2, \ldots, n,$$

then we assume that $\boldsymbol{\Gamma} = (\Gamma_0, \Gamma_1)'$ is a bivariate Gaussian random variable whose marginal components $\Gamma_j \sim N(\gamma_j, \sigma_j^2)$, for $j = 0, 1$, and $\text{Cov}(\Gamma_0, \Gamma_1) = \sigma_{01}$. This model requires the estimation of the parameter vector $\boldsymbol{\theta} = (\gamma_0, \gamma_1, \sigma_0^2, \sigma_1^2, \sigma_{01})'$ to be correctly specified. As in the standard multilevel approach, HCUB models correct for the biases resulting from clustering both in the parameter estimates and in standard errors (DeLeeuw and Meijer, 2008; Gelman and Hill, 2007). Thus, the increased complexity is more than compensated by accuracy and efficiency. In this respect, further evidence is provided by Iannario (2010d).

- *Ranking analysis via CUB models.* Ranks are ordinal data in which we assume that a set of m issues is presented to subjects who are instructed to select and rank them. This task deserves a peculiar approach since ranks are generated by arranging objects/items in increasing or decreasing order. Thus, they associate a permutation of integers to the ordering expressed by each customer. Rank modelling issues and related problems are discussed by Fligner and Verducci (1988, 1993) and Marden (1995).

  For exploiting the CUB model paradigm, we consider this mental assessment as the product of a joint sequential choice of preferences expressed for a list of items. Then, it is convenient to introduce a multi-stage approach, in the vein of ranking models by Plackett (1975) and Xu (2000), with the constraint to maintain the CUB marginal univariate random variables of ranks in order to obtain immediate interpretation and effective parsimony.

- *Imputation of data in CUB models.* This topic includes treatment of missing values, computation of profiles and the measurement of predictive power of such models. Indeed, all these problems are related to the imputation of an ordinal response (absent, future, function of covariates) when such information is not available but a CUB model has been effectively estimated from sample data.

- *Model-based analysis and visualization of ordinal data.* This class of models has been effectively applied in multivariate analysis to perform a model-based classification (Corduas 2008a,b), and for covariate selection by using tree methods (Cappelli and D'Elia (2006). In addition, when marketing researchers collect thousands of ratings in a regularly manner (by panel interviews, email, internet, etc.), CUB models allow huge amounts of variable information to be synthesized into a few parameters and above all to be visualized in a sharp manner. In this respect, CUB models may be routinely estimated and shown on dynamic plots within the framework of data mining (Iannario and Piccolo (2011a) to assess variation in time, space and circumstances.

13.10   Concluding remarks

CUB models have been introduced as a probability tool for adequately fitting ordinal data. Such structure turns out to be more parsimonious and easier for interpreting survey information data with respect to classical approaches (as proportional odds models, for instance). However, previous considerations and reported empirical evidence should confirm that the approach deserves more general considerations both for univariate and multivariate rating analyses. A benefit of these models is the direct estimation and visualization of parameters related to unobserved components which drive consumer choices in expressing preferences; this exploratory resource is located within an identifiable probability model and asymptotic inference may be correctly pursued.

Finally, some refinements are perhaps necessary both in qualifying the role of uncertainty in specific contexts and also to improve numerical routines, especially in multivariate analyses. Moreover, measures of predictive ability are currently under scrutiny in order to assess the real usefulness of ordinal models in motivating effective solutions in marketing decisions.

Acknowledgements

This work has been partly supported by a MIUR grant (code 2008WKHJPK-PRIN2008) for the project 'Modelli per variabili latenti basati su dati ordinali: metodi statistici ed evidenze empiriche' (CUP number E61J10000020001) within the Research Unit of the University of Naples Federico II.

References

Agresti, A. (2002) *Categorical Data Analysis*, 2nd edn. New York: John Wiley & Sons, Inc.

Anderson, E.W. and Fornell, C. (2000) Foundations of the American Customer Satisfaction Index. *Total Quality Management*, 11, 869–882.

Arboretti Giancristofaro, R., Bonnini, S., Grossule, E., Ragazzi, S. and Salmaso, L. (2011) Statistical cognitive survey on Passito wine in Veneto region (Italy) from the consumer's point of view. *Book of Abstracts, Proceedings of 2010 GfKl-CLADAG Joint Meeting*, University of Florence, pp. 335–336.

Cappelli, C. and D'Elia, A. (2006) A tree-based method for variable selection in models for ordinal data. *Quaderni di Statistica*, 8, 125–135.

Corduas, M. (2008a) Clustering CUB models by Kullback-Liebler divergence. *Proceedings of 2008 SCF-CLAFAG Meeting*, pp. 245–248. Edizioni Scientifiche Italiane.

Corduas, M. (2008b) Statistical procedures for clustering ordinal data. *Quaderni di Statistica*, 10, 177–189.

Corduas, M., Iannario, M. and Piccolo, D. (2009) A class of statistical models for evaluating services and performances. In M. Bini *et al.* (eds), *Statistical Methods for the Evaluation of Educational Services and Quality of Products*, pp. 99–117. Berlin: Physica-Verlag.

Critchlow, D.E. 1985 *Metric Methods for Analyzing Partially Ranked Data*. Berlin: Springer-Verlag.

Critchlow, D.E., Fligner, M.A. and Verducci, J.S. (1991) Probability models on rankings. *Journal of Mathematical Psychology*, 35, 294–318.

Deldossi, L. and Paroli, R. (2011) Inference on the CUB model: A MCMC approach. *Book of Abstracts, Proceedings of 2010 GfKl-CLADAG Joint Meeting*, University of Florence, pp. 323–324.

De Leeuw, J .and Meijer, E. (eds) (2008) *Handbook of Multilevel Analysis*. New York: Springer.

D'Elia, A. (1999) A proposal for ranks statistical modelling. In H. Friedl, A. Berghold and G. Kauermann (eds.) *Statistical Modelling: Proceedings of the 14th International Workshop*, Graz, pp. 468–471.

D'Elia, A. (2000a) Il meccanismo dei confronti appaiati nella modellistica per graduatorie: sviluppi statistici ed aspetti critici. *Quaderni di Statistica*, 2, 173–203.

D'Elia, A. (2000b) A shifted binomial model for rankings. In V. Nunez-Anton and E. Ferreira (eds), *Statistical Modelling*, XV International Workshop on Statistical Modelling, Servicio Editorial de la Universidad del Pais Vasco, pp. 412–416.

D'Elia, A. and Piccolo, D. (2005) A mixture model for preference data analysis. *Computational Statistics & Data Analysis*, 49, 917–934.

Di Iorio, F and Piccolo, D. (2009) Generalized residuals in CUB models. *Quaderni di Statistica*, 11, 73–88.

Dobson, A.J. and Barnett, A.G. (2008) *An Introduction to Generalized Linear Models*, 3rd edn. Boca Raton, FL: Chapman & Hall/CRC.

Feigin, P.D. (1993) Modelling and analysing paired ranking data. In M.A. Fligner and J.S. Verducci (eds), *Probability Models and Statistical Analysis for Ranking Data*, Lecture Notes in Statistics 80, pp. 75–91. New York: Springer-Verlag.

Feigin, P.D. and Cohen, A. (1978) On a model of concordance between judges. *Journal of the Royal Statistical Society, Series B*, 40, 203–213.

Fligner, M.A. and Verducci, J.S. (1988) Multistage ranking models. *Journal of the American Statistical Association*, 83, 892–901.

Fligner, M.A. and Verducci, J.S. (eds) (1993) *Probability Models and Statistical Analysis for Ranking Data*, Lecture Notes in Statistics 80. New York: Springer-Verlag.

Fornell, C., Johnson, M.D., Anderson, E.W., Cha, J. and Bryant, B.E. (1996) The American Customer Satisfaction Index: Nature, purpose, and findings. *Journal of Marketing*, 60(4), 7–18.

Franses, P.H. and Paap, R. (2001) *Quantitative Models in Marketing Research*. Cambridge: Cambridge University Press.

Gelman, A. and Hill, J. (2007) *Data Analysis Using Regression and Multilevel/Hierarchical Models*. Cambridge: Cambridge University Press.

Gouriéroux, C., Monfort, A., Renault, E. and Trognon, A. (1987) Generalized residuals. *Journal of Econometrics*, 34, 5–32.

Iannario, M. (2008a) A class of models for ordinal variables with covariates effects. *Quaderni di Statistica*, 10, 53–72.

Iannario, M. (2009a) Fitting measures for ordinal data models. *Quaderni di Statistica*, 11, 39–72.

Iannario, M. (2009b) Selecting feeling covariates in rating surveys. *Italian Journal of Applied Statistics*, 20, 121–134.

Iannario, M. (2010a) On the identifiability of a mixture model for ordinal data. *Metron*, LXVIII, 87–94.

Iannario, M. (2010b) Preliminary estimators for a mixture model of ordinal data. *Submitted*.

Iannario, M. (2010c) Modelling shelter choices in a class of mixture models for ordinal responses. *Submitted*.

Iannario, M. (2010d) Hierarchical CUB models for ordinal variables. *Submitted*.

Iannario, M. and Piccolo, D. (2010a) Statistical modelling of subjective survival probabilities. *GENUS*, LXVI, 17–42.

Iannario, M. and Piccolo, D. (2010b) A new statistical model for the analysis of customer satisfaction. *Quality Technology & Quantitative Management*, 7, 149–168.

Iannario, M. and Piccolo, D. (2011a) A model-based approach for qualitative assessment in opinion mining. *Book of Abstracts, Proceedings of 2010 GfKl-CLADAG Joint Meeting*, University of Florence, pp. 57–58.

Iannario, M. and Piccolo, D. (2011b) University teaching and students' perception: Models of the evaluation process. In M. Attanasio and V. Capursi (eds), *Statistical Methods for the Evaluation of University Systems*, pp. 93–112. Berlin: Physica-Verlag.

Iannario, M. and Piccolo, D. (2011c) A program in R for CUB models inference, Version 3.0, available at *http://www.dipstat.unina.it/cub/cubmodels1.htm/*

King, G., Tomz, M. and Wittenberg, J. (2000) Making the most of statistical analyses: improving interpretation and presentation. *American Journal of Political Science*, 44, 341–355.

Louviere, J.J., Hensher, D.A. and Swait, J.D. (2000) *Stated Choice Methods. Analysis and Applications*, Cambridge University Press.

Marden, J.I. (1995) *Analyzing and Modeling Rank Data*. Chapman and Hall.

McCullagh, P. (1980) Regression models for ordinal data (with discussion). *Journal of the Royal Statistical Society, Series B*, 42, 109–142.

McCullagh, P. and Nelder, J.A. (1989) *Generalized linear models*. 2nd edn. Chapman and Hall.

McLachlan, G. and Krishnan, G.J. (2008) *The EM Algorithm and Extensions*. 2nd edn. J. Wiley & Sons

Molenberghs, G. and Verbeke, G. (2007) Likelihood ratio, score, and Wald tests in a constrained parameter space. *The American Statistician*, 61, 22–27.

Nelder, J.A. and Wedderburn, R.W.M. (1972) Generalized linear models, *Journal of the Royal Statistical Society, Series A*, 135, 370–384.

Pawitan, Y. (2001) *In All Likelihood: Statistical Modelling and Inference Using Likelihood*. Oxford: Oxford University Press.

Piccolo, D. (2003) On the moments of a mixture of uniform and shifted binomial random variables. *Quaderni di Statistica*, 5, 85–104.

Piccolo, D. (2006) Observed information matrix for MUB models. *Quaderni di Statistica*, 8, 33–78.

Piccolo, D. and D'Elia, A. (2008) A new approach for modelling consumers' preferences. *Food Quality and Preference*, 19, 247–259.

Plackett, R.L. (1975) The analysis of permutations. *Applied Statistics*, 24, 193–202.

Train, K.E. (2003) *Discrete Choice Methods with Simulation*. Cambridge: Cambridge University Press.

Tubbs, J.D. (1993) *Assessing changes in repeated ranked lists.* Department of Mathematical Science, University of Arkansas, USA.

van Blokland-Vogelsang, R. (1993) A non parametric distance model for unidimensional unfolding. In M.A. Fligner and J.S. Verducci (eds), *Probability Models and Statistical Analysis for Ranking Data*, Lecture Notes in Statistics 80, pp. 241–276. New York: Springer-Verlag.

Xu, L. 2000 A multistage ranking model. *Psychometrika*, 65, 217–231.

Appendix

A program in R for CUB models

This Appendix documents the essential steps for performing statistical inference by using an R program designed for the estimation and testing of CUB models. Several functions are available for further refinements and simulation analyses. The current version significantly improves initial value estimation, fitting measures, plotting facilities for several items and also deals with estimation within the multi-object framework.

A.1   Main structure of the program

The main structure of the program was first implemented around 2003 in the GAUSS language. Here, we present version 3.0 (released in summer 2011) available at *www.dipstat.unina.it/cub/cubmodels1.htm*. The website provides information about previous and current versions, data sets and a paper (Iannario and Piccolo, 2011c) which explains the main steps of the program, with some examples. In addition, all the published papers and/or reports on CUB models are mentioned.

We assume that a vector of ordinal data or a data matrix of several ordinal data related to a set of items is available (without missing values) in a text (ASCII) file or a file readable by R. In the following, we denote by `ordinal` a single vector and `dati` a matrix of ordinal data.

```
> dati=read.table(''filename.txt'',header=T)
> ordinal=dati[,j]  # if ordinal data are in the jth column
```

The program runs a function `CUB(.)` which in turn is related to several other modular functions used for more specific analysis. The main command is:

```
> source(''CUB.R'')
> m=number_of_categories
```

Explicit definition of m is fundamental for a correct analysis since an automatic definition of m (as the maximum integer of the vector `ordinal`) would fail if no respondents chose the highest value of the support.

A.2   Inference on CUB models

The general instruction to perform statistical inference on a general CUB model is:

```
> CUB(ordinal,Y=paicov, W=csicov, shelter=c)
```

where we denote by `Y`, `W` the matrices of data whose columns contain subjects' covariates for uncertainty and feeling, respectively. Since default values are set to `Y=0, W=0, shelter=0`, it is quite simple to run specific CUB models, as exemplified in Table A.1.

Sometimes responses are available as aggregated frequencies $(n_1, n_2, \ldots, n_m)'$ in a vector `frequencies`. Then, we expand them to generate a vector of length $n = n_1 + n_2 + \ldots + n_m$ by means of:

```
> ordinal=rep(1:m,frequencies)
> CUB(ordinal)
```

Table A.1   Instructions for CUB models estimation

Model	Covariates	Instruction
CUB $(0, 0)$	====	`CUB(ordinal)`
CUB $(p, 0)$	`paicov`	`CUB(ordinal,Y=paicov)`
CUB $(0, q)$	`csicov`	`CUB(ordinal,W=csicov)`
CUB (p, q)	`paicov,csicov`	`CUB(ordinal,Y=paicov,W=csicov)`
CUB +*shelter*	====	`CUB(ordinal,shelter=c)`

Notice that `frequencies` must be a vector of length m even if some observed frequencies are 0.

A.3   Output of CUB models estimation program

The program output varies as a function of the presence/absence of covariates and of their nature (dichotomous or not). Specifically, in addition to standard statistical output (estimates, standard errors, p-values, fitting measures), the program offers some graphical displays according to the following criteria:

- $CUB(0, 0)$ model: a plot of the observed frequency distribution for ordinal categories and related estimated probabilities is shown.

- $CUB(1, 0)$ model: *if* a single covariate for π is *dichotomous* and *if it is coded* as 0, 1, plots of estimated CUB distributions for the two subgroups are automatically generated.

- $CUB(0, 1)$ model: *if* a single covariate for ξ is *dichotomous* and *if it is coded* as 0, 1, plots of estimated CUB distributions for the two subgroups are automatically generated.

- $CUB(p, q)$ model: currently, no plot is automatically generated. Users are encouraged to study expected responses as functions of covariates (if continuous) or conditioned responses (if dichotomous or polytomous).

- $CUB + shelter$ model: a plot of the observed frequency distribution for ordinal categories and related estimated probabilities is shown.

The program allows estimated or computed quantities to be saved for further use (comparison, test, plots, and so on). The available variables are listed in Table A.2 (the symbols have obvious meanings).

Table A.2   List of available variables after running CUB models

Model	Variables
$CUB(0, 0)$	`pai,csi,varmat,loglik,effe2,n,AICCUB00,BICCUB00`
$CUB(p, 0)$	`bet,csi,varmat,loglik,effe2,n,AICCUBp0,BICCUBp0`
$CUB(0, q)$	`pai,gama,varmat,loglik,effe2,n,AICCUB0q,BICCUB0q`
$CUB(p, q)$	`bet,gama,varmat,loglik,effe2,n,AICCUBpq,BICCUBpq`
$CUB + shelter$	`pai1,pai2,csi,varmat,loglik,effe2`

A.4   Visualization of several CUB models in the parameter space

As an example of the facilities offered by the program, we list the commands[1] used to obtain Figure 13.3 (left panel) and assume ordinal data are available in columns $j = 10, \ldots, 15$ of the matrix `dati`.

```
dati=read.table("...\\Wiley2011.txt",header=T)
source("CUB.R"); m=5;
   ### sink("output.out")
nomi=c("Overall","Equipment","Sales","Technical","Supplies","Terms")
nk=length(nomi)
vettpai=rep(NA,nk); vettcsi=rep(NA,nk);  ### vmat=vector("list",nk)
for(j in 1:nk){
CUB(dati[,j+1])
vettpai[j]=pai; vettcsi[j]=csi;          ### vmat[[j]]=varmat
}
   ### sink()
vettpos=c(2,2,1,1,3,3)
plot(vettpai,vettcsi,xlim=c(0.,1),ylim=c(0.,1),
       xlab=expression(pi),ylab=expression(xi),,pch=19,lwd=5,
       main="CUB models for rating of Satisfaction")
text(vettpai,vettcsi,labels=nomi,pos=vettpos, offset=0.5,
       cex=0.8,font=4)
arrows(0.05,0.0,0.75,0.0,code=1,lwd=1.5)
text(0.9,0.0,"Uncertainty",cex=1.25,font=3)
arrows(0.0,0.0,0.0,0.7,code=1,lwd=1.5)
text(0.0,0.85,"Feeling",srt=90,cex=1.25,font=3)
```

If we uncomment the code related to `vmat` in the previous commands, it is possible to draw asymptotic confidence ellipses around estimates plotted in the parameter space by means of:

```
> library(ellipse)
> for(jj in 1:nk){
    lines(ellipse(vmat[[jj]],centre=c(vettpai[jj],vettcsi[jj])),
    lwd=2)
  }
```

A.5   Inference on CUB models in a multi-object framework

Assume that ratings are in the vectors **item1**, **item2**, ... and are vectorized into **ITEM**. In addition, subjects' covariates for uncertainty and feeling and objects' covariates are expanded into **Ytilde**, **Wtilde**, **Xtilde**, respectively. Then, the following R command:

```
> CUB(ITEM,Y=cbind(Ytilde,Xtilde),W=cbind(Wtilde,Xtilde))
```

will generate ML estimates of θ parameters and related statistics.

---

[1] If lines containing `sink` command are uncommented, the output of all estimated CUB model will be directed to a specified file.

It is important to notice that such command are similar to the previous ones but for the composition of the matrices involved.

A.6   Advanced software support for CUB models

The current version 3.0 of the program includes several functions that may be used for specific computations (expectation, variance, log-likelihood functions, simulation of CUB random variables, and so on). For methodological studies, we also present marginal likelihood computation and plots. In addition, given the parameters of a CUB model, an effective function for simulating a sample of data has been included in the program.

14

The Rasch model

Francesca De Battisti, Giovanna Nicolini and Silvia Salini

This chapter introduces the Rasch model and its use in the context of customer satisfaction surveys. The Rasch analysis provides effective measures of customer satisfaction with specific items related to product or service quality. In general, it is more difficult to measure satisfaction and quality levels of a service than it is for a product, since a product's quality is directly observable and may be different from the level of satisfaction it provides. When dealing with a service, the perceived quality and level of customer satisfaction are the results of a complex process. The Rasch analysis supplies two sets of coefficients which provide a simultaneous evaluation of subjective features related to the degree of satisfaction and objective features related to intrinsic quality. Instead of deriving a measurement of these two aspects, the Rasch model assigns a score to each individual and to each questionnaire item along a continuum. These scores make it possible to carry out an effective analysis of the survey responses. The model is introduced from a theoretical perspective, listing relevant literature and websites. A discussion of Rasch models in practice is given, along with a detailed application to the ABC annual customer satisfaction survey. Software packages implementing the Rasch model are described.

14.1 An overview of the Rasch model

14.1.1 The origins and the properties of the model

The Rasch model (RM) was first proposed in the 1960s to evaluate ability tests (Rasch, 1960). These tests are based on a set of items, and the assessment of a subject's ability depends on two factors: his relative *ability* and the item's intrinsic *difficulty*. Later the RM was used to evaluate behaviours or attitudes. In this context, the two factors become the subject's *characteristic* and the item's *intensity*, respectively. In recent years, the model has been employed in the evaluation of service quality (De Battisti *et al.*, 2005). Here the two factors become: the subject's (i.e. the customer's) *satisfaction* and the item's *quality*. In the RM,

Modern Analysis of Customer Surveys: with applications using R, First Edition. Edited by Ron S. Kenett and Silvia Salini.
© 2012 John Wiley & Sons, Ltd. Published 2012 by John Wiley & Sons, Ltd.

these two factors are measured by the parameters θ_i, referring to subject i, and β_j, referring to item j. It is then possible to compare these parameters because they belong to the same continuum. Their interaction is expressed by the difference $\theta_i - \beta_j$. In a deterministic sense, a positive difference means that the subject's abilities are superior to the item's difficulty. From a probabilistic perspective, this is not necessarily observed since a subject who is intrinsically capable of giving a correct response ($\theta_i > \beta_j$) may instead, in negative circumstances, give an incorrect response. Likewise, it is possible that a subject lacking in ability can accidentally give a correct response. The difference $\theta_i - \beta_j$, in the RM, determines the probability of a response. In particular, in the dichotomous case, the probability of a correct response $x_{ij} = 1$ by subject i, of ability θ_i, to item j, of difficulty β_j, is

$$P\left\{x_{ij} = 1 | \theta_i, \beta_j\right\} = \frac{\exp\left(\theta_i - \beta_j\right)}{1 + \exp\left(\theta_i - \beta_j\right)} = p_{ij}. \tag{14.1}$$

The RM framework is not the one typically used in statistical modelling for describing data. When the RM is applied in ability testing, the aim is to obtain data that fits the model (Andrich, 2004). This has some similarity to statistical process control, where a process out of control is reset with data fitting a stable process (see Chapter 20).

In the dichotomous model, data is gathered into a *raw score matrix*, with n rows (one for each subject) and J columns (one for each item); each cell of the matrix contains a 0 or a 1. The sum of each row, $r_i = \sum_{j=1}^{J} x_{ij}$, is the total score of subject i for all the items, while the sum of each column, $s_j = \sum_{i=1}^{n} x_{ij}$, represents the score given by all subjects to item j. These scores are given according to a metric that, being nonlinear, produces some conceptual distortion when comparing the row and column totals. In this case, it is necessary to replace these scores according to a metric based on the conceptual distances between subjects and items (Wright and Masters, 1982). The transformation is based on a logit:

$$\log \frac{p_{ij}}{1 - p_{ij}}. \tag{14.2}$$

By substituting equation (14.1), with the numerator and the denominator of (14.2), respectively, it is possible to define the parameters θ_i and β_j in the same measurement unit as an interval scale. Consequently, even the difference $\theta_i - \beta_j$ is graduated according to the same measurement unit.

The RM possesses some important properties. The first is that the items measure only one latent feature (*one-dimensionality*). This constitutes a limitation in applications to customer satisfaction surveys where there are usually several independent dimensions. Another important characteristic of the RM is that the responses to an item are independent of responses to other items (*local independence*). In the customer satisfaction survey context, this is an advantage. For parameters where no assumptions are made, by applying the logit transformation (14.2), θ_i and β_j can be expressed by a common measurement unit on the same continuum (*linearity of parameters*); the estimates of θ_i and β_j are respectively test and sample free (*separability of parameters*); and the row and column totals on the raw score matrix are sufficient statistics for the estimation of θ_i and β_j (*sufficient statistics*). For more on these properties, see Andrich (1988).

There are three main approaches to estimating a RM: *joint maximum likelihood* (JML), *conditional maximum likelihood* (CML) and *marginal maximum likelihood* (MML). In JML and CML subject-specific parameters are considered fixed effects, whereas in MML they are assumed to be random and independent variables drawn from a density distribution that describes the population. For an overview of parameter estimation techniques in the logistic models with one, two and three parameters, see Baker (1987). For the examination of theoretical features linked to the existence and uniqueness of the maximum likelihood estimates for the Rasch model, see Bertoli-Barsotti (2003, 2005).

The problem of evaluating the position of subjects/items with only correct or incorrect answers is solved in the MML case for the item parameters see Wright and Stone (1979) and in the Warm (1989) approach for the subject parameters.

The Rasch dichotomous model has been extended to the case of more than two ordered categories. The innovation of this approach lies in the assumption that, between one category and the next, there is a threshold that qualifies the item's position and determines the β_j as a function of the difficulty presented by every response category. A threshold is a point at which two adjacent categories have the same probability of being chosen. So, for example, the probability of choosing the first category is the probability of not exceeding the first threshold. Thus, the response to every threshold h of item j depends on the value $\beta_j + \tau_h$, where the second term represents the hth threshold of item j. The thresholds are ordered ($\tau_{h-1} < \tau_h$), because they reflect the category order.

Different polytomous models have been proposed. One of these is the *rating scale model* (RSM). A fundamental condition of the RSM, and also its limitation, is the equality of the threshold values for all the items; that is, even if the distance between thresholds can differ, the pattern of these distances is constant for all the items (Andrich, 1978). In this case the probability that subject i gives response x_{ij} to item j is

$$P(X = x_{ij}) = \frac{\exp\left[\kappa_x + x_{ij}(\theta_i - \beta_j)\right]}{\sum_{h=0}^{m} \exp(\kappa_h + h(\theta_i - \beta_j)},\tag{14.3}$$

where X is the random variable which describes the response of subject i to item j; $x_{ij} = 0, 1, \ldots, m$ is the number of thresholds exceeded; $\kappa_0 = 0$, $\kappa_x = \sum_{h=1}^{x} \tau_h$ and $\kappa_m = 0$.

The other polytomous model is the *partial credit model* (PCM). In this model the difficulty levels differ item by item and the subject receives a partial credit (score for each item) equivalent to the relative level of difficulty of the performance achieved (Masters, 1982). The thresholds m_j can differ freely in the same item or from one item to another. The probability that subject i gives response x_{ij} to item j is

$$P(X = x_{ij}) = \frac{\exp \sum_{h=0}^{x} (\theta_i - \beta_{jh})}{\sum_{l=0}^{m_j} \exp \sum_{h=0}^{l} (\theta_i - \beta_{jh})}, \quad x = 0, 1, \ldots, m_j,\tag{14.4}$$

where $\sum_{h=0}^{0} (\theta_i - \beta_{jh}) = 0$. The observation x_{ij} is a count of the successfully completed item thresholds, and only the difficulties of these x_{ij} completed thresholds appear in the numerator of the model.

The RM requires a specific structure in the response data, namely a probabilistic Guttman structure. A Guttman scale is a psychological instrument developed using the scaling technique proposed by Louis Guttman in 1944. An important purpose of the Guttman scale is to ensure that the instrument measures only a single trait (one-dimensionality). For more on measurement scales of ordinal variables see Chapter 4. Guttman's insight was that for one-dimensional scales, those who agree with a more extreme test item will also agree with all less extreme items that precede it. In the RM, the Guttman response pattern is the most probable response pattern for a subject when items are ordered from least difficult to most difficult. Therefore, as mentioned, the RM is a *model* in the sense that it represents the structure that data should exhibit in order to obtain measurements from it. In this sense, the RM is a useful tool for calibrating questionnaires. In particular, Rasch diagnostics test the independence between the subject parameters and item parameters, compare the estimated probabilities and the observed proportions for the items, and analyse the subject and item residual structures.

Various techniques have been fine-tuned in order to control and to test the congruency of the RM. However, none of them specifies the necessary and sufficient conditions for conformity of the data to the model. Therefore, these techniques test specific departures from theoretical expectations. The techniques can be classified in three ways:

1. Techniques that consider the independence between the subject and item parameters. These techniques seek to control the invariance of estimates through the different classes of individual. Subjects are divided into groups which can have a different dimension according to criteria that can be theoretical or empirical (supplied by data exploration). The next step consists of estimating the item parameters in every group and comparing the resulting estimates. The Andersen likelihood ratio (LR) test (Andersen, 1973) is an example of this approach. See Glas and Verhelst (1995) for a review of RM tests.

2. Techniques that compare the estimated probabilities and observed proportions, both on the group of subjects with a total score. This type of technique is similar to that of χ^2 and is affected by sample size.

3. Techniques that are based on response layouts and refer only to items or subjects. The starting point is the comparison between observed and expected responses (Wright and Stone, 1979; Masters 1982; Wright and Masters, 1982). In formal terms,

$$z_{ij} = \frac{x_{ij} - E(X_{ij})}{\sqrt{\mathrm{Var}(X_{ij})}} = \frac{x_{ij} - \pi_{ij}}{\sqrt{\pi_{ij}(1 - \pi_{ij})}}, \tag{14.5}$$

where z_{ij} is the standardized response or residual, obtained as the difference between the observed response x_{ij} of subject i to item j and the expected response $E(X_{ij})$, divided by the standard deviation of the observations; π_{ij} is the probability that subject i gives the response x_{ij} to item j. In terms of the data approach of the model, the distribution of the residuals z_{ij} is approximately standard normal (Wright and Stone, 1979). Once the standardized differences have been found, it is necessary to summarize them in order to have a global measure of the difference between observations and expectations. If we examine the raw score matrix horizontally (for each subject) or vertically (for each item), we have, respectively

$$U_i = \sum_{j=1}^{k} z_{ij}^2 / k, \quad U_j = \sum_{i=1}^{n} z_{ij}^2 / n.$$

This third approach, with an analysis of the single response vectors, allows for the discovery and possible deletion of the subjects or items that appear inconsistent with the expectations of the models. This is an advantage because it allows us to understand whether it is the item that does not work, and therefore has to be deleted or substituted, or whether the subject has given totally unexpected answers and therefore has to be excluded from the analysis. The *outfit* and *infit* indices (Linacre and Wright, 1994) are commonly used as indices of fit of the items and subjects, based on the U statistics (see also Bacci, 2006; Hardouin, 2007). Recently a Monte

Carlo algorithm performing a family of nonparametric tests for the RM has been introduced (Pononcy, 2001).

14.1.2   Rasch model for hierarchical and longitudinal data

The RM belongs to the larger family of item response theory (IRT) models. As reported in Kamata (2001), the multilevel formulation of IRT models, and in particular of RMs, is not a new idea. The approach can be described from three different perspectives.

In a sense, a regular IRT model can simultaneously be viewed as a multilevel model. This perspective is characterized by the treatment of subject ability parameters as random parameters in an IRT model, a treatment originally intended to facilitate the marginal maximum likelihood estimation of item parameters. The subject abilities are the residuals, or random effects, obtained by the integration of a likelihood function over subject parameters, as found in MML estimation. In this approach, the IRT model is interpreted as a multilevel model because item parameters are fixed and subject abilities are random. When subject parameters are treated as random parameters, they can be decomposed into a linear combination of fixed and random effects. So, it is also possible to introduce a latent regression analysis with subject-level predictors and decompose subject abilities from the simple RM into two parts, one representing the adjusted ability estimates, that is, the amount of ability conditional on the covariates (Zwindermann, 1991).

A second perspective on the multilevel formulation of RMs is the decomposition of an item parameter into more than one parameter. Fischer's (1973, 1983, 1995a) linear logistic test model (LLTM) can be classified under this perspective. Fischer generalized the standard binary RM by decomposing an item difficulty parameter into linear combination of more than one item-varying parameter. This modelling approach allows the inclusion of multiple item characteristic variables. In particular, the LLTM has its origins in a paper by Scheiblechner (1972) on regression analysis of the item parameters in the RM. The integration of the linear regression structure into the RM (Fischer, 1973) led to the formulation of the LLTM. Most applications of the LLTM aim to explain the difficulty of items in terms of the underlying cognitive operations (such as negation, disjunction, conjunction, sequence, intermediate result, permutation). The LLTM is obtained by setting the following linear constraints on the item parameters β_j in the RM (14.1):

$$\beta_j = \sum_w q_{jw} \eta_w + c \quad \text{for } j = 1, \ldots, J, w = 1, \ldots, r, r < J, \tag{14.6}$$

where η_w is the difficulty of the cognitive operation w, q_{jw} is the hypothetical minimum number of times the operation w has to be used in solving item j (these weights have to be fixed *a priori*), and c is the usual additive normalization constant in the RM.

From this perspective, the LLTM is more parsimonious than the RM, because the linear condition (14.6) typically reduces the number of free parameters. From (14.6) it follows immediately that

$$\beta_j - \beta_z = \sum_w (q_{jw} - q_{zw}) \eta_w, \tag{14.7}$$

which means that the difference between each pair of item difficulty parameters is explained as a sum of the difficulty parameters of those cognitive operations which have to be performed in solving item j but not in solving item z, and vice versa.

The $J \times r$ matrix $\mathbf{Q} = [q_{jw}]$ is called the *design matrix* of the model. The parameters η_w may represent any experimental factor or aspect influencing the easiness of the items. In educational or clinical applications, some η_w might be effects or treatments given with dosages q_{ijw} to subjects S_i, or represent the easiness of the respective items when given under standard conditions prior to treatments. In longitudinal studies, some η_w might even measure the interaction of certain observable characteristics of the subjects, such as gender or age, with treatment effects, given that the design matrix $\mathbf{Q}$ is redefined accordingly. The linear decomposition of the β_j is sufficiently general to allow for a large diversity of interpretations of the basic parameters.

Many applications of the LLTM have indicated that learning occurs during test taking. The difficulty of an item, therefore, is not a fixed constant but depends on the item's position in the test. Hence the validity of the results of individual tests may be considered questionable (Hohensinn *et al.*, 2008). The singular suitability of the RM for detecting item bias or for the realization of meaningful cross-cultural comparisons has been recognized by many authors. Such studies are typically based on the comparison of the item parameters in two or more populations. The LLTM can go one step further and investigate the reasons for such bias in terms of the underlying cognitive processes.

Another field of application for linearly constrained logistic test models is the measurement of change as induced, for example, by educational or therapeutic treatments. In this case, it is assumed that the reactions of the subjects on a set of J criteria (items, clinical symptoms or other behaviours) are observed at several points in time (before, during and after a treatment period). The response probability is assumed to be of logistic form; the probabilities of correct responses depend on parameters pertaining to the criteria and to effects of the treatments as well as certain observables, namely individual doses and the duration of the treatments given. A distinctive feature of this model is that there is not just one latent dimension, but rather a set of J latent traits or abilities that are assumed to be sensitive to the treatments; this is known as linear logistic model with relaxed assumptions (LLRA). For more on this model, see Fischer (1983).

Fischer (1974) proposed using the LLTM as a generalization of the RM for repeated measurements (see also Fischer, 1995b). When operating with longitudinal models, the main research question is whether an individual's test performance changes over time. In Fischer's theory, the subject parameters are fixed over time and the item parameters change. The basic idea is that one item I_j is presented at two different times to the same subject S_i and is regarded as a virtual item. So, any change in θ_i occurring between the measurement points can be described without loss of generality as a change of the item parameters, instead of describing change in terms of the subject parameters. Thus, with only two measurement points, I_j with the corresponding parameters β_j generates two virtual items I_w and I_u with associated item parameters β_w^* and β_u^*. For the first measurement point $\beta_w^* = \beta_j$, whereas for the second $\beta_u^* = \beta_j + \gamma$. In this linear combination, the β^* parameters are composed additively by means of the real item parameters β and the treatment effects γ. This concept extends to an arbitrary number of measurement points.

Correspondingly, for each measurement point t, we have a vector of *virtual item parameters* $\beta^{*(t)}$. These are linear reparameterizations of the original $\beta^{(t)}$, and thus the CML approach can be used for estimation. Using the virtual-items concept, any repeated measurement design may be treated as an incomplete design of the model (where a complete design is when all subjects take all items). When using models with linear extensions it is possible to impose group contrasts. In this way, one allows for the item difficulties to be different across subgroups. But this is possible only for models with repeated measurements and virtual items since otherwise the introduction of a group contrast leads to overparameterization and the group effect cannot be estimated by using CML.

A few years later, Fischer and Parzer (1991) extended the RSM in such a way that the item parameters are linearly decomposed into certain basic parameters; this extended model is called the linear rating scale model (LRSM). Glas and Verhelst (1989) and Fischer and Ponocny (1994) extended the PCM by imposing a linear structure on the item parameters to obtain the linear partial credit model (LPCM). Wright and Masters (1984) identified several models that can be derived from the PCM by imposing linear restrictions on the item parameters.

The attempts to investigate rater effects in rating scales, such as performance assessments, represent a third perspective on the multilevel formulation of RMs. Linacre's (1989) many-faceted RM is an example of this well-established approach. In this model there is an additional indicator variable specifically for raters. In particular, the model manages situations where several judges rate subjects on different tests. Observations are the result of the interaction of three elements: subject ability, test difficulty and judge severity. The many-faceted RM is considered to be a three-factorial logit model.

Bacci (2006) introduces the Rasch multilevel model, suggesting that the simplest approach is followed by Kamata (2002), who combines the multilevel structure pertaining the hierarchical nature of the data with the multilevel structure of the Rasch models. Although the literature provides extensive discussions on modelling, it may not provide sufficient information on how to conduct such a data analysis for practitioners. Kamata (2002) provides detailed descriptions of practical procedures to conduct item response data analysis using HLM software (Raudenbush et al., 2000). Bacci (2008) proposes an alternative model to analyse longitudinal data in the context of Rasch measurement models: given the multilevel structure of longitudinal data (measurement points are first-level units and individuals are second-level units), a latent regression model with random intercept is estimated to take into account the variability among individuals. This model is different from other multilevel linear models because the response variable is not observed, but has to be estimated by means of an RM. Bartolucci et al. (2011) describe an extension of the latent Markov RM for the analysis of binary longitudinal data with covariates when subjects are collected in clusters.

14.1.3   Rasch model applications in customer satisfaction surveys

As previously stated, the RM is a dichotomous or polytomous model which, through the parameters θ_i and β_j, measures subject ability and item difficulty, respectively. In the case of customer satisfaction survey data analysis, we interpret these coefficients as the subject's satisfaction (θ_i) and the items' intrinsic quality (β_j) of a service. In the original RM context, the scale of the β_j parameters is interpreted as follows: the smallest values of the β_j parameter are associated with items of low difficulty (so the subjects have a high probability of exceeding the item's difficulty), while the highest values are associated with the more difficult items (the probability of exceeding the item's difficulty is lower). On the other hand, in a quality context, the scale has to be read in the opposite way: the smallest values of the β_j parameter identifies items of greater intrinsic quality (because the subject satisfaction probabilities are high), whilst the highest values of the item parameters correspond to items of poor quality (lower subject satisfaction probabilities). For the scale of the parameters θ_i, the interpretation is the same in both cases: the smallest values of the parameter, which identified subjects of low ability, now identifies subjects with low levels of satisfaction, and the greatest values, which previously corresponded to subjects with a high degree of ability, now correspond to subjects with a high level of satisfaction.

Therefore the subject's response to each item depends on the quality of item β_j, the subject's satisfaction θ_i and the thresholds between the categories. Nevertheless, the use of the RM in a service quality/satisfaction context is restricted to the analysis of a single dimension. Since the theoretical framework of quality and customer satisfaction is usually a mixture of K dimensions, in order to apply the RM we need to analyse each dimension separately or consider only one dimension made up of a set of items, each of them defined as a dimension in its own right (Nicolini and Salini, 2006). These two solutions are not always satisfactory: it may be, for example, that we are interested in studying first the single dimensions and then defining an overall individual satisfaction measure – that is, a measure that is able to summarize the satisfaction of subject i with the K dimensions of the service (Nicolini and De Battisti, 2008). Given that the subjects are the same for every dimension, in order to obtain an overall satisfaction coefficient, we apply the RM for each dimension. Thus we obtain K continuous variables θ_k, each with n sets of individual coefficients θ_{ik} ($i = 1, \ldots, n; k = 1, \ldots, K$). Either

1. the K variables are uncorrelated, and to define an overall satisfaction measure we have to consider a function of the linear combination of the variables θ_k; or

2. the K variables are correlated, in which situation there are two possible scenarios:

   (a) they are perfectly correlated;

   (b) they are not perfectly correlated, but there is positive correlation between them. This is the most often the case and the overall satisfaction measure is again a linear combination of the variables θ_k.

In case 2(a) it is possible to run a unique RM on all the items, because in this scenario there is only one latent variable to be investigated, and therefore we have one-dimensionality. In case 1 and 2(b) we need to determine weights of the linear combination. In order to define the weights, we suggest two methods.

Method A is based on a function of the item coefficients β_{jk} ($j = 1, 2, \ldots J; k = 1, 2, \ldots, K$). The theoretical reason for this method is that the satisfaction level of each subject does not depend on the response to each item, but on the global score that the subject assigns to all the items of the same dimension. Two subjects with the same global scores have the same satisfaction level. So we propose to evaluate the variability of the item coefficient β_{jk}; in particular, we consider the standard deviation of these coefficients for each dimension $\sigma_{\beta k}$ and we use the normalized reciprocals of these variability indices as weights in the linear combination:

$$w_k = \frac{1/\sigma_{\beta k}}{\sum_{k=1}^{K} (1/\sigma_{\beta k})}. \tag{14.8}$$

In this way, the dimensions with more homogeneity have a greater importance than the others and the difference schemes of subject response play an important role. So the linear combination is

$$_A\tilde{O}_i = \sum_{k=1}^{K} w_k \theta_{ik}.$$ (14.9)

We suggest the following function as an overall individual standardized satisfaction measure:

$$_A O_i = \left[\frac{_A\tilde{O}_i - M(_A\tilde{O})}{\sqrt{\text{Var}(_A\tilde{O})}} \right].$$ (14.10)

Method B is based on a factor analysis among the K variables θ_k; the weights of the linear combination are

$$w_k = \frac{\rho_{f_1,\theta_k}}{\lambda},$$ (14.11)

that is, the correlation coefficients between the variables θ_k and the first factor f_1 divided by the first eigenvalue, and the score of the first factor for each subject becomes the overall individual satisfaction measure:

$$_B O_i = f_{1i} = \sum_{k=1}^{K} w_k \theta_{ik}^*,$$ (14.12)

where θ_{ik}^* is the standardized original coefficients θ_{ik}. It is important to note that method B can only be used in case 2(b).

14.2 The Rasch model in practice

In order to present a practical application of the Rasch model, we use the ABC 2010 annual customer satisfaction survey (ACSS) presented in Chapter 2 (see also De Battisti *et al.*, 2010). In this application, we consider six dimensions: equipment; sales support; technical support; supplies and orders; purchasing support; and contracts and pricing. There are 266 records. Here, we consider three different methods of analysis: the first (*single model*) involves the application of the RM to the items of all the dimensions, excluding the overall items, so that the model considers 37 items. The second method (*overall model*) examines only the six overall satisfaction items. The third and final method (*dimension model*) calls for the estimation of an RM for each dimension and then the combining of the obtained results.

14.2.1 Single model

As mentioned above, one possible method is to consider all the items together (excluding the overall items) and then to apply the RM to the 37 single items. However, this method presents methodological and practical problems in customer satisfaction surveys. First of all, the hypothesis of one-dimensionality is unlikely. In fact the theoretical framework of customer satisfaction consists of many different dimensions, each with differing satisfaction levels. Therefore, considering them all together is not necessarily a good solution. This is not the case for the RM applied in the classical IRT context, where the items can have differing levels of difficulty and a capable subject is generally able to answer more items correctly than a less able subject. In our specific case, the dimensions appear independent: for example, a subject could be satisfied with sales support and dissatisfied with technical support and so on. On the other hand, we are not interested in having a ranking of the single items but we are interested in knowing which dimension is more important for customers and which is the one with the best quality and the highest evaluation. Moreover, looking at customer satisfaction survey data, it is not possible to consider such dimensions and this generates many item non-responses. If all dimensions are considered, it is then possible to verify many non-responses for the same customer or for the same item and thus many invalid records may be present. In fact, this is the case in the ABC 2010 ACSS when we consider the 37 items. There are only 30 valid records out of the original 266. Therefore, the RM cannot be estimated due to convergence problems.

14.2.2   Overall model

We now consider the model which deals with the overall satisfaction level with the six dimensions mentioned above. We consider only the full records. Not available (NA) cells are excluded from the analysis. There are 215 valid records with at least one non-response in each of the other records.

Customers can give a satisfaction score ranging from '1' (very low) to '5' (very high). Figure 14.1 shows the frequency distribution of raw scores. The raw scores r_i are obtained, for each respondent, by calculating the sum of the answers from item 1 to item J ($J = 6$) which are sufficient statistics (see Section 14.1.1). The majority of respondents returned a raw score around 21 (21 is the mode and also the median), with a minimum of 8 and a maximum of 30. A few subjects returned a raw score less than 18 or greater than 25, respectively the first and third quartiles.

We use the partial credit model (14.4) available in the R library eRm. Table 14.1 shows item location and thresholds. If we sort the items by location parameter we obtain a ranking of

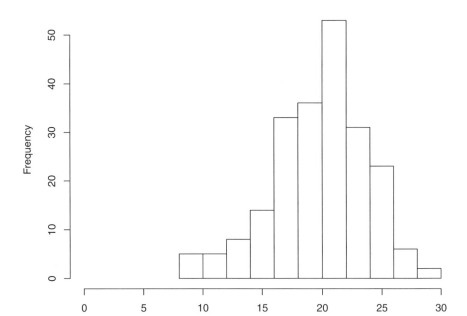

Figure 14.1   Frequency distribution of raw scores in the overall model

Table 14.1   Item location and thresholds for the overall model

Item	Location	Threshold 1	Threshold 2	Threshold 3	Threshold 4
Technical support (TS)	0.33	−0.93	0.37	−0.05	1.93
Equipment (E)	0.57	−1.55	−0.71	0.41	4.12
Supplies and orders (SO)	0.66	−2.03	−1.14	1.31	4.52
Purchasing support (PS)	0.76	−1.43	−0.03	0.87	3.62
Sales support (SS)	0.85	−0.49	0.25	1.07	2.58
Contracts and pricing (CP)	1.38	−0.50	−0.18	1.94	4.25

items with quality rating from best to worst, according to the interpretation of the scale given in the previous paragraph. We observe that the item with the best quality rating is *technical support* and the item with the worst quality rating is *contracts and pricing*.

In order to verify the goodness of fit of the model we calculate the Andersen LR statistic (Andersen, 1973), which tests the assumption that the estimates of the difficulty parameters are equal, whatever the level of the latent trait. The underlying principle of this test statistic is the subgroup homogeneity in RMs. For arbitrary disjoint subgroups the parameter estimates have to be the same. LR is asymptotically distributed as chi-square with degrees of freedom (df) equal to the number of parameters estimated in the subgroups minus the number of parameters in the total data set. In our case, the Andersen LR statistic is 33 with 38 df and *p*-value 0.70, so we fail to reject the null hypothesis.

Table 14.2 highlights the test-of-fit statistics. Rasch *misfit* values flag items which do not share the same construct with the others. As mentioned, the most widely used statistics to evaluate the overall fit of an item to the Rasch model are the *outfit* and *infit* statistics. These are based on a comparison between observed responses for each individual to each item of the questionnaire and responses expected on the basis of the estimated Rasch model. Items with a very high misfit should be removed. *Equipment*, *supplier orders* and *contracts and pricing* have slight misfit: usually outfit and infit values are outside the range −3 to +3, nevertheless their chi-square values suggest that these items could remain in the model.

Figure 14.2 shows the classical 'Rasch ruler' (also called the 'item map') obtained from the ABC 2010 ACSS data. The horizontal line represents the ideal less-to-more continuum of 'satisfaction' and 'quality'; this is the implicit latent dimension of the model. Items and customers share the same linear measurement units (logits). The figure shows both item and thresholds. It can be seen also that the thresholds for technical support are reversed; the third category overlaps with the second, and this is marked with an asterisk.

Table 14.2   Item fit: chi-square, outfit and infit

Item	Chi-sq	df	*p*-value	Outfit *t*	Infit *t*
Equipment (E)	164.166	214	0.995	−3.37	−3.64
Sales support (SS)	193.758	214	0.836	−1.53	−2.27
Technical support (TS)	191.830	214	0.860	−1.47	−0.67
Supplies and orders (SO)	170.194	214	0.988	−3.28	−3.05
Purchasing support (PS)	190.865	214	0.871	−1.76	−1.14
Contracts and pricing (CP)	162.732	214	0.996	−3.93	−3.76

Person-Item Map

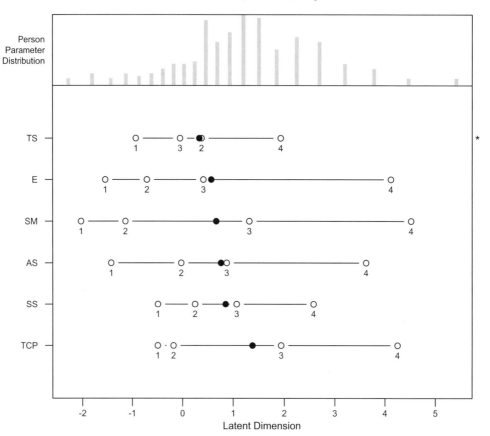

Figure 14.2   – Subject/Item Map for the Overall Model

Figure 14.3 again shows the item map, but now the continuum latent dimension is on the vertical axis, and both items and subjects are plotted according to the infit *t* statistic. Customer scores range from –3.3 to 6.4 logits (customers achieving extreme scores are excluded), while item locations (considering the thresholds as well) range from –2 to 4.5. Thus, we observe a spread of more than 6.5 units for quality and almost 9 for satisfaction. The measurement of satisfaction obtained from this set of items seems reliable, with the range being sufficiently wide. If, on the other hand, all the items had the same characteristics, then the probabilities associated to the answer profiles would be concentrated at one point, and not along a continuum, as observed. The range of items is quite small with respect to the range of satisfaction. There are subjects both at the upper end and the lower end of the scale. Furthermore, only one subject (no. 97) has a level of satisfaction higher than the item with the worst quality rating (see Figure 14.3) and no item thresholds have a quality score greater than the least satisfied subject. Thus, it would seem that the items (quality) are appropriately targeted to the subjects (satisfaction). Furthermore, the item thresholds are well spanned and spaced throughout the continuum. This can be taken as an indicator of a high level of accuracy. For a given increase in the satisfaction

Item/Person Map

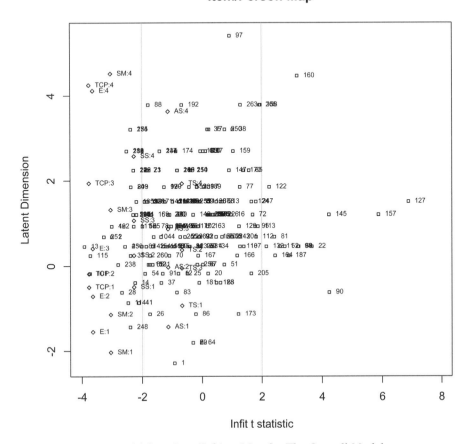

Figure 14.3 – Item/Subject Map for The Overall Model

level there is the same increase in the total raw score. In fact this is not completely true since there is a *potential redundancy* where many item thresholds are on the same horizontal line in Figure 14.3. This means that, when a particular level of satisfaction is achieved, an additional 4–5 marks (as many item thresholds on the same tick) could be present in the total raw score.

Figures 14.4 and 14.5 show the item characteristic curves for *technical support* and *contracts and pricing*, respectively. The subject location is shown on the horizontal axis and the probability related to each response category on the vertical axis. Figure 14.4 shows the items with the best effective quality rating. In this case, we see that higher scoring responses are more likely to be achieved regardless of a subject's location (and therefore the satisfaction rating achieved). On the other hand, Figure 14.5 deals with the items with the worst effective quality rating. In this case, again regardless of a subject's location, we see the opposite effect whereby the lowest scoring response categories are more probable.

The category probability curves for *technical support* reveal the problem of reversed thresholds mentioned above. As we can clearly see in Figure 14.4, the third category overlaps with the second and is probably unnecessary; the probability associated with it is always less than the probability associated to the curves of the other categories.

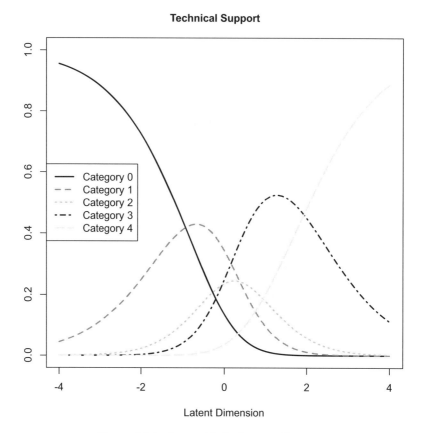

Figure 14.4   Item with the best quality score

We now analyse the parameters related to the respondents (satisfaction) using the θ_i, $i = 1, \ldots, n$, individual subject parameters. Figure 14.6 plots the distribution of raw scores and subject parameters; as expected, the relationship between them is not linear.

The misfit analysis can be used also for subjects. It turns out that about 5% of the parameter residuals exceed the ± 2 limits of the infit t statistic (see Figure 14.3). Moreover, for each respondent, the residuals can be analysed. By the Kolmogorov–Smirnov normality test ($D = 0.0458$, p-value 0.6539), we fail to reject the null hypothesis of normal distribution; see also the quantile–quantile plot (Figure 14.7).

If we consider the residuals item by item, we are able to verify that the residuals are not correlated. In addition, factor analysis (principal component analysis, PCA) brings to light the presence of five relevant factors (Figure 14.8) so that the five residual dimensions appear to be independent (Figure 14.9).

14.2.3   Dimension model

We apply the RM for each dimension. Therefore we have $K = 6$ variables, each consisting of the customer coefficients θ_{ik} ($i = 1, \ldots, n; k = 1, \ldots, 6$), and a ranking of the items for each dimension. We see that the items are different in each dimension, while the subjects are

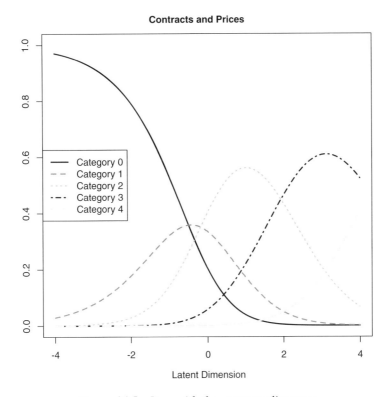

Figure 14.5   Item with the worst quality score

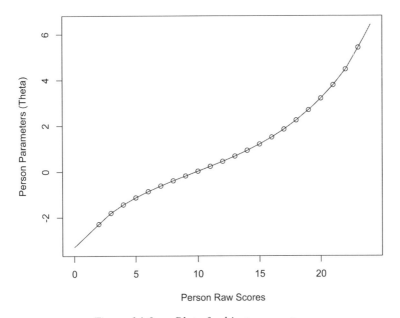

Figure 14.6   – Plot of subject parameters

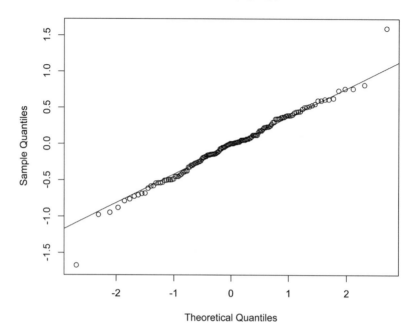

Figure 14.7   Normal quantile–quantile plot

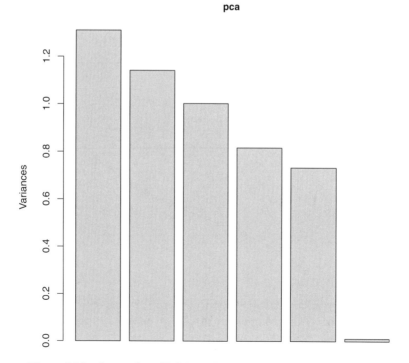

Figure 14.8   Scree plot of PCA analysis of residuals – overall model

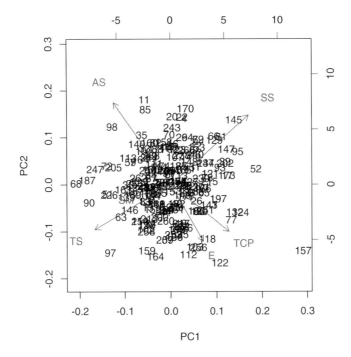

Figure 14.9   Biplot PCA analysis of residuals – overall model

the same for each dimension. Our goal is to define an overall individual satisfaction measure obtained as a linear combination of the $K = 6$ variables θ_k.

Using method B presented in Section 14.1.3 (Nicolini and De Battisti, 2008), we apply the PCM to each component and then apply PCA to the θ_k, thus obtaining for each subject an overall individual satisfaction measure $_B O_i$. Nevertheless, before using the one-dimensional solution as a feasible indicator of satisfaction, it is necessary to evaluate its validity. For this purpose, we need to address two requirements. First, is the first eigenvalue much larger than the others? Second, do all the weights have the same sign?

In our case, both these requirements are satisfied. The scree plot in Figure 14.10 shows that the first eigenvalue is effectively much larger than the others, while the biplot in Figure 14.11 shows that all the weights have the same sign – high values for the first component scores mean high positive values for all the dimensions.

To check the coherence of the two indices, we created two new variables, the rank variables of the measures O_i and of the overall model parameters (see the scatter plot in Figure 14.12). The Spearman correlation between these two variables is 0.88. So the two methods maintain the same ranking, which is an encouraging result.

Another important advantage of the dimension model is that we are able to obtain a detailed analysis of each dimension. As an example, Table 14.3 shows the item location and thresholds, the χ^2 values and p-values for *purchasing support*. The questions are ranked in decreasing order of quality rating from (question 54) 'Complaints are handled promptly' to (question 55) 'Administrative personnel are friendly and courteous'. Question 52 – 'Invoices are clear and easy to understand' – presents some problems because it has a low p-value; if this question

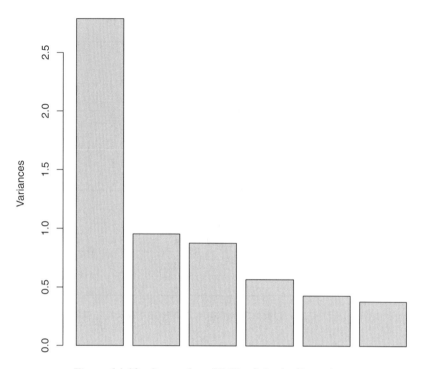

Figure 14.10   Scree plot of PCA of single dimensions

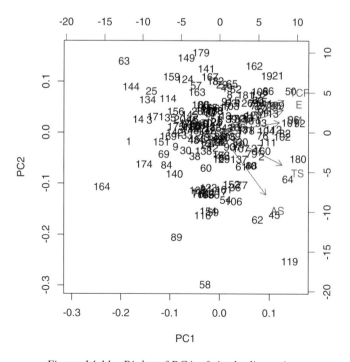

Figure 14.11   Biplot of PCA of single dimensions

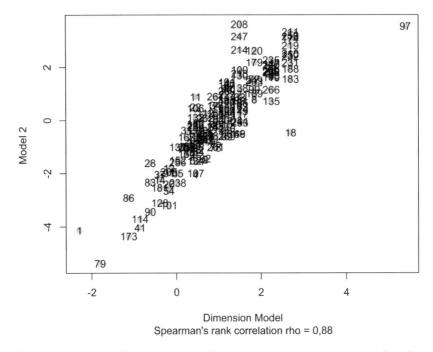

Figure 14.12   Overall model (model 2) ranking versus dimension model ranking

were deleted, the overall χ^2 would decrease. The same analysis can be performed for each dimension. In this sense the RM is a useful tool for calibrating a questionnaire.

14.3   Rasch model software

A list of software programs that implement the Rasch model can be found at http://www.rasch.org/software.htm. Popular software packages for Rasch applications are WINSTEPS (1998, MESA Press, Chicago; http://www.winsteps.com/index.htm), RUMM 2030 (RUMM Laboratory, Duncraig, Western Australia; http://www.rummlab.com.au/) and

Table 14.3   Location, thresholds and chi-square for items of purchasing support

Item	Location	Threshold 1	Threshold 2	Threshold 3	Threshold 4	Chi-sq.	p-value
q54	1.28	0.45	−0.22	0.89	4.00	191.41	0.79
q53	0.88	−0.28	0.00	0.62	3.18	127.13	1.00
q56	0.76	−0.42	−0.22	0.75	2.92	198.80	0.92
q52	0.41	−0.60	−0.59	0.44	2.41	250.68	0.01
q51	0.24	−0.44	−1.08	0.13	2.35	165.47	0.90
q50	0.04	0.66	−2.03	−0.25	1.77	247.64	0.20
q55	−0.06	−1.33	−1.18	−0.24	2.51	224.16	0.47

ConQuest (Wu *et al.*, 2007; http://www.assess.com). For a comparison between WINSTEPS and RUMM, see De Battisti *et al.* (2010). ConQuest produces estimates for the parameters of a wide variety of IRT models, including muldimesional models. General statistical software packages, such as STATA, SAS, JMP and PWStatistics (SPSS) estimate parameters of IRT models with generalized linear mixed models. In STATA, one can estimate the parameters of the Rasch model with the *clogit, xtlogit* or *gllamm* commands. These commands require special data preparation. The *raschtest* command provides estimates of the model parameters and fits the resulting model (Hardouin, 2007).

In the R programming language, Mair and Hatzinger (2007) developed the eRm package which is used in this chapter. With eRm it is possible to compute the ordinary Rasch model for dichotomous items, the LLTM, RSM, LRSM, PCM and LPCM. The eRm package uses a design matrix approach which allows the user to impose repeated measures designs as well as group contrasts. By combining these types of contrasts, one can have item parameters that vary over time with respect to certain subgroups. It is not possible, however, to have group contrasts without applying repeated measurement because the Rasch model assumes 'subgroup invariance'. In order to examine only longitudinal hypotheses, one can impose any number of time contrasts without regard to group differences.

The eRm package applies the CML method for parameter estimation. The advantage of CML is that it is based on the concept of *specific objectivity*.

Among the features and recent developments implemented in eRm are the following:

- Missing values are allowed. If missing values occur, they are coded as NA. Because of the NA structure, the likelihood value is computed separately for each subgroup. The corresponding theoretical treatment can be found in Fischer and Ponocny (1994).

- Parametric and nonparametric goodness-of-fit tests for the dichotomous case such as the Andersen LR test, Wald test and Martin–Löf test (Ponocny, 2001).

- Stepwise selection of item for the dicothomous case.

- A design matrix approach for basic parameters.

- Some utility functions for data simulations.

- Various plots: ICC, goodness-of-fit, subject–item maps and pathway maps.

The same authors have also developed the *RaschSampler* package which implements a Markov chain Monte Carlo algorithm for sampling binary matrices with fixed margins and complying with the Rasch model. For more details, see the package documentation at http://www.r-project.org and Verhelst *et al.* (2007).

Another author, Rizopoulos (2006), developed the *ltm* packagefor the analysis of multivariate dichotomous and polytomous data using latent variable models, under IRT approach. Parameter estimates are obtained by MML using Gauss–Hermite quadrature.

14.4   Summary

The Rasch model, originally proposed by Georg Rasch in 1960 in the context of ability tests, has been further developed over the past 50 years. As mentioned in Section 14.1.2, it belongs to the larger family of IRT models, albeit with a different conceptual approach. Indeed, IRT assumes that the model should fit the data observed in research, while the Rasch perspective requires that the data should fit the model. This particular perspective follows a confirmatory approach, where the model is hypothesized prior to data collection and data–model fit is used to confirm the research hypotheses.

The application of Rasch modelling in the context of customer satisfaction is rather recent, dating back as little as 10 years. Many applications have focused on the evaluation of university teaching, while others have been concerned with the calibration of scales and survey questionnaires. In this chapter, the Rasch model is presented as a promising tool in customer satisfaction studies. However, as far as customer satisfaction studies are concerned, the Rasch model has several competitors. In particular, the CUB models, presented in Chapter 13, propose different statistical models for ordinal data, and in Chapter 17 the Rasch model approach is compared to the non-linear PCA method. All these models generate important insights from the analysis of data collected in customer satisfaction surveys.

References

Andersen, E.B. (1973) A goodness of fit test for the Rasch model. *Psychometrika*, 38, 123–140.

Andrich, D. (1978) A rating formulation for ordered response categories. *Psychometrika*, 43, 561–573.

Andrich, D. (1988) *Rasch Models for Measurement*. Beverly Hills, CA: Sage.

Andrich, D. (2004) Controversy and the Rasch model: A characteristic of incompatible paradigms? *Medical Care*, 42, 1–16.

Bacci, S. (2006) I modelli di Rasch nella valutazione della didattica universitaria. *Statistica Applicata*, 18(1).

Bacci, S. (2008) Analysis of longitudinal health related quality of life using latent regression in the context of Rasch modelling. In C. Huber, N. Limnios, M. Mesbah and M. Nikuline (eds), *Mathematical Methods for Survival Analysis, Reliability and Quality of Life*, pp. 275–290. Hoboken, NJ: John Wiley & Sons, Inc.

Baker F.B. (1987) Methodology review: Item parameter estimation under the one, two and three-parameter logistic models. *Applied Psychological Measurement*, 11(2), 111–141.

Bartolucci, F., Pennoni, F. and Vittadini, G. (2011) Assessment of school performance through a multilevel latent Markov Rasch model. *Journal of Educational and Behavioral Statistics*. doi: 10.3102/1076998610381396. http://arxiv.org/abs/0909.4961.

Bertoli-Barsotti, L. (2003) An order-preserving property of the maximum likelihood estimates for the Rasch model. *Statistics and Probability Letters*, 61, 91–96.

Bertoli-Barsotti, L. (2005) On the existence and uniqueness of JLM estimates for the partial credit model. *Psychometrika*, 70, 517–531.

De Battisti, F., Nicolini, G. and Salini, S. (2005) The Rasch Model to measure service quality. *ICFAI Journal of Services Marketing*, 3(3), 58–80.

De Battisti, F. , Nicolini, G. and Salini, S. (2010) The Rasch model in customer satisfaction survey data, *Quality Technology & Quantitative Management*, 7(1), 15–34.

Fischer G.H. (1973) The linear logistic test model as an instrument in educational research. *Acta Psychologica*, 37, 359–374.

Fischer G.H. (1974) *Einführung in die Theorie psychologischer Tests [Introduction to Mental Test Theory]*. Bern: Huber.

Fischer, G.H. (1983) Logistic latent trait models with linear constraints. *Psychometrika*, 48(1), 3–26.

Fischer, G.H. (1995a) Linear logistic test model. In G.H. Fisher and I.W. Molenaar I.W., *Rasch Models: Foundations, Recent Developments and Applications*, pp. 131–155. New York: Springer.

Fischer, G.H. (1995b) Linear logistic models for change. In G.H. Fisher and I.W. Molenaar I.W., *Rasch Models: Foundations, Recent Developments and Applications*, pp. 157–180. New York: Springer.

Fischer, G.H. and Parzer, P. (1991) An extension of the rating scale model with an application to the measurement of change. *Psychometrika*, 56, 637–651.

Fischer, G.H. and Ponocny I. (1994) An extension of the partial credit model with an application to the measurement of change. *Psychometrika*, 59, 177–192.

Glas, C.A.W. and Verhelst, N.D. (1989) Extensions of the partial credit model. *Psychometrika*, 54, 635–659.

Glas, C.A.W. and Verhelst, N.D. (1995) Testing the Rasch model. In G.H. Fisher and I.W. Molenaar I.W., *Rasch Models: Foundations, Recent Developments and Applications*, pp. 69–95. New York: Springer.

Hardouin, J.B. (2007) Rasch analysis: Estimation and tests with raschtest. *Stata Journal*, 7(1), 1–23.

Hohensinn, C., Kubinger, K.D., Reif, M., Holocher-Ertl, S., Khorramdel, L. and Frebort, M. (2008) Examining item-position effects in large-scale assessment using the linear logistic test model. *Psychology Science Quarterly*, 50, 391–402.

Kamata, A. (2001) Item analysis by the hierarchical generalized linear model. *Journal of Educational Measurement*, 38(1), 79–93.

Kamata, A. (2002) Procedure to perform item response analysis by hierarchical generalized linear model. Paper presented to the annual meeting of the American Educational Research Association, New Orleans, April. http://mailer.fsu.edu/~akamata/AERA_2002.pdf

Linacre, J.M. (1989) *Many-faceted Rasch measurement*, Chicago, MESA Press.

Linacre, J.M. and Wright B.D. (1994) Dichotomous mean-square chi-square fit statistics. *Rasch Measurement Transactions*, 8, 360.

Mair, P. and Hatzinger, R. (2007) Extended Rasch model: The eRm package for the application of IRT models in R. *Journal of Statistical Software*, 20(9).

Masters, G.N. (1982) A Rasch model for partial credit scoring. *Psychometrika*, 47, 149–174.

Nicolini, G. and De Battisti, F. (2008) Methods for summarizing the Rasch model coefficients. In L. D'Ambra, P. Rostirolla and M. Squillante (eds), *Metodi, Modelli e Tecnologie dell'Informazione a Supporto delle Decisioni. Parte prima: Metodologie*. Milan: Franco Angeli.

Nicolini, G. and Salini, S. (2006) Customer satisfaction in the airline industry: The case of British Airways. *Quality and Reliability Engineering International*, 22, 1–9.

Ponocny, I. (2001) Nonparametric goodness-of-fit tests for the Rasch model. *Psychometrika*, 66, 437–460.

Rasch G. (1960) *Probabilistic Models for Some Intelligence and Attainment Tests*. Copenhagen: Danish Institute for Educational Research. Expanded edition (1980) with foreword and afterword by B.D. Wright. Chicago: University of Chicago Press.

Raudenbush, S.W., Bryk A.S., Cheong Y.F. and Congdon R. (2000) *HLM5: Hierarchical linear and nonlinear modeling* [computer program]. Chicago: Scientific Software International.

Rizopoulos, D. (2006) ltm: An R Package for latent variable modeling and item response analysis. *Journal of Statistical Software*, 17(5).

Scheiblechner, H. (1972) Das Lernen und Lösen komplexer Denkaufgaben [The learning and solving of complex reasoning items]. *Zeitschrift für Experimentelle und Angewandte Psychologie*, 3, 456–506.

Verhelst, D.V., Hatzinger, R. and Mair, P. (2007) The Rasch sampler. *Journal of Statistical Software*, 20(4).

Warm, T.A. (1989) Weighted likelihood estimation of ability in item response theory. *Psychometrika*, 54, 427–450.

Wright, B.D. and Masters, G.N. (1982) *Rating scale analysis*. Chicago: MESA Press.

Wright, B.D. and Masters, G.N. (1984) The essential process in a family of measurement models. *Psychometrika*, 49, 529–544.

Wright, B.D. and Stone, M.H. (1979) *Best Test Design*. Chicago: MESA Press.

Wu, M.L., Adams, R.J., Wilson, M.R. and Haldane, S. (2007) *ACER ConQuest 2.0: General item response modelling software* [computer program manual]. Camberwell, Vic.: ACER Press.

Zwinderman, A.H. (1991) A generalized Rasch model for manifest predictors. *Psychometrika*, 56, 589–600.

15

Tree-based methods
and decision trees

Giuliano Galimberti and Gabriele Soffritti

This chapter reviews a selection of tree-based methodologies that can be used for customer satisfaction evaluation. After a short introduction of some basic concepts, three popular methods – CART, CHAID and PARTY– are described and compared; particular attention is given to the treatment of missing values and to specific solutions suitable for dependent variables measured on an ordinal scale. Results obtained by analysing the ABC ACSS data set are illustrated and discussed. The chapter concludes by focusing on the main drawbacks of tree-based methods and discusses ways to overcome them.

15.1 An overview of tree-based methods and decision trees

15.1.1 The origins of tree-based methods

Tree-based methods are data-driven tools based on sequential procedures that recursively partition the data. They provide conceptually simple ways of understanding and summarizing the main features of the data; in particular, they exploit tree graphs to provide visual representations of the rules underlying a given data set. Automatic construction of such rules has been developed in the fields of statistics, engineering and decision theory, in the context of regression and classification analysis, pattern recognition and decision table programming, respectively. Recently, several contributions to this topic can be found in the literature on machine learning and neural networks. As a result, many real-world situations can be successfully analysed through tree-based methods. These methods can be used for both data exploration and prediction rule construction (Siciliano *et al.*, 2008). Prediction rules obtained by tree-based methods are also referred to as decision trees.

One of the first works on tree-based methods in statistics was motivated by the need for matching population samples (Belson, 1959). Morgan and Sonquist (1963a) proposed a

---

Modern Analysis of Customer Surveys: with applications using R, First Edition. Edited by Ron S. Kenett and Silvia Salini.
© 2012 John Wiley & Sons, Ltd. Published 2012 by John Wiley & Sons, Ltd.

tree-based method for exploring survey data by means of a regression analysis, that is, in order to study the influence of some explanatory factors (also referred to as independent variables or predictors) on a continuous variable Y (i.e., the dependent variable, the response or target). The automatic interaction detection (AID) technique (Morgan and Sonquist, 1963a,b) is able to take account of possible interaction effects between the predictors in regression analysis. Further important work in this field was done by Kass (1980): in this work the same problem was addressed within a classification framework, that is, for a categorical dependent variable Y. The chi-square automatic interaction detection (CHAID) algorithm proposed by Kass was the first approach to tree-based classification to use statistical significance tests. However, the popularity of tree-based methods in statistics is mainly due to a unified framework embedding tree-structured methodology in regression and classification, also known as classification and regression trees or CART (Breiman *et al.*, 1984). Since then, several papers on this topic have been published in which improvements on previous methods or solutions applicable to special situations are provided, such as the cases of a censored, ordinal or multiple response (see, for example, Segal (1988), LeBlanc and Crowley (1992), Zhang (1998), Hothorn *et al.* (2006), Zhang and Ye (2008)) and multilabel or multiway data (Noh *et al.*, 2004; Tutore *et al.*, 2007). Other relevant work has been done by Clark and Niblett (1989) and Quinlan (1993). General overviews of tree-based methods and decision trees can be found, for example, in Murthy (1998), Siciliano and Conversano (2009), Zhang and Singer (2010) and Loh (2011). Some applications of tree-based methods to customer satisfaction evaluation are described in Nicolini and Salini (2006) and Tutore *et al.* (2007).

In this chapter a selection of tree-based methodologies that can be useful for evaluating customer satisfaction is presented. In particular, we focus on three methods: CART (Section 15.1.3), CHAID (Section 15.1.4), and a recent proposal due to Hothorn *et al.* (2006) for recursive binary partitioning (PARTY; Section 15.1.5). These three methods are briefly compared in Section 15.1.6. Section 15.1.7 shows how tree-based methods can deal with missing values. Section 15.1.8 describes some specific solutions suitable for applications in customer satisfaction surveys. In Section 15.2.1 we present an illustrative example in which tree-based methods are applied to the ABC ACSS data set described in Chapter 2. Finally, Section 15.3 addresses some limitations of tree-based methods and recent solutions that allow us to overcome them.

15.1.2   Tree graphs, tree-based methods and decision trees

A tree graph (hereafter tree) $\mathcal{T}$ is a graph consisting of one or more nodes connected by arcs. A node is said to be internal if it is associated with a split (or a splitting rule), which is a rule based on logical conditions on the values of the predictors. Each internal node is connected to two or more child nodes, one for each distinct outcome of the corresponding splitting rule. If all the splitting rules have two possible outcomes, then the resulting tree will be binary; otherwise, it will be a multiway tree. Nodes that do not have child nodes are referred to as leaves or terminal nodes. Usually, a value of the dependent variable is associated with each leaf node. The size of the tree is given by the number of its leaves.

An example of a binary tree is shown in Figure 15.1. The dependent variable considered in the example is the satisfaction of a customer with a given service provided by an international company (Satisfaction, with categories 'no' and 'yes') and the predictors are the country in which the customer lives (Country, with categories Benelux, Germany, Israel and United Kingdom) and how many years the customer has used the service (Seniority). The tree is composed of two internal nodes (nodes 1 and 3) and three leaf nodes (nodes 2, 4 and 5). Node

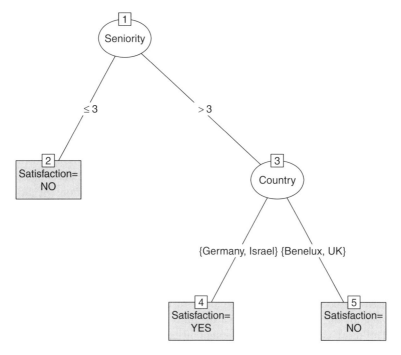

Figure 15.1   An example of a binary tree

1 is associated with a splitting rule defined with respect to the predictor Seniority, leading to the child nodes 2 and 3. This splitting rule separates customers who have been using the service for three years or less (node 2) from those using the service for more than three years (node 3). The splitting rule for node 3 is based on the predictor Country and distinguishes customers who live in Germany or Israel (node 4) from those living in Benelux or the UK (node 5). The value of Satisfaction associated with leaf nodes 2 and 5 is No, while the one associated with leaf node 4 is Yes. The size of this tree is three.

Trees are generally constructed from a data set (also referred to as training set or learning set) which contains information about a random sample of objects. The learning set is composed of data on a dependent variable Y and K $(K \geq 1)$ predictors $X_1, \ldots, X_K$ for n $(n > 1)$ sample units: $\mathcal{L}_n = \{(y_i, x_{i1}, \ldots, x_{iK}); \ i = 1, \ldots, n\}$. According to the properties of $\mathcal{Y}$, which is the set of the possible values of Y, the resulting tree is commonly referred to as a regression tree (if Y is numerical) or a classification tree (when Y is categorical). In both cases, predictors can be numerical and/or categorical.

The process of constructing a tree from the learning set is called tree building or tree growing. Most of the procedures for growing trees start from the entire learning set, which is graphically represented as a tree composed of a single leaf, also called a root node. The learning set is partitioned into two or more subsets according to a given splitting rule; this produces a new tree with an increased size. Then, the partitioning procedure is recursively applied by splitting the leaves of the tree constructed in the previous step. Thus, tree growing is performed through a top-down procedure. Tree-based methodology is also known as recursive partitioning methodology. The result of the recursive partition process is a sequence of nested trees having an increasing number of leaves.

The crucial points with regard to tree growing are: the definition of the admissible splitting rules for a given leaf; the choice of a suitable goodness-of-split criterion, which is used to score all possible splitting rules for a given leaf node and select the best one; and the definition of criteria to stop the recursive partitioning. Many recursive partitioning methods have been proposed in the literature, all of which are characterized by specific choices for these three crucial points. In particular, in CART methodology the selection of the best split is based on the notion of node impurity with respect to the dependent variable Y; CHAID and PARTY exploit tests for the association between Y and each predictor.

Once a tree is obtained from a learning set, each unit of this set can be assigned to one of the leaf nodes. Since it can belong to only one leaf, the tree induces a partition of the learning set into as many subsets as there are leaves. Each element of such a partition is defined by specific values of the predictors. The comparison of the conditional distributions of the dependent variable given leaf node membership allows for the exploration of the relationships between the dependent variable and the predictors as well as the interactions among the predictors. These conditional distributions are also used to assign the value of the dependent variable to each leaf node. For example, the conditional mean value and the conditional modal category are generally computed if Y is continuous or categorical, respectively. In this perspective, the tree obtained can be used as a data-driven tool for data exploration.

Tree-based methods can also be employed to define decision rules for predicting the unknown value of the dependent variable for units on which only the predictor values are observed. This result can be obtained by assigning each of these units to one of the leaf nodes, according to the splitting rules in the tree. The predicted value of Y for each of these units is given by the value of Y associated with the leaf the unit has been assigned to. For categorical dependent variables this prediction rule is known as the majority rule. It is important to note that, in order to obtain trees with optimal predictive performances, tree-based methods should take into account a suitable measure of prediction accuracy, which must be employed in order to select the most accurate tree. Trees obtained in this way are usually referred to as decision trees (Siciliano and Conversano, 2009; Siciliano et al., 2008). When the dependent variable is categorical, tree-based methods can also be used for ranking units according to the probability of their belonging to a given category of Y; see, for example, Fierens et al. (2010) and Clémencon et al. (2011).

Tree-based methods have many advantages. The first is the interpretability of their results. A tree identifies a sequence of simple rules that are easy to understand even without a sound statistical background. The graphical representation of such a sequence allows a straightforward assignment of the predicted values of Y for any sample unit, without performing any computation on the observed predictor values. A second relevant feature of trees is their ability to reveal interaction effects between the predictors. A third advantage lies in the fact that, unlike other statistical methods for regression and classification analysis, classification and regression trees are non-parametric methods, that is, they do not rely on any assumption about the probability distribution of Y nor on the type of relationship (e.g., linearity assumption) between Y and the predictors. A further important feature of tree-based methods is their ability to deal with missing predictor values, without resorting to any preliminary imputation of the missing data (some solutions are described in Section 15.1.7). Finally, every tree growing procedure also performs an automatic selection of the best predictors for the dependent variable.

In the following the attention is focused mainly on binary trees, which are the most widely employed in practical applications. While trees with multiway splits can sometimes be useful, in general they could provide poor solutions. The main problem with multiway trees is that data are fragmented too quickly, leaving insufficient information for further growing the tree.

Furthermore, since a suitable sequence of binary splits can reproduce any multiway split, the former are preferred (Hastie *et al.*, 2009).

15.1.3 CART

Admissible splitting rules for growing binary trees

As described in Section 15.1.2, the first crucial point with regard to tree growing is the selection of the splits of any leaf node t in a tree $\mathcal{T}$. This selection is performed with respect to a set S_t composed of admissible splits for leaf t.

A simple way to define admissible splits consists of examining each single predictor X_k, one at a time (univariate splits), and considering all the binary partitions of $\mathcal{X}_k$, that is, the set of possible values of X_k. The admissible splits depend on the measurement scale of X_k. Consider the first step of the tree growing procedure: the selection of the splitting rule for the root node. If X_k is numerical (such as Seniority in the previous example), admissible binary splits are obtained by choosing a threshold x_k among the observed values of X_k, and by defining the two logical conditions $(X_k \leq x_k)$ and $(X_k > x_k)$. The admissible values of the threshold x_k are all the observed values of X_k aside from the maximum one; thus, for a numerical predictor the number of admissible splits is equal to the number of observed values of X_k minus one.

In the case of a categorical predictor X_k with D_k $(D_k \geq 2)$ observed categories $\{u_1, \ldots, u_{D_k}\}$, the admissible splits depend on the properties of its categories. If the categories of X_k are ordered (e.g., overall satisfaction with the technical support provided by a company, with categories 'very low', 'low', 'intermediate', 'high', 'very high'), admissible binary splits are obtained by choosing a threshold u_k among the categories of X_k, and by defining the logical conditions $(X_k \leq u_k)$ and $(X_k > u_k)$. Thus, for ordered categorical predictors the number of admissible splits is equal to $D_k - 1$. If the categories of X_k are unordered, as for the variable Country in the example, the admissible splits are obtained by listing all the partitions of the D_k categories into two non-empty classes. The total number of admissible splits is given by $2^{D_k - 1} - 1$. For the predictor Country used in the example, the $2^{4-1} - 1 = 7$ admissible splits are $s_1 = \{(\text{Germany, Israel}), (\text{Benelux, UK})\}$, $s_2 = \{(\text{Germany, Benelux}),$ $(\text{Israel, UK})\}$, $s_3 = \{(\text{Benelux, Israel}), (\text{Germany, UK})\}$, $s_4 = \{(\text{Germany, Israel, Benelux}),$ $(\text{UK})\}$, $s_5 = \{(\text{Germany, Israel, UK}), (\text{Benelux})\}$, $s_6 = \{(\text{Germany, Benelux, UK}), (\text{Israel})\}$, $s_7 = \{(\text{Benelux, Israel, UK}), (\text{Germany})\}$. In the special case of a binary categorical predictor, that is, a categorical predictor with $D_k = 2$ categories (e.g., gender), there is only one admissible split. Admissible splits for the partition of a node t different from the root node can be derived in a similar way. The main difference is that only the values of the predictors observed on units assigned to node t are considered. For example, for node 4 in Figure 15.1 the admissible splits are $s_1 = \{(\text{Germany}), (\text{Israel})\}$, and $s_2 = \{(\text{Seniority} \leq 4), (\text{Seniority} > 4)\}$.

When all the predictors are numerical, another strategy for defining admissible splitting rules is to consider only dyadic splits, which are obtained by considering only one threshold for each predictor, given by the midpoint of $\mathcal{X}_k$ for all k. Trees grown using admissible splitting rules of this kind are referred to as binary dyadic trees; see, for example, Donoho (1997) and Hush and Porter (2010).

More complex ways of defining admissible splits can be based on more than one predictor at a time (multivariate splits). For example, with respect to numerical predictors, further splits can be defined by choosing thresholds on the possible values of linear combinations of two or more predictors. However, this can reduce the interpretability of the resulting tree. Most of the software that implements CART and other tree-based methods only allows for univariate splits.

Goodness-of-split criteria based on impurity measures

CART methodology uses criteria for measuring the goodness of a split which exploit impurity measures. A node in a tree is said to be pure if all the sample units assigned to it have the same value of the dependent variable Y. Thus, any variability or heterogeneity measure can be theoretically used to measure the impurity of a node with respect to a numeric or categorical dependent variable, respectively. In practice, because of their appealing properties, the following measures are generally employed: the variance when Y is numeric, the Gini heterogeneity index or the cross-entropy when Y is categorical.

In order to illustrate the definition of these measures with respect to a given node t of a tree, it is useful to introduce the following notation: let $\mathbf{w}(t)$ be an n-dimensional vector of 0–1 elements that allow selection of the units of the learning set that have been assigned to node t. That is, $w_i(t) = 1$ if the ith sample unit has been assigned to node t, and $w_i(t) = 0$ otherwise, for $i = 1, \ldots, n$. For a numeric dependent variable, let $\bar{y}(t)$ be the mean value of Y in node t, that is, the arithmetic mean of the values of Y for the sample units that have been assigned to t:

$$\bar{y}(t) = \frac{\sum_{i=1}^{n} y_i w_i(t)}{\sum_{i=1}^{n} w_i(t)}.$$

Then, the variance of Y in node t can be defined as

$$I_V(t) = \frac{\sum_{i=1}^{n} (y_i - \bar{y}(t))^2 w_i(t)}{\sum_{i=1}^{n} w_i(t)}. \tag{15.1}$$

In the case of a categorical dependent variable with M categories $\{c_1, \ldots, c_M\}$, let $p_m(t)$ be the proportion of sample units belonging to the mth category of Y in node t, for $m = 1, \ldots, M$:

$$p_m(t) = \frac{\sum_{i=1}^{n} 1_{\{c_m\}}(y_i) w_i(t)}{\sum_{i=1}^{n} w_i(t)},$$

where $1_{\{c_m\}}(y_i) = 1$ if the ith sample unit belongs to the mth category of Y, and 0 otherwise. Then the Gini heterogeneity index and the cross-entropy in node t can respectively be defined as

$$I_G(t) = 1 - \sum_{m=1}^{M} [p_m(t)]^2 \tag{15.2}$$

and

$$I_E(t) = - \sum_{m=1}^{M} p_m(t) \log[p_m(t)]. \tag{15.3}$$

Whenever a node t is pure, each of the measures illustrated above is equal to 0. The measures defined in equations (15.2) and (15.3) reach their maximum value, equal to $(M - 1)/M$ and $\log(M)$ respectively, when $p_m(t) = 1/M$ for $m = 1, \ldots, M$, that is, when all the categories of the dependent variable are present in node t with the same proportion of sample units.

In order to illustrate the computation of measures $I_G(t)$ and $I_E(t)$, consider the tree in Figure 15.2, which shows the partition of a learning set of $n = 209$ customers obtained according to the binary tree of Figure 15.1 (the cross-classifications of the customers with respect to Y and each of the two predictors are reported in Table 15.1). For each node t the figure provides information about the number of customers (n) and the proportions of those

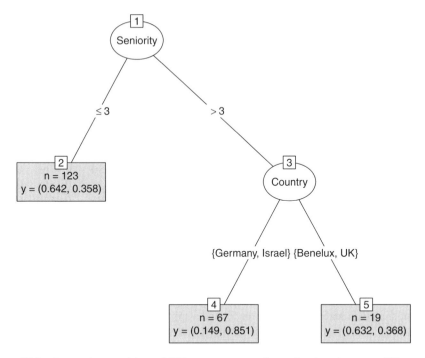

Figure 15.2 Recursive partition of 209 customers performed using the tree of Figure 15.1

belonging to the categories 'no' and 'yes' of Satisfaction, namely $\mathbf{y} = (p_1(t), p_2(t))$. Thus, the impurity of node 1 evaluated using the Gini index is equal to 0.499, while according to the cross-entropy it is equal to 0.693.

Given an impurity measure $I(t)$, it is possible to compute the total tree impurity as

$$I(T) = \sum_{t \in \tilde{T}} I(t)p(t), \tag{15.4}$$

where $\tilde{T}$ is the set of the leaf nodes of the tree T, and $p(t) = \frac{1}{n}\sum_{i=1}^{n} w_i(t)$ is the proportion of units assigned to leaf node t. For example, considering the Gini index, the total impurity of the tree illustrated in Figure 15.2 is equal to the weighted sum of the impurities of nodes 2, 4, and 5: $I(T) = 0.460\frac{123}{209} + 0.254\frac{67}{209} + 0.465\frac{19}{209} = 0.394$.

Table 15.1 Cross-classifications of Satisfaction (Y) with respect to Country and Seniority ($n = 209$ customers)

| | Country | | | | Seniority | | | | |
	Benelux	Germany	Israel	UK	1	2	3	4	5
$Y = $ no	20	39	14	28	24	35	20	9	13
$Y = $ yes	5	72	9	22	15	18	11	16	48

Table 15.2   Admissible splits of node 1 in the tree of Figure 15.2 and decreases in the total tree impurity

Admissible splits s of node 1	$p(t_l(s))$	$I(t_l(s))$	$p(t_r(s))$	$I(t_r(s))$	$\Delta I(1, s)$
$s_1 =\{(\text{Germany, Israel}), (\text{Benelux, UK})\}$	0.359	0.4608	0.641	0.4782	0.027503
$s_2 =\{(\text{Germany, Benelux}), (\text{Israel, UK})\}$	0.349	0.4886	0.651	0.4912	0.009104
$s_3 =\{(\text{Benelux, Israel}), (\text{Germany, UK})\}$	0.770	0.4859	0.230	0.4132	0.030208
$s_4 =\{(\text{Germany, Israel, Benelux}), (\text{UK})\}$	0.761	0.4967	0.239	0.4928	0.003704
$s_5 =\{(\text{Germany, Israel, UK}), (\text{Benelux})\}$	0.880	0.4929	0.120	0.3200	0.027263
$s_6 =\{(\text{Germany, Benelux, UK}), (\text{Israel})\}$	0.890	0.4979	0.110	0.4764	0.003892
$s_7 =\{(\text{Benelux, Israel, UK}), (\text{Germany})\}$	0.469	0.4648	0.531	0.4558	0.039412
$s_8 =\{(\text{Seniority} \le 1), (\text{Seniority} > 1)\}$	0.813	0.4956	0.187	0.4734	0.008010
$s_9 =\{(\text{Seniority} \le 2), (\text{Seniority} > 2)\}$	0.560	0.4602	0.440	0.4601	0.039295
$s_{10} =\{(\text{Seniority} \le 3), (\text{Seniority} > 3)\}$	0.411	0.3807	0.589	0.4595	0.072336
$s_{11} =\{(\text{Seniority} \le 4), (\text{Seniority} > 4)\}$	0.292	0.3354	0.708	0.4821	0.060155

The Gini heterogeneity index $I_G(t)$ has been used.

Consider a given leaf node t^* of a tree $\mathcal{T}$, and a splitting rule $s \in S_{t^*}$, which generates two child nodes $t_l(s)$ and $t_r(s)$ of t^*. This splitting rule leads to a new tree $\mathcal{T}'$ whose set of leaf nodes contains the two child nodes $t_l(s)$ and $t_r(s)$ instead of t^*, which becomes an internal node of $\mathcal{T}'$. Thus, the total tree impurity of $\mathcal{T}'$ is equal to

$$I(\mathcal{T}') = I(\mathcal{T}) - I(t^*)p(t^*) + I(t_l(s))p(t_l(s)) + I(t_r(s))p(t_r(s)).$$

Brieman *et al.* (1984) proved that, for any of the impurity measures mentioned above,

$$I(\mathcal{T}') \le I(\mathcal{T}), \quad \forall t^* \in \tilde{T}, \ \forall s \in S_{t^*}.$$

Thus, the decrease in the total tree impurity induced by the partition of t^* according to the splitting rule $s \in S_{t^*}$ is given by

$$\Delta I(t^*, s) = I(t^*)p(t^*) - I(t_l(s))p(t_l(s)) - I(t_r(s))p(t_r(s)). \tag{15.5}$$

Note that $\Delta I(t^*, s)$ only depends on the units assigned to t^* and not on those assigned to the other leaf nodes of tree $\mathcal{T}$. At each step of the tree growing procedure, an exhaustive search for the pair (t, s) which corresponds to the maximum value of $\Delta I(t, s)$ is performed by considering all the admissible splits for all leaf nodes. The results of this search with respect to the splitting of node 1 in the tree of Figure 15.2 using the Gini index are summarized in Table 15.2. The information about the learning set that is necessary for the computation of the decreases in the total tree impurity can be found in Table 15.1. At the first step of the tree growing procedure, the splitting rule with the largest decrease in the total tree impurity is $s_{10} =\{(\text{Seniority} \le 3), (\text{Seniority} > 3)\}$. Mola and Siciliano (1997) provided a method that allows for the reduction of the computational complexity of the search of the best split when the Gini index is used to compute $\Delta I(t, s)$.

Breiman *et al.* (1984) also introduced an alternative procedure for selecting the best splitting rule when the dependent variable is categorical with more than two categories, also known as the twoing method. The basic idea of this method is to evaluate each splitting rule not with respect to the distribution of Y but with respect to all the possible dichotomizations of its M categories. In other words, the M-class problem is transformed into several two-class problems, by examining all $2^{M-1} - 1$ partitions of the set $\{c_1, \ldots, c_M\}$ of the categories of Y into two non-empty sets. In particular, for each split s of a given node t the decrease $\Delta I(t, s)$ has to be evaluated with respect to each possible partition of $\{c_1, \ldots, c_M\}$, and the greatest decrease is selected. Breiman *et al.* (1984) proved that, if the Gini heterogeneity index is used in these two-class problems, then the best twoing splitting rule for node t is the one that maximizes

$$\Delta I_{\text{TW}}(t, s) = \frac{p(t_l(s)) \cdot p(t_r(s))}{4} \left[\sum_{m=1}^{M} |p_m(t_l) - p_m(t_r)| \right]^2 ; \qquad (15.6)$$

they also proved that the partition of $\{c_1, \ldots, c_M\}$ associated with this best twoing splitting rule is $C_1(t) = \{c_m : p_m(t_l) \geq p_m(t_r)\}$, $C_2(t) = \{c_m : p_m(t_l) < p_m(t_r)\}$. Since the optimal twoing partition of $\{c_1, \ldots, c_M\}$ may change from node to node, the twoing splitting process does not fit into the general CART framework described above, based on the decrease in the total tree impurity. A theoretical comparison among splitting criteria used within the CART methodology for a categorical dependent variable can be found in Breiman (1996c).

Stopping criteria and pruning

Determining when to stop the recursive partitioning procedure in tree-based methods is a crucial issue. Several criteria have been suggested in the literature. The simplest introduces a restriction on the minimum number of units in a node (node size): a node cannot be further split if it has less than $n_{\min_1}$ units, where $n_{\min_1} \geq 1$ has to be specified in advance by the researcher. A similar restriction has been proposed with respect to the admissible splitting rules: splitting rules which generate at least one child node with less than $n_{\min_2}$ units are not allowed, where $n_{\min_2} \leq n_{\min_1}$ has to be specified in advance by the researcher. These two solutions do not depend on the goodness-of-split criterion for growing a tree, thus they can be applied to any recursive partitioning procedure.

It is evident that different choices for $n_{\min_1}$ and $n_{\min_2}$ can lead to trees with different sizes. While, on the one hand, trees having a small number of leaves may not detect some of the interesting features of the relationship between the dependent variable and the predictors, on the other hand, trees with many leaves can be prone to overfitting, that is, they may reflect irrelevant features that are specific to the learning set used to grow the tree. The overfitting problem is particularly relevant when tree-based methods are used to build a prediction rule: predicted values of Y can be severely biased when obtained from trees with too few leaves, or they can have high variability when computed using a tree with too many leaves; see Hastie *et al.* (2009) for a detailed discussion of the so-called bias–variance trade-off. Within the CART methodology, Breiman *et al.* (1984) introduced a procedure which is known as cost-complexity pruning in order to obtain decision trees with optimal predictive performances. This procedure involves three main steps.

1. Given a learning set $\mathcal{L}_n$, a tree $\mathcal{T}_0$ is grown using one of the stopping criteria described above concerning the minimum node size. It is important that $\mathcal{T}_0$ has a large number of leaves so that all the relevant splitting rules are discovered. Thus, a small value of $n_{\min_1}$ (relatively to the sample size n) should be selected.

2. A sequence of subtrees of T_0 is created by minimizing a cost-complexity criterion. A tree T is a subtree of T_0 ($T \subset T_0$) if it can be obtained by pruning T_0, that is, by collapsing all the leaf nodes that originate from a given internal node. The cost-complexity criterion is defined as follows:

$$R_\lambda(T) = \frac{1}{n} \sum_{t \in \tilde{T}} R(t) + \lambda \cdot |\tilde{T}|, \qquad (15.7)$$

where $R(t) = \sum_{i=1}^{n} w_i(t) L(y_i, \hat{y}(t))$ summarizes the prediction error for units assigned to node t, according to a loss function L. Usually, for numerical dependent variables, $\hat{y}(t) = \bar{y}(t)$ and $L(y_i, \hat{y}(t)) = (y_i - \bar{y}(t))^2$ (quadratic loss); while for categorical dependent variables, $\hat{y}(t) = c(t)$, where $c(t)$ is the modal category of Y in node t, and $L(y_i, \hat{y}(t)) = 1 - 1_{\{c(t)\}}(y_i)$ (0–1 loss). Equation (15.7) also depends on a tuning parameter $\lambda \geq 0$ which governs the trade-off between the prediction accuracy of the tree T on the training set and its complexity (measured by $|\tilde{T}|$, the number of leaf nodes). The sequence of subtrees of T_0 is created by finding for each possible value of λ the subtree $T_\lambda \subseteq T_0$ that minimizes equation (15.7). As the notation suggests, T_0 is the solution with $\lambda = 0$. For $\lambda \to \infty$, T_λ is given by the root node. Breiman *et al.* (1984) proved that for any given value of λ, T_λ is unique. Furthermore, even if λ can take an infinite number of values, the sequence of subtrees is finite. Thus, a finite sequence $(\lambda_1 = 0, \ldots, \lambda_L)$ of increasing values of the tuning parameter λ is selected. Finally, the sequence of subtrees is nested: for any $\lambda_l > \lambda_{l'}$, $T_{\lambda_l} \subseteq T_{\lambda_{l'}}$.

3. The prediction performance of each subtree of the sequence is evaluated using a validation set, which is a set $V_u = \{(y_j, x_{j1}, \ldots, x_{jK}); j = 1, \ldots, u\}$ independent from the learning set and composed of u sample units. Usually, learning and validation sets are obtained by randomly partitioning a data set composed of $n+u$ sample units. The prediction performance of a subtree T_λ of the sequence based on V_u is measured by

$$R_V(T_\lambda) = \frac{1}{u} \sum_{t \in \tilde{T}_\lambda} \left[\sum_{j=1}^{u} z_j(t) L(y_j, \hat{y}(t)) \right], \qquad (15.8)$$

where $z_j(t) = 1$ if the jth validation sample unit is assigned to node t, and $z_j(t) = 0$ otherwise. The subtree of the sequence for which (15.8) is minimum is selected as optimal (in the literature this criterion is usually referred to as the 0-SE rule).

As an alternative to the 0-SE rule described above, Breiman *et al.* (1984) suggested selecting the smallest subtree of the sequence whose value of $R_V(T_\lambda)$ is not larger than $\min_\lambda \{R_V(T_\lambda)\}$ plus its standard error (1-SE rule). The introduction of this rule was motivated by the need to reduce the instability of the results obtained with the 0-SE rule. Furthermore, the 1-SE rule also allows the simplest subtree to be selected whose prediction accuracy is comparable to $\min_\lambda \{R_V(T_\lambda)\}$.

The requirement of an independent validation set might be problematic, especially when small data sets are involved in the analysis. Breiman *et al.* (1984) proposed a cross-validation procedure that avoids reserving a part of the data for pruning: the learning set is randomly split into V roughly equally-sized subsets; the vth subset (for $v = 1, \ldots, V$) is used as a validation set to evaluate the prediction performances of the sequence of trees $T_{\lambda_l}^v$ (for $l = 1, \ldots, L$) grown using the other $V - 1$ subsets as a learning set; finally, for each λ_l, the cross-validated prediction performance $R_{CV}(\mathcal{T}_{\lambda_l})$ is obtained by aggregating the prediction performances of the V trees $\mathcal{T}_{\lambda_l}^v$. Usually $V = 5$ or 10; when $V = n$ the cross-validation procedure is called leave-one-out cross-validation. Once the values of $R_{CV}(\mathcal{T}_{\lambda_l})$ are obtained, the 0-SE or 1-SE rule can be applied to select the optimal tree. Clearly, this procedure increases the computational burden of the pruning. Further details on the cost-complexity pruning can be found, for example, in Breiman *et al.* (1984).

Besides cost-complexity pruning, other pruning procedures have been proposed in the literature; see, for example, Niblett and Bratko (1986), Mingers (1987), Quinlan (1993), Cappelli *et al.* (2002) and Zhong *et al.* (2008). A comparison of some of these pruning procedures can be found in Esposito *et al.* (1997).

15.1.4 CHAID

The CHAID algorithm has been specifically proposed to deal with a categorical dependent variable and categorical (unordered or ordered) predictors, and uses the χ^2 statistic as a measure of goodness of split. A special feature of this algorithm is that it also considers multiway splits among the admissible splits for a given node.

Consider the first step of the tree growing procedure. For a given predictor X_k with $D_k \geq 2$ categories, the algorithm tries to reduce the $M \times D_k$ contingency table obtained from the cross-classification of the learning set with respect to Y and X_k to an $M \times J_k$ table whose χ^2 has the lowest p-value, where $2 \leq J_k \leq D_k$ (i.e., a reduced table with the strongest association between Y and a grouping of the categories of X_k). This table is obtained by merging (in an allowable manner) categories of X_k with similar conditional distributions of Y through a stepwise procedure. The type of predictor determines the allowable groupings of its categories: if X_k is an unordered predictor, any grouping of its categories is admissible; if X_k is ordered, only contiguous categories can be grouped together. This procedure leads to a splitting rule with respect to X_k with as many outcomes as the number J_k of optimally merged categories of X_k. Once this merging has been performed for all the K predictors, the p-values $\{pval_1, \ldots, pval_K\}$ of the K optimally reduced contingency tables are compared, and the lowest is selected: if it is lower than a given significance level, the corresponding splitting rule is selected, otherwise the tree growing stops.

The stepwise procedure is composed of merging and splitting steps. The first merging step searches for the pair of categories of the predictor X_k whose $M \times 2$ sub-table shows the χ^2 statistic with the highest p-value. If this p-value is larger than a given significance level, the two corresponding conditional distributions of Y can be considered similar and, hence, these two categories are merged and considered as a single compound category. Merging steps are repeated as long as pairs of (possibly compound) categories with similar conditional distributions of Y are found. Then a splitting step is performed with respect to each compound category consisting of at least three of the original categories. The binary split of these categories which identifies the $M \times 2$ sub-table whose χ^2 statistic has the lowest p-value is searched. If this p-value is lower than a given significance level, the corresponding conditional distributions of Y can be considered significantly different and, hence, the compound category is resolved according to this split. If at least one compound category is resolved, further merging steps are performed. Note that, unlike CART, CHAID does not perform an exhaustive search among all the admissible splitting rules.

Since the CHAID algorithm performs multiple tests, the p-values $\{pval_1, \ldots, pval_K\}$ of the χ^2 statistics computed with respect to the K optimally reduced contingency tables have to

Table 15.3   Results of the stepwise procedure in CHAID applied to node 1 of the tree in Figure 15.2

Country		Seniority	
Pairs of categories	*p*-values	Pairs of categories	*p*-values
Merging step 1		Merging step 1	
Benelux, Germany	0.0000	1, 2	0.6566
Benelux, Israel	0.1452	**2, 3**	0.8874
Benelux, UK	0.0412	3, 4	0.0338
Germany, Israel	0.0216	4, 5	0.1563
Germany, UK	0.0129	Merging step 2	
Israel, UK	0.6958	**1, (2, 3)**	0.6716
Merging step 2		(2, 3), 4	0.0086
Benelux, Germany	0.0000	4, 5	0.1563
Benelux, (Israel, UK)	0.0443	Merging step 3	
Germany, (Israel, UK)	0.0028	**(1, 2, 3), 4**	0.0088
		4, 5	0.1563
		Splitting step	
		(1, 2), 3	0.9691
		1, (2, 3)	0.6716

The merged pairs at the end of each merging step are in bold.

be adjusted before selecting the lowest. In particular, in CHAID the Bonferroni adjustment for dependent tests is used: $pval_k^* = B_k \cdot pval_k$, where B_k is the number of ways the D_k categories of X_k can be merged into J_k compound categories. When X_k is unordered,

$$B_k = \sum_{l=0}^{J_k-1}(-1)^l \frac{(J_k - l)^{D_k}}{l!(J_k - l)!};$$

when X_k is ordered,

$$B_k = \frac{(D_k - 1)!}{(J_k - 1)!(D_k - J_k)!}.$$

Table 15.3 gives the details of the steps performed by the stepwise procedure within the CHAID algorithm for selecting the first splitting rule based on the learning set described in Table 15.1 and using a significance level $\alpha = 0.05$. According to these merging steps, the categories of the two predictors are merged as follows: {Benelux, Germany, (Israel, UK)} for Country, and {(Seniority $\leq$ 3), (Seniority $>$ 3)} for Seniority. The adjusted *p*-values for the corresponding optimally merged contingency tables are $6 \cdot 3.99 \cdot 10^{-5} = 2.40 \cdot 10^{-4}$ for Country and $4 \cdot 3.76 \cdot 10^{-8} = 1.50 \cdot 10^{-7}$ for Seniority. Thus, at the first step of the tree growing procedure, the splitting rule {(Seniority $\leq$ 3), (Seniority $>$ 3)} shows the lowest adjusted *p*-value, which is also lower than $\alpha = 0.05$.

The best split for any node t different from the root node is derived using a strategy similar to that described above. The main difference is that only the values of the predictors observed on units assigned to node t are considered in the definition of the contingency tables.

Note that the significance level α, specified in advance by the researcher, provides an implicit stopping criterion: CHAID does not further split nodes for which the lowest adjusted p-value of the corresponding optimally reduced contingency tables is larger than α.

15.1.5 PARTY

PARTY has been devised in a general framework based on the use of suitable linear statistics which allow the user to deal not only with categorical dependent variables but also with continuous, censored, ordinal or multivariate responses. In PARTY the optimal split for a given node t is established using a two-step procedure: in the first step the predictor with the highest association with the dependent variable in node t is selected, provided that this association is significant; in the second step only admissible splits defined according to the selected predictor are examined.

The general form for the linear statistics used in PARTY for measuring the association between Y and the predictor X_k in node t is

$$\mathbf{A}_k(t) = \text{vec}\left(\sum_{i=1}^{n} w_i(t) g_k(x_{ik}) h(y_i, (y_1, \ldots, y_n))^{\top}\right) \in \mathbb{R}^{G_k \cdot H}, \qquad (15.9)$$

where $g_k : \mathcal{X}_k \to \mathbb{R}^{G_k}$ is a non-random, G_k-dimensional ($G_k \geq 1$) transformation of the predictor X_k, and $h : \mathcal{Y} \times \mathcal{Y}^n \to \mathbb{R}^H$ is an H-dimensional ($H \geq 1$) function which depends on the values of $(y_1, \ldots, y_n)$ in a permutation symmetric way, that is, the value of function h must not depend on the order in which $(y_1, \ldots, y_n)$ appear. The vec operator converts a $G_k \times H$ matrix into a $G_k \cdot H \times 1$ vector by stacking. The choice of functions g_k and h depends on the type of X_k and Y, respectively. Natural choices for g_k are $g_k(x) = x$ for numerical predictors and $g_k(u_d) = e_{D_k}(d)$ for categorical predictors, where $e_{D_k}(d)$ is a D_k-dimensional vector with dth element equal to 1 and zeros elsewhere. For example, with respect to the predictor Country, with categories {Benelux, Germany, Israel, UK}, $g_k(\text{Germany}) = (0, 1, 0, 0)^{\top}$. If a vector $\gamma = (\gamma_1, \ldots, \gamma_{D_k})^{\top}$ of numerical scores is assigned to the categories of an ordered categorical predictor, an alternative choice for function g_k is $g_k(u_d) = \gamma^{\top} e_{D_k}(d) = \gamma_d$. Similar choices can be used for h. When both Y and X_k are categorical, the linear statistics $\mathbf{A}_k(t)$ are given by the $D_k \cdot M$ frequencies of the contingency table obtained from the cross-classification of the units assigned to node t with respect to Y and X_k. In the cases of numerical or ordered predictors with numerical scores, the linear statistics $\mathbf{A}_k(t)$ are the M sums of the numerical values of the scores of X_k computed within the M categories of Y for the units assigned to node t. With respect to the learning set described in Table 15.1, the values of the linear statistics obtained from equation (15.9) at the first step of the tree growing procedure are $\mathbf{A}_1(1) = (20, 39, 14, 28, 5, 72, 9, 22)^{\top}$ for Country, and $\mathbf{A}_2(1) = (255, 388)^{\top}$ for Seniority.

The distribution of $\mathbf{A}_k(t)$ under the hypothesis of independence between Y and X_k is unknown in almost all practical circumstances. However, Strasser and Weber (1999) derived the conditional distribution of $\mathbf{A}_k(t)$ in the permutation test framework (Good, 2000). A univariate test statistic $v_k(t)$ can be obtained by standardizing $\mathbf{A}_k(t)$ using its conditional expectation and covariance, and by mapping the result into the real line. For example,

$$v_k(t) = \max_{l=1,\ldots,G_k \cdot H} \left| \frac{A_{kl}(t) - \mu_{kl}(t)}{\sigma_{kl}(t)} \right|, \qquad (15.10)$$

where $A_{kl}(t)$ is the lth element of $\mathbf{A}_k(t)$, and $\mu_{kl}(t)$ and $\sigma_{kl}(t)$ are the corresponding conditional expectation and standard deviation, respectively; for further details, see Hothorn *et al.* (2006). The p-values $\{pval_1, \ldots, pval_K\}$ of the K test statistics $v_1(t), \ldots, v_K(t)$ are then computed and the Bonferroni adjustment for independent tests is applied: $\widetilde{pval}_k = 1 - (1 - pval_k)^K$ for $k = 1, \ldots, K$. The lowest adjusted p-value is selected: if it is lower than a given significance level, the corresponding predictor is selected for defining admissible splits for node t; otherwise, no splitting rule is associated with node t, which is declared a leaf node. With respect to the learning set described in Table 15.1, the values of the univariate test statistic obtained from equation (15.10) at the first step of the tree growing procedure are $v_1(1) = 4.0514$ for Country, and $v_2(1) = 5.0954$ for Seniority; the corresponding adjusted p-values are $\widetilde{pval}_1 = 3.91 \cdot 10^{-4}$ and $\widetilde{pval}_2 = 6.96 \cdot 10^{-7}$. The predictor Seniority shows the highest association with Y, and this association is significant at levels $\alpha = 0.01$ or larger. Hence, this predictor is selected for defining admissible splits of node 1.

Once a predictor X_{k^*} has been selected, PARTY exploits the permutation test framework also to select the optimal binary splitting rule for node t. Let $S_t^{k^*}$ be the set of the admissible binary splits for node t defined according to X_{k^*} (see Section 15.1.3); for any $s \in S_t^{k^*}$, the following linear statistics are computed:

$$\mathbf{A}_{k^*}^s(t) = \mathrm{vec}\left(\sum_{i=1}^n w_i(t_l(s))h(y_i, (y_1, \ldots, y_n))^\top\right) \in \mathbb{R}^H. \tag{15.11}$$

$\mathbf{A}_{k^*}^s(t)$ induces a two-sample statistic measuring the discrepancy between the distributions of Y in the two child nodes of t defined by the splitting rule s. A univariate test statistic $v_{k^*}^s(t)$ can be obtained by standardizing $\mathbf{A}_{k^*}^s(t)$ using its conditional expectation and covariance, and by mapping the result into the real line; further details can be found in Hothorn *et al.* (2006). After computing $v_{k^*}^s(t)$ for all $s \in S_t^{k^*}$, the optimal splitting rule s^* for node t is the one with the largest value of $v_{k^*}^s(t)$:

$$s^* = \arg\max_{s \in S_t^{k^*}} v_{k^*}^s(t).$$

Table 15.4 reports the admissible splits based on the predictor Seniority and the corresponding values of $v_2^s(1)$: according to these results, in the first step of the tree growing procedure the optimal splitting rule is $\{(\text{Seniority} \leq 3), (\text{Seniority} > 3)\}$.

Table 15.4   Admissible splits of node 1 in the tree of Figure 15.2 and values of the two-sample test statistic $v_2^s(1)$

Admissible splits s of node 1	$v_2^s(1)$
$\{(\text{Seniority} \leq 1), (\text{Seniority} > 1)\}$	1.8265
$\{(\text{Seniority} \leq 2), (\text{Seniority} > 2)\}$	4.0449
$\{(\text{Seniority} \leq 3), (\text{Seniority} > 3)\}$	5.4887
$\{(\text{Seniority} \leq 4), (\text{Seniority} > 4)\}$	5.0053

Only splits based on Seniority are examined.

Similarly to CHAID, in PARTY no splitting rule is associated with a given node if the lowest adjusted p-value of the corresponding association tests is larger than α, which provides a stopping criterion for tree growing.

15.1.6   A comparison of CART, CHAID and PARTY

With respect to the learning set used in the previous examples, all of the methods examined select the same splitting rule in the first step of the tree growing. However, it should be stressed that this is not always the case. Some authors have argued that the splitting rules selected by CART are biased towards predictors with many admissible splits or many missing values; see, for example, Kass (1980), Kim and Loh (2001) and Hothorn *et al.* (2006). This bias is related to the use of an exhaustive search of the best split over all the possible predictors. The splitting rules selected by CHAID and PARTY do not share this systematic tendency. This result is achieved by separating the predictor selection from the splitting rule selection, and by resorting to comparisons among p-values. The idea of separating the selection of the predictor from the selection of the splitting rule was introduced in Loh and Vanichsetakul (1988). This approach was also employed in the two-stage algorithm described in Siciliano and Mola (2000).

Another feature that distinguishes CHAID and PARTY from CART is the absence of a pruning step after tree growing. However, using some benchmark data sets, Hothorn *et al.* (2006) showed that prediction rules obtained using PARTY trees achieve a prediction accuracy which is similar to that of the pruned trees obtained with CART. They argued that the use of association tests in the tree growing procedure avoids problems of overfitting.

15.1.7   Missing values

The methods described above require a data set without missing values. When the value of the dependent variable Y for some units is missing, these units should be discarded before growing the tree, or a preliminary data imputation should be performed by using, for example, the tree-based method for data imputation proposed by Conversano and Siciliano (2009).

In contrast, given missing values on the predictors, tree-based methods can be applied by suitably modifying the recursive partitioning procedure without requiring a preliminary data imputation or discarding units. This represents one of the main advantages of tree-based methods. Several solutions have been proposed in the literature. A first approach involves defining admissible splitting rules which explicitly take into account missing values on the predictors. For unordered categorical predictors, this can be achieved by simply adding a new category which is 'missing' and by deriving the admissible splitting rules as described in Section 15.1.3. For ordered categorical and numerical predictors, the definition of admissible splitting rules able to deal with the new 'missing' category is more complex, since the position of this category on the predictor scale is unknown. A possible solution (Kass, 1980) is to consider splitting rules in which the 'missing' category may be isolated or may be added to one of the two elements of the binary partition of $\mathcal{X}_k$ defined by a given threshold. For example, for node 4 in Figure 15.1 the admissible splits according to this approach are $s_1 = \{(\text{Germany, missing}), (\text{Israel})\}$, $s_2 = \{(\text{Germany}), (\text{Israel, missing})\}$, $s_3 = \{(\text{missing}), (\text{Germany, Israel})\}$, $s_4 = \{(\text{Seniority} \leq 4, \text{missing}), (\text{Seniority} > 4)\}$, $s_5 = \{(\text{Seniority} \leq 4), (\text{Seniority} > 4, \text{missing})\}$, and $s_6 = \{(\text{Seniority} = \text{missing}), (\text{Seniority} \neq \text{missing})\}$. This approach provides clear paths as to where the units with missing values on the predictors end up in the tree structure, and hence it allows for the discovery of relationships between the presence of missing values on some predictors and the dependent variable (Gentle *et al.*, 2004).

A second approach relies on the definition of surrogate splits to be assigned to each optimal split (Breiman *et al.*, 1984). According to this approach, when considering a predictor for a split at each step of the recursive partitioning procedure, only units for which that predictor is not missing are used to compute any of the goodness-of-split criteria described hitherto in this section. Having chosen the best splitting rule for a given node t, a list of surrogate splits is selected. The surrogate splits are ranked according to their ability to mimic the optimal split. More specifically, a provisional binary dependent variable is created by considering t_l or t_r node membership; for each predictor (excluding the one that defines the optimal split) the splitting rule with the lowest prediction error for this provisional dependent variable is selected. The $K - 1$ surrogate splits thus obtained are then ranked according to their prediction errors and compared to the blind splitting rule which assigns units to the largest child node; surrogate splits with a prediction error larger than that of the blind splitting rule are discarded. Any sample unit in node t which has a missing value for the predictor which defines the optimal splitting rule is assigned to one child node of t using the first surrogate split; if it has a missing value for the predictor defining the first surrogate split, it is assigned using the second surrogate split, and so on. If a sample unit is missing for all the predictors defining surrogate splits, the blind splitting rule is employed. An example of the use of surrogate splits is reported in Section 15.2.1. It should be noted that this second approach exploits the relationships among the predictors; however, there is no guarantee that it can find satisfactory surrogate splits, that is, surrogate splits that are better than the blind splitting rule (Gentle *et al.*, 2004).

15.1.8   Tree-based methods for applications in customer satisfaction surveys

Tree-based methods can be used to analyse customer satisfaction survey data in order to discover the aspects of a product or service that most influence overall satisfaction, recommendation levels or repurchasing intentions. This goal can be achieved by using any of these aspects as a dependent variable Y, and the opinions on specific aspects of the product or the service as predictors. Examples can be found in Nicolini and Salini (2006) and Tutore *et al.* (2007) in the context of the airline industry and public transport, respectively. In particular, in Tutore *et al.* (2007) a two-stage method allowing stratifying factors to be accounted for when growing trees is also described.

Usually, customers are asked to express their opinions on an ordinal scale with more than two levels, using categories such as 'very unsatisfied', 'somewhat unsatisfied', 'somewhat satisfied', and 'very satisfied', or 'very unlikely', 'somewhat unlikely', 'somewhat likely', and 'very likely'. The use of ordered categorical predictors is described in Section 15.1.3. The methods illustrated in Section 15.1.3 to evaluate the goodness of a split are suitable for numerical or unordered categorical dependent variables, and hence it is necessary to modify them in order to explicitly take into account the ordering of the categories of Y.

As far as the CART methodology is concerned, Piccarreta (2008) suggested the use of the following heterogeneity measure for ordinal categorical variables:

$$I_{\text{GO}}(t) = \sum_{m=1}^{M} P_m(t)[1 - P_m(t)], \tag{15.12}$$

where $P_m(t) = \sum_{j=1}^{m} p_j(t)$, $m = 1, \ldots, M$, are the cumulative proportions of Y in node t.

It is worth noting that both measures defined in equations (15.2) and (15.12) are special cases of a general mutability measure originally proposed by Gini (1954),

$$I_{GG}(t) = \sum_{m=1}^{M} \sum_{m'=1}^{M} \delta_{m,m'} p_m(t) p_{m'}(t), \qquad (15.13)$$

where $\delta_{m,m'}$ is an appropriately chosen measure of the dissimilarity between the categories c_m and $c_{m'}$ of Y. In particular, equations (15.2) and (15.12) are obtained by choosing $\delta_{m,m'} = 1 - 1_{\{c_m\}}(c_{m'})$ (0–1 loss) and $\delta_{m,m'} = \frac{|m-m'|}{2}$, respectively.

The mutability measure $I_{GG}(t)$ defined in equation (15.13) has an interesting interpretation (Breiman *et al.*, 1984). Suppose that, instead of using the majority rule for predicting a categorical dependent variable, units within node t are randomly assigned to category c_m with probability $p_m(t)$, for $m = 1, \ldots, M$. Using this alternative rule and assuming that $p_{m'}(t)$ is the estimated probability that a unit randomly selected within node t belongs to category $c_{m'}$, the estimated probability of wrongly predicting category c_m when the actual category is $c_{m'}$ is equal to $p_m(t)p_{m'}(t)$. The overall estimated probability of any prediction error in node t is given by

$$\sum_{m} \sum_{m' \neq m} p_m(t) p_{m'}(t), \qquad (15.14)$$

which coincides with the Gini heterogeneity index (15.2). Equation (15.14) implicitly assumes that all prediction errors are equally serious, that is, each prediction error is associated with a constant loss $L(c_{m'}, c_m) = \delta_{m,m'} = 1 - 1_{\{c_m\}}(c_{m'})$ (0–1 loss). Thus, $\delta_{m,m'}$ in equation (15.13) can be interpreted as a misclassification cost. For ordinal categorical dependent variables, a reasonable choice for the misclassification costs can be based on the number of categories lying between $c_{m'}$ and c_m on the corresponding ordinal scale, that is, considering costs proportional to the absolute difference between m' and m. Thus $I_{GG}(t)$ can be interpreted as an estimate of the overall cost of any prediction error at node t.

The ordinal nature of Y should also be considered in the pruning procedure, in particular when defining the loss function used in equations (15.7) and (15.8). Furthermore, any misclassification cost structure that takes into account the ordering of the categories of Y can be exploited in equation (15.13) in order to obtain further impurity measures different from $I_{GO}(t)$, and in equations (15.7) and (15.8) in order to perform tree pruning.

Another strategy for dealing with the ordering of the categories of Y within the CART methodology can be obtained by modifying the twoing method. In particular, only the $M - 1$ partitions $\{c_j : j \leq m\}$, $\{c_j : j > m\}$, for $j = 1, \ldots, M - 1$, of the set $\{c_1, \ldots, c_M\}$ are examined. Piccarreta (2008) showed that the best ordered twoing splitting rule at a node t is that which maximizes

$$\Delta I_{OTW}(t, s) = p(t_l(s)) \cdot p(t_r(s)) \max_{m \in \{1, \ldots, M-1\}} [P_m(t_l) - P_m(t_r)]^2. \qquad (15.15)$$

A comparison of these and other approaches for dealing with ordinal dependent variables within the CART methodology can be found in Piccarreta (2008).

As far as PARTY is concerned, the ordering of the categories of Y can be explicitly accounted for by the appropriate choice of the function h in equations (15.9) and (15.11). In particular, if a vector $\boldsymbol{\xi} = (\xi_1, \ldots, \xi_M)^\top$ of M numerical scores is assigned to the M categories of Y, a natural choice is $h(c_m) = \boldsymbol{\xi}^\top e_M(m) = \xi_m$, where $e_M(m)$ is an M-dimensional vector with mth element 1 and all other elements 0. As Hothorn *et al.* (2006) pointed out, when both Y and X_k are ordinal, the test resulting from equation (15.9) corresponds to the linear-by-linear association test (Agresti, 2002).

15.2   Tree-based methods and decision trees in practice

In this section an application of tree-based methods and decision trees to customer satisfaction survey data is described. In particular, data from the ABC annual customer satisfaction survey are analysed. Detailed information about this data set can be found in Chapter 2. The analysis was carried out using R (R Development Core Team, 2008). In particular, the add-on packages `rpart` (Therneau and Atkinson, 1997) and `party` were used, both being freely available from CRAN (`http://CRAN.R-project.org/`). The `rpart` package implements the CART methodology for unordered categorical and numerical dependent variables, using cross-validation techniques in the third step of the pruning procedure; `party` implements the general PARTY methodology. Another R package, `rpartOrdinal` (Archer, 2010), implements some of the goodness-of-split criteria described in Section 15.1.8 that have to be used along with `rpart`. However, the results obtained using this package are not shown in Section 15.2.1 since some of the functions in this package returned errors and warning messages when applied to the ABC data. Furthermore, this package does not allow automatic pruning.

15.2.1   ABC ACSS data analysis with tree-based methods

The aim of this application is to show how tree-based methods can be used to discover the aspects of the service provided by ABC that most influence overall satisfaction. Thus, the dependent variable considered is the overall satisfaction level with ABC (q_1), while the predictors considered are the overall satisfaction levels with: equipment (q_{11}), sales support (q_{17}), technical support (q_{25}), training (q_{31}), supplies and orders (q_{38}), software solutions (q_{43}), the customer website (q_{49}), purchasing support (q_{57}) and contracts and pricing (q_{65}). Each of these variables is measured on an ordinal scale with five levels, ranging from 'very low' (coded '1') to 'very high' (coded '5'). Since satisfaction may also be influenced by the country in which customers are based and the number of years for which customers have been using the service, in the analysis we also considered Country and Seniority as predictors. Thus, $K = 11$. Further variables in the ABC data set which could be used as dependent variables are recommendation level (q_4) and repurchasing intention (q_5).

Customers with missing values on q_1 are excluded from the analysis, while missing values on the predictors are handled by surrogate splits (see Section 15.1.7). The learning set is thus composed of $n = 262$ customers. In growing trees, no restrictions on $n_{\min_1}$ and $n_{\min_2}$ are considered (so that $n_{\min_1} = n_{\min_2} = 1$, thus allowing us to initialize the pruning step in CART from the maximally expanded tree), while for PARTY trees α is set equal to 0.01.

CART methodology is applied first, using the Gini heterogeneity index (15.2) as the impurity measure, and considering as dependent variable a dichotomization of q_1 obtained by aggregating the two highest levels on the one hand, and the three lowest ones on the other hand. The categories thus obtained are labelled 'yes' and 'no', respectively. The frequencies of these two categories in the learning set are equal to 156 and 106, respectively. Clearly, this transformation may discard some relevant information about customer satisfaction, thus it is important to be aware of this limitation.

Table 15.5 Cross-validated prediction performance of the nested trees obtained using CART

λ	0.000	0.002	0.003	0.005	0.006	0.008	0.009	0.019	0.038	0.052	0.415
$|\tilde{T}_\lambda|$	53	49	46	38	35	21	11	6	4	2	1
$r_{CV}(T_\lambda)$	0.500	0.491	0.434	0.472	0.500	0.443	0.547	0.453	0.491	0.585	1.000

The main results of the pruning procedure are summarized in Table 15.5. This procedure starts with a tree T_0 consisting of 53 leaves, and creates a sequence of 11 subtrees. Prediction performance is evaluated using leave-one-out cross-validation ($V = 262$). For each subtree, the relative prediction error $r_{CV}(T_\lambda) = \frac{R_{CV}(T_\lambda)}{0.405}$ is reported in Table 15.5, where $0.405 = \frac{106}{262}$ is the so-called root node error. The subtree with the lowest value for $r_{CV}(T_\lambda)$ is the one with 46 leaves, corresponding to $\lambda = 0.003$. The estimated standard error for $r_{CV}(T_{0.003})$ is 0.058. According to the 1-SE rule, the subtree with four leaves is selected. This tree is shown in Figure 15.3, where the distribution of the dichotomized overall satisfaction with ABC within each terminal node is also graphically represented. This tree exploits the overall levels of satisfaction with equipment (q_{11}) and technical support (q_{25}), as well as Country, to partition the customers into four groups. Node 7 contains customers with high or very high levels of satisfaction with both equipment and technical support; in this group the percentage of

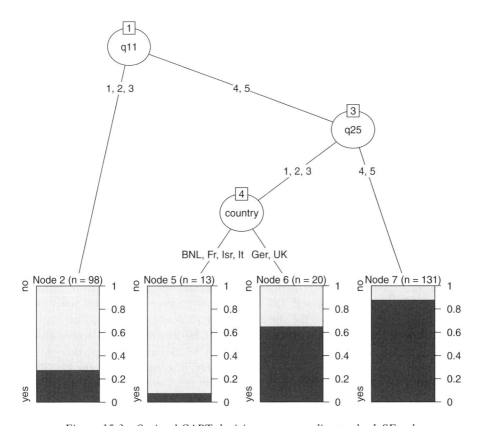

Figure 15.3   Optimal CART decision tree according to the 1-SE rule

Table 15.6   Surrogate splits for the splitting rule $\{(q_{11} \leq 3), (q_{11} > 3)\}$ selected using CART, ranked by prediction error

Splitting rule	Prediction error	Usage
$\{(q_{65} \leq 2), (q_{65} > 2)\}$	0.278	6
$\{(q_{25} \leq 2), (q_{25} > 2)\}$	0.306	1
$\{(q_{17} \leq 2), (q_{17} > 2)\}$	0.365	0
{(France, Germany, Israel, Italy, UK), (Benelux)}	0.365	0
$\{(q_{38} \leq 2), (q_{38} > 2)\}$	0.369	0
$\{(q_{57} \leq 2), (q_{57} > 2)\}$	0.373	0
Blind splitting rule	0.376	0

customers satisfied with ABC is 87.8% (against 59.5% in the learning set as a whole). Nodes 5 and 6 are both composed of customers having high or very high levels of satisfaction with equipment but very low, low or intermediate levels of satisfaction with technical support. These two nodes differ according to the Country in which customers are based: Benelux, France, Israel and Italy for node 5, Germany and UK for node 6. It is worth noting that the percentages of customers satisfied with ABC in these two nodes are very different: 7.7% and 65.0%, respectively. This highlights an interesting interaction effect of q_{11}, q_{25} and Country on the overall satisfaction with ABC. Finally, customers with very low, low or intermediate levels of satisfaction with equipment are assigned to node 2; only 27.6% of them are satisfied with ABC.

In order to deal with missing values on the predictors, rpart implements the surrogate split approach (see Section 15.1.7). For example, the value of q_{11} is missing for seven customers; therefore, they are assigned to nodes 2 or 3 using the surrogate splits. Table 15.6 reports the surrogate splitting rules identified by rpart and the blind splitting rule, along with their prediction errors (with respect to the optimal splitting rule) and the number of customers for whom they are used. Six out of the seven customers with missing values for q_{11} are assigned to the leaf nodes according to the overall satisfaction level with contracts and pricing (q_{65}), while for the assignment of the remaining customer whose values of both q_{11} and q_{65} are missing, the overall satisfaction level with technical support (q_{25}) is used (see the third column of Table 15.6, labelled 'Usage'). The same approach (with different surrogate splits) is also used at each internal node of the CART tree.

In order to take into account the ordinal nature of the dependent variable without losing any relevant information about overall satisfaction available in the data, the PARTY methodology is applied. This second analysis is carried out by assigning scores ranging from 1 to 5 to the categories of the dependent variable and of each of the ordered categorical predictors, and using equation (15.10) to derive the univariate test statistics. The resulting tree is shown in Figure 15.4. This figure also contains the bar charts for the overall satisfaction with ABC within each terminal node and the adjusted p-value for the univariate test statistic for the predictor used to define the split at each internal node. This tree exploits the overall levels of satisfaction with technical support (q_{25}), equipment (q_{11}), and contracts and pricing (q_{65}) to partition the customers into five groups. These groups show very different distributions of the overall satisfaction with ABC. In particular, nodes 3 and 4 show the highest percentages of customers whose satisfaction with ABC is very low or low (69.3%, 26.5%, respectively, against 13.7% in the learning set as a whole). In contrast, the largest percentages of customers

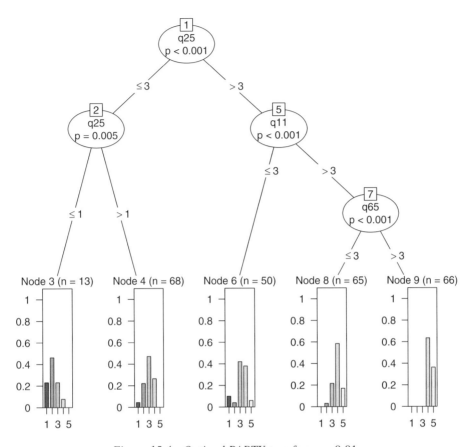

Figure 15.4   Optimal PARTY tree for $\alpha = 0.01$

who are highly or very highly satisfied with ABC are observed in nodes 8 and 9 (75.4% and 100.0%, against 59.5% in the learning set as a whole). It is interesting to note that nodes 3 and 4 are characterized by splitting rules which only involve q_{25}: node 3 contains customers whose satisfaction with technical support is very low, while customers with low or intermediate levels of satisfaction with technical support are assigned to node 4. Furthermore, the predictors q_{11} and q_{65} are only relevant for defining nodes 6, 8 and 9. In particular, customers in node 6 are highly or very highly satisfied with technical support but have low or very low levels of satisfaction with equipment. Finally, customers with high or very high levels of satisfaction with technical support, equipment, and contracts and pricing are assigned to node 9. Note that the terminal node 7 in the CART tree corresponds to the internal node 7 in the PARTY tree (which is further split into nodes 8 and 9). Note also that if this optimal tree is used to predict overall satisfaction with ABC, according to the majority rule neither the very low nor the very high levels will be assigned to any new observation.

15.2.2   Packages and software implementing tree-based methods

Tree-based methods are implemented in many commercial statistical programs, such as IBM SPSS Statistics, SAS Enterprise Miner, Salford Systems CART® and S-Plus. As far as ordered categorical dependent variables are concerned, trees based on the ordered twoing splitting method can be obtained using IBM SPSS Statistics 19 and Salford Systems CART®. The general Gini mutability measure defined in equation (15.13) is also available in CART®. In SAS Enterprise Miner the ordered nature of Y can be taken into account by specifying a suitable misclassification cost structure.

15.3   Further developments

Due to their flexibility and intuitive structure, tree-based methods are useful tools for analysing the relationship between a dependent variable and a set of predictors. However, they also have some drawbacks. The major problem with trees is their instability: even a small change in the learning set can lead to a very different sequence of optimal splitting rules and, hence, to a tree with a very different structure and very different predicted values for the same units. This issue should be taken into account in the interpretation of the results. In particular, predictors which do not define any split in a tree are not necessarily irrelevant for predicting the dependent variable, especially when predictors are highly correlated. Breiman *et al.* (1984) proposed a method to evaluate predictor importance based on surrogate splits.

The main reason for the instability of trees is the stepwise nature of the tree growing process: the selection of the split at a given step is optimal only conditionally on all the splits selected at the previous steps. Several methods have been suggested for reducing the variability in the predicted values caused by tree instability and for improving prediction performance, such as bagging (Breiman, 1996a), boosting (Breiman, 1998; Bühlmann and Hothorn, 2010; Friedman, 2001; Freund and Schapire, 1996; Shalev-Shwartz and Singer, 2010) and random forests (Breiman, 2001; Zhang and Singer, 2010). These methods involve constructing a large number of trees on the same learning set and aggregating them according to a suitable criterion. Each of these trees is obtained after introducing some random perturbation in the data or in the tree growing process (Breiman, 1996b). Several studies have shown that these methods not only outperform single trees in prediction accuracy, but also improve the evaluation of predictor importance (Sandri and Zuccolotto, 2010). A survey of these methods can be found, for example, in Hastie *et al.* (2009). One of their main drawbacks is that the aggregation of the trees they produce is no longer representable as a tree, thus losing the intuitive structure of a single tree. Alternative approaches which preserve the tree structure use resampling techniques within each node to stabilize the selection of the best splitting rule (Briand *et al.*, 2009; Dannegger, 2000) or combine a large number of trees into a consensus tree (Miglio and Soffritti 2003, 2004; Shannon and Banks, 1999). The use of these methods for analysing ordered categorical dependent variables has not yet been explored in depth and thus represents an interesting direction for future research.

References

Agresti, A. (2002) *Categorical Data Analysis*, 2nd edn. New York: John Wiley & Sons, Inc.

Archer, K.J. (2010) rpartOrdinal: An R package for deriving a classification tree for predicting an ordinal response. *Journal of Statistical Software*, 34, 1–17.

Belson, W.A. (1959) Matching and prediction on the principle of biological classification. *Applied Statistics*, 8, 65–75.

Breiman, L. (1996a) Bagging predictors. *Machine Learning*, 24, 123–140.

Breiman, L. (1996b) Heuristics of instability and stabilization in model selection. *Annals of Statistics*, 24, 2350–2383.

Breiman, L. (1996c) Technical note: some properties of splitting criteria. *Machine Learning*, 24, 41–47.

Breiman, L. (1998) Arcing classifiers. *Annals of Statistics*, 26, 801–849.

Breiman, L. (2001) Random forests. *Machine Learning*, 45, 5–32.

Breiman, L., Friedman, J.H., Olshen, R.A. and Stone, C.J. (1984) *Classification and Regression Trees*. Belmont, CA: Wadsworth.

Briand, B., Ducharme, G.R., Parache, V. and Mercat-Rommens, C. (2009) A similarity measure to assess the stability of classification trees. *Computational Statistics & Data Analysis*, 53, 1208–1217.

Bühlmann, P. and Hothorn, T. (2010) Twin boosting: Improved feature selection and prediction. *Statistics and Computing*, 20, 119–138.

Cappelli, C., Mola, F. and Siciliano, R. (2002) A statistical approach to growing a reliable honest tree. *Computational Statistics & Data Analysis*, 38, 285–299.

Clark, P. and Niblett, T. (1989) The CN2 induction algorithm. *Machine Learning*, 3, 261–283.

Clémençon, S., Depecker, M. and Vayatis, N. (2011) Adaptive partitioning schemes for bipartite ranking. How to grow and prune a ranking tree. *Machine Learning*, 83, 31–69.

Conversano, C. and Siciliano, R. (2009) Incremental tree-base imputation with lexicographic ordering. *Journal of Classification*, 26, 361–379.

Dannegger, F. (2000) Tree stability diagnostics and some remedies for instability. *Statistics in Medicine*, 19, 475–491.

Donoho, D.L. (1997) CART and best-ortho-basis: a connection. *Annals of Statistics*, 25, 1870–1911.

Esposito, F., Malerba, D. and Semeraro, G. (1997) A comparative analysis of methods for pruning decision trees. *IEEE Transactions on Pattern Analysis and Machine Intelligence*, 19, 476–491.

Fierens, D., Ramon, J., Blockeel, H. and Bruynooghe, M. (2010) A comparison of pruning criteria for probability trees. *Machine Learning*, 78, 251–285.

Friedman, J.H. (2001) Greedy function approximation: A gradient boosting machine. *Annals of Statistics*, 29, 1189–1232.

Freund, Y. and Schapire, R. (1996) Experiments with a new boosting algorithm. In L. Saitta (ed.), *Machine Learning: Proceedings of the Thirteenth International Conference*, pp. 148–156. San Francisco: Morgan Kaufmann.

Gentle, J.E., Härdle, W. and Mori, Y. (eds) (2004) *Handbook of Computational Statistics: Concepts and Methods*. Berlin: Springer.

Gini, C. (1954) *Variabilità e Concentrazione*. Roma: Veschi, Italy.

Good, P. (2000) *Permutation Tests: A Practical Guide to Resampling Methods for Testing Hypotheses*, 2nd edn. New York: Springer.

Hastie, T., Tibshirani, R. and Friedman, J.H. (2009) *The Elements of Statistical Learning*, 2nd edn. New York: Springer.

Hothorn, T., Hornik, K. and Zeileis, A. (2006) Unbiased recursive partitioning: A conditional inference framework. *Journal of Computational and Graphical Statistics*, 15, 651–674.

Hush, D. and Porter, R. (2010) Algorithms for optimal dyadic decision trees. *Machine Learning*, 80, 85–107.

Kass, G.V. (1980) An exploratory technique for investigating large quantities of categorical data. *Applied Statistics*, 29, 119–127.

Kim, H. and Loh, W.Y. (2001) Classification trees with unbiased multiway splits. *Journal of the American Statistical Association*, 96, 589–604.

LeBlanc, M. and Crowley, J. (1992) Relative risk trees for censored survival data. *Biometrics*, 48, 411–425.

Loh, W.Y. (2011) Classification and regression trees. *Wiley Interdisciplinary Reviews: Data Mining and Knowledge Discovery*, 1, 14-23.

Loh, W.Y. and Vanichsetakul, N. (1988) Tree-structured classification via generalized discriminant analysis. *Journal of the American Statistical Association*, 83, 715–725.

Miglio, R. and Soffritti, G. (2003) Methods to combine classification trees. In M. Schader, W. Gaul and M. Vichi (eds), *Between Data Science and Applied Data Analysis*, pp. 65–73. Berlin: Springer.

Miglio, R. and Soffritti, G. (2004) The comparison between classification trees through proximity measures. *Computational Statistics & Data Analysis*, 45, 577–593.

Mingers, J. (1987) Expert systems – rule induction with statistical data. *Journal of the Operational Research Society*, 28, 39–47.

Mola, F. and Siciliano, R. (1997) A fast splitting procedure for classification trees. *Statistics and Computing*, 7, 209–216.

Morgan, J.N. and Sonquist, J.A. (1963a) Problems in the analysis of survey data, and a proposal. *Journal of the American Statistical Association*, 58, 415–434.

Morgan, J.N. and Sonquist, J.A. (1963b) Some results from a non-symmetrical branching process that looks for interaction effects. *Proceedings of the Social Statistics Section of the ASA*, 58, 40–53.

Murthy, S.K. (1998) Automatic construction of decision trees from data: A multi-disciplinary survey. *Data Mining and Knowledge Discovery*, 2, 345–389.

Niblett, T. and Bratko, I. (1986) Learning decision rules in noisy domains. In *Proceedings in Expert Systems 86*, pp. 25–34. Cambridge: Cambridge University Press.

Nicolini, G. and Salini, S. (2006) Customer satisfaction in the airline industry: The case of British Airways. *Quality and Reliability Engineering International*, 22, 581–589.

Noh, H.G., Song, M.S. and Park, S.H. (2004) An unbiased method for constructing multilabel classification trees. *Computational Statistics & Data Analysis*, 47, 149–164.

Piccarreta, R. (2008) Classification trees for ordinal variables. *Computational Statistics*, 23, 407–427.

Quinlan, J.R. (1993) *C4.5: Programs for Machine Learning*. San Mateo, CA: Morgan Kaufmann.

R Development Core Team (2008) *R: A Language and Environment for Statistical Computing*. Vienna: R Foundation for Statistical Computing.

Sandri, M. and Zuccolotto, P. (2010) Analysis and correction of bias in total decrease in node impurity measures for tree-based algorithms. *Statistics and Computing*, 20, 393–407.

Segal, M.R. (1988) Regression trees for censored data. *Biometrics*, 44, 35–47.

Shalev-Shwartz, S. and Singer, Y. (2010) On the equivalence of weak learnability and linear separability: New relaxations and efficient boosting algorithms. *Machine Learning*, 80, 141–163.

Shannon, W.D. and Banks, D. (1999) Combining classification trees using MLE. *Statistics in Medicine*, 18, 727–740.

Siciliano, R. and Mola, F. (2000) Multivariate data analysis and modeling through classification and regression trees. *Computational Statistics & Data Analysis*, 32, 285–301.

Siciliano, R. and Conversano, C. (2009) Decision tree induction. In J. Wang (ed.), *Encyclopedia of Data Warehousing and Data Mining*, pp. 624–630. Hershey, PA: Information Science Reference.

Siciliano, R., Aria, M. and D'Ambrosio, A. (2008) Posterior prediction modelling of optimal trees. In P. Brito (ed.), *Proceedings in Computational Statistics (COMPSTAT 2008)*, pp. 323–334. Heidelberg: Physica-Verlag.

Strasser, H. and Weber, C. (1999) On the asymptotic theory of permutation statistics. *Mathematical Methods of Statistics*, 8, 220–250.

Therneau, T.M. and Atkinson, E.J. (1997) An introduction to recursive partitioning using the rpart routines. Technical report 61, Mayo Foundation.

Tutore, V.A., Siciliano, R. and Aria, M. (2007) Conditional classification trees using instrumental variables. In M.R. Berthold, J. Shawe-Taylor J. and N. Lavrač (eds), *Advances in Intelligent Data Analysis VII*, Lecture Notes in Computer Science 4723, pp. 163–173. Berlin: Springer.

Zhang, H. (1998) Classification trees for multiple binary responses. *Journal of the American Statistical Association*, 93, 180–193.

Zhang, H. and Singer, B.H. (2010) *Recursive Partitioning and Applications*, 2nd edn. New York: Springer.

Zhang, H. and Ye, Y. (2008) A tree-based method for modeling a multivariate ordinal response. *Statistics and Its Interface*, 1, 169–178.

Zhong, M., Georgiopoulos, M. and Anagnostopoulos, G.C. (2008). A k-norm pruning algorithm for decision tree classifiers based on error rate estimation. *Machine Learning*, 71, 55–88.

16

PLS models

Giuseppe Boari and Gabriele Cantaluppi

This chapter deals with the partial least squares (PLS) estimation algorithm and its use in the context of structural equation models with latent variables (SEM-LV). After a short description of the general structure of SEM-LV models, the PLS algorithm is introduced; then, statistical and geometrical interpretations of PLS are given. A detailed application example based on the ABC annual customer satisfaction survey data concludes the chapter.

16.1 Introduction

The partial least squares (PLS) algorithm, developed by Wold (1982, 1985), was first introduced in the context of multiple linear regression models, in order to overcome the problems arising in the estimation procedures when overparameterization and multicollinearity occur. The former, often encountered in chemiometric analyses, is due to the presence of more explanatory variables than there are observations available, so that ordinary least squares (OLS) procedures cannot operate correctly. The latter derives from subsets of explanatory variables being highly correlated, which dramatically reduces estimator efficiency.

In these cases, the so-called PLS regression algorithm may be adopted: in order to summarize the original explanatory variables, a new set of latent (non-observable) auxiliary variables is generated, which, in a fashion similar to latent factors, reduces the complexity of the model. These factors are then used as the regressors of the dependent variables in the original model. The values assumed by the latent variables (latent scores) are estimated by means of the alternating least squares (ALS) algorithm (Wold, 1966) which maximizes the correlation between these variables and the original explanatory variables, as well as that with the dependent variables. This is the main characteristic of all PLS-based algorithms (see below for a more detailed presentation of the procedure).

More recently (Lohmöller, 1989; Wold, 1985), the PLS estimation algorithm has been adopted within the framework of structural equation models with latent variables (SEM-LV)

Modern Analysis of Customer Surveys: with applications using R, First Edition. Edited by Ron S. Kenett and Silvia Salini.
© 2012 John Wiley & Sons, Ltd. Published 2012 by John Wiley & Sons, Ltd.

as an alternative to the so-called covariance structure estimation procedures (Bollen, 1989), the latter ones making use of the celebrated LISREL procedure.[1] SEM-LV models, also called path analysis models, are concerned with the analysis of the relationships existing between latent variables, not directly observable, which are also defined latent constructs.

These models are typical in socio-economic analysis (e.g., those concerning consumer behaviour (Chin, 1998)), psychometric analysis, and the analysis of all complex systems described by a multi-equation model of the form

$$\eta = f(\eta, \xi) + \zeta, \tag{16.1}$$

where η represents the vector of the so-called endogenous variables (which depend at least on one other variable internal to the model), ξ is the vector of the exogenous variables (explanatory variables or input variables that might be explained by variables external to the model), $f(\cdot)$ represents a vector function, usually linear, describing the relationship between all the components of the endogenous η and the exogenous ξ and possibly the other components of η, and ζ represents an error component, a random variable with mean zero, uncorrelated with ξ, describing random errors on η and explaining factors not considered in the model. Observe that mean independence of ζ conditional on ξ, also called the pseudo-isolation (Bollen, 1989), is a condition for achieving good properties of the estimators when OLS procedures are applied; moreover, recall that under the stated conditions model (16.1) cannot be interpreted as being of a strictly equation error type, since the error terms can only be interpreted as representing possible exogenous variables being excluded from the model, but uncorrelated with all the explanatory variables considered in the model.

16.2   The general formulation of a structural equation model

A linear SEM-LV consists of two sets of equations: the structural or inner model (see (16.1)) describing the path of the relationships among the latent variables, and the measurement or outer model, representing the relationships among the latent variables and appropriately corresponding measurable variables (called manifest variables or proxy variables) used to reconstruct the unobservable latent ones. The direction of causality should be known in advance, on the basis of the theory explaining the phenomena under investigation.

16.2.1   The inner model

Following the conventional notation of the LISREL community, the structural model is represented by the following linear relation:

$$\eta = \mathbf{B}\eta + \mathbf{\Gamma}\xi + \zeta, \tag{16.2}$$

where η is an $(m \times 1)$ vector of latent endogenous random variables (dependent variables) whose levels are conditional on those of the other latent variables in the model; ξ is an $(n \times 1)$ vector of latent exogenous random variables (explanatory variables, which may possibly depend on variables external to the model); ζ is an $(m \times 1)$ vector of error components, zero-mean random variables, summarizing the measurement errors on the η components and/or

---

[1] LISREL is an acronym for 'Linear Structural Relationships', and the name is also used for the computer software developed by Jöreskog and Sörbom (2006).

other explanatory variables, independent of the $\boldsymbol{\xi}$, not directly included in the model. $\mathbf{B}$ and $\boldsymbol{\Gamma}$ are respectively $(m \times m)$ and $(m \times n)$ matrices containing the so-called structural parameters. In particular, the matrix $\mathbf{B}$ contains information concerning the linear relationships among the latent endogenous variables: their elements represent the direct casual effect on each η_i $(i = 1, \ldots, m)$ of the remaining η_j $(j \neq i)$. The matrix $\boldsymbol{\Gamma}$ contains the coefficients explaining the relationships among the latent exogenous and the endogenous variables; their elements represent the direct causal effects of the $\boldsymbol{\xi}$ components on the η_i variables.

Relation (16.2) is called a structural equation model since it describes the causal structure of the process under investigation. The presence of zeros in $\mathbf{B}$ or $\boldsymbol{\Gamma}$ means the absence of causal relationships between the corresponding latent variables. The main diagonal of $\mathbf{B}$ has necessarily null elements, since an endogenous variable cannot cause itself.

If the matrix $\mathbf{I} - \mathbf{B}$ is not singular, which means that none of the m equations in (16.2) is redundant, the model may be written in the alternative reduced form

$$\boldsymbol{\eta} = (\mathbf{I} - \mathbf{B})^{-1}\boldsymbol{\Gamma}\boldsymbol{\xi} + (\mathbf{I} - \mathbf{B})^{-1}\boldsymbol{\zeta}, \tag{16.3}$$

showing that the endogenous variables are functions only of the exogenous variables and the error components.

Finally, the structural model also defines covariance matrices $\boldsymbol{\Phi}$ $(n \times n)$ and $\boldsymbol{\Psi}$ $(m \times m)$ respectively of the exogenous latent variables $\boldsymbol{\xi}$ and of the errors $\boldsymbol{\zeta}$. It is assumed that $E(\boldsymbol{\xi}) = \mathbf{0}$ and $E(\boldsymbol{\zeta}) = \mathbf{0}$, so that $\boldsymbol{\Phi} = E(\boldsymbol{\xi}\boldsymbol{\xi}')$ and $\boldsymbol{\Psi} = E(\boldsymbol{\zeta}\boldsymbol{\zeta}')$.

When the matrix $\mathbf{B}$ is lower triangular or can be recast as lower triangular by changing the order of the elements in $\boldsymbol{\eta}$ – which is possible if $\mathbf{B}$ has all zero eigenvalues (Faliva, 1992) – and $\boldsymbol{\Psi}$ is diagonal, then model (16.2) is said to be of recursive type, which excludes feedback effects. In the sequel we will assume $\mathbf{B}$ to be lower triangular.

A simple example of structural model is shown in Figure 16.1, where the latent variables are represented by ellipses. The path diagram refers to a model containing two exogenous latent variables, ξ_1 and ξ_2, and two endogenous variables, η_1 and η_2. The corresponding structural model, written in a non-concise form, is

$$\begin{bmatrix} \eta_1 \\ \eta_2 \end{bmatrix} = \begin{bmatrix} 0 & 0 \\ \beta_{21} & 0 \end{bmatrix} \begin{bmatrix} \eta_1 \\ \eta_2 \end{bmatrix} + \begin{bmatrix} \gamma_{11} & \gamma_{12} \\ 0 & 0 \end{bmatrix} \begin{bmatrix} \xi_1 \\ \xi_2 \end{bmatrix} + \begin{bmatrix} \zeta_1 \\ \zeta_2 \end{bmatrix},$$

where

$$\mathbf{B} = \begin{bmatrix} 0 & 0 \\ \beta_{21} & 0 \end{bmatrix} \quad \text{and} \quad \boldsymbol{\Gamma} = \begin{bmatrix} \gamma_{11} & \gamma_{12} \\ 0 & 0 \end{bmatrix}.$$

Under the hypothesis that the covariance matrix $\boldsymbol{\Psi}$ is diagonal, the model is of recursive type.

The recursive model is the one most commonly employed in applications, where the main interest of the experimenter is focused on the estimation of the non-null parameters contained in $\mathbf{B}$ and $\boldsymbol{\Gamma}$ and of the levels (scores) of the latent variables (in particular, the endogenous ones).

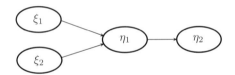

Figure 16.1   A path diagram example

Those inference procedures are made possible by the availability of the observable variables (manifest variables) linked to the latent ones according to the measurement model, also called the outer model.

16.2.2 The outer model

In order to make inference on the structural model (16.2) it is necessary to identify appropriate observable measures linked to the latent variables. The relationships between manifest and latent variables are described by the measurement (outer) model, and may assume, according to the nature of the problem at hand, one of the following forms:

$$\mathbf{x} = \mathbf{\Lambda_x \xi} + \mathbf{\delta_x}, \tag{16.4}$$

$$\mathbf{y} = \mathbf{\Lambda_y \eta} + \mathbf{\varepsilon_y}, \tag{16.5}$$

or

$$\mathbf{\xi} = \mathbf{\Pi_x x} + \mathbf{\delta_\xi}, \tag{16.6}$$

$$\mathbf{\eta} = \mathbf{\Pi_y y} + \mathbf{\varepsilon_\eta}, \tag{16.7}$$

or a mixture thereof, where the vector random variables $\mathbf{x}$ ($q \times 1$) and $\mathbf{y}$ ($p \times 1$) represent the indicators for the latent variables $\mathbf{\xi}$ and $\mathbf{\eta}$, respectively. Relations (16.4) and (16.5) describe the so-called measurement model of reflective type, where the proxy variables $\mathbf{x}$ and $\mathbf{y}$ are functions of the corresponding latent ones. Relations (16.6) and (16.7) constitute a formative measurement model, the latent variables being functions of the observed/manifest ones (Diamantopoulos *et al.*, 2008).

For a recursive model to be identified by means of the PLS-based estimation procedures, it is necessary that $q \geq n$ and $p \geq m$; moreover, there must be at least one proxy for each latent variable (the covariance structure approach requires more stringent conditions for model identifiability).

The matrices $\mathbf{\Lambda_x}$ and $\mathbf{\Lambda_y}$ contain the linear coefficients expressing the linear relationships between the latent and the corresponding manifest variables, while $\mathbf{\Pi_x}$ and $\mathbf{\Pi_y}$ contain the linear regression coefficients representing the reverse relationships.

The vectors $\mathbf{\delta_x}$, $\mathbf{\varepsilon_y}$, $\mathbf{\delta_\xi}$, $\mathbf{\varepsilon_\eta}$ are the error components, zero-mean random variables, mutually independent and independent of the explanatory components considered in the linear models.

It is also assumed, for the reflective measurement models, that they possess the so-called factor complexity of order 1, which means that each manifest variable is linked to one and only one latent variable.

Finally, as detailed in Bayol *et al.* (2000), the sets of exclusive proxy variables linked to the latent variables included in the model should possess a high level of reliability, also called 'internal consistency' of the scale. The reflective proxy variables making up the set of measures pertaining each individual latent variable should in fact be positively correlated[2] both with the latent factor and, as a consequence, with each other, because they are measuring, to a certain extent, the same common entity. Cronbach's α coefficient is obtained as an average of the correlations between every pair of proxy variables; as a rule of thumb (Nunnally, 1978), values of α greater than 0.7 are interpreted as indicating good internal consistency of the scale.

---

[2] If one manifest variable has negative correlation with other variables, we define it to be a 'reverse item', in which case a score reversing procedure is appropriate.

In Van Zyl *et al.* (2000) the construction of a confidence interval for α is presented; in Zanella and Cantaluppi (2006) the assumptions for applying Cronbach's α are reported and a suggestion is made for the construction of a test procedure aimed at establishing the existence and the validity of the latent construct.

16.3   The PLS algorithm

The PLS algorithm proceeds in a manner substantially different from the more traditional estimating techniques. In the so-called covariance-based procedures, such as LISREL, the model parameter estimates are first obtained, using an approach similar to the method of moments, and then the latent scores may be estimated (Bollen, 1989). In the PLS procedure the sequence is reversed: a first iterative phase is carried out to obtain the latent score estimates, and then the values of the vector $\boldsymbol{\theta}$, containing all the unknown parameters in the model (**B**, **Γ**, etc.), are estimated, by applying the OLS method to all the linear multiple regression sub-problems into which the inner and outer models are decomposed. Since we have assumed that the inner model is of the recursive type, the OLS procedure is always appropriate. Figure 16.2 shows the flow diagram for the PLS algorithm.

For the sake of simplicity, the following notation is conventionally used by PLS community. The set of all the latent variables (always measured as differences from their respective average

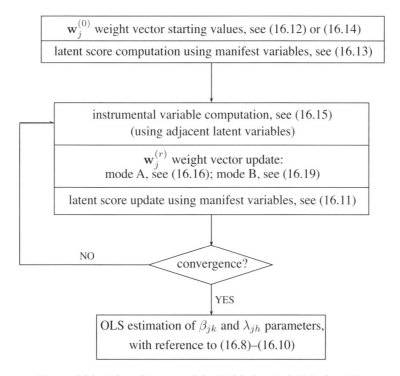

Figure 16.2   Flow diagram of the Wold classical PLS algorithm

values) is represented by the vector

$$\mathbf{Y} = [Y_1, \ldots, Y_n, Y_{n+1}, \ldots, Y_{n+m}]'$$

where the first n elements are of exogenous type and the remaining m endogenous. The p_j manifest variables corresponding to every latent Y_j, $j = 1, \ldots, n + m$, are gathered together in the vector $\mathbf{X}_j$, with elements X_{jh}, $h = 1, \ldots, p_j$. Every manifest variable is therefore defined as $X_{jh} = X_{jh}^* - \bar{x}_{jh}^*$, in which X_{jh}^* is the original observed variable whose mean value $\bar{x}_{jh}^*$ is computed by the average of the available data x_{jhi}^*, where $i = 1, 2, \ldots, N$ refers to the generic statistical unit. Moreover, the $\mathbf{B}$ and $\boldsymbol{\Gamma}$ matrices are combined into the matrix $\mathbf{B}^* = [\boldsymbol{\Gamma}|\mathbf{B}] = \{\beta_{jk}^*\}$, where

$$\beta_{jk}^* = \begin{cases} \gamma_{jk}, & k \le n, \\ \beta_{j,k-n}, & k > n, \end{cases}$$

giving (16.2) the form

$$\begin{bmatrix} Y_{n+1} \\ \vdots \\ Y_{n+m} \end{bmatrix} = \mathbf{B}^* \mathbf{Y} + \boldsymbol{\zeta} = [\boldsymbol{\Gamma}|\mathbf{B}] \begin{bmatrix} \boldsymbol{\xi} \\ \boldsymbol{\eta} \end{bmatrix} + \boldsymbol{\zeta},$$

and $\boldsymbol{\Lambda}^* = \{\lambda_{jh}\}$ may be defined by applying the direct sum operator to the matrices $\boldsymbol{\Lambda}_{\mathbf{X}}$ and $\boldsymbol{\Lambda}_{\mathbf{Y}}$,

$$\boldsymbol{\Lambda}^* = \{\lambda_{jh}\} = \boldsymbol{\Lambda}_{\mathbf{X}} \oplus \boldsymbol{\Lambda}_{\mathbf{Y}} = \begin{bmatrix} \boldsymbol{\Lambda}_{\mathbf{X}} & \mathbf{0} \\ \mathbf{0} & \boldsymbol{\Lambda}_{\mathbf{Y}} \end{bmatrix}.$$

The error terms in the measurement model are analogously combined into the vector ε.

Consequently, the SEM-LV model (recursive in the inner part) may be rewritten in the following extended manner:

$$Y_j = \sum_{k=1}^{j-1} \beta_{jk}^* Y_k + \zeta_j, \qquad \text{(inner part)} \qquad j = n+1, \ldots, n+m, \qquad (16.8)$$

$$X_{jh} = \lambda_{jh} Y_j + \varepsilon_{jh}, \qquad \text{(outer mode A)} \qquad j = 1, \ldots, n+m, \qquad (16.9)$$

$$Y_j = \sum_{h=1}^{p_j} \pi_{jh} X_{jh} + \delta_j, \qquad \text{(outer mode B)} \qquad j = 1, \ldots, n+m, \qquad (16.10)$$

where mode A refers to a model of the reflective type (such as factor analysis) and mode B to a formative model (such as multiple regression analysis). Observe that, since $\mathbf{B}$ is assumed to be lower triangular, the generic endogenous variable Y_j, $j = n + 1, \ldots, n + m$, may only depend on the exogenous variables $Y_1, \ldots, Y_n$ and on a subset of its preceding endogenous latent variables. Moreover, it is also assumed that $E(\zeta_j|Y_1, \ldots, Y_{j-1}) = 0$ ($j = n + 1, \ldots, n + m$), $E(\varepsilon_{jh}|Y_j) = 0$ and $E(\delta_j|X_{j1}, \ldots, X_{jp_j}) = 0$ ($j = 1, \ldots, n + m$), also called predictor specification.

With reference to the inner model, we may define the following square matrix $\mathbf{T}$, of order $n + m$, indicating the structural relationships among the latent variables:

$$
\begin{array}{cc}
\text{exogenous} \left\{ \begin{array}{c} Y_1 \\ \vdots \\ Y_n \end{array} \right. &
\begin{array}{|cccccc|}
\hline
0 & 0 & 0 & 0 & 0 & 0 \\
0 & 0 & 0 & 0 & 0 & 0 \\
0 & 0 & 0 & 0 & 0 & 0 \\
\end{array} \\
\text{endogenous} \left\{ \begin{array}{c} Y_{n+1} \\ \vdots \\ Y_{n+m} \end{array} \right. &
\begin{array}{|cccccc|}
 & & & 0 & 0 & 0 \\
 & & & & 0 & 0 \\
 & & & & & 0 \\
\hline
\end{array}
\end{array}
$$

The generic element t_{jk} of $\mathbf{T}$ is given unit value if the endogenous Y_j is linked to Y_k (i.e., those variables are adjacent in the path diagram); t_{jk} is null otherwise. Observe that the first n rows of $\mathbf{T}$, containing only zeros, correspond to the exogenous variables; furthermore, the upper triangular part of this matrix contains null elements and the last m rows correspond to the matrix $\mathbf{B}^* = [\mathbf{\Gamma}|\mathbf{B}]$ previously defined.

Once causal relationships pertaining the inner model have been established, the PLS algorithm follows the iterative structure described in Figure 16.2, which defines, at the rth generic step, the scores of each latent variable Y_j as a linear combination of the manifest variables corresponding to Y_j,

$$\hat{Y}_j = \sum_{h=1}^{p_j} w_{jh}^{(r)} X_{jh}, \quad j = 1, 2, \ldots, n + m, \tag{16.11}$$

with appropriate weights $w_{j1}^{(r)}, \ldots, w_{jp_j}^{(r)}$ summing to 1. The relationships defining the estimates $\hat{Y}_j$ of the latent scores are called 'outside approximations' (Lohmöller, 1989).

The starting step of the algorithm uses an arbitrarily defined weighting system that, for the sake of simplicity, may take the form

$$\mathbf{w}_j^{(0)} = [1, 0, \ldots, 0], \quad j = 1, 2, \ldots, n + m, \tag{16.12}$$

to which the following initial latent score estimates correspond:

$$\hat{Y}_j = \sum_{h=1}^{p_j} w_{jh}^{(0)} X_{jh}, \quad j = 1, 2, \ldots, n + m. \tag{16.13}$$

Dealing with missing values

When any missing values are encountered, the following approaches are available to deal with the problem: case deletion; mean replacement by variable (column) or by group mean; or all the available data can be used. Since the problem of missing data is commonly encountered in practice, we give a short description of these approaches.

Denote the state of a generic data value x_{jhi}, $j = 1, 2, \ldots, n + m$, $h = 1, 2, \ldots, p_j$, $i = 1, 2, \ldots, N$, for the ith statistical unit and the jth observed variable belonging to the group of the p_j manifest variables linked to the latent variable Y_j, by means of the indicator function

$$I_{jhi} = \begin{cases} 1, & \text{if } x_{jhi} \text{ is a valid case,} \\ 0, & \text{if } x_{jhi} \text{ is missing,} \end{cases}$$

where N is the total number of respondents. Observe that in this case the x_{jhi} represent the original values, on which no centring transformation has been carried out.

1. *Case deletion.* When there is at least one missing value over the whole set of manifest variables of a data record, then the complete case is deleted from the data set, so that it is not considered in the subsequent analyses (with a non-negligible loss of information when several deletions are done).

2. *Mean replacement.* Two alternative procedures may be adopted:

    (a) by variable (by column),

    $$\text{if } I_{jhi} = 0 \text{ then } x_{jhi} = \frac{1}{\sum_{i=1}^{N} I_{jhi}} \sum_{i=1}^{N} x_{jhi} I_{jhi},$$

    which replaces the missing value with the average computed all over the available data for the X_{jh} variable;

    (b) by group mean (within the same block of manifest variables),

    $$\text{if } I_{jhi} = 0 \text{ then } x_{jhi} = \frac{1}{\sum_{h=1}^{p_j} I_{jhi}} \sum_{h=1}^{p_j} x_{jhi} I_{jhi},$$

    which replaces the missing value with the average computed over the group of the remaining X_{jh} variables pertaining to the latent Y_j.

3. *Use all the available data.* When the generic case, for every block of manifest variables, has at least one non-missing value, impute the missing x_{jhi} with the $\tilde{y}_{ji}$ defined by (16.20) below; otherwise if all values in a block are missing, the entire case is deleted.

Adaptation of the PLS algorithm when there are missing values

Due to the possibility of missing values, we define the vectors of initial PLS weights as

$$\mathbf{w}_j^{(0)} = [1/p_j, \ 1/p_j, \dots, 1/p_j]', \quad j = 1, 2, \dots, n + m, \tag{16.14}$$

in order to prevent possible inconsistent computation of the latent variable scores when using (16.12), which may also give rise to non-convergence of the iterative procedure.

Let $\bar{x}_{jh}$, $h = 1, 2, \dots, p_j$, denote the mean of the manifest variables X_{jh} associated with the latent variable Y_j:

$$\bar{x}_{jh} = \frac{1}{\sum_{i=1}^{N} I_{jhi}} \sum_{i=1}^{N} x_{jhi} I_{jhi},$$

where I_{jhi} is redefined to have a unit value too, when the specific missing value has been replaced; N now becomes the number of cases remaining after possible deletions.

In procedure 3 (use all available data), the uncentred scores of the latent variables Y_j, $j = 1, 2, \ldots, n + m$, are initially defined as linear combinations of the values of the corresponding manifest variables,

$$\tilde{y}_{ji} = \frac{1}{\sum_{h=1}^{p_j} w_{jh}^{(0)} I_{jhi}} \sum_{h=1}^{p_j} w_{jh}^{(0)} x_{jhi} I_{jhi};$$

then all missing data values of the generic case $i = 1, 2, \ldots, N$ of the manifest variables X_{jh}, $h = 1, 2, \ldots, p_j$, are replaced with the previous values $\tilde{y}_{ji}$, that is, $x_{jhi} = \tilde{y}_{ji}$ (if $I_{jhi} = 0$); in this way, missing values are imputed by group *weighted* mean. In this case the zero value of the corresponding indicator I_{jhi} does not need to be redefined, since the procedure of imputation is carried out at every step of the PLS algorithm.

In every procedure for dealing with missing data, the initial scores of the latent variables Y_j, $j = 1, 2, \ldots, n + m$, are defined as linear combinations of the centred values of the corresponding manifest variables $X_{jh}, h = 1, 2, \ldots, p_j$:

$$\hat{y}_{ji} = \frac{1}{\sum_{h=1}^{p_j} w_{jh}^{(0)}} \sum_{h=1}^{p_j} w_{jh}^{(0)} (x_{jhi} - \bar{x}_{jh})$$

where $\bar{x}_{jh} = \sum_{i=1}^{N} x_{jhi}/N$ if any imputation of missing data values has been done. Observe that the previous procedure, also reported in Boari *et al.* (2007), differs slightly from that proposed by Lohmöller (1989) and described in Tenenhaus *et al.* (2005, pp. 171–172): with the imputation of the missing data, at every algorithm iteration (see step 3 below), the average values $\bar{x}_{jh}$ are updated and then the latent variables centred.

Iterative phases of the PLS algorithm

Step 1. For each latent variable Y_j, define an instrumental variable Z_j as a linear combination of the estimates of the latent variables Y_k linked to Y_j in the path diagram:

$$Z_j = \sum_{k=1}^{n+m} \tau_{jk} \hat{Y}_k, \tag{16.15}$$

where $\tau_{jk} = \max(t_{jk}, t_{kj})\text{sign}[\text{Cov}(\hat{Y}_j, \hat{Y}_k)]$ (remember that t_{jk} is the generic element of the matrix $\mathbf{T}$, used to formalize the relationships in the inner model: $t_{jk} = 1$ if the latent variable Y_j is connected with Y_k in the path model representation, $t_{jk} = 0$ otherwise) and

$$\text{Cov}(\hat{Y}_j, \hat{Y}_k) = \frac{1}{N-1} \sum_{i=1}^{N} \hat{y}_{ji} \hat{y}_{ki}$$

where $\hat{Y}_j$ has zero mean.

Step 2. For the so-called mode A (reflective outer model) model, at every stage r of the iteration ($r = 1, 2, \ldots$) update the vectors of the weights $w_j^{(r)}$ as follows:

$$w_{jh}^{(r)} = \pm C_{jh} / \sum_{h=1}^{p_j} C_{jh}, \quad j = 1, 2, \ldots, n + m, \ h = 1, 2, \ldots, p_j, \tag{16.16}$$

where

$$C_{jh} = \text{Cov}(X_{jh}, Z_j) = \frac{1}{N-1} \sum_{i=1}^{N} (x_{jhi} - \bar{x}_{jh})(z_{ji} - \bar{z}_{jh}), \tag{16.17}$$

$$\bar{z}_{jh} = \frac{1}{N} \sum_{i=1}^{N} z_{ji}, \quad \pm = \text{sign} \left(\sum_{h=1}^{p_j} \text{sign}[\text{Cov}(X_{jh}, \hat{Y}_j)] \right), \tag{16.18}$$

in which

$$\text{Cov}(X_{jh}, \hat{Y}_j) = \frac{1}{N-1} \sum_{i=1}^{N} (x_{jhi} - \bar{x}_{jh})(\hat{y}_{ji} - \bar{y}_j)$$

and

$$\bar{y}_j = \frac{1}{N} \sum_{i=1}^{N} \hat{y}_{ji}.$$

For the formative outer model (Mode B) the weights are obtained by normalizing the solutions of the following multiple OLS problem:

$$Z_j = \sum_{h=1}^{p_j} w_{jh}^{(r)} (x_{jhi} - \bar{x}_{jh}) + \delta_j, \quad j = 1, 2, \dots, n+m, \ h = 1, 2, \dots, p_j. \tag{16.19}$$

Step 3. Compute the provisional uncentred latent scores with the updated weights:

$$\tilde{y}_{ji} = \frac{1}{\sum_{h=1}^{p_j} w_{jh}^{(r)} I_{jhi}} \sum_{h=1}^{p_j} w_{jh}^{(r)} x_{jhi} I_{jhi}. \tag{16.20}$$

Then, after replacement of the missing data x_{jhi} ($I_{jhi} = 0$) with the values $\tilde{y}_{ji}$, calculate

$$\hat{y}_{ji} = \frac{1}{\sum_{h=1}^{p_j} w_{jh}^{(r)}} \sum_{h=1}^{p_j} w_{jh}^{(r)} (x_{jhi} - \bar{x}_{jh}), \tag{16.21}$$

where

$$\bar{x}_{jh} = \frac{1}{N} \sum_{i=1}^{N} x_{jhi}.$$

Observe that, as already mentioned for procedure 3 (use all available data), at every iteration of the PLS algorithm the $\tilde{y}_{ji}$ values are temporarily computed only by means of the available manifest values; then a group weighted mean procedure is used for imputation of the missing data with the latent variable value, calculated on available data.

Looping. Loop to step 1 until the convergence criterion

$$\left[\sum_j \sum_h \left(w_{jh}^{(r)} - w_{jh}^{(r-1)} \right)^2 \right]^{1/2} \leq \varepsilon.$$

is attained, where ε is an appropriately chosen positive convergence tolerance value.

Ending phase of the PLS algorithm

By using the final scores, one can carry out OLS estimation of the β_{jk} coefficients linking Y_k to Y_j (for every inner submodel), the λ_{jh} parameters (outer models, mode A), the π_{jh} parameters (outer models, mode B), specifying the linear relations between the latent Y_j and the corresponding manifest X_{jh} and finally the residual variances. Observe that when the missing-data imputation is carried out all covariances are computed using the available data together with the reconstructed data.

16.4   Statistical interpretation of PLS

In general, all the PLS estimation procedures introduced in the literature (Esposito Vinzi *et al.*, 2010) are based on the construction of latent scores as appropriate linear combinations of the observed variables, each method employing a different weighting system. Moreover, under the assumption of factor complexity of order 1, every latent variable Y_j, $j = 1, \ldots, n + m$, is related to a unique set of p_j manifest variables grouped into the vector $\mathbf{X}_j = [X_{j1}, X_{j2}, \ldots, X_{jp_j}]'$.

In the sampling framework, for each latent Y_j there exists an $N \times p_j$ matrix $\mathbf{X}_j$ of the available observations; let $\hat{\mathbf{y}}_j$ be the corresponding $N \times 1$ vector of score estimates.

The PLS iterative estimation algorithm defines at the rth iteration the so-called 'outer approximation' of the jth latent variable as

$$\hat{y}_{ji} = \sum_{h=1}^{p_j} w_{jh}^{(r)} x_{jhi}, \quad i = 1, \ldots, N,$$

or, in matrix notation,

$$\hat{\mathbf{y}}_j = \mathbf{X}_j \mathbf{w}_j^{(r)}, \tag{16.22}$$

where $\hat{y}_{ij}$ is the generic element of the vector $\hat{\mathbf{y}}_j$, $\mathbf{w}_j^{(r)}$ the weighting vector defined at the rth step of the iteration, and x_{jhi} the (i, h)th element of the matrix $\mathbf{X}_j$. Observe that the entire set of observations constitutes a matrix $\mathbf{X}$ of order $(N, \sum_{j=1}^{n+m} p_j)$ obtained from the $(n + m)$ matrices $\mathbf{X}_j$ just defined.

The so-called 'inner' estimate of the latent Y_j is defined upon the definition of the so-called instrumental variables Z_j according to (16.15).

We now wish to clarify the meaning of the updating phase of the weights $w_{jh}^{(r)}$, supporting the existence of the convergence of the PLS algorithm. Once the instrumental variables have been defined the procedure assumes, in the case of the reflective measurement models, the following relationship between the observed and instrumental variables:

$$X_{jh} = \gamma_{jh}^{(r)} Z_j + E_{jh}, \quad h = 1, \ldots, p_j, \tag{16.23}$$

where the $\gamma_{jh}^{(r)}$ are the regression coefficients and E_{jh} the error components such that $\mathrm{Cov}(Z_j, E_{jh}) = 0$. The weights $w_{jh}^{(r)}$ are then obtained as the solution of an OLS problem; for this reason their values will be proportional to the correlation between X_{jh} and Z_j. Therefore, relationship (16.22) becomes (Tenenhaus, 1999),

$$\hat{Y}_j \propto \sum_h \mathrm{Corr}(X_{jh}, Z_j) X_{jh}.$$

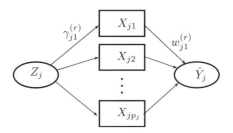

Figure 16.3   Main phases of the PLS recursive algorithm

In the reflective model (mode A), in fact, each manifest variable X_{jh} is linked to the corresponding instrumental Z_j by (16.23). In order to estimate the unknown $\gamma_{jh}^{(r)}$ coefficients we need to minimize $\mathrm{Var}(E_{jh})$, that is, to maximize the explained variance $\mathrm{Var}(\gamma_{jh}^{(r)} Z_j)$, which also corresponds to the maximization of $\mathrm{Var}(w_{jh}^{(r)} X_{jh})$ for each X_{jh} ($h = 1, \ldots, p_j$), when the weighting system $w_{jh}^{(r)}$ defined by (16.16) is used. Recalling that $\hat{Y}_j$ is obtained as a linear combination of the observed X_{jh}, the main phases of the iterative algorithm may be represented by the diagram in Figure 16.3.

Should the observations of the Y_j actually be available, then the right-hand side of the diagram will represent a multiple regression model. If not, the $\hat{Y}_j$ are defined as a linear combination of the X_{jh}, with the weights proportional to the correlations between Z_j and X_{jh}.

However, since $\hat{Y}_j$ is an approximation of Z_j, the use of the weights $w_{jh}^{(r)}$ corresponds to the choice/selection of the coefficients of the linear model minimizing the distance between $\hat{Y}_j$ and the X_{jh} by means of the OLS criterion.

Therefore, the procedure stops when $\hat{Y}_j$ becomes as close as possible to Z_j, that is, when the correlation between $\hat{Y}_j$ and Z_j attains its maximum value.

Recalling (16.16), where the weights are appropriately rescaled, it can be observed that the use of the corresponding $\hat{Y}_j$ estimates, obtained with (16.22) (or (16.11) when no missing values are present), seems to be a possible solution to the problem of the indeterminacy of the latent scores of Y_j; this problem occurs since, for instance, every transformation $c\hat{Y}_j$ represents an equivalent estimate of Y_j.

A possible remedy could arise by pre-defining the length of the vector of the estimated scores $\hat{Y}_j$ (i.e., $[\sum_h \hat{y}_{jh}^2]^{1/2} = constant$), which corresponds to fixing the variance of $\hat{Y}_j$; this condition is also used in order to ensure model identifiability.

16.5   Geometrical interpretation of PLS

The statistical interpretation of the PLS algorithm, set out in the previous section, may be complemented by the following geometrical interpretation: the values of the estimated scores for each latent variable form a vector $\hat{\mathbf{y}}_j$, in the domain $\mathfrak{R}^n$ of the observed vectors, obtained as a linear combination of the columns of the matrix $\mathbf{X}_j$ pertaining to Y_j. Since, in general, all the elements in the weighting system combining the observed proxy variables linked to Y_j are positive and have unit sum, $\hat{\mathbf{y}}_j$ belongs to the convex cone of the linear space spanned by the p_j vectors of those manifest variables.

Moreover, recall that the $\mathbf{w}_j$ weights are proportional to the $\boldsymbol{\gamma}_j$ coefficients attaining the minimum distances between the Z_j and the observed variables X_{jh} (or equivalently,

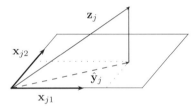

Figure 16.4   Geometrical interpretation of the PLS solution in $\Re^2$

between $\hat{Y}_j$ and Z_j). This also corresponds to the maximization of the correlation between $\hat{Y}_j$ and Z_j.

Figure 16.4 shows the particular case where $p_j = 2$: it can be observed that the solution to the problem is given by that linear (convex) combination corresponding to the orthogonal projection of the vector $\mathbf{z}_j$, the realization of the random variable Z_j, into the subset of the linear space spanned by the two vectors $\mathbf{x}_{j1}$ and $\mathbf{x}_{j2}$.

Finally, observe that the vector $\hat{\mathbf{y}}_j$ so identified is the nearest to the vectors $\mathbf{x}_{jh}$ (the columns of the matrix $\mathbf{X}_j$) that are close to the adjacent latent variables that are represented by Z_j, so giving a path analysis structure (the coefficients of the matrix $\mathbf{B}^*$) showing the best linear relationship.

16.6   Comparison of the properties of PLS and LISREL procedures

In this section we compare the characteristics of the models to which PLS and LISREL refer and consequently the properties of the corresponding parameter estimators.

Firstly, we recall that, under the normality assumption, the LISREL estimates are invariant with respect to scale transformations, are consistent, asymptotically unbiased and efficient within the class of consistent estimators. Moreover, they have an asymptotic multivariate normal distribution. Nevertheless, they may suffer from a drawback, referred to in the literature as the improper solution problem, due in general to the limited number of available observations and possibly the presence of multicollinearity among the observed variables (Grapentine, 2000; Grewal *et al.*, 2004; Hulland *et al.*, 2010). Moreover, the LISREL procedure typically makes use of the maximum likelihood method, requiring explicit knowledge of the data distribution (usually assumed to be multivariate normal). The PLS approach imposes different conditions (authors refer to 'soft modelling'); nevertheless it provides robust estimators, though they are not in general scale invariant and unbiased.

Concerning the parameters in the PLS and LISREL models, we make the following remarks with reference to the noteworthy paper by Schneeweiss (1993). The LISREL approach takes into consideration the covariance structure of the model and refers to the bijective relationship existing between the space of the unknown parameters and that of the covariance matrices, whose elements can be expressed as a function of the parameters, according to the specified statistical model (16.2) and (16.4)–(16.7); The PLS algorithm, as observed above, makes reference to the Wold causality structure and all parameters are obtained, having defined the vectors of the latent scores $\hat{Y}_j$ as a linear convex combination of the columns of the matrices $\mathbf{X}_j$, such that maximum closeness to the instrumental variables Z_j (which also best summarize all the adjacent latent variables of each $\hat{Y}_j$) is achieved. The reference model for the PLS procedure is to be considered slightly different from that for LISREL, and theoretical relationships between the sets of outer and inner parameters defining the two different models are given. In particular, one may expect that the distributions of the inner model parameter estimators are stochastically lower for PLS than for LISREL (Dijkstra, 1983). As a consequence PLS will, on average, underestimate the inner model parameters compared to the LISREL approach. Thus, in the perspective of testing the hypothesis that the above-mentioned coefficients are zero, statistical significance of a β_{jk} coefficient is expected to be confirmed also within the LISREL approach. However, the PLS estimators do not possess all the other asymptotic properties previously mentioned.

In Schneeweiss (1993) it is shown that every entity (parameter or latent score) estimated by means of the PLS algorithm may be expressed as a function of the corresponding theoretical (i.e., actual) entity describing the structural model; the author also states the condition for the PLS entities to converge towards the LISREL ones. Briefly, with reference to the generic block of p_j manifest variables $\mathbf{X}_j$, corresponding to the latent Y_j, we recall (16.9) which, for the sake of simplicity, we rewrite, omitting the indices, in the following matrix form:

$$\mathbf{X} = \lambda Y + \mathbf{E},$$

where the components of $\mathbf{E}$ are orthogonal to Y. Following Wold (1985), we also assume that $\sigma_Y^2 = 1$, that is, the latent variable considered is standardized. It then follows that

$$\boldsymbol{\Sigma}_{XX} = \lambda\lambda' + \boldsymbol{\Sigma}_{EE}.$$

Now define the quantities

$$k_1 = a^2/\lambda'\lambda, \quad k_2 = a^4/\lambda'\lambda, \tag{16.24}$$

where a represents the largest eigenvalue of $\boldsymbol{\Sigma}_{EE}$. Schneeweiss (1993) shows that, as k_1 vanishes, the PLS latent variables converge in probability to the corresponding LISREL ones. Moreover, denoting by $\hat{\lambda}$ the PLS estimate of λ, we have

$$\lim_{k_2 \to 0} (\hat{\lambda} - \lambda)'(\hat{\lambda} - \lambda) = 0,$$

which implies that $(\hat{\lambda} - \lambda) \to 0$ with vanishing k_2, where the convergence is uniform with respect to all the elements of $(\hat{\lambda} - \lambda)$, and for finite values of p_j.

In a similar fashion it is shown that

$$\lim_{k_1 \to 0} [(\hat{\lambda} - \lambda)'(\hat{\lambda} - \lambda)]/(\lambda'\lambda) = 0.$$

The convergence of the PLS latent variables to the corresponding LISREL values, for vanishing k_1 and k_2, is called 'consistency at large'. It occurs, for example, when p_j is large, but it may also happen that both k_1 and k_2 are small when the latent variable Y_j is not linked to a large number of proxy variables.

In other words, the quantity controlling the closeness of the two models is not only the number of observed variables linked to each latent variable (just sufficient condition), but also the ratio between the maximum eigenvalue of the error covariance matrix and the sum of squares of the weights. The smaller the ratio, the most profitable the PLS method becomes (and, in addition, PLS is computationally simpler than LISREL). However, it must be observed that the quantities k_1 and k_2 are defined as functions of unknown entities; so it seems reasonable to wonder when they may be considered negligible. For this purpose, it may be underlined that a^2 will assume, in general, a small value, since the proxy variables belonging to a single block should show high correlation. In particular, it is reasonable to expect that

$a^2 \simeq O(1/p_j)$, recalling that p_j is the number of observable variables linked to the generic latent Y_j. Furthermore, the correlation between the manifest and latent variables generally being high, the variances of the proxy and latent variables are finite and of the same order of magnitude; then $k_1 \simeq o(1/p_j^2)$, leading in practice to small values of the ratios previously considered.

16.7   Available software for PLS estimation

There are several packages available for carrying out PLS estimation, some of which are confined to the PLS regression algorithm. For models including (structural equation models with) latent variables we may refer to the software review presented in Temme *et al.* (2006). A list of the available software dealing substantially with all the features described in the preceding paragraphs must begin with the Fortran executable program LVPLS by Lohmöller (1984), followed by PLSPATH (Sellin, 1989). PLS-GUI (Li, 2005) and VisualPLS (Fu, 2006) are graphical interfaces to LVPLS. Also available are PLS-Graph (Chin, 2001), SmartPLS (Ringle *et al.*, 2005) and the PLS module in the Spad system (Test&Go, 2006). XLSTAT-PLSPM is an Excel add-in running within XLSTAT (Addinsoft, 2007). These have recently been joined by the stand-alone WarpPLS 1.0 (Kock, 2010), developed in Matlab.

Two procedures for PLS estimation are also available in the R libraries *semPLS* (Monecke, 2010) and *plspm* (Sanchez and Trinchera, 2010), but they do not provide tools for dealing with missing values other than case deletion.

Matlab scripts, reproducing the PLS-VB Excel add-in developed by Boari and Cantaluppi (2005), were used for the analysis of the application discussed in the next section, where missing data values are present.

We conclude with a brief list of the most common available software dealing with the covariance-based procedures, previously mentioned as an alternative approach to structural equation modelling with latent variables: LISREL (Jöreskog and Sörbom, 2006), Mplus (Muthén and Muthén, 2010), EQS (Bentler, 1995), Statistica-SEPATH (StatSoft, 2011), IBM SPSS Amos (IBM, 2011), which are commercial software, and the R libraries *sem* (Fox, 2010) and *lavaan* (Rosseel, 2011).

16.8   Application to real data: Customer satisfaction analysis

The most celebrated models for the analysis of customer satisfaction concern the definition of the so-called National Customer Satisfaction Indices, which started with the construction of the well-known Swedish Barometer (Fornell, 1992), followed by the analogous ACSI and ECSI models referring respectively to the US and European markets (Cassel and Eklöf, 2001; ECSI, 1998; Fornell *et al.*, 1996).

The aim of National Customer Satisfaction models is to measure the satisfaction of consumers of goods offered by the most important industries (e.g., automotive and consumer electronics) and of services (e.g., retail banking, telecommunications, insurance and transportation) provided by public or private organizations. The corresponding indices are embedded within a system of cause and effect relationships which has evolved in recent decades (Hackl and Westlund, 2000; Johnson *et al.*, 2001); in the latter work the possible application of the model is suggested at the company level (see also Kanji and Wallace, 2000).

The ACSI and ECSI models are slightly different, though both involve the following constructs:

- customer expectations,

- perceived quality,

- perceived value,

- overall satisfaction, (CS Index)

- customer loyalty,

where customer expectations, perceived quality and perceived value act as antecedents of overall satisfaction. The image latent variable is present only in the ECSI model, playing the role of unique exogenous latent variable; with regard to the latent variables depending on overall satisfaction, both models define customer loyalty as the final endogenous variable, while the ACSI model considers also the presence of the mediating effect of the customer complaints between overall satisfaction and customer loyalty.

The PLS approach has been adopted for the estimation of the parameters for all the customer satisfaction models mentioned above, as well as the subsequent estimation of the latent scores (e.g., those measuring the perceived overall satisfaction and loyalty), both at the national, industry and company levels.

As an example of the application of the PLS algorithm for the analysis of customer satisfaction at the micro level we considered the analysis of the 2010 annual customer satisfaction survey administered to a sample of ABC company customers.

In particular, only the 65 items referring to the following main aspects of the services supplied by ABC were included in the analysis:

- overall satisfaction with ABC (Y_7, Y_8),

- equipment,

- sales support (Y_3),

- technical support (Y_1),

- training (Y_5),

- supplies (Y_6),

- purchasing support (Y_2),

- contracts and pricing (Y_4).

The items pertaining to software solutions and customer website were excluded from the analysis due to the presence of a filtering question, which dramatically reduced the availability of complete data.

A preliminary factor analysis was performed on the chosen variables, as also suggested by Parasuraman *et al.* (1991) in their seminal paper on customer satisfaction measurement for services. Table 16.1 reports the output of the procedure; as usual, negligible loadings

Table 16.1   Factor analysis for the first 65 statements in the ABC 2010 ACSS

Statement	F1	F2	F3	F4	F5	F6	F7
s1						0.6151	
s2						0.6024	
s4						0.7125	
s5						0.6878	
s9				0.4072			
s11				0.5315			
s12			0.6747				
s13			0.8750				
s14			0.9296				
s15			0.7757				
s16			0.7978				
s17			0.9227				
s18	0.8482						
s19	0.8775						
s20	0.8008						
s21	0.6730						
s23	0.8916						
s24	0.6864						
s25	0.9611						
s26					0.9196		
s27					0.8681		
s28					0.8813		
s29					0.7338		
s30					0.7192		
s32						0.5792	0.5001
s33						0.5054	0.4620
s36							0.8683
s37							0.9986
s38							0.6582
s50		0.5507					
s51		0.8530					
s52		0.6847					
s53		0.7967					
s54		0.7168					
s55		0.8064					
s56		0.7092					
s57		0.9539					
s58				0.4637			
s59				0.7616			
s61				0.6142			
s63				0.6541			
s64				0.8307			
s65				0.8956			

Factors are shown in descending order of eigenvalues. A promax rotation was performed.

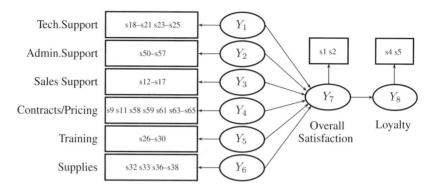

Figure 16.5   SEM-LV model initially considered

(with absolute values less than 0.4) are not presented. Observe that the grouping of items obtained by rotating the initial principal component solution is coherent with the main aspects of the questionnaire listed above. In our opinion, overall satisfaction with ABC can be split into the following two aspects: overall satisfaction (s1, s2) and loyalty (s4, s5). Moreover, items s32 and s33 pertaining to supplies have properly been assigned to factor F7 (Y_6), while items s9 and s11 have been classified within F4 (Y_4), instead of within the group labelled 'equipment'.

In line with the foregoing discussion the SEM model shown in Figure 16.5 is proposed; the symbols previously used for the factors have been appropriately redefined for the six exogenous and two endogenous latent variables considered in the model. Observe that the measurement model is of reflective type.

The analysis of the reliability of the latent constructs is reported in Table 16.2. The values of Cronbach's α are all greater than 0.7 – usually taken as a good reliability level (Nunnally, 1978) – providing evidence that the proxies considered were appropriately selected: they possess high internal consistency, thus measure common entities. Moreover, adapting the results given, for example, in Zanella and Cantaluppi (2006), the 95% confidence limits, reported in the last columns, were obtained.

Table 16.2   Reliability analysis

Latent variable	Proxies	Cronbach's α	95% conf. limits	
Y_1	7	0.9190	0.8701	0.9541
Y_2	8	0.8902	0.8252	0.9376
Y_3	6	0.8716	0.7922	0.9277
Y_4	8	0.8879	0.8215	0.9363
Y_5	5	0.8517	0.7567	0.9172
Y_6	5	0.8103	0.6889	0.8941
Y_7	2	0.7493	0.5033	0.8735
Y_8	2	0.8755	0.7533	0.9371

Table 16.3   Inner model estimates for the initial solution

	Coefficient	t	p-value
β_{71}	0.3518	6.5191	0.0000
β_{72}	0.1683	2.1270	0.0344
β_{73}	0.1593	2.2195	0.0274
β_{74}	0.3669	4.6278	0.0000
β_{75}	0.0232	0.4122	0.6806
β_{76}	−0.0104	−0.1242	0.9012
β_{87}	0.8822	19.0740	0.0000

The analysis then continued with the estimation of the customer satisfaction model using the PLS procedure. Missing values were imputed using the procedure 3 (using all available data). Table 16.3 shows the output from a typical computer program implementing the PLS algorithm. Observe that the regression coefficients linking Y_5 and Y_6 to Y_7 are not significant: the model has consequently been reduced by taking into account only four exogenous variables (see Table 16.4 for the final model parameter estimates). The variables Y_1 (technical support) and Y_4 (contracts and pricing) show the highest regression coefficients with the endogenous variable Y_7: as expected, the level of overall satisfaction expressed by the customers of an IT services company mainly depends on those two aspects. It has also been confirmed, as well known in the literature, that the relationship between overall satisfaction and loyalty is characterized by a very high level of dependence ($\beta_{87} = 0.8834$).

Furthermore, by comparing the average values of the estimated latent scores (see Table 16.5) with the estimated regression coefficients of the final inner model, it may be stated that, in order to improve the average level of overall customer satisfaction (and thus also of customer loyalty), ABC should first intervene by improving the level of Y_4, which is characterized by the lowest average score (performance) and a high level of the coefficient β_{74} (impact), as suggested also by examining the impact–performance matrix shown in Figure 16.6.

In this way the most successful intervention is defined since by moving from the lowest average value there is a greater room to improve. Moreover, the notable value of the impact coefficient (β_{74}) will allow the Y_7 level to be increased in a more effective manner by intervening on Y_4 than on the other latent exogenous variables.

Table 16.4   Inner model estimates for the final solution

	Coefficient	t	p-value
β_{71}	0.3412	6.9136	0.0000
β_{72}	0.1746	2.5943	0.0100
β_{73}	0.1751	2.5517	0.0113
β_{74}	0.3676	4.8840	0.0000
β_{87}	0.8834	19.5550	0.0000

Table 16.5 Estimated latent scores expressed as percentages

	Y_1	Y_2	Y_3	Y_4	Y_7	Y_8
1	27.70	39.97	10.24	6.52	0.00	0.00
2	40.81	39.97	59.71	45.76	50.00	75.00
3	64.87	63.29	43.32	52.79	38.88	38.06
4	32.79	60.57	59.79	41.09	50.00	75.00
5	89.20	68.63	92.09	69.83	75.92	75.00
6	91.45	37.74	75.00	73.47	62.96	75.00
7	87.18	60.02	54.27	51.03	62.96	63.06
8	75.00	75.00	75.00	57.48	62.04	75.00
9	58.52	35.03	57.28	57.39	62.04	75.00
10	68.07	75.00	62.08	50.44	62.96	63.06
11	25.89	54.46	46.10	41.73	37.04	50.00
12	59.23	58.17	45.86	50.00	50.00	61.94
⋮	⋮	⋮	⋮	⋮	⋮	⋮
average	71.11	63.70	61.17	55.29	59.07	68.91

Only the first dozen records and the scores averages are reported.

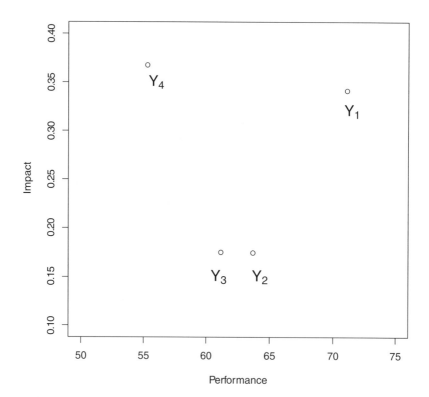

Figure 16.6 Impact–performance matrix

References

Addinsoft (2007) *XLStat-PLSPM*. Available at http://www.xlstat.com/en/products/xlstat-plspm.

Bayol, M.P., de la Foye, A., Tellier, C. and Tenenhaus, M. (2000) Use of the PLS path modelling to estimate the European Customer Satisfaction Index (ECSI) model. *Statistica Applicata*, 12(3), 361–375.

Bentler, P.M. (1995) *EQS Program Manual*. Encinso, CA: Multivariate Software Inc.

Boari, G. and Cantaluppi, G. (2005) Selection of structural equation models with the PLS-VB programme. In M. Vichi, P. Monari, S. Mignani and A. Montanari (eds), *New Developments in Classification and Data Analysis*, pp. 105–112. Berlin: Springer.

Boari, G., Cantaluppi, G. and De Lauri, A. (2007) Handling missing values in PLS path modelling: Comparison of alternative procedures. In *Classification and Data Analysis 2007, Book of Short Papers, Meeting of the Classification and Data Analysis Group of the Italian Statistical Society*, pp. 641–644. Università degli Studi di Macerata, EUM.

Bollen, K.A. (1989) *Structural Equations with Latent Variables*. New York: John Wiley & Sons, Inc.

Cassel, C. and Eklöf, J.A. (2001) Modelling customer satisfaction and loyalty on aggregate levels: Experience from the ECSI pilot study. *Total Quality Management*, 12(7), 834–841.

Chin, W.W. (1998) The partial least squares approach for structural equation modelling. In G.A. Marcoulides (ed.), *Modern Methods for Business Research*, pp. 295–336. Mahwah, NJ: Lawrence Erlbaum Associates.

Chin, W.W. (2001) *PLS-Graph Users Guide Version 3.0*. Houston, TX: C.T. Bauer College of Business, University of Houston.

Diamantopoulos, A., Riefler, P. and Roth, K.P. (2008) Advancing formative measurement models. *Journal of Business Research*, 61, 1203–1218.

Dijkstra, T. (1983) Some comments on maximum likelihood and partial least squares methods. *Journal of Econometrics*, 22, 67–90.

ECSI (1998) European Customer Satisfaction Index – Foundation and structure for harmonised national pilot projects. Report prepared by ECSI Technical Committee, ECSI Document No. 005 ed. 1 20-1 1-98.

Esposito Vinzi, V., Trinchera, L. and Amato, S. (2010) PLS path modeling: From foundations to recent developments and open issues for model assessment and improvement. In V. Esposito Vinzi, W.W. Chin, J. Henseler and H. Wang (eds), *Handbook of Partial Least Squares*, pp. 47–82. New York: Springer.

Faliva, M. (1992) Recursiveness vs. interdependence in econometric models: A comprehensive analysis for the linear case. *Journal of the Italian Statistical Society*, 1(3), 335–357.

Fornell, C. (1992) A National Customer Satisfaction Barometer: The Swedish experience. *Journal of Marketing*, 56(1), 6–21.

Fornell, C., Johnson, M.D., Anderson, E.W., Cha, J. and Bryant, B.E. (1996) The American Customer Satisfaction Index: Nature, purpose, and findings. *Journal of Marketing*, 60(4), 7–18.

Fox, J., with contributions from Kramer, A. and Friendly, M. (2010) sem: Structural equation models. R package version 0.9-21. Available at http://CRAN.R-project.org/package=sem.

Fu, J.-R. (2006) VisualPLS – partial least square (PLS) regression – an enhanced GUI for Lvpls (PLS 1.8 PC) Version 1.04b1. Available at http://www2.kuas.edu.tw/prof/fred/vpls/index.html.

Grapentine, T. (2000) Path analysis vs. structural equation modeling. *Marketing Research*, 12(3), 12–20.

Grewal, R., Cote, J.A. and Baumgartner, H. (2004) Multicollinearity and measurement error in structural equation models: Implications for theory testing. *Marketing Science*, 23(4), 519–529.

Hackl, P. and Westlund, A.H. (2000) On structural equation modeling for customer satisfaction measurement. *Total Quality Management*, 11(4), 820–825.

Hulland, J., Ryan, M.J. and Rayner, R.K. (2010) Modeling customer satisfaction: A comparative performance evaluation of covariance structure analysis versus partial least squares. In V. Esposito Vinzi, W.W. Chin, J. Henseler and H. Wang (eds), *Handbook of Partial Least Squares*, pp. 307–325. New York: Springer.

IBM SPSS Statistics (2011) *IBM SPSS Amos*. Available at http://www.spss.com/amos.

Johnson, M.D., Gustafsson, A., Andreassen, T.W., Lervik, L. and Cha, J. (2001) The evolution and future of national customer satisfaction index models. *Journal of Economic Psychology*, 22, 217–245.

Jöreskog, K.G. and Sörbom, D. (2006)*LISREL 8.8 for Windows* [computer software]. Lincolnwood, IL: Scientific Software International.

Kanji, G.K. and Wallace, W. (2000) Business excellence through customer satisfaction. *Total Quality Management*, 11(7), 979–998.

Kock, N. (2010) *WarpPLS 1.0*. Laredo, TX: ScriptWarp Systems.

Li, Y. (2005) *PLS-GUI Graphic User Interface for Partial Least Squares (PLS-PC 1.8) Version 2.0.1 beta*. University of South Carolina, Columbia, SC.

Lohmöller, J.-B. (1984) *LVPLS Program Manual - Version 1.6*. Zentralarchiv für Empirische Sozialforschung, Universität zu Köln.

Lohmöller, J.-B. (1989) *Latent Variable Path Modeling with Partial Least Squares*. Heidelberg: Physica-Verlag.

Monecke, A. (2010) semPLS: Structural equation modeling using partial least squares. R package version 0.8-7. Available at http://CRAN.R-project.org/package=semPLS.

Muthén, L.K. and Muthén, B.O. (2010) *Mplus User's Guide*, 6th edn. Los Angeles: Muthén & Muthén.

Nunnally, J.C. (1978) *Psychometric Theory*, 2nd edn. New York: McGraw-Hill.

Parasuraman, A., Berry, L.L. and Zeithaml, V.A. (1991) Refinement and reassessment of the SERVQUAL scale. *Journal of Retailing*, 67(4), 420–450.

Ringle, C.M., Wende, S. and Will, A. (2005) *SmartPLS Version 2.0*. Universität Hamburg.

Rosseel, Y. (2011) lavaan: Latent variable analysis. R package version 0.4-8. Available at http://CRAN.R-project.org/package=lavaan.

Sanchez, G. and Trinchera, L. (2010) plspm: Partial least squares data analysis methods. R package version 0.1-11. Available at http://CRAN.R-project.org/package=plspm.

Schneeweiss, H. (1993) Consistency at large in models with latent variables. In K. Haagen, D.J. Bartholomew and M. Deistler (eds), *Statistical Modelling and Latent Variables*, pp. 299–320. Amsterdam: North-Holland.

Sellin, N. (1989) *PLSPATH - Version 3.01. Application Manual*. Universität Hamburg.

StatSoft (2011) *Electronic Statistics Textbook*. StatSoft. Available on http://www.statsoft.com/textbook.

Temme, D., Kreis, H. and Hildebrandt, L. (2006) PLS path modeling – a software review. SFB 649 Discussion Papers SFB649DP2006-084, Sonderforschungsbereich 649, Humboldt University, Berlin. Available at http://sfb649.wiwi.hu-berlin.de/papers/pdf/SFB649DP2006-084.pdf.

Tenenhaus, M. (1999) L'approche PLS. *Revue de Statistique Appliquée*, 47(2), 5–40.

Tenenhaus, M., Esposito Vinzi, V., Chatelin, Y.-M. and Lauro, C. (2005) PLS path modeling. *Computational Statistics & Data Analysis*, 48, 159–205.

Test&Go (2006) *Spad Version 6.0.0*. Paris.

Van Zyl, J.M., Neudecker, H. and Nel, D.G. (2000) On the distribution of the maximum likelihood estimator of Cronbach's alpha. *Psychometrika*, 65, 271–280.

Wold, H. (1966) Estimation of principal component and related models by iterative least squares. In P.R. Krishnaiah (ed.), *Multivariate Analysis*, pp. 391–420. New York: Academic Press.

Wold, H. (1982) Soft modeling: The basic design and some extensions. In K.G. Jöreskog and H. Wold (eds), *Systems under Indirect Observation: Causality, Structure, Prediction*, Vol. 2, pp. 1–54. Amsterdam: North-Holland.

Wold, H. (1985) Partial least squares. In S. Kotz and N.L. Johnson (eds), *Encyclopedia of Statistical Sciences*, Vol. 6, pp. 581–591. New York: John Wiley & Sons, Inc.

Zanella, A. and Cantaluppi, G. (2006) Some remarks on a test for assessing whether Cronbach's coefficient α exceeds a given theoretical value. *Statistica Applicata*, 18, 251–275.

17

Nonlinear principal component analysis

Pier Alda Ferrari and Alessandro Barbiero

In this chapter we present a technique for the analysis of customer satisfaction based on a dimensionality reduction approach. This technique, usually referred to as nonlinear principal component analysis, assumes that the observed ordinal variables can be mapped into a one-dimensional quantitative variable, but, unlike linear principal component analysis, does not require the adoption of an a priori difference between classification categories and does not presuppose a linear relation among the observed variables. So, neither the weights of the variables nor the differences between their categories are assumed, and both are suitably determined through the data as the solution of an optimization problem. The main features of nonlinear principal component analysis are illustrated, the problem of missing data is dealt with, and several methods for their treatment are described and assessed. Nonlinear principal component analysis is also compared with the Rasch model (Chapter 14) and linear principal component analysis, two direct competitors for studying customer satisfaction data. Finally, the method is applied to the ABC ACSS data, and its applicability and findings are discussed.

17.1 Introduction

As discussed in previous chapters, customer satisfaction with a service or a product is not easily and directly quantifiable, but can be assessed through observed variables linked by different aspects of the service or product under evaluation. Data on customer satisfaction are usually collected by survey questionnaires where respondents are asked to declare their degree of satisfaction with those different aspects.

The statistical analysis of such data is challenging for various reasons. The first reason is due to the cognitive aspects of the phenomenon, since the level of satisfaction is generally dependent on both expectation and individual features of respondents. Furthermore, surveys

Modern Analysis of Customer Surveys: with applications using R, First Edition. Edited by Ron S. Kenett and Silvia Salini.
© 2012 John Wiley & Sons, Ltd. Published 2012 by John Wiley & Sons, Ltd.

include questions about subjective variables that express only what people declare (Freeman, 1978). Measurement errors can emerge from this cognitive dissonance and affect data with consequences on the efficacy of the results, as addressed, for example, by Bertrand and Mullainathan (2001).

There are also several statistical issues related to the data handling. Typically, customer satisfaction is measured by means of a set of related variables or items, whose relevance and/or weight need to be determined. In addition, these variables are categorical and usually measured on an ordinal-level scale, making their treatment more complex. Even if ordinal categories are coded with numbers, as with a Likert scale (Likert, 1932), it does not solve the problem. The numerical labels usually indicate the rank of the categories but not their absolute values. Consequently, the numerical distance between adjacent labels does not reflect the real distance on a numerical scale. It is then evident that the data need to be suitably treated if the findings are not to be affected by unrealistic assumptions. For this purpose, many different statistical methods have been produced, following different approaches. Here, we assume that the observed variables can be mapped into a one-dimensional quantitative variable, the 'level of satisfaction', and that neither the weights of the variables nor the differences between their categories are assumed *a priori*; but both must be suitably determined from the data. A method used to reach this goal is nonlinear principal component analysis (NLPCA: see Gifi, 1990; Michailidis and de Leeuw, 1998), which allows for the setting-up of a numerical indicator of the level of satisfaction, starting with the ordinal responses.

The chapter is organized as follows. In Section 17.2 homogeneity analysis and NLPCA are described. Section 17.3 discusses why and when NLPCA is feasible for the setting-up of an indicator of the level of satisfaction. Problems connected to the presence of missing data and their treatment are presented in Section 17.4. Section 17.5 deals with the comparison of NLPCA with two other popular competitors. In Section 17.6, NLPCA is applied to the ABC ACSS data and the main aspects and findings of this application are described. Finally, Section 17.7 presents a summary and some conclusions.

17.2   Homogeneity analysis and nonlinear principal component analysis

17.2.1   Homogeneity analysis

Homogeneity analysis belongs to the so-called Gifi system (Gifi, 1990). It was introduced by the Data Theory Group of the University of Leiden in the 1980s and developed in the years that followed. For the sake of simplicity, we start with the one-dimensional case and then move on to the m-dimensional case.

Method

In the one-dimensional case, we aim to reduce m categorical variables, observed without missing values, into one-dimensional numerical latent variable, called 'component', retaining as much of the original information as possible.

More specifically, let $\mathbf{H}$ be the $n \times m$ matrix containing the observations of the m variables on n objects, $\mathbf{h}_j$ the jth column of the matrix $\mathbf{H}$, $\mathbf{c}_j$ the k_j-dimensional vector containing the categories of the jth variable, $j = 1, 2, \ldots, m$, and $\mathbf{G}_j$ the $n \times k_j$ indicator matrix such that

$\mathbf{G}_j \mathbf{c}_j = \mathbf{h}_j$. The goal of the analysis is to find the vector $\mathbf{x}$ of dimension $n \times 1$ and the m vectors $\mathbf{y}_j$ of dimension $k_j \times 1$ that minimize the following loss function:

$$\sigma^2(\mathbf{x}; \mathbf{y}_1, \ldots, \mathbf{y}_m) = \frac{1}{m} \sum_{j=1}^{m} \left(\mathbf{x} - \mathbf{G}_j \mathbf{y}_j\right)^T \left(\mathbf{x} - \mathbf{G}_j \mathbf{y}_j\right), \tag{17.1}$$

where $\mathbf{x}$ contains the 'object scores' of a one-dimensional numerical variable, and $\mathbf{y}_j$ the categories of variable j appropriately quantified.

In other words, the goal is to find the best vector $\mathbf{x}$ of values of a one-dimensional variable replacing all the m variables. It is the best because it minimizes the loss of information due to representing m variables by only one component; specifically, it minimizes the distance between the observations of the m variables appropriately quantified and the scores of the component.

In order to avoid trivial solutions, some conditions are imposed. Usually, object scores and transformed variables are standardized by the following restrictions:

$$\mathbf{x}^T \mathbf{x} = n, \quad \mathbf{u}_n^T \mathbf{x} = 0, \tag{17.2}$$

where $\mathbf{u}_n^T$ is an n-vector of 1s; and

$$\mathbf{y}_j^T \mathbf{D}_j \mathbf{y}_j = n, \quad \mathbf{u}_{k_j}^T \mathbf{D}_j \mathbf{y}_j = 0, \tag{17.3}$$

where $\mathbf{D}_j = \mathbf{G}_j^T \mathbf{G}_j$ is a diagonal matrix containing the frequencies of categories of variable j. The two conditions in (17.2) jointly ensure that the object scores have zero mean and unit variance; while the conditions in (17.3) ensure the same features for quantifications $\mathbf{y}_j$, properly weighted by the frequencies of the categories of variable j. Other normalization methods can, however, be adopted.

The above-mentioned method of computing variable transformations is called 'optimal scaling' because the transformations (*scaling*) are chosen so as to minimize (*optimal*) the loss function.

ALS algorithm

The minimization of (17.1), with restrictions (17.2) and (17.3) jointly applied over object scores x and quantifications $\mathbf{y}_j$, is derived by means of an algorithm called alternating least squares (ALS) (Gifi, 1990; Michailidis and de Leeuw, 1998). ALS is an iterative process and produces a decreasing sequence of loss function values, which under some mild regularity conditions converges to a stationary value of the loss function. It starts at iteration $s = 0$ with arbitrary object scores $\mathbf{x}^{(0)}$ and each iteration s consists of three steps:

1. Update category quantifications: $\mathbf{y}_j^{(s)} = \mathbf{D}_j^{-1} \mathbf{G}_j \mathbf{x}^{(s)}$.

2. Update object scores: $\tilde{\mathbf{x}}^{(s)} = m^{-1} \sum_{j=1}^{m} \mathbf{G}_j \mathbf{y}_j^{(s)}$.

3. Normalization: $\mathbf{x}^{(s+1)} = \mathrm{orth}(\tilde{\mathbf{x}}^{(s)})$, where orth is any technique that computes an orthonormal basis for the column space of a matrix.

Note that in some cases the ALS algorithm might converge to a local minimum (Gifi, 1990).

The goodness of the solution depends on the minimization of the sum of the squared distances between the scores obtained and the data. In order to evaluate the goodness of the procedure it is thus possible to use the variance accounted for by the one-dimensional variable. This can be measured by λ_1, i.e. the maximum eigenvalue of the correlation matrix of transformed data, or, better, by the percentage of variance accounted for by the first dimension (given by $100\lambda_1/m$), which expresses it in relative terms; the larger the percentage, the better the solution.

p-dimensional case

More generally, if we are interested in $p \leq m$ components, we seek an $n \times p$ matrix $\mathbf{X}$ of scores, and formula (17.1) takes the form

$$\sigma^2(\mathbf{X}; \mathbf{Y}_1, \ldots, \mathbf{Y}_m) = \frac{1}{m} \sum_{j=1}^{m} \text{tr} \left(\mathbf{X} - \mathbf{G}_j \mathbf{Y}_j\right)^T \left(\mathbf{X} - \mathbf{G}_j \mathbf{Y}_j\right), \tag{17.4}$$

where $\mathbf{Y}_j$ is the $k_j \times p$ matrix of quantifications. Conditions in (17.2) and (17.3) become

$$\mathbf{X}^T \mathbf{X} = n\mathbf{I}_p, \quad \mathbf{u}_n^T \mathbf{X} = \mathbf{0}, \tag{17.5}$$

$$\mathbf{Y}_j^T \mathbf{D}_j \mathbf{Y}_j = n\mathbf{I}_p, \quad \mathbf{u}_{k_j}^T \mathbf{D}_j \mathbf{Y}_j = \mathbf{0}, \tag{17.6}$$

whose meaning is analogous to the previous case. The percentage of total variance accounted for by the first p dimensions is then given by $100 \left(\sum_{j=1}^{p} \lambda_j\right) / m$, where $\lambda_1, \lambda_2, \ldots, \lambda_p$ are the ordered eigenvalues of the correlation matrix of the transformed data.

17.2.2   Nonlinear principal component analysis

NLPCA is a special case of homogeneity analysis that assumes and preserves the order of the categories of the observed variables. For this reason, it is specifically suitable for ordinal variables.

Method

We suppose now that the observed variables are ordinal, and their order is required to be preserved. The loss function in equation (17.1) can then be written as

$$\sigma^2(\mathbf{x}; \mathbf{q}_1, \ldots, \mathbf{q}_m; \beta_1, \ldots, \beta_m) = \frac{1}{m} \sum_{j=1}^{m} \left(\mathbf{x} - \mathbf{G}_j \mathbf{y}_j\right)^T \left(\mathbf{x} - \mathbf{G}_j \mathbf{y}_j\right)$$

$$= \frac{1}{m} \sum_{j=1}^{m} \left(\mathbf{x} - \mathbf{G}_j \mathbf{q}_j \beta_j\right)^T \left(\mathbf{x} - \mathbf{G}_j \mathbf{q}_j \beta_j\right)$$

where the optimal quantifications $\mathbf{y}_j$ are decomposed as $\mathbf{y}_j = \beta_j \mathbf{q}_j$, where $\mathbf{q}_j, j = 1, \ldots, m$, is meant as the $k_j \times 1$ vector that contains the optimal category quantifications for variable j and β_j is a scalar and represents the component loading for variable j.

   In order to incorporate the measurement level of the variables into the analysis – that is, to satisfy the relationship:

$$_j h_1 \leq \ldots \leq _j h_{k_j} \Rightarrow _j q_1 \leq \ldots \leq _j q_{k_j} \quad j = 1, \ldots, m, \tag{17.7}$$

– a restriction is needed on $\mathbf{q}_j$. Specifically, $\mathbf{q}_j$ is taken to belong to C_j, where $C_j = \left\{q_j |_1 q_j \leq _2 q_j \leq \ldots _{k_j} q_j\right\}$ is the convex cone of all vectors having non-decreasing elements.

Hence, conditions (17.3) become

$$\mathbf{q}_j^T \mathbf{D}_j \mathbf{q}_j = n, \quad \mathbf{u}_{k_j}^T \mathbf{D}_j \mathbf{q}_j = 0, \qquad (17.8)$$

and ensure that the category quantifications, properly weighted by the frequencies of the categories of variable j, have zero mean and unit variance.

Algorithm

The solution is still obtained by the ALS algorithm, conveniently adapted. Since it is necessary to take into account the measurement level of the variables, ALS is modified in order to ensure also the order of quantifications: a monotone regression, $\mathbf{x} = \beta_j \mathbf{G}_j \mathbf{q}_j + \text{error}$, is performed at each iteration step to reorder the estimated quantifications of each variable (Gifi, 1990).

Solution

The one-dimensional solution obtained provides object scores $\mathbf{x}$, quantifications $\mathbf{q}_j$ and loadings β_j; they are linked by the relationship

$$\mathbf{x} = \frac{1}{m} \sum_j \mathbf{G}_j \mathbf{q}_j \beta_j = \frac{1}{m} \sum_j \mathbf{t}_j \beta_j, \qquad (17.9)$$

where $\mathbf{t}_j$ is the $n \times 1$ vector of the values of transformed variable j on the n objects and

$$\mathbf{q}_j = \mathbf{D}_j^{-1} \mathbf{G}_j \mathbf{x}. \qquad (17.10)$$

Equation (17.10) expresses the so-called first centroid principle (Michailidis and De Leeuw, 1998) according to which 'a category quantification is in the centroid of the object scores that belong to it', while equation (17.9) shows that 'an object score is the average of the quantifications of the categories it belongs to'. Equation (17.9) connects object scores to the $\mathbf{t}_j$ and β_j and highlights that the a latent component can be meant as a linear combination of the observed variables, optimally quantified and weighted. All this allows us to replace the matrix $\mathbf{H}$ of ordinal data with the matrix of numerical data $\mathbf{T} = [\mathbf{t}_1| \ldots |\mathbf{t}_m] = [\mathbf{G}_1 \mathbf{q}_1| \ldots |\mathbf{G}_m \mathbf{q}_m]$

If no missing data affect the data set, because of standardization requirements (17.2) and (17.3) and because we set $\mathbf{y}_j = \beta_j \mathbf{q}_j$, the component loadings β_j are the correlation coefficients between object scores and quantified variables; hence, they can be nicely interpreted as 'ordinary' component loadings.

p-dimensional case

If, more generally, $p \leq m$ components are required, formula (17.1) takes the form

$$\sigma^2(\mathbf{X}; \mathbf{q}_1, \ldots, \mathbf{q}_m; \boldsymbol{\beta}_1, \ldots, \boldsymbol{\beta}_m) = \frac{1}{m} \sum_{j=1}^m \text{tr} \left(\mathbf{X} - \mathbf{G}_j \mathbf{q}_j \boldsymbol{\beta}_j^T \right)^T \left(\mathbf{X} - \mathbf{G}_j \mathbf{q}_j \boldsymbol{\beta}_j^T \right), \qquad (17.11)$$

where $\mathbf{X}$ is a $n \times p$ matrix of scores, $\mathbf{q}_j$ is the $k_j \times 1$ vector of optimal category quantifications and $\boldsymbol{\beta}_j$ the p-dimensional vector of component loadings. Conditions (17.2) and (17.3) become

$$\mathbf{X}^T \mathbf{X} = n \mathbf{I}_p, \quad \mathbf{u}_n^T \mathbf{X} = \mathbf{0}, \qquad (17.12)$$

$$\mathbf{q}_j^T \mathbf{D}_j \mathbf{q}_j = n \mathbf{I}_p, \quad \mathbf{u}_{k_j}^T \mathbf{D}_j \mathbf{q}_j = \mathbf{0}. \qquad (17.13)$$

In the case of NLPCA, a so-called rank-1 restriction of the form $\mathbf{Y}_j = \mathbf{q}_j \boldsymbol{\beta}_j^T$ is imposed – each matrix $\mathbf{Y}_j$ is restricted to be of rank 1, which implies that its columns are proportional to each other, and, hence, so too are the category quantifications.

The solution is again obtained by the ALS algorithm and provides the following matrix of object scores:

$$\mathbf{X} = \frac{1}{m} \sum_j \mathbf{G}_j \mathbf{q}_j \boldsymbol{\beta}_j^T = \frac{1}{m} \sum_j \mathbf{t}_j \boldsymbol{\beta}_j^T. \tag{17.14}$$

Each column of $\mathbf{X}$ contains the object scores of one of the p latent variables and for each of them relationship (17.9) holds. From equation (17.14) it is worth noting that when moving among the p latent variables the weights of the variables change, whereas the p quantifications for each variable are the same due to the rank-1 condition, so that actually the categorical data are transformed into 'numerical' data and $\mathbf{T}$ contains these 'numerical observations'.

NLPCA is implemented in various statistical packages. Here we refer to the R environment and its package *homals* (de Leeuw and Mair, 2009), which implements NLPCA as a special case of homogeneity analysis (see Section 17.6.2).

17.3   Analysis of customer satisfaction

17.3.1   The setting up of indicator

If we assume the basic hypothesis that a measure of customer satisfaction can be obtained by reducing the dimensionality of multiple items representing the various aspects of the product or service, then NLPCA is a particularly suitable method to use. In fact, it allows for synthesizing observed variables in a reduced space, preserving measurement levels of qualitative ordinal data without assuming an *a priori* difference between adjacent categories. The latent dimension is derived through equation (17.9) as a linear combination of the observed items after an optimal quantification of their ordinal categories and their weights in the linear combination.

In this new context, each object is a unit (customer or respondent), each variable j is an item of the questionnaire, the jth column of the data matrix $\mathbf{H}$ contains the answers of customers to the jth item, and the scores of the latent variable by NLPCA may be used as measures of customer satisfaction. In fact, the level of satisfaction of respondent i is obtained by formula (17.9) as a weighted mean of transformed items with loadings β_j as weights.

Nevertheless, before using the one-dimensional solution as a feasible indicator of satisfaction, it is necessary to evaluate its validity. With this aim, the following requirements need to be ascertained:

1. *The first eigenvalue is high.* This means that the solution fits the data well; as stated, the first eigenvalue constitutes a measure of goodness of the procedure. Alternatively, Cronbach's α (Cronbach, 1951) can be determined. This coefficient, introduced as a way to assess the reliability of scales, is related to λ_1 as follows (Heiser and Meulman, 1994):

$$\alpha = \frac{m(\lambda_1 - 1)}{(m-1)\lambda_1}.$$

The closer α is to unity, the better the indicator. If condition 1 is not satisfied, or we are interested in extracting more than one latent dimension, we cannot stop the analysis at the first component but continue to extract a number $p > 1$ of components.

2. *All the weights have the same sign.* Since the vector of object scores $\mathbf{x}$ is constructed as a simple linear combination of quantified categories, once the ordering of item answers is coherent with the level of satisfaction, we expect that the higher the rank of observed variables, the higher the value of the satisfaction indicator. This requires that the weights of combinations have the same sign for each variable – specifically, positive if orientation is increasing. If, after checking the orientation, the loadings do not have the same sign, the latent variable obtained cannot be used to assess the level of satisfaction, but the item(s) with negative loading should be removed from the analysis.

3. *The solution is stable.* Finally, it is also important to evaluate the stability of the outputs produced (i.e., eigenvalues, component loadings, category quantifications and scores). This is a topic still under active research by scholars and no definitive solution has been presented so far. One proposed solution is the evaluation of stability by means of resampling methods, which, in general, avoid troublesome theoretical proofs. Among the various resampling methods, the bootstrap method is an empirical and easy way to verify whether the indicators resulting from NLPCA are more or less stable. The bootstrap technique, introduced in order to evaluate the bias and variance of an estimator of a parameter of interest (Efron, 1979; Efron and Tibshirani, 1993), can be used in the NLPCA context to check the stability of each outcome. In Section 17.6, to verify the component loadings and λ_1 stability, an algorithm that draws samples with replacement from the data set and performs NLPCA on them is used. The bootstrap confidence intervals for component loadings and λ_1 are produced and the coefficient of variation given by the ratio of the standard deviation of the bootstrap estimates to the corresponding actual value is also calculated. If the outcomes are not stable, the results are strictly connected to the specific data set and do not have a general validity.

Once the optimal quantifications and weights are determined and validated, the scores and the numerical $n \times m$ matrix $\mathbf{T}$ of transformed data, containing each respondent's answers appropriately quantified, may be considered for further statistical analysis; for example, they could be used to investigate the level of satisfaction according to specific factors.

Thus, if one is interested in comparing the level of satisfaction among different groups, for example, by countries and types of customers and so on, an indicator of the group level can be computed in the following way. Let g be the group index, n_g the number of respondents within group g and $\sum_g n_g = n$ the total sample size. The conditional mean of scores $\bar{x}_g$ for group g is given by

$$\bar{x}_g = \frac{1}{n_g} \mathbf{u}_g^T \mathbf{x},$$

where $\mathbf{u}_g^T$ is an indicator vector of order n with a block of n_g 1s in the position corresponding to group g and zeros elsewhere, and $\mathbf{x}$ is the $n \times 1$ vector of scores.

Values of $\bar{x}_g$ can be directly used for comparison. In fact, since the grand population mean is

$$\bar{x} = \frac{1}{n} \sum_g n_g \bar{x}_g = 0,$$

then $\bar{x}_g \geq 0$ (< 0) denotes equal or greater (lower) average satisfaction of the group with respect to the grand mean and measures the group effect on satisfaction.

Furthermore, we can use submatrices of $\mathbf{T}$ to analyse the correlation between different items or groups of items describing specific topics of products or service and so on.

17.3.2   Additional analysis

Sometimes, in addition to the items connected to specific aspects of the service or product, the questionnaire contains other items representing ordinal variables that are rough overall satisfaction indicators. Each of the latter variables can be thought of as a proxy for the continuous indicator of satisfaction under analysis or as its ordinal measurement.

In this case, it would be interesting to detect which of these variables better expresses the continuous indicator or to state the relationships between each of them and the different aspects of the service or product. For this purpose, we propose using NLPCA on the overall satisfaction variables as well.

Let G_A be the group containing the manifest variables describing overall satisfaction and G_B the group of manifest variables describing specific aspects of the service or product. The idea is to carry out an NLPCA on each of the two groups, thereby obtaining two indicators, I_A and I_B, with the vector of scores

$$\mathbf{x}_A = \sum_i {}_A\beta_{iA}\mathbf{t}_i,$$

$$\mathbf{x}_B = \sum_i {}_B\beta_{iB}\mathbf{t}_i,$$

and matrices $\mathbf{T}_A$ and $\mathbf{T}_B$ of data appropriately quantified. Then, the coefficient ${}_B\beta_j$ expresses the weight of the jth aspect of the service or product on the indicator I_B, while, if missing data are not present or if they are imputed (see Section 17.4 on imputation), coefficient ${}_A\beta_i$ can be interpreted as the correlation coefficient between the ith variable of group G_A and I_A.

Besides, if the correlation coefficient between I_A and I_B (calculated through their values $\mathbf{x}_A$ and $\mathbf{x}_B$) is high, a joint interpretation of the two analyses is possible, since the correlations between each variable of group A and each variable of group B, appropriately quantified, become meaningful. An example is provided in Section 17.6.3.

Finally, in evaluating the magnitude of the correlation coefficient, we need to take into account that when the variables are discrete these coefficients are less extended than in the case of continuous variables and might not be able to reach the extreme values -1 and $+1$ (see, for example, Guilford, 1965), so expect their range to be small.

17.4   Dealing with missing data

When statistical data are collected with survey questionnaires, many missing values are often recorded. The problem of missing data is well known in the statistical literature, with many methods available for dealing with them (Little and Rubin, 2002; Chapter 8 of this book), but not all are applicable with NLPCA. Three options have been proposed for handling missing data in NLPCA (Gifi, 1990; Michailidis and de Leeuw, 1998). All three options involve the matrix $\mathbf{G}_j$, albeit in different ways. Specifically, $\mathbf{G}_j$ is left incomplete (*missing data passive*) or is completed by adding one column for each variable with missing data (*missing data single category*) or multiple columns (*missing data multiple category*). With the first option, the ith row of $\mathbf{G}_j$ is a zero row if the ith object has a missing observation for variable j. For the second option, an extra column (single extra category) is added to $\mathbf{G}_j$ with entry '1' for each object having a missing value on the jth variable, while the third option (multiple extra category) adds to $\mathbf{G}_j$ as many extra columns as there are objects with missing data on the jth variable. The first option discards missing observations from the computation, while the others require strong assumptions regarding the pattern of missing data.

The R package *homals* (see Section 17.6.2), which performs NLPCA as a special case of homogeneity analysis, implements only the 'passive treatment', which will be discussed in more detail.

Formally, in a situation with missing data, if the passive method is adopted, an incomplete indicator matrix $\mathbf{G}_j$ and a binary matrix $\mathbf{M}_j$ are included (Michailidis and de Leeuw, 1998) in the loss function (17.1) as follows:

$$\sigma^2(\mathbf{x}; \mathbf{q}_1, \ldots, \mathbf{q}_m; \beta_1, \ldots, \beta_m) = \frac{1}{m} \sum_{j=1}^{m} \left(\mathbf{x} - \mathbf{G}_j \mathbf{q}_j \beta_j\right)^T \mathbf{M}_j \left(\mathbf{x} - \mathbf{G}_j \mathbf{q}_j \beta_j\right),$$

where $\mathbf{M}_j$, of order $n \times n$, has the role of binary indicator of missing observations for variable j, while the normalization restrictions (17.2) on object scores become

$$\mathbf{x}^T \mathbf{M}_* \mathbf{x} = m \cdot n, \quad \mathbf{u}_n^T \mathbf{M}_* \mathbf{x} = 0,$$

with

$$M_* = \sum_{j=1}^{m} \mathbf{M}_j.$$

In this case, the score of the ith unit is actually computed as

$$x_i = \frac{1}{\#V_i} \sum_{j \in V_i} \beta_j \mathbf{g}_{ij} \mathbf{q}_j = \frac{1}{\#V_i} \sum_{j \in V_i} t_{ij} \beta_j$$

where V_i is the subset of variables really observed on unit i, $\#V_i < m$ is the cardinality of V_i, $\mathbf{g}_{ij}$ is the ith row of matrix $\mathbf{G}_j$, t_{ij} is the ith element of vector $\mathbf{t}_j$ and equation (9) can be rewritten (Ferrari *et al.*, 2006) as

$$x_i = \frac{1}{m} \left(\#V_i \frac{1}{\#V_i} \sum_{j \in V_i} \beta_j \mathbf{g}_{ij} \mathbf{q}_j + (m - \#V_i) \frac{1}{\#V_i} \sum_{j \in V_i} \beta_j \mathbf{g}_{ij} \mathbf{q}_j \right).$$

This expression highlights the fact that, under passive treatment, unobserved variables contribute to the object score with a quantity that is exactly the average of the terms $\beta_j \mathbf{G}_j \mathbf{q}_j$, actually observed, completely ignoring the others. This seems infeasible since variables play a different role in setting up the indicator. Moreover, in this way, observed outliers have, *ceteris paribus*, more relevance to the final score if they are observed on units presenting more missing values rather than having few or no missing observations. This fact could lead to unsatisfactory classification and/or ordering of the units based on scores.

Recently, Ferrari *et al.* (2011) discussed the problem of missing data in NLPCA. They considered four available techniques that can be adapted to qualitative data for missingness at random or completely at random: listwise deletion (Lw), mode (Mo) imputation, median (Me) imputation, and the passive method (Pa).

Lw eliminates all objects with at least one missing datum. It is a complete-case analysis, confined to the set of cases with no missing values.

In Mo/Me imputation, the mode/median of the variable is substituted for every missing value of the same variable. These methods are simply the modification of the classical mean substitution technique that are suitable for categorical data. Median imputation is more feasible when the variable is ordinal.

If missing data are few and sparse, no substantial impact is expected to affect results. In contrast, if missing data are significantly present and/or show a specific pattern, results can be significantly influenced by the specific option that is adopted to treat them.

None of the previous proposals seems to be completely satisfactory. The very popular option of complete subsets or Lw is inefficient, since it does not use the overall information contained in the data. The reduced set of cases may be smaller, sometimes much smaller, than the initial data set. In fact, in multivariate analysis it is relatively common to find a missing value on any one of the m variables; this leads to elimination of the related unit, quickly reducing the effective sample size.

The imputation of all missing values of a variable with its mode or median or with any other category transforms the incomplete data set into a complete one. However, this method of assigning the same category to all the units with the missing observation on that variable artificially diminishes the variance of the variable and, when an extra category is adopted, increases the number of categories. Moreover, data from surveys and questionnaires usually show high levels of association between variables. Consequently, mode/median imputation, ignoring values observed on the same unit for the other variables, does not take into account important auxiliary information and works as if all the variables were independent (unconditional imputation).

Nonetheless, the three methods have the advantage of using a complete data set for the construction of the indicator. They preserve all the characteristics of NLPCA, specifically, relationship (17.9), and the coincidence between weights and correlation coefficients. This does not occur for the passive method where loadings 'lose' their useful meaning and object scores have the pitfalls just highlighted.

Ferrari *et al.* (2011) propose a new missing data treatment method ('Forward imputation', Fo) overcoming these drawbacks; it seems particularly suitable for situations where NLPCA is used to obtain a numerical indicator. This method is based on an iterative algorithm that alternates the application of NLPCA on a subset of the data with no missing data (complete matrix), and the imputation of missing data cells with the corresponding cells of the nearest unit in the complete matrix. This sequential process starts from the unit with the lowest number of missing values and ends with the unit with the highest number of missing values. The Fo imputation possesses some nice theoretical properties, and seems to work well. An in-depth study of the performance of the methods shows Fo imputation as having the best performance among the above missing-data treatment methods; Me comes next, followed by Mo imputation and Pa, which vary in their respective position depending on experimental conditions.

17.5   Nonlinear principal component analysis versus two competitors

In this section we briefly discuss two other methods, PCA and the Rasch model (RM), which are different approaches to analysing the latent trait underlying a multiple-item scale. The first is connected to an algorithmic procedure according to which no data generating process is assumed, but the best representation of the data is searched. The second assumes a model entirely known except for the values of parameters, which have to be estimated. The RM is presented in detail in Chapter 14.

PCA, like NLPCA, is a reduction process that constructs the indicator as a linear combination of manifest variables, but these are required to be numerical. So, in the case of ordinal variables, the categories of the variables must be coded numerically before the analysis. In contrast to NLPCA, this implies the adoption of an *a priori* difference between adjacent categories. Furthermore, PCA assumes a linear relation among the observed variables, which is unrealistic. This very popular method thus appears not to be particularly suitable when these conditions are not met, and NLPCA could lead to an appreciable improvement in the analysis. In order to evaluate this improvement, here we will consider the eigenvalue ratio, given by

$$I_{NL} = \frac{\lambda_1^{NLPCA}}{\lambda_1^{PCA}}. \tag{17.15}$$

This is a global index that expresses in relative terms the improvement obtained by using NLPCA rather than PCA; the larger the index, the greater the improvement. Other indices for detecting the nonlinearity of single items can be adopted (see, for example, Carpita and Manisera, 2011).

Incidentally, for both analyses, if required, a convenient number $p \geq 1$ of components can be chosen, but, in contrast to PCA, in NLPCA the solutions for $1, 2, \ldots, p$ dimensions are not nested, even if, given p, optimal quantifications are the same in each of the p dimensions.

In Section 17.6.3 we compare the two analyses and present some results with regard to the ABC survey.

The Rasch model was introduced by Rasch as a psychometric tool to measure both item difficulty and subject ability along a shared continuum, based on responses to the items of a questionnaire (Rasch, 1980). Originally proposed for dichotomous responses, it was later extended to polytomous ordinal responses, (see Andrich, 1978; Masters, 1982). Since this model is specifically presented in Chapter 14 we do not discuss it any further, confining ourselves to comparing it with NLPCA.

We observe (see, for example, Ferrari *et al.*, 2005) that although based on different premises, both methods give scores for every respondent, which can be assumed to be measures of their satisfaction as well as distances between categories of items. Nevertheless, NLPCA appears more suitable in some contexts for two reasons. First, since NLPCA gives the level of satisfaction as a linear combination of quantified questionnaire answers, it is immediately possible to compute the score for a new customer. Secondly, while for the RM unidimensionality is a fundamental assumption and, hence, only one dimension can be considered, for the NLPCA it is possible to consider more than one dimension when one dimension does not synthesize the data properly or in order to enrich the analysis.

However, Ferrari et al. (2005) suggested a complementary use of the two techniques, providing a joint interpretation of RM parameters and NLPCA component loadings.

17.6   Application to the ABC ACSS data

Previous sections have discussed the possibility of adopting NLPCA to assess customer satisfaction. In this section, its main characteristics and its application are examined in the case of the ABC annual customer satisfaction survey (ACSS) data introduced in Chapter 2. On these data, two types of analysis are carried out. In the first (Section 17.6.3), the subset with no missing data (complete subset) is considered with the aim of showing how to set up and evaluate the indicator, how to extract more than one component, how to identify a good proxy variable, and (Section 17.6.4) how to compare NLPCA with two competitors: PCA and the RM. In the second (Section 17.6.5) the entire data set including missing data is considered, with the aim of comparing different ways of handling missing data. Specifically, the methods Fo, Mo, Me and Pa already discussed are applied. Therefore, since the analysis of the complete subset can be classified as a listwise treatment of missing data, the results of the previous analysis on the complete subset are also compared.

We have derived many findings, but, for the sake of brevity, we confine our attention to only some of them. We do not discuss the results in depth, but rather show the kind of outcomes one can achieve with NLPCA and present them in figures or tables. Again, for brevity, we do not describe the data already presented in Chapter 2, but discuss the selection of variables and preparation of the data set. We provide only the essential information with regard to the *homals* package as well.

17.6.1   Data preparation

Given our focus on the global indicator of satisfaction, in the first stage the 11 variables connected with 'overall satisfaction' were selected and clustered into two groups:

A. Overall satisfaction with ABC

   Q_1: Overall satisfaction level with ABC

   Q_2: Overall satisfaction level with ABC's improvements during 2010

   Q_3: Is ABC your best supplier?

   Q_4: Would you recommend ABC to other companies?

   Q_5: If you were in the market to buy a product, how likely would it be for you to purchase an ABC product again?

B. Specific satisfaction with ABC

   Q_{11}: Overall satisfaction level with equipment and system

   Q_{17}: Overall satisfaction level with sales support

   Q_{25}: Overall satisfaction level with technical support

Table 17.1   Number of missing values for each variable

Variable	Q_1	Q_2	Q_3	Q_4	Q_5	Q_{11}	Q_{17}	Q_{25}	Q_{38}	Q_{57}	Q_{65}
No. of missing values	2	7	13	3	2	7	16	0	12	21	11

$Q_{31:}$ Overall satisfaction level with ABC training

Q_{38}: Overall satisfaction level with ABC's supplies and orders

Q_{43}: Overall satisfaction level with overall software add-on solution

Q_{49}: Overall satisfaction level with the customer website

Q_{57}: Overall satisfaction level with purchasing support

Q_{65}: Overall satisfaction level with contracts and pricing

Some of these variables (i.e., Q_{31}, Q_{43} and Q_{49}) have a huge number of missing values, which might not be random, because they are connected to particular aspects that have not been experienced by the customer (e.g., use of the customer website). For this reason, these variables have been excluded and only variables Q_1, Q_2, Q_3, Q_4, Q_5, Q_{11}, Q_{17}, Q_{25}, Q_{38}, Q_{57} and Q_{65} considered for the analysis. So we have a data set of 264 units by 11 variables, henceforth referred to as the 'entire data set'. The number of missing data for each variable is given in Table 17.1, while the number of missing values by number of units is shown in Table 17.2. We select from this incomplete data set the subset of 208 units with no missing data, which we will refer to as the 'complete subset'. On each of the two ('incomplete' and 'complete') data sets an analysis is carried out with different purposes.

Before starting the analysis, we need to check that the orientation of the categories of each variable is coherent with the orientation of the level of satisfaction. Since the answer to question Q_3, 'Is ABC your best supplier?' presents 'yes' as the first category and 'no' as the second, an inversion of these categories is required for coherence with the other variables, where high values of categories correspond to high levels of customer satisfaction.

17.6.2   The *homals* package

NLPCA can be implemented using different packages. Here we discuss the *homals* package in R, which is freely available on CRAN. The core routine of the package is *homals*. This function performs a homogeneity analysis, also known as a multiple correspondence analysis but with many additional options. Variables must be grouped into sets for emulating regression analysis and canonical analysis. For each variable there are, in addition, rank constraints on the category quantifications and level constraints, which allow one to treat a variable as nominal, ordinal or numerical. By combining the various types of restrictions we obtain generalizations

Table 17.2   Number of units by number of missing values

No. of missing values	0	1	2	3	6
No. of units	208	34	12	8	2

of principal component analysis, canonical analysis, discriminant analysis and regression analysis.

In order to perform NLPCA by extracting p dimensions, it is sufficient to set the parameters $ndim = p$, $rank = 1$ and $level = $ '$ordinal$'. The meaning of $rank = 1$ and $level = $ '$ordinal$' follows directly from the content of Section 17.2.2. The function allows the user to set a maximum number of iterations for the ALS algorithm $itermax$ and a maximum admissible error eps. If some data are missing, the $missing$ $data$ $passive$ treatment is automatically applied.

The $homals$ package considers a set of normalization conditions different from those described in Section 17.2; for $p = 1$ they are:

$$\mathbf{x}^T\mathbf{x} = \frac{1}{m}, \mathbf{u}_n^T\mathbf{x} = 0,$$

$$\mathbf{q}_j^T\mathbf{D}_j\mathbf{q}_j = \beta_j^2, \quad \mathbf{u}_{k_j}^T\mathbf{D}_j\mathbf{q}_j = 0,$$

so that the weight β_j does not stand for the correlation coefficient ρ_j between the object scores and the transformed variable j, the latter being obtained by the relationship $\rho_j = \beta_j\sqrt{m}$.

17.6.3   Analysis on the 'complete subset'

In Section 17.3 an indicator of customer satisfaction meant a linear combination of the observed items describing single aspects of satisfaction. For this purpose the six group B (specific satisfaction) variables Q_{11}, Q_{17}, Q_{25}, Q_{38}, Q_{57} and Q_{65} that actually express these different aspects are selected. The analysis is done on the complete data set (208 units) and the first component is extracted. The latent variable obtained is characterized by $\lambda_1 = 2.99$ (49.8% of the total variance) and $\alpha = 0.798$. In Figure 17.1, the component loadings (left) and the correlation coefficients (right) of the six manifest variables with respect to the first latent variable are displayed. As expected, the component loadings (and correlation coefficients) all have the same sign (positive). For testing the stability of the loadings, a bootstrap procedure has been carried out, drawing $B = 2000$ samples of size 208 with replacement from the complete

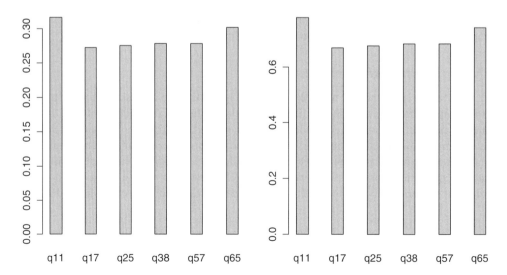

Figure 17.1 A1: Component loadings (left) and correlation coefficients (right) of the six manifest variables with respect to the first latent variable

Table 17.3   Percentile bootstrap confidence intervals and coefficients of variation (CV) for maximum eigenvalue and component loadings

	λ_1	Q_{11}	Q_{17}	Q_{25}	Q_{38}	Q_{57}	Q_{65}
2.5% bound	2.665	0.275	0.236	0.225	0.235	0.241	0.271
97.5% bound	3.451	0.346	0.308	0.321	0.325	0.316	0.337
CV	0.036	0.058	0.077	0.092	0.088	0.070	0.055

subset and performing an NLPCA on each of them. The 2.5% and 97.5% quantiles of the λ_1 and loading distributions (i.e., percentile bootstrap confidence intervals) are computed and reported in Table 17.3, along with the corresponding coefficients of variation, given by the ratio between the bootstrap standard deviation and the true value obtained on the complete subset. The stability analysis stresses the robustness of the outcomes. Since the one-dimensional component satisfies all the three conditions explained in Section 17.3.1, the latent variable may effectively be assumed to be a measure of the level of satisfaction and the findings can be discussed and used. The component loadings can then be construed as weights of the different aspects in assessing the level of satisfaction; Q_{11} (equipment) and Q_{25} (technical support) seem to point to the most and the least important aspects, respectively.

Since NLPCA also provides the optimal quantifications of the ordinal categories of the variables, for the categories of each item, the quantifications resulting from NLPCA are compared in Figure 17.2 with the integer values assigned to the response categories by the

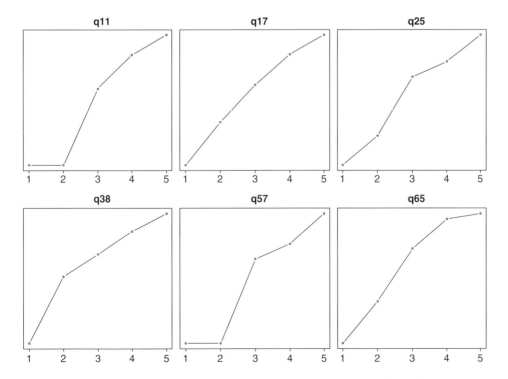

Figure 17.2   Category quantifications of NLPCA versus integer values of responses

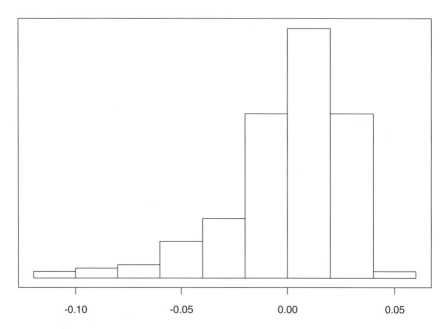

Figure 17.3   Histogram of the scores of the first latent variable

questionnaire. Note that for several items the categories appear quite equidistant, and only for a few items does the assumption of equal distance not seem to hold. We will return to this issue when we discuss the comparison between NLPCA and PCA.

Finally, scores are calculated and their distribution is represented. Figure 17.3 shows that the frequency distribution of the scores is quite asymmetrical, highlighting the presence of some units actually unsatisfied, which might need more attention.

Interestingly, scores can also be analysed or clustered according to a specific grouping factor (e.g., country). Figure 17.4 displays how the satisfaction level changes by country, recording the most satisfaction in Israel ($\bar{x}_I = 0.711$) and the least in Benelux ($\bar{x}_B = -0.603$).

If a richer analysis and more components are required, the extraction could be continued. For example, for $p = 2$ components, NLPCA provides $\lambda_1 = 2.95$, $\lambda_2 = 0.84$ and the percentage of variance accounted for reaches 3.79 (63.2%), an increase of 14.4%. The loadings for the two latent dimensions are reported in Table 17.4. Notably, on this second dimension, of course, coefficients do not have the same sign because the new latent variable does not measure the overall level of satisfaction; rather, it highlights a specific characteristic of satisfaction. On this dimension, equipment, sales support and contracts and pricing are opposed to technical support, supplies and orders and purchasing support. Figure 17.5 shows the plot of scores with regard to both dimensions, and the average of the two latent variables for each country is shown in black. The purpose is to check on the effect of the country factor. The figure shows that the overall level of satisfaction varies with the country, as already underlined, but no effect of country emerges with regard to the second latent variable; the country averages are quite similar.

As yet, only the six manifest variables describing specific aspects of ABC's service have been considered because our purpose has been to obtain an indicator of the level of

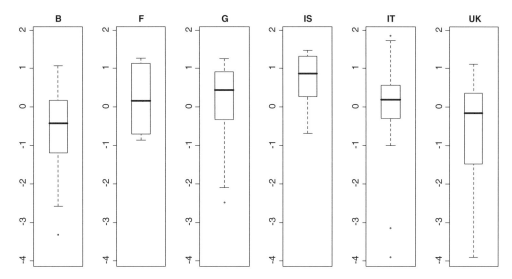

Figure 17.4 Boxplot of the normalized scores by country (B = Benelux, F = France, G = Germany, IS = Israel, IT = Italy, UK = United Kingdom)

satisfaction. Nevertheless, the presence of five manifest variables for assessing the overall level of satisfaction with ABC through an ordinal scale suggests that we carry out a parallel NLPCA on these variables, according to the aims set out in Section 17.3.2. The latent variable I_A obtained in this way is characterized by $\lambda_{max} = 3.5252$ (70.5%) and $\alpha = 0.8954$. Table 17.5 displays the correlation coefficients between each of the variables of group A and the indicator I_A. As we can see, the highest values of these coefficients correspond to variables Q_4 and Q_1, which thus best express the global level of satisfaction; the lowest value is related to the variable Q_3, as expected, being a dichotomous variable.

The large value of corr(I_A, I_B) = 0.94 allows for an integrated reading of the two analyses. To do that, the correlation coefficients between each optimally quantified variable of group A and each optimally quantified variable of group B have been calculated and are shown in Table 17.6. They can be read across rows or columns. So, by comparing the values within the ith row, we can detect which aspects better contribute to that row variable. We see that, in almost all rows, variables Q_{11} and Q_{65} have the highest values, and this result confirms, on the one hand, the relevant weight of these variables in determining the overall level of satisfaction and, on the other, the soundness of the variables of group A in approximating the global level of satisfaction. If we compare the coefficients within the columns, we can investigate which group A variable best expresses the single aspect. Specifically, variable Q_{57} (overall satisfaction level with purchasing support) is better represented by Q_2 (overall satisfaction

Table 17.4 NLPCA loadings for the two dimensions

	Q_{11}	Q_{17}	Q_{25}	Q_{38}	Q_{57}	Q_{65}
1st dim	0.321	0.262	0.276	0.275	0.278	0.303
2nd dim	−0.008	−0.247	0.115	0.189	0.099	−0.145

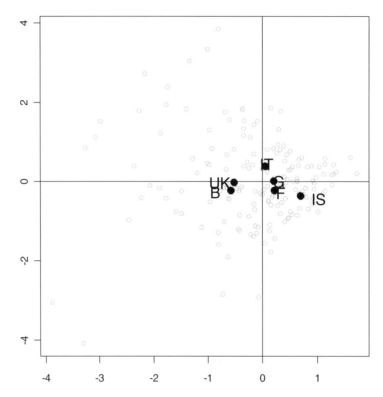

Figure 17.5   Plot of the scores of the two latent variables and country averages (in black)

level with ABC's improvements during 2010), while Q_{17} (overall satisfaction level with sales support) is better expressed by Q_4 (Would you recommend ABC to other companies?) and Q_5 (If you were in the market to buy a product, how likely would it be for you to purchase an ABC product again?). All this enriches the analysis and provides more information on single aspects or on the global level of satisfaction, even if we know only the five ordinal variables.

17.6.4   Comparison of NLPCA with PCA and Rasch analysis

It is of interest to compare the findings obtained on the complete data set by NLPCA with those obtained by two popular competitors: PCA and Rasch analysis.

The results obtained by NLPCA for the one-dimensional case are first compared with those obtained by PCA. The percentage of variance accounted for is 47.8% in PCA versus 49.8% in

Table 17.5   Correlation coefficients between the manifest variables and I_A

	Q_1	Q_2	Q_3	Q_4	Q_5
I_A	0.894	0.814	0.707	0.911	0.857

Table 17.6   Correlation coefficients between the optimally quantified variables of group A (rows) and group B (columns)

	Q_{11}	Q_{17}	Q_{25}	Q_{38}	Q_{57}	Q_{65}
Q_1	0.552	0.394	0.5611	0.421	0.350	0.512
Q_2	0.443	0.414	0.4204	0.363	0.406	0.491
Q_3	0.367	0.173	0.3650	0.282	0.201	0.361
Q_4	0.520	0.447	0.4545	0.399	0.297	0.551
Q_5	0.526	0.466	0.3666	0.366	0.269	0.575

NLPCA, resulting in a slight improvement in fit when NLPCA is applied. This highlights the fact that the relationship among the data is almost linear as confirmed by the global index of nonlinearity, defined in (17.15) and taking the value $I_{NL} = 1.0362$, which is extremely low and signals a small improvement in using NLPCA rather than PCA. Some inkling of this fact came from the comparison between scale points (used in PCA) and optimal quantifications (used in NLPCA), as stressed in Figure 17.2, already discussed. Finally, NLPCA and PCA loadings of the six group B variables are sketched in Table 17.7 and compared, after the NLPCA loadings have been appropriately normalized so as to have a norm equal to 1. These values are pretty close in both NLPCA and PCA, confirming the previous findings. The largest difference in loadings (0.386 versus 0.406), related to variable Q_{17} (sales support), is in fact very small.

The NLPCA results are now compared with those obtained from the RM (see Chapter 14). Since the RM results have been widely discussed and those of NLPCA discussed above, here we only examine the possibility of a joint reading of the item parameters of RM and component loadings of NLPCA. Following the suggestion of Ferrari and Salini (2011), for each item the parameter estimated by the RM is intended as a measure of the 'quality' of that item, while the NLPCA loading represents its 'relevance' to customer satisfaction, being the weight of that manifest variable in defining the indicator of satisfaction. It is then possible to obtain one ranking of the items according to their quality and one according to their relevance or represent each item in a Euclidean plane through its component loading and item parameter

Table 17.8 shows, for each item, the item parameter and component loading values. With regard to quality, we can observe that the item with the highest quality is technical support and the item with the lowest quality is terms, conditions and prices. At the same time we can analyse the relevance of each item for the composite indicator of customer satisfaction; a small weight means low importance of the corresponding item in determining the level of satisfaction. The component loadings show that sales support has the lowest relevance while equipment has the highest.

In Figure 17.6 each item is represented by its component loading (horizontal axis) and item parameter (vertical axis). The origin is arbitrary but helpful in clarifying the interpretation.

Table 17.7   Comparison of NLPCA and PCA loadings

	Q_{11}	Q_{17}	Q_{25}	Q_{38}	Q_{57}	Q_{65}
PCA	0.433	0.406	0.388	0.405	0.378	0.436
NLPCA	0.449	0.386	0.391	0.395	0.396	0.429

Table 17.8 Item parameter and component loading for each item

	Equipment	Sales support	Technical support	Supplies and orders	Purchasing support	Contracts and pricing
Item parameter	0.566	0.849	0.330	0.664	0.758	1.376
Component loading	0.776	0.667	0.676	0.683	0.684	0.741

This representation allows us to identify items with low/high quality and low/high weight. For example, contracts and pricing (quadrant I) is relevant for the level of customer satisfaction, but its quality seems low. Hence, action to improve that aspect could produce a large increase in satisfaction, while equipment (quadrant IV) has a high weight but, since its quality is already good, no action would be necessary. The reading of the other items is straighforward. An analysis of this kind, based on 'quality' and 'relevance', may provide useful support for decision makers.

17.6.5   Analysis of 'entire data set' for the comparison of missing data treatments

The high rate of units with missing data (56 out of 264, equal to 21% of units) requires a missing-data treatment so that a large amount of information is not lost.

The problem of missing data in NLPCA was specifically examined by Ferrari *et al.* (2011) with the simulation study already mentioned. Following their work, NLPCA was carried out

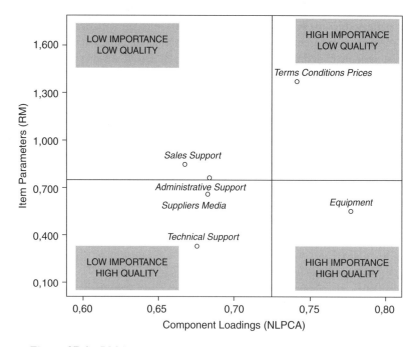

Figure 17.6   RM item parameters versus NLPCA component loadings

on the entire data set (264 units) containing missing data using five treatment methods: Fo, Mo, Me, Pa and Lw. The Fo imputation and the comparison procedure were carried out using an R script developed by those authors.

In order to compare the results from these methods, two indices were considered:

$$\cos(\beta_t, \beta_{t'}) = \frac{\beta_t^T \beta_{t'}}{|\beta_t||\beta_{t'}|}$$

and

$$d(\mathbf{x}_t, \mathbf{x}_{t'}) = \sqrt{\frac{1}{n} \sum_{i=1}^{n} (x_i - x_i')^2}.$$

The first index is the cosine of the angle between the two vectors of loadings obtained by applying the missing-data treatment methods t and t': the closer the cosine is to unity, the closer the vectors; hence, the weights assigned by the two methods to the variables are also close. The second index is the Euclidean distance between the vector of the scores obtained by methods t and t' divided by $\sqrt{n}$; the smaller this distance, the closer the indicators, in the sense that they assign more similar scores to the units.

Two remarks are necessary here. In the comparison of loadings, the listwise method is also considered. In the comparison of scores, by contrast, the listwise method is excluded, because it is based on a different number (208) of units, working only on the complete subset.

To carry out a reliable study on the performance of the different methods, we would need to know the true values of the missing data; since they are not, of course, available, we adopt as a reference the results of the simulation study quoted, where the outcomes from Fo are shown to be the closest to the 'true' outcomes.

In any case, in a first step all the methods are compared to each other. The values of the cosines between vectors of loadings under the different missing-data treatments seem similar; the smallest cosine (0.9996) is recorded between Fo and Pa. A more detailed comparison is given in Figure 17.7, where plots of Fo loadings versus Me, Mo, Pa and Lw loadings are displayed. They confirm the more general findings of Ferrari *et al.* (2011). In fact, Fo and Me provide closer results, while Pa and Lw provide the more varied ones. That might be due to the presence of outliers that penalizes Pa and to the high rate of units with missing values that

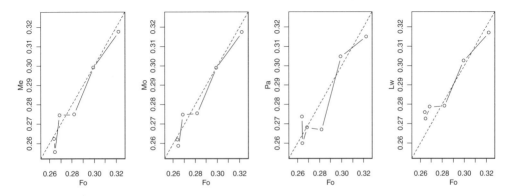

Figure 17.7   Plot of Fo loadings versus Me, Mo, Pa, Lw loadings

Table 17.9   Comparison of missing-data treatments: distance between score vectors

Method	Me	Mo	Pa
Fo	0.136	0.142	0.203
Me		0.035	0.263
Mo			0.259

causes Lw to lose information, because of the deletion of those units from the analysis. For these specific data, Mo works pretty well; it is not far from Me or Fo.

Turning to the normalized object scores, the results for $d(\mathbf{x}_t, \mathbf{x}'_t)$ are shown in Table 17.9 and confirm that the scores provided by the Me method are the closest to the Fo scores. With Fo as reference method, the empirical cumulative distribution functions of difference in scores between Fo and each of the other methods, $|x_i^{Fo} - x_i^t|$, are plotted in Figure 17.8. The largest differences in scores are observed with regard to Pa; in detail, the percentage of units that differ more than 0.15 from Fo is 15.5% for Pa and 7.7% for Me and Mo.

Finally, the percentages of variance accounted for are computed. These are 48.5% (Fo), 47.7% (Mo), 47.5% (Me), 47.8% (Pa) and 49.8% (Lw). However, the last two figures cannot be directly compared with the others, since for Pa the percentage of variance accounted for is not computed upon the correlation matrix of the transformed variables, while Lw is performed on a different data set. For these data, Mo seems to work slightly better than Me with respect to the reference method Fo.

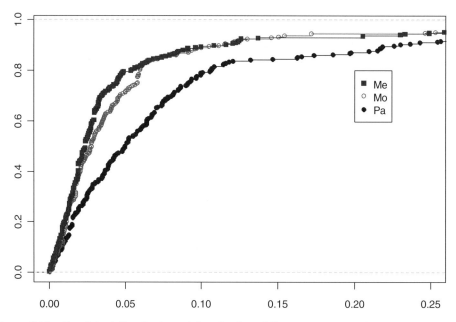

Figure 17.8   Empirical distribution of the absolute difference in scores between Fo and Mo, Me and Pa missing-data methods.

In the case of the ABC ACSS data, the different approaches to dealing wih missing data do not affect the results. This is because missing data are sparse, and the majority of incomplete units have only one missing value (see Table 17.2). It is important to underline that also in this case, where the different methods offer similar results, their relative performance is coherent with the quoted simulation study, confirms the analysis findings, and suggests the use of the forward imputation method for the treatment of missing data in NLPCA.

17.7   Summary

In this chapter we have shown how NLPCA can be applied in assessing customer satisfaction levels. Customer satisfaction is a complex phenomenon, not directly observable, but identifiable through several observable variables that represent different aspects of satisfaction and whose ordinal categories indicate different levels.

The latent dimension of the phenomenon is derived as a linear combination of observed variables through an optimal quantification of their ordinal categories and an optimization of the weights to be assigned to each variable.

NLPCA appears suitable for this purpose as it allows an indicator of satisfaction to be obtained that takes into account the nonlinear relationships between manifest variables and their weights. By assigning a real number to each unit (customer), it is possible to carry out an analysis on the basis of these values, creating customer profiles, forecasting new customer 'types', detecting environmental factors affecting the level of satisfaction and so on. If necessary, it is also possible to consider more than one latent dimension, thus enriching the analysis.

Moreover, if there are manifest variables intended as proxies for the level of satisfaction, a double application of NLPCA allows us to identify which of these best represents the global level of satisfaction and which of them best reflects each aspect of satisfaction. Furthermore, the analysis can also be carried out when there are missing data: some popular methods for handling missing data and a recent proposal specifically conceived for NLPCA are briefly described and their pros and cons discussed.

The application of NLPCA to the ABC survey data showed the advantages of this method in constructing and evaluating an indicator and in identifying a proxy for satisfaction. It also provided the possibility of examining and comparing different methods of missing-data treatment.

References

Andrich, D. (1978) A rating formulation for ordered response categories. *Psycometrika*, 43, 561–573.

Bertrand, M. and Mullainathan, S. (2001) Do people mean what they say? Implications for subjective survey data. *American Economic Review*, 91(2), 67–72.

Carpita M. and Manisera M. (2011) On the nonlinearity of homogeneous ordinal variables. In S. Ingrassia, R. Rocci and M. Vichi (eds), *New Perspectives in Statistical Modeling and Data Analysis*. Berlin: Springer.

Cronbach, L.J. (1951) Coefficient alpha and the internal structure of tests. *Psychometrika*, 16, 297–334.

de Leeuw, J. and Mair, P. (2009) Gifi methods for optimal scaling in R: The package homals. *Journal of Statistical Software*, 31(4).

Efron, B. (1979) Bootstrap method: Another look at the jackknife. *Annals of Statistics*, 7, 1–26.

Efron, B. and Tibshirani, R.J. (1993) *An Introduction to the Bootstrap*. New York: Chapman & Hall.

Ferrari, P.A. and Salini, S. (2011), Complementary use of Rasch models and nonlinear principal compo-nents analysis in the assessment of the opinion of Europeans about utilities. *Journal of Classification*, 28(1), 53–69.

Ferrari P.A., Annoni P. and Salini S. (2005) A comparison between alternative models for environmental ordinal data: Nonlinear PCA vs Rasch analysis. In A.R. Francis, K.M. Matawie, A. Oshlack and G.K. Smyth (eds), *Statistical Solutions to Modern Problems: Proceedings of the 20th International Workshop on Statistical Modelling*, pp. 173–177. Sydney: University of Western Sydney.

Ferrari, P.A., Annoni, P. and Urbisci, S. (2006) A proposal for setting-up indicators in the presence of missing data: The case of vulnerability indicators. *Statistica e Applicazioni* 4(1), 73–88.

Ferrari, P.A., Annoni, P., Barbiero, A. and Manzi, G. (2011) An imputation method for categorical variables with application to nonlinear principal component analysis. *Computational Statistics and Data Analysis*, 55(7), 2410–2420.

Freeman, R.B. (1978) Job satisfaction as an economic variable. *American Economic Review*, 68(2), 135–141.

Gifi, A. (1990) *Nonlinear Multivariate Analysis*. New York: John Wiley & Sons, Inc.

Guilford, J.P. (1965) *Fundamental Statistics in Psychology and Education*. New York: McGraw-Hill.

Heiser, W.J. and Meulman, J.J. (1994) Homogeneity analysis: Exploring the distribution of variables and their nonlinear relationship. In M.J. Greenacre and J. Blasius (eds), *Correspondence Analysis in the Social Sciences: Recent Developments and Applications*, pp. 179–209. New York: Academic Press.

Likert, R. (1932) A technique for the measurement of attitudes. *Archives of Psychology*, 140, 595–639.

Little, R.J.A. and Rubin, D.B. (2002) *Statistical Analysis with Missing Data*, 2nd edn. New York: Wiley Interscience.

Masters, G.B. (1982) A Rasch model for partial credit scoring. *Psychometrika*, 47, 149–174.

Michailidis, G. and De Leeuw, J. (1998) The Gifi system of descriptive multivariate analysis. *Statistical Science*, 13, 307–336.

Rasch G. (1980) *Probabilistic Models for Some Intelligence and Attainment Tests*. Chicago: University of Chicago Press.

18

Multidimensional scaling

Nadia Solaro

This chapter presents multidimensional scaling (MDS) methods and their application to customer satisfaction surveys. MDS methods are multivariate statistical analysis techniques of particular relevance to survey data analysis. In fact, despite some criticism, such applications are gaining in popularity, especially in market research studies. The chapter begins by presenting the theory of MDS, including theoretical results on so-called proximity data, the basic input data of MDS. An overview is given of the most widely applied MDS models, the classical, least squares and nonmetric MDS. Several reserach topics in MDS are also considered, i.e. the problems of assessing goodness of fit, comparing two different MDS solutions, and diagnosing anomalous results that could derive from analyses. The chapter then goes on to deal with the application of metric MDS models to the ABC annual customer satisfaction survey. It concludes by outlining some future directions for MDS research.

18.1 An overview of multidimensional scaling techniques

Multidimensional scaling (MDS) is the name given to a large family of multivariate data analysis techniques for dealing with dimensionality reduction and data visualization problems in situations where the reproduction of 'closeness' of data from an observational space is of primary concern. The common, basic idea of these methods is to represent a given set of proximities, that is, measures of pairwise similarity/dissimilarity between objects (subjects, units, variables, stimuli, etc.) in a multidimensional space, typically Euclidean, with few dimensions, typically two or three. Hence, 'distance-like' data are the basic input structure. MDS techniques are recognized as flexible analysis tools, as they can handle different types of proximities.

The literature on MDS methods is extremely rich, with several milestone results in multivariate statistical analysis of data. Recent comprehensive references on MDS are the monographs by Cox and Cox (2001) and Borg and Groenen (2005), where classical and recent

Modern Analysis of Customer Surveys: with applications using R, First Edition. Edited by Ron S. Kenett and Silvia Salini.
© 2012 John Wiley & Sons, Ltd. Published 2012 by John Wiley & Sons, Ltd.

results are gathered together and discussed. Since the first steps in MDS, when Eckart and Young (1936) and Young and Householder (1938) laid the mathematical foundations for the *par excellence* MDS method, that is, the classical MDS, many methodological advances have been achieved. Moreover, many implementations of MDS models are now available in statistical software, such as SPSS and SAS. This has contributed to the wide spread and popularity of MDS in several fields of application, such as ecological and environmental sciences, as well as the growing interest in MDS in many other areas, such as medical applications, where typically alternative statistical techniques, for example factor analysis or principal component analysis, have been often used. Borg and Groenen (2005, Chapter 1) review four main purposes of MDS:

1. *MDS techniques are tools for data exploration.* Investigators having no explicit theory about the structure of a data set can use MDS to find systematic rules in the data. A common approach is to describe data as points in a multidimensional space, visualizing data through geometrical representations.

2. *MDS techniques are methods for testing structural hypotheses.* This is the confirmatory approach to MDS. Researchers formulate hypotheses when they feel confident that they have a good knowledge of the phenomenon under study. They verify whether available data support assumptions that classify the original items into different semantic fields.

3. *MDS techniques are methods for exploring psychological structures*, when the objective is to find the underlying dimensions explaining systematic trends in psychological constructions.

4. *MDS can be interpreted as a methodology for modelling similarity judgements*, when subjects are asked to express their own judgement of how close pairs of objects are according to 'distances' they perceive in their 'psychological space'.

This section is organized into eight parts covering the origins of MDS, proximity data, MDS models, goodness-of-fit assessment, Procrustes analysis for comparing MDS solutions, robustness analysis, and dealing with missing values. Final remarks pertain to the use of MDS in customer satisfaction surveys and market research.

18.1.1   The origins of MDS models

MDS has its roots in psychometrics, where the primary interest was in understanding the psychological dimensions that explain similarity judgements expressed by subjects on certain objects. Pioneering methodological contributions were published by the journal *Psychometrika*, in particular the work by Eckart and Young (1936), and Young and Householder (1938), who established the mathematical background for MDS. Torgerson (1952) developed the first metric MDS method, later known as Torgerson's or classical MDS. Subsequently, Gower (1966) extended Torgerson's results and highlighted the relationship between classical MDS and principal component analysis. His method is known as *principal coordinate analysis*. Around the same time, Shepard (1962a, 1962b) and Kruskal (1964a, 1964b) proposed a non-metric MDS method in which proximities are monotonically related to distances in the representation space. Kruskal's breakthrough contribution was to introduce the concept of the loss function in the MDS framework. This opened up the field to further proposals for MDS models based on the specification and optimization of a loss function, such as Sammon's non-linear mapping (Sammon, 1969). Later contributions led to further generalizations of MDS. These include the unfolding model (Coombs, 1964), which deals with data derived from two distinct sets of objects; the individual differences models (Carroll and Chang, 1970) and three-way MDS (Carroll and Wish, 1974), which extended MDS models to three-way proximity data – proximities collected or computed on more than one source of information, such as different points in time. In the 1970s, much work was also done on implementing MDS models in computationally efficient algorithms. For instance, Takane *et al.* (1977) developed the ALSCAL algorithm, which is included in SPSS. More recently, Ramsay (1982) introduced the maximum likelihood approach to MDS, while Meulman (1992, 1993) included MDS in Gifi's (1990) descriptive system of multivariate analysis by reformulating it as an optimal scaling approach.

18.1.2   MDS input data

In all MDS applications, information on the closeness of points in a multidimensional space plays the role of input data. These are generally termed *proximities* and are measures of the degree to which pairs of objects (individuals, items, stimuli, etc.) are alike (similarity measures) or different (dissimilarity measures). Proximity measures can be of a very diverse nature. A first distinction arises in the way proximities are collected, namely directly or indirectly. Proximities are called *direct* if they consist of quantitative appraisals or qualitative judgements that are directly expressed by subjects who are asked to state how (dis)similar are elements in a set of objects. These judgement calls can take the form of preference rankings when subjects are asked to sort objects according to their opinion, perception or preference. Proximities are called *indirect* if they are computed from certain types of available information, usually variables in a data matrix, by applying a specific (dis)similarity or distance measure. Information in a data matrix format could be either quantitative or qualitative, or both, in which case the data is said to be of mixed type. Therefore, a crucial task in setting up indirect proximities is to choose the most suitable measure for handling variables consistently with their metric or non-metric characteristics.

Proximities are organized into proximity matrices. These represent basic input data of any MDS application. They result in square symmetric matrices when measures are derived from pairwise comparisons performed within the same set of objects (e.g., subjects with themselves, or stimuli with themselves). Alternatively, when proximities involve comparisons between two distinct sets of entities, for instance subjects' preference rankings towards a set of objects, proximity matrices are asymmetric, and they may be rectangular or square.

In this chapter we restrict our attention to proximities with specific characteristics, namely indirect proximities forming square and symmetric matrices. Our choice is driven by the applications discussed in Section 18.2, where we consider data derived from a customer satisfaction study. For more details on proximities, see Davison (1983, Chapter 3), Everitt and Rabe-Hesketh (1997, Chapter 2), Cox and Cox (2001, Chapter 1), Borg and Groenen (2005, Chapter 6), and Gower (2004).

Moreover, since MDS techniques are usually applied to dissimilarities rather than similarities, in what follows we mostly refer to dissimilarities. This does not, however, imply any loss of generality, given that similarities can be obtained from dissimilarities, and vice versa, through appropriate transformations. We briefly deal with this point in the last part of this subsection.

A function $\delta(u_i, u_j) = \delta_{ij}$ is a dissimilarity function if it is non-negative ($\delta_{ij} \geq 0$), symmetric ($\delta_{ij} = \delta_{ji}$) and satisfies the identity property ($\delta_{ii} = 0$), for every pair (i, j) of objects u_i and u_j ($i, j = 1, \ldots, n$). Authors such as Gower and Legendre (1986) and Cailliez and Kuntz (1996) stress that the above definition of dissimilarity does not contain enough conditions to prevent inconsistent situations where, even though $\delta_{ij} = 0$ holds for two distinct objects u_i and u_j, it may occur that $\delta_{ik} \neq \delta_{kj}$ for a third object u_k. This inconvenience does not arise, however, if the dissimilarity satisfies at least one of the following properties: (a) $\delta_{ij} = 0$ if, and only if, u_i and u_j coincide (definiteness property), or (b) $\delta_{ij} \leq \delta_{ik} + \delta_{kj}$ (triangle inequality), for every i, j and k. If both conditions (a) and (b) occur, the dissimilarity becomes a metric or, equivalently, a distance (see Cailliez and Kuntz, 1996).

There is a large variety of dissimilarity and distance measures in the literature. Various treatments of proximity data, such as Everitt and Rabe-Hesketh (1997) or Cox and Cox (2001), report on the best-known measures and give helpful advice on their use. Again, for more details we refer the reader to these more specialized texts. Here, we give more emphasis to results that are useful for the MDS applications proposed in Section 18.2, with special attention to mixed-type data handling.

The choice between using a dissimilarity measure or metric depends on the measurement level of variables. As a matter of fact, dissimilarities are generally reserved for qualitative variables and distances for interval or ratio-scaled variables. This does not mean, however, that dissimilarity measures are always at ordinal level, and measures for quantitative variables are always at interval or ratio-scale level. There are instances of non-metric measures for quantitative variables, while most measures for qualitative variables preserve metric properties.

Traditionally, dichotomous (or binary) variables are handled in terms of similarities rather than dissimilarities. A key distinction among similarity measures is the concept of co-absence, which denotes the situation where a category or an attribute of a binary variable is absent on the objects u_i and u_j at the same time. The opposite concept is that of co-presence, where a category or an attribute is present on the objects u_i and u_j at the same time. If co-absences are deemed as an integral part of a similarity evaluation between objects, then they should be taken into account in the measure computation (e.g., as in the simple matching coefficient), otherwise they should be discarded, and co-presences only should be considered (as in Jaccard's coefficient).

Polytomous qualitative variables are treated like binary variables. Similarity coefficients can be generalized to cover co-presences arising from variables with multiple categories. Gower (2004) evokes the two possible approaches to introduce co-absences in this case. Usually, it is preferable not to include them in the similarity evaluation, in order to avoid assigning spurious weights in comparisons.

For quantitative variables, one often applies the well-known Minkowski family of distances. Given a set of p quantitative variables Y_l with n observations y_{il}, the general parametric form is

$$_r\delta_{ij} = \left\{ \sum_{l=1}^{p} \left| y_{il} - y_{jl} \right|^r \right\}^{1/r}, \quad r \geq 1,$$

which includes the city-block or Manhattan ($r = 1$), Euclidean ($r = 2$), and Chebyshev or Lagrange ($r \to \infty$) distances as special cases. As Euclidean distance,

$$_2\delta_{ij} = \sqrt{\sum_{l=1}^{p} \left(y_{il} - y_{jl} \right)^2},$$

is particularly important in MDS methods, we dedicate a subsection to describing its main properties. In addition, such a family can be generalized to include differentiated weights for variables. This gives rise to the so-called generalized (or weighted) Minkowski family of distances, which has the general parametric form

$$_r\delta_{ij}^w = \left\{ \sum_{l=1}^{p} w_l \left| y_{il} - y_{jl} \right|^r \right\}^{1/r}, \quad r \geq 1,$$

where w_l is a non-negative weight for Y_l ($l = 1, \ldots, p$).

The Minkowski family does not obviously exhaust all the possibilities. Other measures for quantitative variables are, for instance, the Canberra and Bray–Curtis measures, particularly popular in environmental fields of application. These two are prominent instances of measures for quantitative variables that are not necessarily metric. In fact, the Canberra measure is metric for positive variable values only, while the Bray–Curtis measure is not a metric.

We are not aware of specific (dis)similarity coefficients for ordinal variables. Many measures for binary variables, such as the simple matching coefficient, can be generalized to ordinal variables. Another possibility is to employ a distance from the Minkowski family, usually the city-block distance, to ordinal variables after they have been transformed into ranks.

Mixed-type data, involving a mixture of qualitative and quantitative variables, can be handled by Gower's (1971) general coefficient of similarity (GCS). This is still the most popular method for such data. The GCS is defined as a weighted average of similarity scores, which are computed by accounting for the different measurement levels of variables. Given a set of p variables Y_l, the general formula for the similarity of objects u_i and u_j is

$$s_{ij} = \frac{\sum_{l=1}^{p} w_{ijl} s_{ijl}}{\sum_{l=1}^{p} w_{ijl}}, \qquad 0 \leq s_{ij} \leq 1; \quad i, j = 1, \ldots, n, \qquad (18.1)$$

where s_{ijl} is the similarity score of u_i and u_j computed on Y_l and w_{ijl} is a non-negative weight.

Weighting is a convenient device for handling missing values. A weight w_{ijl} can be set to zero whenever either observation y_{il} or y_{jl} of Y_l is missing, or set to unity if both y_{il} and y_{jl} are known. Weighting is also used to discern between co-absences and co-presences in the case of binary variables. If u_i and u_j do not coincide on variable Y_l, that is, $y_{il} \neq y_{jl}$, then $w_{ijl} = 1$ and $s_{ijl} = 0$. Otherwise, when they share the same category of interest, that is, $y_{il} = y_{jl} = 1$, then $w_{ijl} = s_{ijl} = 1$. The case $y_{il} = y_{jl} = 0$ requires a choice between including co-absences ($w_{ijl} = s_{ijl} = 1$) and excluding them ($w_{ijl} = s_{ijl} = 0$) from comparisons. For nominal variables, the similarity score is $s_{ijl} = 1$ provided that $y_{il} = y_{jl}$, otherwise $s_{ijl} = 0$. For interval or ratio-scaled variables, the similarity score is $s_{ijl} = 1 - \left| y_{il} - y_{jl} \right| / \{\max(Y_l) - \min(Y_l)\}$, which is a normalized value due to the division by the range. Although ordinal variables were not explicitly considered by Gower (1971), they can be included in the GCS computation nonetheless. It suffices to transform them into ranks (average ranks should be considered in the presence of ties), and then treat them as ratio-scaled variables. Podani (1999) proves that this procedure gives rise to a metric version of the GCS. He also provides a non-metric version of the GCS for use with ordinal variables, which is not considered in this chapter. Finally, the GCS can be re-expressed in terms of a dissimilarity measure by applying standard transformations (Gower, 1971; Gower and Legendre, 1986).

As a final remark, GCS tends to give too much weight to qualitative variables, especially when their frequency distribution is strongly heterogeneous, as measured by one of the diversity indices, such as Shannon's or Gini's heterogeneity measures (Solaro, 2010). Gower (1971) does, however, suggest a way to design weights that account for the importance of variables by introducing a more general formulation than (18.1). For example, by putting $w_l = w_{ijl}$ for all (i, j) in (18.1) it is possible to give a constant weight to variable Y_l over all possible comparisons between objects. Cox and Cox (2000) suggest an extension to GCS for computing pairwise similarities for objects, which also simultaneously produces a set of 'similarities' for variables. The weights assigned to both objects and variables are thus learned directly from data, through an iterative procedure.

Proximities and Euclidean properties

One of the main aspects of concern in MDS applications is the fact that dissimilarities are embeddable in a Euclidean space, since, with some exceptions, this is the geometrical structure of the representation space. Gower and Legendre (1986, pp. 7–10) assemble many results on metric and Euclidean properties of dissimilarity coefficients. As far as MDS is concerned, it is worth listing these properties. Let $\mathbf{\Delta}$ be an $n \times n$ dissimilarity matrix with δ_{ij} as generic element. Such a matrix is said to be non-metric if its elements satisfy the requirements of non-negativity, symmetry and identity; it is metric if its elements also satisfy the definiteness property and triangle inequality. Moreover, $\mathbf{\Delta}$ is a Euclidean matrix if its elements are derived as Euclidean distances or can be assumed as such. Then, the following facts hold:

(i) If $\mathbf{\Delta}$ is a non-metric matrix, it is always possible to obtain a metric dissimilarity matrix $\mathbf{\Delta}^*$ by adding a proper constant c to the off-diagonal elements δ_{ij} ($i \neq j$). Such a constant must be no smaller than $\max_{k,s,m} |\delta_{ks} + \delta_{km} - \delta_{sm}|$. This result is derived from the triangle inequality.

(ii) Let $\mathbf{A}$ be the $n \times n$ matrix with elements $-\frac{1}{2}\delta_{ij}^2$ and $\mathbf{w}$ a n-dimensional vector such that $\mathbf{w}^t\mathbf{1} = 1$, where the vector $\mathbf{1}$ contains n ones. Then, a sufficient and necessary condition for $\mathbf{\Delta}$ to be Euclidean is that the matrix $(\mathbf{I} - \mathbf{1w}^t)\mathbf{A}(\mathbf{I} - \mathbf{w1}^t)$ is positive semi-definite (p.s.d.), where $\mathbf{I}$ is the $n \times n$ identity matrix. This is the fundamental result on which classical MDS is grounded.

(iii) Let $\mathbf{S}$ be a similarity matrix with elements $0 \leq s_{ij} \leq 1$ and $s_{ii} = 1$. If $\mathbf{S}$ is p.s.d., then $\mathbf{\Delta}$ with elements $\delta_{ij} = \sqrt{1 - s_{ij}}$ is Euclidean.

(iv) Given a dissimilarity matrix $\mathbf{\Delta}$, there exist constants h and k such that the matrix with elements $(\delta_{ij}^2 + h)^{1/2}$ (Lingoes, 1971) or $\delta_{ij} + k$ (Cailliez, 1983) is Euclidean. This is the additive constant problem.

With specific reference to GCS, Gower (1971) proved that the matrix $\mathbf{S}$ formed with the similarities (18.1) turns out to be p.s.d. provided that there are no missing values. Therefore, due to result (iii), the Gower dissimilarity matrix, that is, the matrix with elements $\delta_{ij} = \sqrt{1 - s_{ij}}$, where s_{ij} is given by formula (18.1), has Euclidean properties.

18.1.3   MDS models

Many MDS models have been proposed in the literature. All of them provide methods for representing the coordinates of observed dissimilarities in a q-dimensional Euclidean space, such that the observed dissimilarities are reproduced 'as well as possible' by Euclidean distances computed on the final MDS configuration of points. Between-objects dissimilarities δ_{ij} of an observational space are thus mapped into between-objects Euclidean distances d_{ij}

of a representation space with the least possible loss of information. Formally, let $\mathbf{X}$ denote an $n \times q$ configuration of points. Dissimilarities are related to distances through a so-called *representation function*, $f : \delta_{ij} \rightarrow d_{ij}(\mathbf{X})$, where the notation $d_{ij}(\mathbf{X})$ stresses that distances d_{ij} are functions of the unknown coordinates $\mathbf{X}$. Defining a specific type of mapping is the same as identifying a specific MDS model. Hence, in every MDS model the equation $d_{ij}(\mathbf{X}) = f(\delta_{ij})$ between dissimilarities and distances is the basic starting point. It turns out, however, that this equality has only theoretical value. From a practical point of view, it is rare for the equality to hold exactly, for a variety of reasons mentioned below. The problem is therefore usually stated in weaker terms that rely on an approximate relation, $d_{ij} \approx f(\delta_{ij})$, and the exact equality applied only to the definition of *disparity*. A disparity, usually denoted by $\hat{d}_{ij}$, is the value actually obtained by applying the function f to the dissimilarity δ_{ij}, that is, $\hat{d}_{ij} = f(\delta_{ij})$, so that the above approximation can be expressed also as $d_{ij} \approx \hat{d}_{ij}$ $(i, j = 1, \ldots, n)$. Commonly used representation functions, and relations defining disparities from dissimilarities, are the identity function $\hat{d}_{ij} = \delta_{ij}$, which gives rise to the *absolute* MDS; the linear functions $\hat{d}_{ij} = b\,\delta_{ij}$, to which the *ratio* MDS corresponds, and $\hat{d}_{ij} = a + b\,\delta_{ij}$, which specifies the *interval* MDS, where the parameters a and b are chosen to guarantee that the above relation between Euclidean distances and observed dissimilarities holds as far as possible. Additional possible choices include the exponential, logarithmic and polynomial functions. Thus, in principle, any kind of function could be considered. It is not even necessary for f to be completely specified. The parameters in f can be derived directly from data via iterative procedures. The main requirement in all these cases is that the function f is continuous, parametric and monotone. A different situation arises when f is required only to preserve the order of dissimilarities. This is the *ordinal* MDS. In this case, f must be a monotone function such that if $\delta_{ij} < \delta_{rs}$, then $\hat{d}_{ij} \leq \hat{d}_{rs}$. The function f represents a weak monotonicity relation if no further requirement is explicitly introduced for the case $\delta_{ij} = \delta_{rs}$. This is the primary approach to ties (i.e., tied dissimilarities). When the equality $\delta_{ij} = \delta_{rs}$ needs $\hat{d}_{ij} = \hat{d}_{rs}$ as its image, the approach followed is known as the secondary approach to ties, and the function f defines a strong monotonicity relation, $(i \neq j,$ $r \neq s, i, j, r, s = 1, \ldots, n)$.

According to a historical view that continues to persist in the literature, the distinction between metric MDS and non-metric MDS is based on the measurement level of dissimilarities. When dissimilarities are at interval or ratio-scale level, metric MDS is involved, whilst non-metric MDS is concerned with ordinal dissimilarities. Some authors argue, however, that the classification of MDS models should be more elaborate, and that it should account for the relation linking observed dissimilarities with distances, such as ordinal, interval or ratio MDS (Heady and Lucas, 2010). Apart from the ordinal MDS already discussed, the ratio MDS tries to find a final configuration of points where the ratio of any two disparities coincides with the ratio of the corresponding dissimilarities, $\hat{d}_{ij}/\hat{d}_{rs} = \delta_{ij}/\delta_{rs}$ or $\hat{d}_{ij}/\delta_{ij} = b$ for every pair (i, j) and (r, s). In the interval MDS, it is the ratio between the difference of any two disparities and the difference between the corresponding two dissimilarities that has to remain constant, that is, $(\hat{d}_{ij} - \hat{d}_{rs})/(\delta_{ij} - \delta_{rs}) = b$ for every (i, j) and (r, s) (Borg and Groenen, 2005, Chapter 9).

As mentioned above, the relation linking distances $d_{ij}(\mathbf{X})$ with observed dissimilarities δ_{ij} should hold only approximately. How acceptable the approximation is depends on several factors, and should be assessed through some optimal criteria such as a goodness-of-fit index or a measure based on a loss function. The choice of the number q of dimensions of the MDS configuration contributes strongly to the accuracy of the final solution. Nonetheless, another fundamental factor that can affect the goodness of an MDS analysis is the whether or not the input dissimilarity matrix is Euclidean. This is a crucial point: if dissimilarities are far from being embeddable in a Euclidean space, because they are non-Euclidean or even non-metric, a large number of dimensions are required to satisfactorily recover the coordinates. A well-known theoretical result establishes that it is always possible to perfectly embed a Euclidean dissimilarity matrix of order n in an $(n - 1)$-dimensional Euclidean space. This is the fundamental theorem of classical MDS, which will be discussed later. In all other situations, as mentioned earlier, appropriate transformations can be applied to observed dissimilarities to render them metric or even Euclidean.

Classical multidimensional scaling

One of the most widespread MDS models is the classical MDS (CMDS). This evolved from the mathematical foundations developed by Eckart and Young (1936) and Young and Householder (1938), and its main theoretical results were set out within the scaling framework by Torgerson (1952) and Gower (1966). This is the reason why CMDS is also known in the literature as Torgerson's MDS or Gower's principal coordinate analysis method. CMDS is the prime example of a metric absolute MDS, which assumes the identity as representation function, $d_{ij}(\mathbf{X}) = \delta_{ij}$, and provides an analytical solution. Let $\mathbf{B} = \mathbf{XX}^t$ be the inner-product matrix with generic element b_{ij}, that is, the product $\mathbf{x}_i^t \mathbf{x}_j$ of the unknown coordinates $\mathbf{x}_i$ and $\mathbf{x}_j$, given by the rows i and j of $\mathbf{X}$. In order to prevent indeterminate solutions due to arbitrary translations, the matrix $\mathbf{X}$ is assumed to be column-centred, that is, its columns (also called dimensions) have zero mean. The basic idea of CMDS is to recover the set of coordinates $\mathbf{X}$ through the spectral decomposition of matrix $\mathbf{B}$. However, $\mathbf{B}$ is unknown, so that a more practicable formulation that links $\mathbf{B}$ with the observed dissimilarities δ_{ij} is required. According to Torgerson (1952), if δ_{ij} are Euclidean distances, then the solution is immediate. Since the square of (observed) Euclidean distances, $\delta_{ij}^2 = (\mathbf{x}_i - \mathbf{x}_j)^t(\mathbf{x}_i - \mathbf{x}_j)$, can be expressed as $\delta_{ij}^2 = \mathbf{x}_i^t\mathbf{x}_i + \mathbf{x}_j^t\mathbf{x}_j - 2\mathbf{x}_i^t\mathbf{x}_j$, straightforward computations lead to the formula

$$b_{ij} = \mathbf{x}_i^t\mathbf{x}_j = a_{ij} - a_{i.} - a_{.j} + a_{..},$$

where

$$a_{ij} = -\tfrac{1}{2}\delta_{ij}^2, \quad a_{i.} = \tfrac{1}{n}\sum_j a_{ij}, \quad a_{.j} = \tfrac{1}{n}\sum_i a_{ij}, \quad a_{..} = \tfrac{1}{n^2}\sum_i\sum_j a_{ij},$$

for any i and j (see Cox and Cox, 2001). Let $\mathbf{A}$ be a matrix with generic element a_{ij}. Then the matrix $\mathbf{B}$ is linked to $\mathbf{A}$ by the relation $\mathbf{B} = \mathbf{HAH}$, where $\mathbf{H}$ is the so-called centring matrix,

$$\mathbf{H} = \mathbf{I} - \frac{\mathbf{11}^t}{n},$$

and $\mathbf{1}$ is a n-vector of ones. The matrix $\mathbf{B}$ so expressed is also referred to as a doubly-centred matrix. To find the configuration $\mathbf{X}$, we then rely on the sufficient and necessary condition (ii) in Section 18.1.2 (Gower and Legendre, 1986), according to which the matrix $\mathbf{B} = \mathbf{HAH}$ turns out to be p.s.d. thanks to $\boldsymbol{\Delta}$ being Euclidean. Such a condition is applied in the present case by putting the vector $\mathbf{w}$ equal to $\tfrac{1}{n}\mathbf{1}$. Finally, given the positive semi-definiteness of $\mathbf{B}$, which implies the non-negativity of all its eigenvalues λ_m, a set of coordinates $\mathbf{X}$ can be recovered from the spectral decomposition of $\mathbf{B}$. We have $\mathbf{X} = \boldsymbol{\Gamma}\boldsymbol{\Lambda}^{1/2}$, where $\boldsymbol{\Lambda}^{1/2}$ is the diagonal matrix with the square roots of the eigenvalues λ_m of $\mathbf{B}$, taken in non-increasing order, and $\boldsymbol{\Gamma}$ is the matrix with the corresponding orthogonal, normalized eigenvectors.

Regarding the number of columns in $\mathbf{X}$, that is, the number of dimensions that can be extracted, it is worth noting that $\mathbf{B}$ admits at least one zero eigenvalue, since $\mathbf{B1} = \mathbf{HAH1} = \mathbf{0}$.

Therefore, the maximum number of dimensions can never exceed $n - 1$. In practice, a large number of zero eigenvalues might be associated with the matrix $\mathbf{B}$, so that the maximum number of dimensions that can be extracted is equal to the total number of its positive eigenvalues, that is, the rank of the matrix $\mathbf{HAH}$. This, however, is not always relevant to the problem of how many dimensions it is appropriate to retain. For this purpose, we rely on a range of criteria for choosing the 'ideal' number of dimensions, which are related to goodness-of-fit appraisals, as mentioned later. For now, we denote by q the number of dimensions retained in the analysis, which must be sufficiently small with respect to the maximum number ($q < n - 1$). By taking the first q positive eigenvalues of $\mathbf{B}$, the solution is $\mathbf{X} = \boldsymbol{\Gamma}_q \boldsymbol{\Lambda}_q^{1/2}$, where $\boldsymbol{\Lambda}_q^{1/2}$ and $\boldsymbol{\Gamma}_q$ have the same meaning as, respectively, $\boldsymbol{\Lambda}^{1/2}$ and $\boldsymbol{\Gamma}$, but limited to the first q eigenvalues.

For the sake of completeness, we remind the reader that by the spectral decomposition theorem, any symmetric $n \times n$ matrix $\mathbf{M}$ can be written as: $\mathbf{M} = \boldsymbol{\Gamma}\boldsymbol{\Lambda}\boldsymbol{\Gamma}^t = \sum_{m=1}^{n} \lambda_m \boldsymbol{\gamma}_m \boldsymbol{\gamma}_m^t$, where $\boldsymbol{\Lambda}$ is a diagonal matrix of the n eigenvalues λ_m of $\mathbf{M}$, and $\boldsymbol{\Gamma}$ is an $n \times n$ orthogonal matrix whose columns $\boldsymbol{\gamma}_m$ are normalized eigenvectors. The reader is referred to textbooks on matrix algebra for all definitions and properties regarding eigenvalues and eigenvectors, quadratic forms and definiteness (see, for example, Mardia *et al.*, 1979, Appendix A). The main results concerning symmetric matrices are as follows:

1. For a symmetric matrix, eigenvectors corresponding to distinct eigenvalues are orthogonal to each other. Then, normalization and orthogonality imply the orthonormality of the matrix $\boldsymbol{\Gamma}$ containing the eigenvectors, that is, $\boldsymbol{\Gamma}\boldsymbol{\Gamma}^t = \boldsymbol{\Gamma}^t\boldsymbol{\Gamma} = \mathbf{I}$, where $\mathbf{I}$ is the identity matrix of order n.

2. The eigenvalues of $\mathbf{M}$ are all positive if and only if $\mathbf{M}$ is positive definite. The eigenvalues of $\mathbf{M}$ are all non-negative if and only if $\mathbf{M}$ is p.s.d.

3. If $\lambda_m > 0$ for $m = 1, \ldots, q$, and $\lambda_r = 0$ for $r = q + 1, \ldots, n$ (i.e., $\mathbf{M}$ is p.s.d.), then the spectral decomposition of $\mathbf{M}$ can be written as $\mathbf{M} = \sum_{m=1}^{q} \lambda_m \boldsymbol{\gamma}_m \boldsymbol{\gamma}_m^t = \boldsymbol{\Gamma}_q \boldsymbol{\Lambda}_q \boldsymbol{\Gamma}_q^t$.

4. If $\lambda_m \geq 0$ for all m, then $\mathbf{M}^{1/2} = \boldsymbol{\Gamma}\boldsymbol{\Lambda}^{1/2}\boldsymbol{\Gamma}^t$, where $\boldsymbol{\Lambda}^{1/2} = \mathrm{diag}(\lambda_1^{1/2}, \ldots, \lambda_n^{1/2})$.

In general situations where the observed dissimilarities δ_{ij} are not computed as Euclidean distances, CMDS can still be applied. Within this framework, two situations may occur. In the first, the matrix $\mathbf{B}$ is p.s.d. In such a case, due to the aforementioned sufficient and necessary condition (ii), the dissimilarity matrix $\boldsymbol{\Delta}$ has Euclidean properties, thus it is perfectly embeddable in a Euclidean space. In the second case, the matrix $\mathbf{B}$ is not p.s.d., so that a more or less large number of eigenvalues may turn out to be negative. Under these circumstances, we can then proceed in two possible directions. The observed matrix $\boldsymbol{\Delta}$ can be rendered Euclidean by appropriate transformations of its elements δ_{ij}, such as those recalled in Section 18.1.2 (Gower and Legendre, 1986). For a collection of results concerning the additive constant problem, see Cox and Cox (2001, Section 2.2.8). As an alternative, if the magnitude of negative eigenvalues is relatively small, these may simply be ignored, and the analysis confined to the first q positive eigenvalues; see Cox and Cox (2001, Section 2.2.3).

Finally, Gower (1966) proved the dual relationship between CMDS and principal component analysis. In particular, he established that the coordinates of a CMDS solution are already referred to their principal axes, that is, the axes that reproduce the maximum possible variation of points in the same sense as principal components. In fact, given $\mathbf{X}$ in q dimensions, we have $\mathbf{X}^t\mathbf{X} = \boldsymbol{\Lambda}_q^{1/2}\boldsymbol{\Gamma}_q^t\boldsymbol{\Gamma}_q\boldsymbol{\Lambda}_q^{1/2} = \boldsymbol{\Lambda}_q$, which is the diagonal matrix of eigenvalues, so that the deviances of the q dimensions are given by the entries of Λ_q. From this interpretation, the name 'principal coordinate analysis', attributed by Gower to CMDS, follows.

Least-squares multidimensional scaling

Least-squares MDS (LSMDS) techniques attempt to find a set of coordinates $\mathbf{X}$ by optimizing some least-squares fit criterion, such as minimizing a least-squares loss function, between the observed dissimilarities and the distances in the final configuration. Using a LS criterion, distances are implicitly expressed as linear functions of the observed dissimilarities plus an error term, $d_{ij}(\mathbf{X}) = f(\delta_{ij}) + \varepsilon_{ij} = \hat{d}_{ij} + \varepsilon_{ij}$, for all (i, j). The term ε_{ij} may represent both measurement errors, possibly affecting the observed dissimilarities, and distortion errors, arising because in general the final distances do not perfectly fit the observed dissimilarities. In the literature, many loss functions have been introduced. Each identifies a specific LSMDS model. One recurring, general version of the loss function has the form

$$S(\mathbf{X}) = \sum_{i=1}^{n} \sum_{j<i}^{n} w_{ij}(\hat{d}_{ij} - d_{ij}(\mathbf{X}))^2, \tag{18.2}$$

where the w_{ij}s are positive weights for, if necessary, strengthening (or weakening) differences between disparities and distances pertaining to objects we intend to emphasize or de-emphasize. In the case of absolute MDS, that is, when $\hat{d}_{ij} = \delta_{ij}$, for all (i, j), formula (18.2) is usually called 'raw stress', where stress stands for *s*tandardized *r*esidual *s*um of *s*quares, and is given by

$$\sigma_r(\mathbf{X}) = \sum_{i=1}^{n} \sum_{j<i}^{n} w_{ij}(\delta_{ij} - d_{ij}(\mathbf{X}))^2. \tag{18.3}$$

One of the main objections raised towards loss functions of types (18.2)–(18.3) is that they are not invariant under scale transformations of coordinates. To guarantee this invariance property, some forms of normalization can be introduced. A natural choice is to express weights as a function of disparities, for instance by putting weights in (18.2) equal to $w_{ij} = \hat{d}_{ij}^{-2}$. Another instance arises when $w_{ij} = \hat{d}_{ij}^{-1} \left(\sum_i \sum_{j<i} \hat{d}_{ij} \right)^{-1}$. In this way, smaller disparities, thus contributing to the stress loss function more heavily than greater disparities, tend to be reproduced more accurately than larger ones. Sammon (1969) was the first to introduce this particular LSMDS model by assuming the relation $\hat{d}_{ij} = \delta_{ij}$. The model is thus known as Sammon's non-linear mapping. Another choice involves expressing weights as a function of the final distances. In particular, if $w_{ij} = \left(\sum_i \sum_{j<i} d_{ij}^2 \right)^{-1}$, and the square root of $S(\mathbf{X})$ is taken, Kruskal's Stress-1 formula $S_1 = \sqrt{\sum_i \sum_{j<i} \left(\hat{d}_{ij} - d_{ij} \right)^2 \Big/ \sum_i \sum_{j<i} d_{ij}^2}$ is obtained. Another normalization is based on putting $w_{ij} = \left\{ \sum_i \sum_{j<i} \left(d_{ij} - \bar{d} \right)^2 \right\}^{-1}$, where $\bar{d}$ is the mean of distances computed for $1 \leq i < j \leq n$. This gives the Stress-2 formula $S_2 = \sqrt{\sum_i \sum_{j<i} \left(\hat{d}_{ij} - d_{ij} \right)^2 \Big/ \sum_i \sum_{j<i} \left(d_{ij} - \bar{d} \right)^2}$. The Stress-1 and Stress-2 formulas are designated in the literature as forms of implicit normalization, in the sense that the size of $\mathbf{X}$ is implicitly taken into account through the assessment of the total size of distances computed from the same coordinates $\mathbf{X}$. Furthermore, the following normalization is usually introduced

for formula (18.3):

$$\sigma_N(\mathbf{X}) = \frac{\sigma_r(\mathbf{X})}{\sum_i \sum_{j<i} w_{ij}\delta_{ij}^2} = \frac{\sum_i \sum_{j<i} w_{ij}(\delta_{ij} - d_{ij}(\mathbf{X}))^2}{\sum_i \sum_{j<i} w_{ij}\delta_{ij}^2}, \tag{18.4}$$

which is termed normalized stress (Borg and Groenen, 2005, Chapter 11).

LSMDS models are used to find a set of coordinates $\mathbf{X}$ through iterative numerical procedures, and in general can be looked upon as a two-stage process that is iteratively repeated. In the first stage, the procedure works by assuming that a set of coordinates $\mathbf{X}$ is given in q dimensions. This permits us to compute the distances d_{ij} and estimate unknown parameters that are possibly involved in the representation function. For instance, in the interval MDS if the parameters a and b defining the disparities $\hat{d}_{ij} = f(\delta_{ij}) = a + b\delta_{ij}$ have not previously been fixed by the researcher, they can be estimated at this stage by applying the least-squares method to the simple linear regression of d_{ij} on δ_{ij}. In the second stage, the loss function enters a minimization procedure the aim of which is to recover a new set of coordinates $\mathbf{X}$. Final distances d_{ij} are updated based on these new coordinates. The two stages are then iteratively repeated until a stopping criterion is satisfied. Usually, a maximum number of algorithm steps or a fixed threshold under which the stress function must lie are considered as criteria for stopping the procedure. In the initial step, the algorithm needs to be initialized by fixing a convenient configuration of points. To be satisfactory, the initial solution should be not too far from the final solution. Usually, the CMDS solution represents a good initial choice, so much so that most statistical software offers it as a default option.

In a series of papers, de Leeuw and colleagues introduce a methodology for minimizing (18.3) under the constraint $\sum_i \sum_{j<i} w_{ij}\delta_{ij}^2 = n(n-1)/2$, assumed without loss of generality (de Leeuw, 1977; de Leeuw and Heiser, 1980; de Leeuw and Mair, 2009). This implies that the observed dissimilarities are normalized before entering the minimization process. This kind of normalization is known as explicit, in that the sum of the weighted squared dissimilarities is explicitly set equal to a constant. This methodology, called SMACOF (i.e., *sc*aling by *ma*jorizing a *co*mplicated *f*unction), is based on the principle of majorization, which requires the construction of a surrogate function that must be majorizing for the loss function (18.3). This majorizing function must be simpler to minimize than (18.3). This approach, which is particularly efficient from a computational point of view, has been extended to many versions of MDS, both metric and non-metric. For more details of the mathematical derivation of SMACOF, see Cox and Cox (2001, Chapter 11), Borg and Groenen (2005, Chapter 8), and de Leeuw and Mair (2009).

CMDS can also be addressed within the framework of loss function minimization. In particular, CMDS is not a true distance-preserving method, as it attempts to preserve pairwise scalar products (i.e., squared dissimilarities), instead of inter-object dissimilarities (Lee and Verleysen, 2007, Chapter 4). In this sense, a classical scaling solution minimizes the S-Stress criterion, given by S-Stress $= \sum_i \sum_{j<i} (\delta_{ij}^2 - d_{ij}^2)^2$, or the STRAIN criterion, $\mathrm{tr}(\mathbf{B} - \hat{\mathbf{B}})^2$, where $\mathbf{B} = \mathbf{HAH}$ is the initial, known matrix given as a function of the observed dissimilarities, and $\hat{\mathbf{B}} = \mathbf{XX}^t$ is the matrix given by employing the unknown coordinates $\mathbf{X}$. This result is derived from the optimal property of CMDS, which was proved in both Euclidean and non-Euclidean cases (Mardia,1978; Meulman, 1992, 1993; Cox and Cox, 2001).

As a historical aside, Kruskal was the first author to define the concept of stress, in particular in his pioneering work on non-metric MDS (Kruskal, 1964a, 1964b). Subsequently, the definition of stress was suitably adapted to also cover metric MDS methods.

Multidimensional scaling with weights for objects or variables

In principle, we can assign different weights to objects in order to strengthen (or weaken) the influence that a subset of them can exert on recovering the coordinates $\mathbf{X}$. Moreover, when an initial data matrix is available, we can handle variables via a similar approach, thus assigning more (or less) importance to a subset of them than to others. Object weighting and variable weighting can be managed in MDS framework by means of two different approaches. With regard to variables, given that MDS is an object-oriented methodology, input data are explicitly set up as between-objects comparisons. In consequence, we can introduce weights for variables only at the stage of computation of dissimilarities. As is known, weighting can also serve to adjust quantitative variables for their different degree of dispersion around the mean, thus allowing them to be directly comparable. A common choice is therefore to use the inverse of their standard deviation in the computation of distances. As far as the Minkowski family of distances is concerned, the weights thus defined can be raised to the rth power and inserted in the generalized parametric form given in Section 18.1.2. This procedure turns out to be the same as computing distances on standardized variables using the unweighted form of the Minkowski family. Within the framework of MDS, Greenacre (2005), besides reviewing the standard case of variable weighting through application of the weighted Euclidean distance, describes a procedure for learning weights for variables directly from data. Moreover, if the objective is also to take into account correlations between quantitative variables, the Mahalanobis distance provides a solution to this problem (Mardia *et al.*, 1979). Given that the inverse of the variance–covariance matrix is involved in its computation, variables are not only standardized, but also corrected for their bivariate linear interrelationships. As just said, for mixed-type data, variable weighting can be inserted into GCS computations by specifying weights that appear in formula (18.1). Analogously to the Minskowski family of distances, the GCS does not account for the possible association/correlation between variables. With regard to this type of data, Cuadras and Fortiana (1995, 1998) propose a variant of MDS called *related metric scaling* in order to allow for the possible association/correlation between variables. However, their approach is not likely to be considered as a pure variable-weighting approach, given that they correct for between-variables relationships at the stage of the MDS procedure, rather than at the step of dissimilarity computation.

Object weighting can be handled instead during the MDS procedure stage. In particular, when the classical approach is applied, this gives rise to the *weighted classical* MDS (WCMDS), also known as *weighted principal coordinates analysis* (Gower, 1982; Gower and Legendre, 1986) or *weighted metric scaling* (Cuadras and Fortiana, 1996). Given an n-vector $\mathbf{w}$ of weights such that $\mathbf{1}^t \mathbf{w} = 1$, the matrix $\mathbf{B}$, which here will be denoted by $\mathbf{B_w}$, can be formed as follows: $\mathbf{B_w} = \mathbf{D}_w^{1/2} \mathbf{H_w} \mathbf{A} \mathbf{H}_w^t \mathbf{D}_w^{1/2}$, where the matrix $\mathbf{H_w}$, analogous to $\mathbf{H}$ in CMDS, encompasses the weights, $\mathbf{H_w} = \mathbf{I} - \mathbf{1}\mathbf{w}^t$, and the matrix $\mathbf{D}_w^{1/2}$ is the diagonal matrix with entries the square roots of the weights. Then, by applying the spectral decomposition theorem to $\mathbf{B_w}$, the coordinates $\mathbf{X}$ are recovered from $\mathbf{X} = \mathbf{D}_w^{1/2} \mathbf{\Gamma} \mathbf{\Lambda}^{1/2}$. In this weighted case, too, the configuration $\mathbf{X}$ preserves the dual interpretation outlined for CMDS, that is, the rows of $\mathbf{X}$ are the principal coordinates of the n points and the columns of $\mathbf{X}$ give their representation in terms of principal axes (Cuadras and Fortiana, 1996). As a final remark, most LSMDS models can handle object weighting in a natural way. It suffices to specify appropriate weights in formulas (18.2) and (18.3). Among others, the SMACOF procedure could also be applied to find a final solution after a system of weights for objects has been specified.

Non-metric multidimensional scaling

Non-metric MDS (Cox and Cox, 2001, Chapter 3; Borg and Groenen, 2005, Chapter 9) is applied when the observed dissimilarities are at the ordinal level of measurement. Nonetheless, even if dissimilarities could be at a higher level of measurement, it might be wiser to handle them at the ordinal level because of potential measurement errors. These might heavily bias any conclusion that should require metric-type information. Non-metric MDS, accounting for the order of dissimilarities instead of their magnitude, could in principle take care of these situations. Breakthrough works on this matter are the articles by Shepard (1962a, 1962b) and Kruskal (1964a, 1964b). The latter author, in particular, established a method to find disparities which satisfy a monotone relationship with the observed dissimilarities and which are as close as possible to the final distances. This method is known as monotonic or isotonic regression (Kruskal, 1964a, 1964b). Space does not permit us to summarize the procedure, so we limit ourselves to remarking that Kruskal's method produces an MDS configuration $\mathbf{X}$ by minimizing Stress-1. This minimization is carried out with respect to $\mathbf{X}$ through a least-squares monotone regression of distances d_{ij} on disparities $\hat{d}_{ij}$. Another version of stress which is sometimes applied in ordinal MDS is based on the Stress-2 formula, which is used to avoid potential degeneracies occurring during the minimization process. However, this approach appears to be much less popular than Stress-1. As another instance of non-metric MDS, it is worth recalling Guttman's (1968) approach, based on the concept of the coefficient of alienation.

18.1.4 Assessing the goodness of MDS solutions

One of the most crucial choices in applications is concerned with the number of dimensions to retain in analysis, or, in other terms, the dimension of the space in which the coordinates $\mathbf{X}$ have to be recovered. This point is strictly connected with the assessment of the goodness of fit of an MDS solution, for which several criteria are known in the literature. If, in particular, the MDS configuration derives from a CMDS application, then a convenient method for deciding on the ideal number of dimensions can be set up using the eigenvalues from the spectral decomposition of the matrix $\mathbf{B}$. As briefly mentioned, it turns out that the total number of positive eigenvalues of $\mathbf{B}$ represents the actual maximum dimension of the space of the MDS solution. This number is generally too large, and the choice of q can be based on the so-called goodness-of-fit (GOF) indices, which give 'the proportion of a distance matrix explained' (Cox and Cox, 2001). Taking into account also the situation where there exist negative eigenvalues, the general form of these indices is

$$\text{GOF}_1 = \frac{\sum_{m=1}^{q} \lambda_m}{\sum_{m=1}^{n-1} |\lambda_m|} \quad \text{and} \quad \text{GOF}_2 = \frac{\sum_{m=1}^{q} \lambda_m}{\sum_{m \in \Lambda^+} \lambda_m^+},$$

where λ_m is the mth eigenvalue and λ_m^+ the mth element of the set $\mathbf{\Lambda}^+$ of positive eigenvalues of $\mathbf{B}$. When the dissimilarity matrix $\mathbf{\Delta}$ is Euclidean, the two GOFs obviously coincide. Moreover, the assessment of the contribution of each single dimension to the overall fit can be evaluated through the relative eigenvalues, which are given, respectively, by

$$\tilde{\lambda}_m^{(1)} = \frac{\lambda_m}{\sum_{m=1}^{n-1} |\lambda_m|} \quad \text{and} \quad \tilde{\lambda}_m^{(2)} = \frac{\lambda_m}{\sum_{m \in \Lambda^+} \lambda_m^+}.$$

Another measure that can naturally be invoked in CMDS is the STRAIN, which, according to Mardia (1978), can be used as a measure of discrepancy between the two configurations of points, one observed and the other fitted.

When a least-squares MDS method is applied, or when the main interest lies in pairwise comparisons between observed dissimilarities and final distances, a more general measure may be required. A natural choice is the normalized stress (18.4). De Leeuw (1977), in particular, proved that when $\mathbf{X}$ represents a stationary point for the raw stress (18.3), the normalized stress is likely to be interpreted as 'the proportion of variation of the dissimilarities not accounted for by the distances', given that it is strictly connected with the square of Tucker's coefficient of congruence r_c. The latter is a correlation coefficient for two variables about their origin and not their mean. In the present context, where dissimilarities and distances are involved, it can be defined as follows:

$$r_c = \frac{\sum_{i=1}^{n} \sum_{j<i}^{n} d_{ij} \delta_{ij}}{\left[\left(\sum_{i=1}^{n} \sum_{j<i}^{n} d_{ij}^2 \right) \left(\sum_{i=1}^{n} \sum_{j<i}^{n} \delta_{ij}^2 \right) \right]^{1/2}}. \tag{18.5}$$

Then, if $\mathbf{X}$ is a stationary point for (18.3) the relation between the two measures is $r_c = \sqrt{1 - \sigma_N(\mathbf{X})}$, which is always well defined since $0 \leq \sigma_N(\mathbf{X}) \leq 1$, as dissimilarities and distances are non-negative quantities. Another strictly related index is the dispersion accounted for (DAF), given by DAF $= 1 - \sigma_N(\mathbf{X})$, which can be interpreted as a coefficient of determination due to the relation between the normalized stress and Tucker's coefficient of congruence. However, many authors, Borg and Groenen (2005) among them, urge some caution when using the normalized stress (18.4) or DAF as goodness-of-fit indices. Observed dissimilarities and final distances being non-negative, it is not infrequent to observe values of Tucker's coefficient close to 1, and consequently values of $\sigma_N(\mathbf{X})$ smaller than 0.1. Furthermore, when the solution $\mathbf{X}$ does not represent a stationary point for the stress loss functions (18.3) and (18.4), the algebraic formula linking DAF with Tucker's coefficient no longer holds. It is therefore important to evaluate them both, since the information content is quite different.

As a final remark, there exist many other fit measures, and the choice among them much depends on the type of loss function involved. For instance, in SMACOF MDS de Leeuw and Mair (2009) use the so-called metric stress in order to evaluate the goodness of fit of metric SMACOF solutions. After a solution $\mathbf{X}$ has been found for the normalized observed dissimilarities, the metric stress is given by the value of the raw stress (18.3) divided by $n(n-1)/2$. It falls therefore within the definition of normalized stress (18.4). As another instance, in Kruskal's non-metric MDS the Stress-1 is employed with the double purpose of assessing goodness of fit and finding the ideal number of dimensions. On this matter, Kruskal (1964a) suggested a sort of 'rule of thumb' for evaluating Stress-1 values. While a zero value denotes a perfect fitting, Stress-1 values in the interval $(0, 0.025]$ denote an excellent fit, $(0.025, 0.05]$ good, $(0.05, 0.1]$ fair, $(0.1, 0.2]$ poor, and values greater than 0.2 a very bad fit. For a full discussion on fit measures, see Borg and Groenen (2005, Chapters 3 and 11).

18.1.5   Comparing two MDS solutions: Procrustes analysis

Once an MDS configuration of points is produced, it turns out to be not unique with respect to translation, rotation, reflection and dilation. If $\mathbf{X}$ is a solution in q dimensions, with $\mathbf{x}_i$ representing the ith row, then $\mathbf{x}_i' = \rho \mathbf{T} \mathbf{x}_i + \mathbf{c}$, for all i, is also a solution, where ρ is a scale factor defining the dilation (shrinking if $\rho < 1$, or stretching if $\rho > 1$). $\mathbf{T}$ is an orthogonal matrix of order q, representing a rotation or reflection, and $\mathbf{c}$ is any q-dimensional vector giving a translation. The non-uniqueness of MDS solutions raises the question of how two configurations can be appropriately compared by taking into account the above transformations. Procrustes analysis offers such a possibility. Let $\mathbf{X}$ and $\mathbf{Z}$ be two configurations for the same set of n points in, respectively, q and q' dimensions, $q' \leq q$. Such matrices might be obtained by different MDS techniques, or by the same MDS technique applied to two different dissimilarity matrices obtained by different dissimilarity measures. Moreover, if $q' < q$, then it is always possible to add $q - q'$ columns of zeros to $\mathbf{Z}$ to make it an $n \times q$ matrix. Then a measure that expresses the matching degree between $\mathbf{X}$ and $\mathbf{Z}$, and that, at the same time, takes into account all possible rotations, reflections, translations and dilations of $\mathbf{X}$, can be found as the solution of the following minimization problem with respect to the matrix $\mathbf{T}$, the vector $\mathbf{c}$ and the scalar ρ:

$$\min_{\rho, \mathbf{c}, \mathbf{T}} R^2 = \min_{\rho, \mathbf{c}, \mathbf{T}} \sum_{i=1}^{n} (\mathbf{z}_i - \rho \mathbf{T} \mathbf{x}_i - \mathbf{c})^t (\mathbf{z}_i - \rho \mathbf{T} \mathbf{x}_i - \mathbf{c}), \qquad (18.6)$$

where R^2 is the sum of squared intra-point distances between the two configurations $\mathbf{Z}$ and $\mathbf{X}$, the latter after transformations. It has been proved that the minimum value of (18.6) is attained at

$$R^2 = \mathrm{tr}(\mathbf{Z}^t \mathbf{Z}) - \{\mathrm{tr}(\mathbf{X}^t \mathbf{Z} \mathbf{Z}^t \mathbf{X})^{1/2}\}^2 / \mathrm{tr}(\mathbf{X}^t \mathbf{X}), \qquad (18.7)$$

while for the optimal solutions for $\mathbf{T}$, $\mathbf{c}$ and ρ we refer the reader to Cox and Cox (2001, Chapter 5). Formula (18.7) is called the Procrustes statistic. It assumes non-negative values, and is equal to zero if $\mathbf{Z}$ and $\mathbf{X}$ match perfectly. Since R^2 is not symmetric in $\mathbf{X}$ and $\mathbf{Z}$, it is necessary to decide which is the target configuration, and consequently which configuration has to be transformed. A possible standardized version of (18.7) produces the normalized Procrustes statistic:

$$\tilde{R}^2 = 1 - \{\mathrm{tr}(\mathbf{X}^t \mathbf{Z} \mathbf{Z}^t \mathbf{X})^{1/2}\}^2 / \mathrm{tr}(\mathbf{X}^t \mathbf{X}) \mathrm{tr}(\mathbf{Z}^t \mathbf{Z}), \qquad (18.8)$$

which is symmetric with respect to $\mathbf{X}$ and $\mathbf{Z}$, normalized to $[0, 1]$, assumes value zero if $\mathbf{Z}$ and $\mathbf{X}$ match perfectly, and is equal to one if $\mathbf{Z}$ and $\mathbf{X}$ are completely mismatched.

All the mathematical derivations underlying Procrustes statistics can be found in Sibson (1978), Mardia *et al.* (1979, Chapter 14), Cox and Cox (2001, Chapter 5), and Borg and Groenen (2005, Chapter 20). Gower and Dijksterhuis's (2004) monograph provides an extensive treatment of Procrustes problems; for instance, the authors consider oblique Procrustes transformations as well.

18.1.6   Robustness issues in the MDS framework

At the time they wrote their monograph, Cox and Cox (2001) claimed that there were very few contributions on robustness and diagnostic methods in MDS. Today, to the best of our knowledge, this situation has hardly changed. The main proposals for robust MDS remain those reported in Cox and Cox's monograph: the work of Spence and Lewandowsky (1989), who developed a substantial modification of the Newton–Raphson method to robustify the coordinate determination in the presence of outliers; and the work of Sibson (1979), who studied the effects of perturbations on the eigenvalues and eigenvectors from the matrix **B** in CMDS. A more recent contribution can be found in House and Banks (2004) specifically addressing the case of huge amounts of data and setting out an algorithm for selecting a 'good' subset, uncontaminated and sufficiently representative of the target set of data. An MDS method based on stress function minimization is then applied to the selected data subset.

With regard to diagnostic techniques, it is worth mentioning the proposal by Chen and Chen (2000) for the construction of interactive plots to investigate how well objects are reproduced in an MDS configuration. Apart from this, however, the main diagnostic techniques are still based on the Shepard diagram, a scatterplot of observed dissimilarities (or disparities) plotted against final distances, which may help visualize the greatest discrepancies between them (Cox and Cox, 2001, Chapter 3 and Borg and Groenen, 2005, Chapter 3).

A slightly different approach pertaining to robustness in MDS and relying on the forward search (FS: Atkinson *et al.*, 2004) is proposed by Solaro and Pagani (2010). Roughly speaking, in the exploratory multivariate data analysis framework, the FS is a recently developed methodology based on a collection of robust diagnostic techniques for detecting multivariate outliers and discovering potential hidden patterns in multidimensional data sets. Especially where outlier detection is concerned, the FS turns out to be particularly effective in disclosing 'masking' and 'swamping' effects. The former refers to the situation where an object is not recognized as an outlier due to the presence of many other outliers, while the latter denotes the situation where an object is wrongly recognized as outlier since one or more groups of outliers are present (Atkinson *et al.*, 2004). The FS starts from an 'outlier-free core' subset of size m_0 ($m_0 < n$), which can be formed according to several criteria, such as the 'robustly centred ellipses' or the 'bivariate boxplots' methods. Then, a sequence of subsets of growing size m ($m = m_0, \ldots, n$) are formed by adding one unit at a time on the basis of the ascending ordering drawn from squared Mahalanobis distances. These are called 'forward subsets', and are usually denoted by $S^{(m)}$. Given an $n \times p$ data matrix **Y** of quantitative variables, whose ith row is $\mathbf{y}_i$, the squared Mahalanobis distance d_{im}^2 is thus computed for each unit which:

- enters the subset $S^{(m)}$: $d_{im}^2 = (\mathbf{y}_i - \boldsymbol{\mu}_m)^t \boldsymbol{\Sigma}_m^{-1}(\mathbf{y}_i - \boldsymbol{\mu}_m)$, $i \in S^{(m)}$, where $\boldsymbol{\mu}_m$ and $\boldsymbol{\Sigma}_m$ are, respectively, the centroid and the covariance matrix computed on the m units in $S^{(m)}$ ($i = 1, \ldots, m$; $m = m_0, \ldots, n$);

- does not enter the subset $S^{(m)}$: $d_{jm}^2 = (\mathbf{y}_j - \boldsymbol{\mu}_m)^t \boldsymbol{\Sigma}_m^{-1}(\mathbf{y}_j - \boldsymbol{\mu}_m)$, $j \notin S^{(m)}$, where, once again, $\boldsymbol{\mu}_m$ and $\boldsymbol{\Sigma}_m$ are computed on the m units in $S^{(m)}$ ($j = m + 1, \ldots, n$; $m = m_0, \ldots, n$).

Once forward subsets have been formed, it is possible to monitor the impact of potential perturbations in data (e.g., outliers), on all statistics of interest obtained from a specific statistical method applied to each subset. A typical graphical tool used to monitor the search is the so-called 'forward plot': Mahalanobis distances, or other quantities of interest, are plotted by means of trajectories as the size m increases from m_0 to n. Such trajectories are then studied to detect outlying units or other possible anomalies. As at the beginning of the search subsets are formed by a few, similar units, the covariance matrix may have small elements. The Mahalanobis distances of the units not included in the subset can take very high values. It would then be more appropriate to carry out the search with the scaled Mahalanobis distance: $\tilde{d}_{im} = d_{im} \sqrt[2p]{\det(\boldsymbol{\Sigma}_m)/\det(\boldsymbol{\Sigma}_n)}$, where d_{im} is the Mahalanobis distance and $\boldsymbol{\Sigma}_n$ the covariance matrix computed on all n units. For a more thorough description of the FS, see Atkinson and Riani (2000), who give an extensive treatment of it in statistical modelling; Atkinson *et al.* (2004), who provide a full account of the FS when exploring multidimensional data sets; and Atkinson *et al.* (2010) with regard to the most recent advances in this area of research.

The extension of the FS in MDS proposed by Solaro and Pagani (2010) is currently confined to CMDS, and assumes that a starting data matrix with quantitative variables is given. We are not aware of further methodological enhancements concerning the FS devised to cover situations in which the original data matrix is qualitative or of mixed type, or in which only a proximity matrix is available, rather than a data matrix. All these topics are open fields for research. Nonetheless, a way to overcome such a lack of methodological results may be to rely on an MDS technique, and then apply the FS directly to the MDS configuration of points. On this matter and within the framework of customer satisfaction studies, Solaro (2010) presents an application where the FS is employed on a CMDS configuration recovered from a Gower dissimilarity matrix, which is derived from mixed-type data. Although this approach does not exploit the full capabilities of the FS, because the starting data matrix is not purely quantitative, it represents the simplest way to apply the FS to mixed-type data.

18.1.7   Handling missing values in MDS framework

As already pointed out, missing values can be handled within the MDS framework using standard methods, which can be divided into to two approaches. If a data matrix is available, then we could assign zero weight to those (dis)similarity scores or distance differences that pertain to pairwise comparisons where at least one missing value is present. We have discussed this same procedure for the GCS (18.1). Otherwise, if only a proximity matrix is available, we could handle missing proximities at the very stage of the MDS procedure by choosing an appropriate MDS method that allows zero weights to be included corresponding missing proximities. For instance, LSMDS methods that consider minimization of raw stress (18.3), such as SMACOF MDS, can absorb missing proximities in a natural way, by assigning zero weights to missing pairwise comparisons. Unfortunately, this option is not available in every MDS method. For instance, as we have already pointed out, WCMDS permits us to specify a vector **w** of weights to give greater (or less) influence on results to certain objects rather than others. These weights, however, remain the same over all pairwise comparisons. Therefore, if a zero weight is included in **w**, any comparison pertaining to such an object would actually be discarded from the analysis.

18.1.8   MDS applications in customer satisfaction surveys

MDS methods have a broad tradition of applications in market research, within which customer satisfaction studies fall. The early instances of MDS applications date back to the late 1950s, when certain psychometricians, like Torgerson, began to employ CMDS in order to study latent dimensions involved in customers' perceptions. At first, this specific objective was considered one of the main advantages of MDS over alternative statistical techniques. Later, in the 1970s, MDS was increasingly applied in its many different variants, such as metric and non-metric MDS, unfolding models and three-way MDS, and used for a broader class of problems, for instance the setting-up of maps in brand positioning problems. However, in the early 1980s many market researchers began to turn to alternative methods of analysis, such as conjoint analysis, because of several weak points they claimed MDS suffers. The most crucial of these concern the identification and interpretability of dimensions, as identification, being strictly related to the rotation of axes, seems arbitrary and dimensions appear to be, in some cases, meaningless. It is amply recognized that such problems mainly arise when MDS is used in exploratory rather than confirmatory analysis. Hence, dimensions should be theoretically identified *a priori*. Another point has to do with the problem of operationalizing the dimensions extracted, in the sense that, even though they are identified and a convincing interpretation is provided, they are not likely to be easily manipulated or controlled in order to achieve practical goals. However, despite all the criticisms, MDS techniques have continued to play an important role, especially in customer satisfaction studies.

Excellent reviews of MDS in this area of research are Cooper (1983), in which a complete overview of marketing applications is provided, and Carroll and Green (1997), where the use of MDS is reconstructed from a historical point of view, along with the debate that took place between supporters and sceptics on MDS in market studies. More recent advances, related to MDS, are the works of Wu and DeSarbo (2005) and Klawonn *et al.* (2009). In particular, the former authors develop a stochastic MDS approach to model the relation between performance, expectation and disconfirmation, and overall customers' satisfaction according to the expectancy–disconfirmation paradigm of the determinants of customer satisfaction judgements. Klawonn *et al.* (2009) give a new method for measuring similarities between customer satisfaction profiles when customer are grouped into different customer segments.

18.2   Multidimensional scaling in practice

This section is concerned with the application of the MDS methodology to customer satisfaction surveys. Specifically, we refer to the ABC annual customer satisfaction survey (ACSS) described in Chapter 2. The main objective is to set up satisfaction-related dimensions that can adequately reproduce differences among customers. For this purpose, we apply a set of methods reviewed in Section 18.1 to the Gower dissimilarity matrix computed by using a specific subset of items from the questionnaire. We adopt a metric approach to MDS in order to exploit the metric properties of the chosen dissimilarity measure. Non-metric MDS will therefore not be considered in the present context. We first apply CMDS, WCMDS and SMACOF to a complete set of data, that is, data without missing values. Having assessed the goodness of fit of configurations and provided the interpretation of their dimensions, we compare the various MDS solutions via Procrustes analysis. This will show whether, for any subset of customers, there are noteworthy differences over the various configurations. Applying different MDS models gives us the opportunity to see whether the extracted dimensions, in particular the first, change their interpretation according to the method used. The idea is that, if satisfaction-related dimensions are important latent aspects underlying data, they should be captured independently of the method used. Next, we perform a robustness analysis through standard application of FS in order to detect potential multivariate outliers or particular hidden patterns in the MDS configurations of points. Finally, we show how missing values can be handled by employing standard solutions available in the MDS framework. All the analyses are carried out in the R environment (R Development Core Team, 2011), using in particular the R libraries *Rfwdmv* for FS (Atkinson *et al.*, 2005), *vegan* for the WCMDS (Oksanen *et al.*, 2011), and *smacof* for the SMACOF MDS (de Leeuw and Mair, 2009).

18.2.1   Data sets analysed

As described in Chapter 2, the ACSS questionnaire was completed by 266 ABC customers who were asked to express their level of satisfaction with selected company and product features. They responded to a series of ordinal items and several binary items. Overall, the structure of the questionnaire is fairly complex: overall satisfaction with a specific aspect is combined with a battery of items aimed at closely examining that aspect. Several filter questions are also included, which inevitably produce structural missing values. Beside these, there are many non-responses, arising especially in the detailed batteries of items. Finally, customers' personal data (e.g., country of origin, seniority, position, and market segmentation) are gathered by other questions with responses at either the nominal or ordinal level of measurement.

We simplify the data set in order to show the potential of the metric methods described in Section 18.1 to extract satisfaction-related dimensions. Therefore, we confine our analyses to the subset of 198 customers who gave complete responses (no missing values) to the following items: the first five in the questionnaire (q_1–q_5), covering overall satisfaction with ABC; overall satisfaction with, equipment and system (q_{11}), sales support (q_{17}), technical support (q_{25}), ABC's supplies and orders (q_{38}), purchasing support (q_{57}), and contracts and pricing (q_{65}). In addition, we consider the two filter questions on the use of ABC software add-on solutions (q_{39}) and ABC's customer website (q_{44}). We have decided not to consider the remaining overall satisfaction items because of large amounts of missing responses, be they structural (e.g., items q_{42} and q_{49}) or not (e.g., items q_{31} and q_{43}). Throughout this chapter, we refer for simplicity to the set of items selected as the overall satisfaction items. Moreover, the batteries of more specific items with extensive missing values were also discarded from the analyses. The selected items, 13 in all, are listed in Table 18.1 along with Shannon's normalized entropy index, whose evaluation will be of help in defining weights for the subsequent WCMDS application.

Next, to show how MDS methods can perform in the presence of missing values, we reconsider the data set formed by the whole set of customers. Once again, MDS analyses involve only the overall satisfaction items. Since pairwise comparisons between complete and incomplete customers' responses give rise to entire rows (and columns) of missing values in the dissimilarity matrix, we had to tackle this problem at the stage of computation of Gower's similarity scores (18.1), by assigning appropriate weights. Proceeding otherwise, by controlling for missing values at the subsequent stage of MDS procedures, would lead to numerically singular systems, which admit no solution. As will be apparent, missing values cause the Gower dissimilarity matrix to be non-Euclidean, the remedy for this is either to add a suitable constant to make the matrix Euclidean or simply to ignore the negative eigenvalues if they are of small magnitude. We show both these procedures to give a clear idea of their consequences.

18.2.2   MDS analyses of overall satisfaction with a set of ABC features: The complete data set

The initial step of all the subsequent MDS analyses is represented by the Gower dissimilarity matrix $\mathbf{\Delta}$, computed on the $n = 198$ customers using the items listed in Table 18.1. Since these items are binary or ordinal, Podani's (1999) metric version of GCS is considered. Our purpose is to find a final configuration $\mathbf{X}$ of the 198 customer points in q dimensions, where q needs a proper assessment. As is commonly done in dimensionality reduction problems,

Table 18.1   Items from the ABC annual customer satisfaction survey

Label	Measurement level	Description	Shannon's normalized entropy index[*]	Item weights
q_1	Ordinal	Overall satisfaction level with ABC	0.8554	1.1690
q_2	Ordinal	Overall satisfaction level with ABC's improvements during 2010	0.8950	1.1173
q_3	Binary	Is ABC your best supplier? (No $= 0$; Yes $= 1$)	0.9497	0.9497
q_4	Ordinal	Would you recommend ABC to other companies?	0.8960	1.1161
q_5	Ordinal	If you were in the market to buy a PRODUCT, how likely would it be for you to purchase an ABC product again?	0.8941	1.1184
q_{11}	Ordinal	Overall satisfaction level with the equipment	0.7276	1.3743
q_{17}	Ordinal	Overall satisfaction level with sales support	0.9462	1.0568
q_{25}	Ordinal	Overall satisfaction level with technical support	0.8770	1.1403
q_{38}	Ordinal	Overall satisfaction level with ABC's supplies and orders	0.7068	1.4148
q_{39}	Binary	Do you use ABC software add-on solutions? (No $= 0$; Yes $= 1$)	0.9911	0.9911
q_{44}	Binary	Do you use ABC's customer website? (No $= 0$; Yes $= 1$)	0.5471	0.5471
q_{57}	Ordinal	Overall satisfaction level with purchasing support	0.8357	1.1966
q_{65}	Ordinal	Overall satisfaction level with contracts and pricing	0.8371	1.1946

[*] Shannon's normalized entropy index is given by $E_N = -\sum_{i=1}^{h} f_i \ln f_i \Big/ \ln h$, with f_i denoting relative frequencies and h the number of categories.

we base this choice on a combination of several criteria: parsimony (few dimensions produce simpler descriptions of data), accuracy (many dimensions fit data better), and interpretability of dimensions (dimensions accounting for larger proportions of observed dissimilarities tend to explain systematic tendencies, while the rest mostly reproduce noise in data). With regard to accuracy, since the choice of goodness-of-fit measure strictly depends on which MDS model is actually applied, for each analysis we adopt the same range of goodness-of-fit indices: DAF, Tucker's coefficient of congruence with formula (18.5), and Stress-1. Obviously, where the MDS method is directly related to eigenvalues, such as CMDS and WCMDS, GOF indices and relative eigenvalues can also be provided. Moreover, as the original items are a mixture of binary and ordinal variables, the interpretation of the dimensions obtained will be based on specific association measures with the original items, namely Spearman's rank correlation coefficient when the items are ordinal, or Pearson's correlation coefficient when the items are binary. In particular, the latter measure stands for the square root of the proportion of

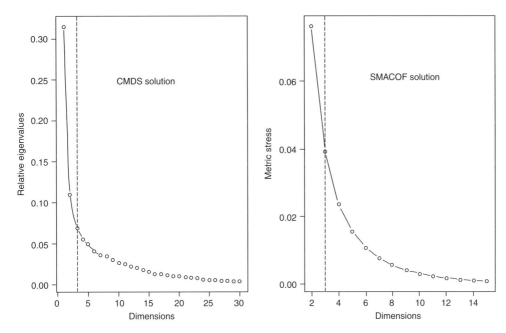

Figure 18.1   Goodness-of-fit indices computed with respect to configurations with an increasing number of dimensions. The dashed vertical line corresponds to a three-dimensional solution. (Left) Relative eigenvalues for each dimension in CMDS. (Right) Metric stress of SMACOF configurations with different dimensionality.

between-category variance of the dimensions extracted which is explained by a categorical variable. By taking into account all the above criteria, we feel it is appropriate to set the ideal number q of dimensions equal to three in all applications. With regard to this, Figure 18.1 exhibits an instance of computation of goodness-of-fit indices carried out with respect to the configurations which are set up with two alternative MDS methods in the presence of an increasing number of dimensions.

Classical MDS

First, we show the results of analyses carried out with CMDS. Solaro (2010) performs the same analysis in a similar customer satisfaction problem. Referring to a three-dimensional configuration **X** of points, the two goodness-of-fit indices turn out to be $GOF_1 = GOF_2 = 0.4914$ (the two GOFs coincide due to the Gower dissimilarity matrix being Euclidean), while $DAF = 0.8638$, $r_c = 0.9666$, $S_1 = 0.5265$ (as a consequence of the arguments outlined in Section 18.1.4, here $r_c \neq \sqrt{DAF}$). In particular, from the evaluation of the relative eigenvalues we have that dimension 1 accounts for 31.39% of the total dispersion, dimension 2 for 10.93% and dimension 3 for 6.82% (Figure 18.1, left). While these results denote a generally poor fit of the CMDS configuration to the observed dissimilarities, extracting a larger number of dimensions would not make much sense in this situation, bearing in mind the need for parsimony and interpretability of results. With regard to this point, Table 18.2 reports Spearman's rank correlation coefficients of the ordinal items, along with Pearson's correlation coefficients of the binary items, and the three dimensions, on which interpretations are mainly based.

Table 18.2   Spearman's rank (ordinal items) and Pearson's (binary items) correlation coefficients of items and dimensions of the CMDS solution

Items	Dim. 1	Dim. 2	Dim. 3
q_1	0.8442	−0.1174	0.0163
q_2	0.7565	−0.0526	−0.1152
q_4	0.8581	−0.0661	−0.0498
q_5	0.8151	−0.0688	−0.0869
q_{11}	0.6608	−0.0889	−0.1514
q_{17}	0.4997	−0.3746	−0.4580
q_{25}	0.6154	−0.2125	0.0238
q_{38}	0.5572	−0.2058	−0.0682
q_{57}	0.3705	−0.4066	−0.2795
q_{65}	0.6781	−0.1469	−0.1821
q_3	0.7461	0.0761	0.5730
q_{39}	0.3858	0.8646	−0.2800
q_{44}	0.2596	0.2757	0.2114

Since all items are positively correlated with dimension 1, this latter can be interpreted as an indicator of the overall satisfaction with the set of ABC features considered. Moreover, as coefficients are all positive, dimension 1 and its items share the same orientation. Its scores (i.e., the values it assumes) thus admit a straightforward interpretation. Higher positive scores tend to identify the most satisfied customers, who also agree in declaring the ABC their best supplier. Lower negative scores characterize the least satisfied customers. Dimension 2 is mostly positively linked with the binary item q_{39} ('Do you use ABC software add-on solutions?') and negatively, though more or less weakly, with the other items, except for the binary item q_{44} ('Do you use ABC's customer website?'). Therefore, dimension 2 seems to identify two types of customers: those who use ABC's products (software add-on solutions and customer website) but tend not to be highly satisfied (positive scores); and those who do not use those specific products, but express satisfaction with the services ABC provides (negative scores). Despite its negative correlation with q_{17} (level of satisfaction with sales support), and its positive correlation with the binary item q_3 ('Is ABC your best supplier?'), dimension 3 is less clearly interpretable. Its positive scores seem, however, to identify customers who, though declaring themselves unsatisfied with sales support, tend to consider ABC as their best supplier. Obviously, the opposite meaning is attached to its negative scores. In order to consider more thoroughly the indicator of satisfaction (dimension 1) we consider Figure 18.2, in which boxplots of its distribution within, respectively, customers' country of origin (left) and profitability level (right) are shown.

With regard to the country of origin, it is worth noting that customers from Benelux and UK appear to be the least satisfied, albeit to a varied extent, while the majority of customers from Germany and Israel appear to be satisfied with ABC. France and Italy occupy an intermediate position, with similar proportions of satisfied and dissatisfied customers. As far as profitability level is concerned, with the exception of missing values, the right-hand panel exhibits a slightly decreasing trend in the overall satisfaction scores when passing

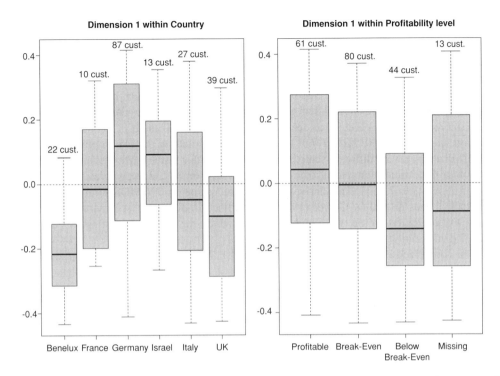

Figure 18.2   Boxplots of (left) the distribution of dimension 1 within customers' country of origin and (right) profitability level.

from profitable to below-break-even customers. The same graphical analysis carried out with respect to the position of the respondent (omitted) reveals the highest scores on dimension 1 for technical staff and operators, while technical managements show mostly positive scores. While managements, administrators, and 'other positions' tend mostly to exhibit negative levels of satisfaction, the satisfaction of owners tends to be distributed uniformly over negative and positive scores. Further, as regards customer seniority, customers of 2–3 years' standing appear to be slightly less satisfied than the most recent or the oldest customers. Finally, regarding market segmentation, there is no evidence of any specific trend – the boxplots of dimension 1 appear quite similar across the segments.

SMACOF MDS

The same kind of analysis carried out with SMACOF MDS leads to substantially analogous results, although the fit to the data is better than for CMDS. The three-dimensional final solution is found in 887 iterations and produces a metric stress value equal to 0.0392 (Figure 18.1, right). Moreover, DAF = 0.9608, r_c = 0.9802, S_1 = 0.1979 (here the relation $r_c = \sqrt{DAF}$ does hold). Spearman's and Pearson's correlation coefficient are very close to those of Table 18.2, especially as regards the first two dimensions. The interpretation of the three dimensions appears therefore to be unchanged. Finally, boxplots of the distribution of dimension 1 within customers' country of origin and profitability level are practically identical to those in Figure 18.2, so are omitted here.

Procrustes analysis

Figure 18.3 shows the two superimposed scatterplots of CMDS and SMACOF configurations (left) and the superimposition plot obtained as a result of Procrustes analysis (right). For simplicity, both graphs are referred to the first two axes.

The first panel shows that the SMACOF procedure produces a much more scattered configuration of points (filled circles) than CMDS (empty circles). In particular, as the variance of the first three dimensions is equal to, respectively, 0.2330, 0.1300 and 0.1150 in the SMACOF solution, and 0.0547, 0.0191 and 0.0119 in the CMDS solution, the two configurations are characterized by a very different scale. It is worth noting that in the CMDS solution the variance of the three dimensions can also be computed as λ_m/n ($m = 1, 2, 3$), where $\lambda_1 = 10.8381$, $\lambda_2 = 3.7753$, $\lambda_3 = 2.3534$. This derives from the interpretation of CMDS solutions in terms of principal axes proved by Gower (1966) and outlined in Section 18.1.3. Moreover, in the SMACOF configuration the two clouds of points, which can be clearly distinguished in the CMDS configuration, seem to have disappeared.

The normalized Procrustes statistic is equal to 0.1170, thus showing that after translation, rotation, reflection and dilation, the two configurations do not match. The right-hand panel in Figure 18.3 displays the results concerning matching after Procrustes transformations have been applied to the SMACOF solution, whereas the CMDS configuration is the target. As this plot is referred to the first two axes, some caution is necessary in interpreting Procrustes analysis results. In fact, all movements of points required to go from the SMACOF to the

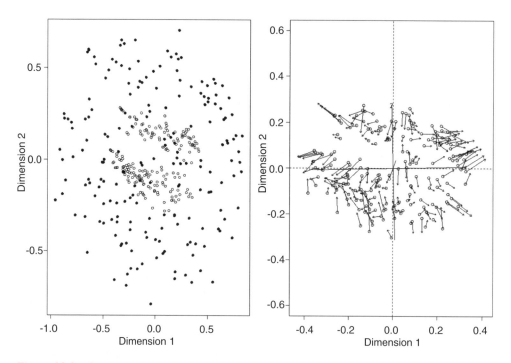

Figure 18.3   Comparing CMDS and SMACOF configurations. (Left) CMDS (empty circles) and SMACOF (filled circles) original configurations. (Right) CMDS and SMACOF configurations after Procrustes transformations.

CMDS solution should more properly be read in a three-dimensional map, which is, however, not of simple interpretation. Circles in the right-hand panel of Figure 18.3 represent points of the transformed SMACOF configuration, while the arrows from each point lead to the corresponding point in the target CMDS configuration. It is apparent that these arrows are characterized by quite different lengths, but we have not been able to find a rule concerning customers' features, such as country of origin, position and market segmentation, that might explain why certain points are very close over the two configurations, while other points are very far from each other.

18.2.3   Weighting objects or items

As argued in Section 18.1.3, in any MDS analysis either objects or variables, or both, could be differently weighted with the objective of increasing (or decreasing) the influence of a subset of objects or variables on the final results. Object weighting can be done at the MDS procedure stage, by employing WCMDS or weighted SMACOF. Variable weighting requires the use of different weights in the computation of Gower's dissimilarity scores. We present here an instance for each case of weighting. We weight customers on the basis of their customer seniority. For items, we set up weights as a function of Shannon's normalized entropy index.

Weighting customers

Given that customers' personal data are available, we employ customer seniority (in years) as weighting variable. In particular, customers of 1 year's standing represent 15.66% of the whole set of data, 2 years' standing 27.27%, 3 years' standing 15.66%, 4 years' standing 15.15%, and 5 years' standing 26.26%. A plausible justification for assigning more weight to the responses of customers of longer standing could lie in the supposition that customers' knowledge of ABC improves with the length of their business relationship, so that their opinion might be considered more reliable than that of new customers. Weights are introduced in the following MDS applications by using the weighted version of CMDS and SMACOF, respectively.

With regard to WCMDS, the analysis is carried out with the *wcmdscale* function from the *vegan* library in R, which implements the theory of WCMDS for weighting objects outlined in Section 18.1.3. Once again, three dimensions are retained in analysis. We obtain $GOF_1 = GOF_2 = 0.4976$, and $DAF = 0.8646$, $r_c = 0.9680$, $S_1 = 0.5266$, while the relative eigenvalues are equal, respectively, to 32.23% for dimension 1, 11.12% for dimension 2 and 6.41% for dimension 3. These values are very close to the previous CMDS analysis. Moreover, with regard to dimension interpretation, Spearman's and Pearson's correlation coefficients are very similar to Table 18.2; no notable change has occurred in the magnitude and signs of coefficients. WCMDS dimensions thus have the same interpretation as before. A more thorough investigation performed with Procrustes analysis reveals that the two CMDS and WCMDS configurations of points are practically identical. In particular, the normalized Procrustes statistic is equal to 0.0105.

The same analysis carried out with the weighted version of SMACOF produces a better fit than WCMDS if DAF and Tucker's coefficient of congruence are taken into account. The three-dimensional solution is found in 1012 iterations with a metric stress value equal to 0.0377, $DAF = 0.9623$, and $r_c = 0.9799$. The value of Stress-1 is $S_1 = 0.6068$, denoting a slightly worse fit than the CMDS solution. The three dimensions thus have the same interpretations as before. Finally, the normalized Procrustes statistic computed for the weighted and unweighted SMACOF solutions is equal to 0.0417, thus indicating that the two configurations are very similar.

Weighting overall satisfaction items

With regard to the weighting of items, we employ the weights reported in the last column of Table 18.1. For binary items, these are given by Shannon's normalized entropy index, while for ordinal items the weights are given as reciprocals of the index. In thus way, we intend to slightly limit the impact of binary variables with respect to ordinal items. On the other hand, by using the reciprocal of Shannon's index, we place more emphasis on items characterized by more homogeneous responses. Unlike objects, variable weights are introduced in the computation of Gower's dissimilarity coefficients. On the matrix thus obtained, we apply CMDS and SMACOF, and retain three dimensions as before. With regard to the goodness of fit of solutions, we obtain $GOF_1 = GOF_2 = 0.4818$, and $DAF = 0.8564$, $r_c = 0.9639$, $S_1 = 0.5459$ for CMDS; metric stress $= 0.0410$ and $DAF = 0.9590$, $r_c = 0.9793$, $S_1 = 0.2026$ for SMACOF. The SMACOF solution remains the one with the better fit. In both cases, as Spearman's and Pearson's correlation coefficients are similar to Table 18.2, the three dimensions retain their interpretations. In order to compare the two solutions, Procrustes analysis is then carried out with CMDS as the target configuration. The normalized Procrustes statistic is equal to 0.1465, showing that CMDS and SMACOF are quite dissimilar. Once again, the graph with the two superimposed scatterplots reveals that SMACOF has produced a much more scattered configuration of points than CMDS. Moreover, as observed earlier, in the SMACOF configuration the two clouds of points, which remain fairly prominent in the CMDS solution, seem absent yet again. This aspect is particularly apparent in the superimposition plot of Procrustes analysis, where it can be clearly seen that most points of the transformed SMACOF configuration have to move in opposite outward directions, with respect to the axis origin, in order to produce the two distinct clouds of points of the CMDS configuration.

18.2.4   Robustness analysis with the forward search

Figure 18.4 presents the forward plots obtained from standard application of the FS to the CMDS and SMACOF configurations in the item weighting case. We consider this case as an important instance of robustness analysis through the FS applied to these data. As outlined in Section 18.1.6, forward plots are particularly convenient diagnostic tools that monitor any statistic of interest over the subsets of growing size that are formed during the search. The graphs presented here show trajectories of the scaled Mahalanobis distances computed for both kinds of customers, that is, those included or not included in the mth forward subset, $m = 4, \ldots, 198$. These trajectories are then studied to identify outlying customers or discover other possible substructures in the data. In Figure 18.4, trajectories are drawn as the subset size grows, one unit at a time, from $m_0 = 4$ to $n = 198$ (horizontal axis).

Two remarks are in order here. First, in the left-hand panel, for CMDS, during the first half of the search trajectories are visibly divided into two well-separated sets. This indicates that the CMDS configuration contains a substructure of data consisting of two groups of customers, as observed before. Then, near the 125th step of the search, these two sets of trajectories begin to unify. The reason is that customers in the group with the highest trajectories are entering the forward subsets. From that point on, the centroid and the covariance matrix are computed by using these units as well, so that their scaled Mahalanobis distance tends to become smaller than

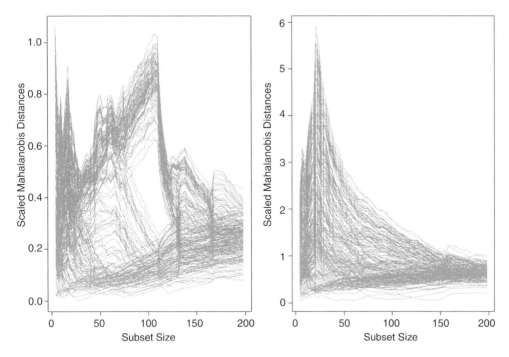

Figure 18.4   Forward search for CMDS and SMACOF in item weighting case. Forward plots of scaled Mahalanobis distances for (left) CMDS and (right) SMACOF configurations.

at the beginning of the search. Trajectories for the SMACOF configuration are characterized by a different pattern. They appear to join in a unique, decreasing sheaf of curves. This gives further support to the previous observation that in the SMACOF configuration points are not separated into two distinct clouds.

Second, we note that multivariate outliers are present in neither configuration. In fact, isolated trajectories, which typically identify outlying units, are completely absent from both graphs.

18.2.5   MDS analyses of overall satisfaction with a set of ABC features: The incomplete data set

In this final part of the analysis, we show how CMDS can be applied in the presence of missing values. For this purpose, in addition to the set of 198 customers with complete responses, we consider 66 customers who respond to at least one overall satisfaction item but not to all of them, giving a total of 264 customers in the data set under analysis. Recall that there were 266 customers in the initial data set; two customers gave no response for any overall satisfaction item and were discarded.

The first step is to set up the Gower dissimilarity matrix. In order to compute dissimilarities for two customers where at least one of them has incomplete responses, we rely on formula (18.1) by introducing zero weights. This produces a non-Euclidean dissimilarity matrix. We then apply CMDS, employing two approaches.

In the first, CMDS is applied without correcting for the non-Euclidean nature of the dissimilarity matrix. This gives rise to a three-dimensional configuration with $\text{GOF}_1 = 0.4480$ and $\text{GOF}_2 = 0.4681$, where, in particular, dimension 1 accounts for 27.34% of total dispersion, dimension 2 for 10.77%, and dimension 3 for 6.69% (computed with the formula for $\tilde{\lambda}_m^{(1)}$ reported in Section 18.1.4). Moreover, Spearman's and Pearson's correlation coefficients are still close to those in Table 18.2. Comparing GOF_1 and GOF_2, negative eigenvalues appear to be small in magnitude, so that they can simply be ignored and the current analysis judged as satisfactory. Now, by considering the 198 customers who appear in both the complete and incomplete data sets, we carry out Procrustes analysis in order to compare the CMDS configuration just recovered with the analogous one derived from the complete data set (see Section 18.2.2). The two configurations turn out to be very similar, as further confirmed by the normalized Procrustes statistic, equal to 0.0169. This may be an argument in favour of the presence of missing values not implying any relevant change in the coordinates of customers with complete responses. Figure 18.5 reports the main results of this analysis. The left-hand panel shows the superimposed scatterplots of the CMDS configuration obtained for the complete data set (empty circles) and for the incomplete data set, where points are marked differently depending on whether customers were present (filled circles) or not (filled triangles) in the complete data set. The right-hand panel shows results of Procrustes analysis confined to the set of customers present in both data sets. The CMDS configuration for incomplete data is depicted with empty circles, while that for complete data, being the target configuration, is represented with arrows. It is immediately evident that the two configurations, after Procrustes transformations, are very similar.

The second approach involves transforming the dissimilarity matrix in order to make it Euclidean before proceeding to the analyses. In the *cmdscale* function in R this problem is solved via the analytical solution proposed by Cailliez (1983). The constant thus found here is equal to 1.010108. Therefore, CMDS is applied to the dissimilarity matrix that has as elements the sums of observed dissimilarities and the constant. This latter CMDS application produces a three-dimensional solution with $\text{GOF}_1 = \text{GOF}_2 = 0.1941$, where dimension 1 accounts for 11.47% of total dispersion, dimension 2 for 4.79%, and dimension 3 for 3.15%. As is apparent, this configuration fits the data much worse than that without any adjustment for the non-Euclidean nature of the matrix. The analysis carried out by ignoring negative eigenvalues gives rise to more satisfactory results.

18.2.6   Package and software for MDS methods

MDS methods are implemented in the vast majority of statistical software. Moreover, there are several programs expressly dedicated to MDS, such as PERMAP (Heady and Lucas, 2010), and some of these can be freely downloaded from the internet. Since in this context it would be impossible to produce an exhaustive list of all programs, we confine our attention to two of the most widely circulated commercial products, SPSS and SAS, and to the R environment, which we have used in our analysis of the ACSS data. SPSS includes three algorithms for MDS: ALSCAL is the implementation of the algorithm described in Takane *et al.* (1977), which tries to fit squared distances to squared dissimilarities by minimizing the S-Stress. It is a very flexible tool, in that it provides MDS models for symmetric and asymmetric proximity data, unfolding and three-way analysis. The PROXSCAL algorithm implements the mathematical theory developed by Commandeur and Heiser (1993), where the loss function is represented by the raw stress (18.3) divided by the total number of distinct pairwise comparisons. Metric

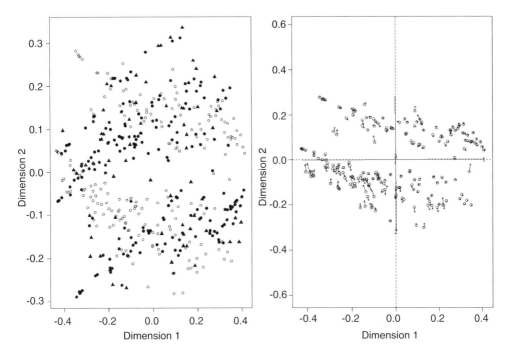

Figure 18.5   Comparing two CMDS configurations for complete and incomplete data. (Left) CMDS for complete data (empty circles) and for incomplete data (filled circles are customers present in the complete data set, triangles are those absent). (Right) CMDS configurations after Procrustes transformations.

and non-metric models are implemented, as well as proximity matrices that derive from multiple sources. The PREFSCAL algorithm performs unfolding analysis of proximity data. The PROXSCAL and PREFSCAL algorithms were developed by the Data Theory Scaling System Group, Leiden University, The Netherlands.

SAS implements MDS through PROC MDS, which fits plenty of metric and non-metric MDS models, as well as three-way analysis. In the case of exploratory MDS, under opportune specification of some options, PROC MDS gives the same results as the ALSCAL procedure.

The R environment base distribution implements CMDS in the function *cmdscale*. As to other MDS models, it is worth pointing out the *MASS* library (Venables and Ripley, 2002), which contains the functions *isoMDS* for Kruskal's non-metric MDS and *sammon* for Sammon's non-linear mapping. The *vegan* library (Oksanen *et al.*, 2011), specifically developed for the analysis of ecological and environmental data, includes the *wcmdscale* function for WCMDS, along with *metaMDS*, which performs non-metric MDS with random starts to achieve much more stable solutions, and *procrustes*, which performs Procrustes analysis, computes values of Procrustes statistics (18.7)–(18.8) and produces Procrustes graphics for comparing two MDS solutions. The *smacof* library (de Leeuw and Mair, 2009) includes *smacofSym* for carrying out LSMDS on symmetric dissimilarity matrices by the SMACOF method. Various other functions in the library fit non-metric and unfolding models, carry out three-way analyses, and implement models with constraints on the configuration.

18.3   Multidimensional scaling in a future perspective

MDS has a long history, and over the years it has witnessed important methodological enhancements. Nevertheless, several open questions remain. In our opinion, the problem of assessing the goodness of fit of MDS configurations deserves further in-depth consideration. As argued, the choice around goodness-of-fit indices much depends on the type of MDS model applied. Broader attention should be given to the problem of identifying a class of indices that could help compare different MDS models in terms of data fit. According to Mardia *et al.* (1979, Chapter 14), Procrustes rotation 'is a goodness of fit measure used to compare two configurations'. Hence, this kind of analysis serves to evaluate the degree of matching between two configurations, and is not expressly concerned with the pure goodness-of-fit problem related to how a configuration fits the original data. Furthermore, with the exception of classical scaling, there is still no method for assessing the contribution of each single extracted dimension to the reproduction of input proximities. Where the spectral decomposition is involved, the assessment of the contribution of single dimensions can be accomplished through relative eigenvalues, given the meaning of eigenvalue in this context, as well as the fact that summing eigenvalues gives precisely 'the total amount' of goodness of fit. For the vast majority of other MDS methods the evaluation is always carried out as a whole, that is, over the set of all extracted dimensions jointly considered. Moreover, most effort should be concentrated on implementing the more recently developed MDS models in widely available statistical software. For instance, de Leeuw and Mair (2009) have implemented in the SMACOF program a method to find an MDS representation for points that lie on a quadratic surface or a sphere. We are not aware of such methodology also being implemented in SPSS or SAS. The same considerations apply to 'boundary' issues such as robustness and sensitivity analyses. The aim of the latter in particular is to study how stable MDS results could be if analyses were repeated on different subsets of objects. De Leeuw and Meulman (1986) presented a method for sensitivity analysis which can be applied to most MDS models, but it is not currently implemented in any of the major statistical software. Finally, we believe that more effort should be given to systematic tools for diagnosing potential outlying proximities, or more generally, particular patterns underlying proximity data, in order to establish the impact that these perturbations may have on MDS results. It seems to us that the forward search could offer a valid solution to such problems (Solaro and Pagani, 2010), but much work still needs to be done, especially to extend its use to the general case where a proximity matrix is available, instead of an initial data matrix.

18.4   Summary

This chapter provides an overview of multidimensional scaling models that are well known in the literature, namely classical MDS, least-squares MDS, and non-metric MDS. We briefly mention several variants of these methods, that is, the weighted version of CMDS, Sammon's non-linear mapping, and the SMACOF method developed to minimize the stress loss function. We discuss in detail proximity data, which represent the input data of any MDS analysis. Various theoretical topics are also discussed, such as the problems of comparing two MDS configurations via Procrustes analysis, performing a robustness analysis for diagnostic purposes, and handling missing data. Moreover, we comment on the role of MDS models in market research and customer satisfaction studies. Finally, we present a case study based on the ABC annual customer satisfaction survey in order to show how the main metric MDS models addressed here can be applied and compared. The chapter concludes with some notes on the main statistical packages where MDS models are implemented, along with a discussion on possible directions for future methodological research on MDS methods.

References

Atkinson, A.C. and Riani, M. (2000) *Robust Diagnostic Regression Analysis*. New York: Springer.

Atkinson, A.C., Riani, M. and Cerioli, A. (2004) *Exploring Multivariate Data with the Forward Search*, Springer-Verlag, New York.

Atkinson, A.C., Cerioli, A. and Riani, M. (2005) Rfwdmv: Forward search for multivariate data. R package version 0.72–2. http://www.riani.it.

Atkinson, A.C., Riani, M. and Cerioli, A. (2010) The Forward Search: Theory and data analysis. *Journal of the Korean Statistical Society*, 39, 117–134.

Borg, I. and Groenen, P.J.F. (2005) *Modern Multidimensional Scaling. Theory and Applications*, 2nd edn. New York: Springer.

Cailliez, F. (1983) The analytical solution to the additive constant problem. *Psychometrika*, 48, 305–308.

Cailliez, F. and Kuntz, P. (1996), A contribution to the study of the metric and Euclidean structures of dissimilarities. *Psychometrika*, 61(2), 241–253.

Carroll, J.D. and Chang, J.J. (1970) Analysis of individual differences in multidimensional scaling via an *N*-way generalization of 'Eckart-Young' decomposition. *Psychometrika*, 35, 283–319.

Carroll, J.D. and Green, P. (1997) Psychometric methods in marketing research: Part II, Multidimensional scaling. *Journal of Marketing Research*, 34(2), 193–204.

Carroll, J.D. and Wish, M. (1974) Models and methods for three-way multidimensional scaling. In D.H. Krantz, R.C. Atkinson, R.D. Luce and P. Suppes (eds), *Contemporary Developments in Mathematical Psychology*, pp. 57–105. San Francisco: Freeman.

Chen, C.H. and Chen, J.A. (2000) Interactive diagnostic plots for multidimensional scaling with applications in psychosis disorder data analysis. *Statistica Sinica*, 10, 665–691.

Commandeur, J.J.F. and Heiser, W.J. (1993) Mathematical derivations in the proximity scaling (PROX-SCAL) of symmetric data matrices. Technical Report, Department of Data Theory, University of Leiden.

Coombs, C.H. (1964) *A theory of data*. New York: John Wiley & Sons, Inc.

Cooper, L.G. (1983) A review of multidimensional scaling in marketing research. *Applied Psychological Measurement*, 7, 427–450.

Cox, T.F. and Cox, M.A.A. (2000) A general weighted two-way dissimilarity coefficient. *Journal of Classification*, 17, 101–121.

Cox, T.F. and Cox, M.A.A. (2001) *Multidimensional Scaling*, 2nd edition, Monograph on Statistics and Applied Probability, 88. Boca Raton, FL: Chapman & Hall/CRC.

Cuadras, C.M. and Fortiana, J. (1995) A continuous metric scaling solution for a random variable. *Journal of Multivariate Analysis*, 52, 1–14.

Cuadras, C.M. and Fortiana, J. (1996) Weighted continuous metric scaling. In A.K. Gupta and V.L. Girko (eds), *Multidimensional Statistical Analysis and Theory of Random Matrices: Proceedings of the Sixth Eugene Lukacs Symposium*, pp. 27–40. Utrecht: VSP.

Cuadras, C.M. and Fortiana, J. (1998) Visualizing categorical data with related metric scaling. In J. Blasius and M. Greenacre (eds), *Visualization of Categorical Data*, pp. 365–376. London: Academic Press.

Davison, M.L. (1983) *Multidimensional Scaling*. New York: John Wiley & Sons, Inc.

de Leeuw, J. (1977) Applications of convex analysis to multidimensional scaling. In J.R. Barra, F. Brodeau, G. Romier and B. van Cutsem (eds), *Recent Developments in Statistics*, pp. 133–145. Amsterdam: North-Holland.

de Leeuw, J. and Heiser, W.J. (1980) Multidimensional scaling with restrictions on the configuration. In P. Krishnaiah (ed.), *Multivariate Analysis, Volume V*, pp. 501–522. Amsterdam: North-Holland.

de Leeuw, J. and Mair, P. (2009) Multidimensional scaling using majorization: SMACOF in R. *Journal of Statistical Software*, 31(3), 1–30.

de Leeuw, J. and Meulman, J. (1986) A special jackknife for multidimensional scaling. *Journal of Classification*, 3(1), 97–112.

Eckart, C. and Young, G. (1936) The approximation of one matrix by another of lower rank. *Psychometrika*, 1, 211–218.

Everitt, B.S. and Rabe-Hesketh, S. (1997) *The Analysis of Proximity Data*, Kendall's Library of Statistics, 4. London: Arnold.

Gifi, A. (1990) *Nonlinear Multivariate Analysis*. Chichester: John Wiley & Sons, Ltd.

Gower, J.C. (1966) Some distance properties of latent root and vector methods used in multivariate analysis. *Biometrika*, 53, 325–338.

Gower, J.C. (1971) A general coefficient of similarity and some of its properties. *Biometrics*, 27(4), 857–871.

Gower, J.C. (1982) Euclidean distance geometry. *Mathematical Scientist*, 7, 1–14.

Gower, J.C. (2004) Similarity, dissimilarity and distance, measures of. In S. Kotz, N. Balakrishnan, C. Read and B. Vidakovic (eds), *Encyclopedia of Statistical Sciences*. Hoboken, NJ: John Wiley & Sons, Inc. http://dx.doi.org/10.1002/0471667196.ess1595.pub2.

Gower, J.C. and Dijksterhuis, G.B. (2004) *Procrustes Problems*, Oxford Statistical Science Series, 30. Oxford: Oxford University Press.

Gower, J.C. and Legendre, P. (1986) Metric and Euclidean properties of dissimilarity coefficients. *Journal of Classification*, 3, 5–48.

Greenacre, M. (2005) Weighted metric multidimensional scaling. In M. Vichi, P. Monari, S. Mignani and A. Montanari (eds), *New Developments in Classification and Data Analysis*, pp. 141–149. Berlin: Springer,

Guttman, L. (1968) A general nonmetric technique for finding the smallest coordinate space for a configuration of points. *Psychometrika*, 33, 469–506.

Heady, R.B. and Lucas, J.L. (2010) *MDS Analysis Using Permap vs. 11.8*. http://www.newmdsx.com/permap/permap.html.

House, L.L. and Banks, D. (2004) Robust multidimensional scaling. In J. Antoch (ed.), *COMPSTAT 2004 – Proceedings in Computational Statistics*, 251–259. Heidelberg: Physica-Verlag.

Klawonn, F., Nauck, D.D. and Tschumitschew, K. (2009) Measuring and visualising similarity of customer satisfaction profiles for different customer segments. In *Hybrid Artificial Intelligence Systems*, Lecture Notes in Computer Science 5572, pp. 60–67. Berlin: Springer.

Kruskal, J.B. (1964a) Multidimensional scaling by optimizing goodness of fit to a nonmetric hypothesis. *Psychometrika*, 29, 1–27.

Kruskal, J.B. (1964b) Nonmetric multidimensional scaling: a numerical method. *Psychometrika*, 29, 28–42.

Lee, J.A. and Verleysen, M. (2007) *Nonlinear Dimensionality Reduction*. New York: Springer.

Lingoes, J.C. (1971) Some boundary conditions for a monotone analysis of symmetric matrices. *Psychometrika*, 36, 195–203.

Mardia, K.V. (1978) Some properties of classical multi-dimensional scaling. *Communications in Statistics – Theory and Methods A*, 7, 1233–1241.

Mardia, K.V., Kent, J.T. and Bibby, J.M. (1979) *Multivariate Analysis*. London: Academic Press.

Meulman, J.J. (1992) The integration of multidimensional scaling and multivariate analysis with optimal transformations. *Psychometrika*, 57, 539–565.

Meulman, J.J. (1993) Principal coordinates analysis with optimal transformation of the variables – minimizing the sum of squares of the smallest eigenvalues. *British Journal of Mathematical and Statistical Psychology*, 46, 287–300.

Oksanen, J., Blanchet, F.G., Kindt, R., Legendre, P., O'Hara, R.B., Simpson, G.L., Solymos, P., Stevens, M.H.H. and Wagner, H. (2011) vegan: Community ecology package. R package version 1.17–0. http://CRAN.R-project.org/package=vegan.

Podani, J. (1999) Extending Gower's general coefficient of similarity to ordinal characters. *Taxon*, 48, 331–340.

R Development Core Team (2011) *R: A Language and Environment for Statistical Computing*. Vienna: R Foundation for Statistical Computing. http://www.R-project.org.

Ramsay, J.O. (1982) Some statistical approaches to multidimensional scaling data. *Journal of the Royal Statistical Society, Series A*, 145, 285–312.

Sammon, J.W. (1969) A nonlinear mapping for data structure analysis. *IEEE Transactions on Computers*, 18, 401–409.

Shepard, R.N. (1962a) The analysis of proximities: Multidimensional scaling with unknown distance function, Part I. *Psychometrika*, 27, 125–140.

Shepard, R.N. (1962b) The analysis of proximities: Multidimensional scaling with unknown distance function, Part II. *Psychometrika*, 27, 219–246.

Sibson, R. (1978) Studies in the robustness of multidimensional scaling: Procrustes statistics. *Journal of the Royal Statistical Society, Series B*, 40, 234–238.

Sibson, R. (1979) Studies in the robustness of multidimensional scaling: Perturbational analysis of classical scaling. *Journal of the Royal Statistical Society, Series B*, 41, 217–229.

Solaro, N. (2010) Sensitivity analysis and robust approach in multidimensional scaling. An evaluation of customer satisfaction, *Quality Technology and Quantitative Management*, 7(2), 169–184.

Solaro, N. and Pagani, M. (2010) The forward search for classical multidimensional scaling when the starting data matrix is known. In C. Lauro, F. Palumbo and M. Greenacre (eds), *Data Analysis and Classification*, pp. 101–109. Berlin: Springer.

Spence, I. and Lewandowsky, S. (1989) Robust multidimensional scaling. *Psychometrika*, 54, 501–513.

Takane, Y., Young, F.W. and de Leeuw, J. (1977) Nonmetric individual differences multidimensional scaling: An alternating least squares method with optimal scaling features. *Psychometrika*, 42, 7–67.

Torgerson, W.S. (1952) Multidimensional scaling: I. Theory and method. *Psychometrika*, 17, 401–419.

Venables, W.N. and Ripley, B.D. (2002) *Modern Applied Statistics with S*. New York: Springer. http://www.stats.ox.ac.uk/pub/MASS4.

Young, G. and Householder, A.S. (1938) Discussion of a set of points in terms of their mutual distances. *Psychometrika*, 3, 19–22.

Wu, J. and DeSarbo, W.S. (2005) Market segmentation for customer satisfaction studies via a new latent structure multidimensional scaling model. *Applied Stochastic Models in Business and Industry*, 21, 303–309.

19

Multilevel models for ordinal data

Leonardo Grilli and Carla Rampichini

This chapter is concerned with regression models for ordinal responses, with special emphasis on random effects models for multilevel or clustered data. After a brief discussion on ordinal variables, it reviews the most common regression models for ordinal responses, focusing on cumulative models, namely models based on cumulative probabilities. It then deals with random effects cumulative models for multilevel data, discussing several issues peculiar to the random effects extension such as the distinction between marginal and conditional effects, the measures of unobserved cluster-level heterogeneity, the consequences of adding covariates, and the main types of predicted probabilities. It also deals with estimation, inference and prediction, with a brief look on available software. Finally, it presents an application of random effects cumulative models to the analysis of student ratings of university courses.

19.1 Ordinal variables

Satisfaction is usually measured using graded scales, also called Likert scales, such as 'Very dissatisfied', 'Dissatisfied', 'Satisfied' and 'Very satisfied'. The resulting statistical variable Y is ordinal, that is, it has ordered categories. Sometimes a score is associated with each label (e.g., 'Very dissatisfied' is 1, 'Dissatisfied' is 2, ...), but even in this case the variable Y is genuinely ordinal: it is not measured on an interval scale since the distances between the categories are unknown and the scoring system is just an arbitrary assumption. For example, the common choice of scoring the categories using the integers $1, 2, 3, \ldots$ amounts to assuming that the categories are evenly spaced (e.g., the difference between 'Very dissatisfied' and 'Dissatisfied' is the same as the difference between 'Dissatisfied' and 'Satisfied').

The statistical methods for ordinal variables avoid the arbitrariness of scoring systems and thus are generally to be preferred. Nonetheless, in the social sciences the use of scoring systems to convert categories into numbers is common practice since the statistical methods

Modern Analysis of Customer Surveys: with applications using R, First Edition. Edited by Ron S. Kenett and Silvia Salini.
© 2012 John Wiley & Sons, Ltd. Published 2012 by John Wiley & Sons, Ltd.

for quantitative variables are more powerful and easier to implement and interpret. The consequences of analysing ordinal variables with methods for continuous variables have been investigated both analytically (Olsson, 1979) and via simulations (Muthén and Kaplan, 1985). In general, the bias depends on the number of categories (five is usually a minimum to get an acceptable level of bias) and the skewness of the distribution: indeed, the bias increases with the degree of skewness and may become large in the case of floor or ceiling effects, that is, when the largest frequency corresponds to a category at the extremes of the scale. The bias may be reduced by using sophisticated scoring systems (Fielding, 1997), but we do not pursue the matter further and later on we focus on appropriate methods for ordinal variables. For more on this topic see Chapter 4 on measurement scales and Chapter 17 on nonlinear principal component analysis (NPLCA).

An ordinal variable is a categorical variable supplemented with information on the ordering of the categories; indeed, the statistical methods for ordinal variables are designed to exploit such information. Formally, a categorical variable Y with categories y_c, $c = 1, \ldots, C$, has a multinomial distribution with probabilities $\pi_c = \Pr(Y = y_c)$. The set of C probabilities $\pi_1, \pi_2, \ldots, \pi_C$ has one redundant probability due to the constraint $\pi_1 + \pi_2 + \ldots + \pi_C = 1$. When the categories are ordered, the cumulative probabilities are defined as $\gamma_c = \Pr(Y \leq y_c) = \pi_1 + \pi_2 + \ldots + \pi_c$. Note that there are $C - 1$ non-redundant cumulative probabilities since the last one is $\gamma_C = 1$.

It is often useful to assume that an ordinal variable Y with C categories is generated by a latent continuous variable Y^* with a set of $C - 1$ thresholds α_c^* such that $Y = y_c$ if and only if $\alpha_{c-1}^* < Y^* \leq \alpha_c^*$. For example, if satisfaction is expressed using a four-point scale (e.g. 'Very dissatisfied', 'Dissatisfied', 'Satisfied', 'Very satisfied'), we can postulate the existence of a latent satisfaction on a continuous scale which is categorized by three thresholds. Figure 19.1 illustrates the density of the underlying satisfaction Y^*, the thresholds α_c^* and the corresponding observed satisfaction Y.

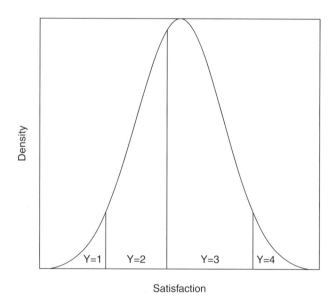

Figure 19.1   Density of underlying satisfaction, thresholds and observed satisfaction.

The existence of an underlying continuous variable cannot be proved or disproved: it corresponds to a different approach useful for both interpretation and development of analytical tools, for example the polychoric correlations (e.g., Agresti, 2010). The representation based on a latent continuous variable is conceptually appealing in settings such as customer satisfaction since it disentangles the process generating the observed satisfaction grade Y into two parts: the underlying satisfaction level Y^* and the measurement process corresponding to the thresholds α_c^*. In this perspective, the different ways of formulating the question and defining the rating scale affect the observed level of satisfaction through the thresholds. Similarly, if the rating scale adopted in the questionnaire is perceived in a different way by a subset of respondents, this affects the observed level of satisfaction through the thresholds. Indeed, the standard assumption that the set of thresholds is the same for all respondents is a measurement invariance assumption, which can be relaxed by allowing heterogeneous thresholds (Johnson, 2003). See also Chapter 14 on Rasch models.

In the next two sections we review the main regression models for an ordinal response: in Section 19.2 we consider the standard, single-level models for independent observations, while in Section 19.3 we consider the multilevel extension to deal with correlated observations. In single-level models (also called *marginal* models) the probabilities are conditioned on the covariates but not on the random effects, while in multilevel models (also called *conditional* models) the probabilities are conditioned both on the covariates and on the random effects. In this chapter the probabilities and parameters for multilevel models have no superscript, while those for single-level models have the superscript $^\circ$ (e.g., the vector of regression coefficients is β in a multilevel model and β° in the corresponding single-level model). In all the models we will denote the statistical units by a subscript with double index ij, where $j = 1, 2, \ldots, J$ is the level 2 (cluster) index and $i = 1, 2, \ldots, n_j$ is the level 1 index (the double index is superfluous in single-level models, but we use it so that the notation is uniform across the chapter).

19.2 Standard models for ordinal data

Suppose that we observe an ordinal response Y_{ij} with C categories for level 1 unit i in cluster j, along with a vector of covariates $\mathbf{x}_{ij}$ (including the constant term). A regression model establishes a relationship between the covariates and the set of probabilities of the categories $\pi_{cij}^\circ = \Pr(Y_{ij} = y_c | \mathbf{x}_{ij})$, $c = 1, \ldots, C$. Since one of the probabilities is redundant, any model must incorporate suitable restrictions to ensure the identification of the parameters. Models for ordinal data also incorporate restrictions to reflect the ordering of the categories.

Models for ordinal data need not be expressed in terms of the set of category probabilities π_{cij}°: they may also refer to convenient one-to-one transformations, such as the set of cumulative probabilities $\gamma_{cij}^\circ = \Pr(Y_{ij} \leq y_c | \mathbf{x}_{ij})$. Indeed, the most popular models for ordinal data are expressed in terms of these cumulative probabilities.

Early papers on regression models for ordinal data include McKelvey and Zavoina (1975), McCullagh (1980), and Winship and Mare (1984). The textbook of Agresti (2010) gives a thorough treatment of ordinal data, while O'Connell (2006) provides applied researchers in the social sciences with accessible and comprehensive coverage of analyses for ordinal outcomes. Other valuable treatments of ordinal outcomes are Johnson and Albert (1999) from a Bayesian perspective, and Greene and Hensher (2010) in the setting of choice theory. Books on statistical modelling often have a chapter on ordinal regression models, for example Skrondal and Rabe-Hesketh (2004) and Hilbe (2009).

19.2.1 Cumulative models

A cumulative regression model for an ordinal response Y_{ij} with C categories is defined by a set of $C - 1$ equations where the cumulative probabilities γ°_{cij} are related to the covariates $\mathbf{x}_{ij}$. We consider *cumulative generalized linear models* where the cumulative probabilities are related to the covariates through a linear predictor $\mathbf{x}'_{ij}\boldsymbol{\beta}^\circ$ and a monotone link function g:

$$g(\gamma^\circ_{cij}) = \alpha^\circ_c - \mathbf{x}'_{ij}\boldsymbol{\beta}^\circ, \quad c = 1, 2, \ldots, C - 1. \tag{19.1}$$

The parameters α°_c, called *thresholds* or *cutpoints*, are in increasing order, $\alpha^\circ_1 < \alpha^\circ_2 < \ldots < \alpha^\circ_{C-1}$. The vector of regression coefficients $\boldsymbol{\beta}^\circ$ (including the intercept β°_0) does not have the category index c, thus the effects of the covariates are constant across response categories, a feature called the *parallel regression assumption*; indeed, plotting $g(\gamma^\circ_{cij})$ against a covariate yields $C - 1$ parallel lines (or parallel curves if the covariate has non-linear terms). Cumulative models are known in psychometrics as *graded response models* (Samejima, 1969) or *difference models* (Thissen and Steinberg, 1986). The latter name indicates that the probabilities of the categories are obtained by difference: $\pi^\circ_{cij} = \gamma^\circ_{cij} - \gamma^\circ_{c-1,ij} = g^{-1}(\alpha^\circ_c - \mathbf{x}'_{ij}\boldsymbol{\beta}^\circ) - g^{-1}(\alpha^\circ_{c-1} - \mathbf{x}'_{ij}\boldsymbol{\beta}^\circ)$.

The minus sign before the linear predictor in model (19.1) implies that increasing a covariate with a positive slope is associated with a shift towards the right-hand end of the response scale, namely a rise of the probabilities of the higher categories. Some authors write the model with a plus sign before the linear predictor: in that case the interpretation of the effects of the covariates is reversed.

In model (19.1) we cannot simultaneously estimate the constant of the linear predictor and all the $C - 1$ thresholds; in fact, adding an arbitrary constant to the linear predictor can be counteracted by adding the same constant to each threshold. This identification problem is usually solved by either omitting the constant from the linear predictor ($\beta^\circ_0 = 0$) or fixing the first threshold to zero ($\alpha^\circ_1 = 0$).

Typical choices of the link function g are *logit*, *probit* and *complementary log-log*. The logit link is widely used (except in the social sciences) mainly because of the connection with odds ratios. In the following we will focus on the logit cumulative model, also known as the *proportional odds model*:

$$\text{logit}(\gamma^\circ_{cij}) = \alpha^\circ_c - \mathbf{x}'_{ij}\boldsymbol{\beta}^\circ, \quad c = 1, 2, \ldots, C - 1, \tag{19.2}$$

where the logit on the left-hand side is a *cumulative logit*, namely the logarithm of the odds of not exceeding the cth category:

$$\text{logit}(\gamma^\circ_{cij}) = \log \frac{\gamma^\circ_{cij}}{1 - \gamma^\circ_{cij}} = \log \frac{\Pr(Y_{ij} \le y_c)}{\Pr(Y_{ij} > y_c)}. \tag{19.3}$$

In model (19.2) the parallel regression assumption implies the proportional odds property; in fact, the ratio of the odds of not exceeding the cth category for units ij and $i'j'$ is $\exp(-(\mathbf{x}'_{ij} - \mathbf{x}'_{i'j'})\boldsymbol{\beta}^\circ)$, which does not depend on c and thus is constant across response categories.

The parallel regression assumption of the cumulative models may be too restrictive (for a test, see Brant, 1990). It can be relaxed by allowing the thresholds to depend on covariates or, alternatively, by allowing covariates to have category-specific slopes (these models are called *partial proportional odds*, after Peterson and Harrell, 1990). Another way to relax the parallel regression assumption is to let the variance of the residual in the underlying linear model (see Section 19.3.1) to depend on covariates (Cox, 1995) or, alternatively, to use a scaled link such as the *scaled probit link* of Skrondal and Rabe-Hesketh (2004). A further approach is to introduce latent classes (Breen and Luijkx, 2010). Models violating the parallel regression assumption should be used with care since they raise identification and interpretation issues (Agresti, 2010).

19.2.2 Other models

The rest of this chapter is devoted to the multilevel extension of cumulative models, but here we briefly mention some non-cumulative models that may be preferable in some contexts and can be extended to handle multilevel data as well.

A wide class of models is obtained by specifying a multinomial logit model for the probabilities of the categories $\pi^{\circ}_{cij} = \Pr(Y_{ij} = y_c | \mathbf{x}_{ij})$ with additional parameter constraints reflecting the ordering of the categories (Skrondal and Rabe-Hesketh, 2004). An example is the *adjacent category logit model* (Agresti, 2010), where the linear predictor is equated to the logarithm of the odds between adjacent categories $\pi^{\circ}_{cij}/\pi^{\circ}_{c-1,ij}$. In fact, such a model can be written as a multinomial logit model with linear predictor $c\mathbf{x}'_{ij}\boldsymbol{\beta}^{\circ}$, which can be seen as either a model with category-specific slopes $c\boldsymbol{\beta}^{\circ}$ and category-invariant covariates $\mathbf{x}_{ij}$, or a model with category-invariant slopes $\boldsymbol{\beta}^{\circ}$ and category-specific covariates $\mathbf{x}^c_{ij} = c\mathbf{x}_{ij}$. The last formulation is known in item response theory as the *partial credit model* (Masters, 1982), which is a generalization of the Rasch model to ordinal items.

A valuable alternative to traditional models for ordinal responses is represented by the CUB models outlined in Chapter 13. While traditional models considered in this chapter are based on a multinomial distribution, CUB models are based on a mixture between a shifted binomial distribution (to be interpreted as *feeling*) and a discrete uniform distribution (to be interpreted as *uncertainty*).

19.3 Multilevel models for ordinal data

Let us now consider the multilevel extension of regression models for ordinal responses. These models are outlined in most books on multilevel analysis. In addition, we recommend the reviews of Agresti and Natarajan (2001) and Hedeker (2008), and Chapter 10 of Agresti (2010). As the field is vast, we focus on the most popular configuration in applications, namely cumulative models (outlined in Section 19.2.1) with a random intercept in a two-level hierarchy:

$$g(\gamma_{cij}) = \alpha_c - (\mathbf{x}'_{ij}\boldsymbol{\beta} + u_j), \quad c = 1, 2, \ldots, C - 1, \tag{19.4}$$

where $j = 1, 2, \ldots, J$ is the level 2 (cluster) index, $i = 1, 2, \ldots, n_j$ is the level 1 index, and γ_{cij} is the cumulative probability up to the cth category for unit i in cluster j. The term u_j is a random effect representing unobserved factors at the cluster level: since it is shared by all the units of the cluster, it induces within-cluster correlated responses. If the overall intercept β_0 is unconstrained, we can view u_j as a random shift of the intercept so that the intercept of cluster j is $\beta_0 + u_j$; otherwise, if the overall intercept is fixed at zero, we can view u_j as a random shift of the thresholds so that the set of thresholds of cluster j is $\alpha_c - u_j, c = 1, 2, \ldots, C - 1$.

The standard assumption on the random effects u_j is that, conditionally on the covariates, they are independent and identically distributed with zero mean and a common cluster variance σ^2_u to be estimated. In contrast, the assumption of common cluster variance can be easily relaxed (Hedeker, 2008, Section 6.7), as can the conventional normality assumption (Agresti and Natarajan, 2001, Section 4.2). In order to obtain unbiased estimates, the key part of the standard assumption is *exogeneity*, that is to say, the mean of the random effects does not depend on the covariates: $E(u_j|\{\mathbf{x}_{ij} : i = 1, 2, \ldots, n_j\}) = 0$. A multilevel model like that in equation (19.4) may be useful in several kinds of applications in customer satisfaction, for example: (i) analysis of a single response from customers clustered in units offering a product or service (firms, schools, hospitals, …) or clustered in geographical regions; (ii) analysis of repeated responses to a given question in a longitudinal survey on a panel of customers; (iii) joint analysis of a set of items of a survey questionnaire on customers. Note that customers are level 1 units in example (i) and level 2 units (clusters) in examples (ii) and (iii).

The sample size required in order reliably to fit a multilevel model for ordinal data depends on several factors, including the complexity of the model, the value of the cluster variance, and the estimation method. Moreover, the requirement is higher for the variances of the random effects than for the regression coefficients. Some guidelines are provided by recent simulation studies on the closely related multilevel logit models for binary responses: Austin (2010) considers a random intercept logit model, whereas Moineddin *et al.* (2007) focus on a logit model where both the intercept and the slope randomly vary across clusters. In the random intercept case, the estimates are reasonably good with most estimation methods even with 10–15 clusters, as long as the average cluster size is at least 10. If the clusters are smaller, more clusters are needed. In the random slope case, the requirement is considerably higher, say 30 clusters of size 30.

19.3.1   Representation as an underlying linear model with thresholds

As noted in Section 19.1, an ordinal response Y_{ij} with C categories can be represented as an underlying continuous response Y_{ij}^* with a set of $C - 1$ thresholds α_c^* such that $Y_{ij} = y_c$ if and only if $\alpha_{c-1}^* < Y_{ij}^* \leq \alpha_c^*$. It follows that a cumulative generalized linear model for an ordinal response is equivalent to a system composed of a set of thresholds α_c^* and a linear regression model for an underlying continuous response,

$$Y_{ij}^* = \mathbf{x}_{ij}'\boldsymbol{\beta}^* + u_j^* + e_{ij}^*, \tag{19.5}$$

where e_{ij}^* is a level 1 error with standard deviation σ_{e^*} and u_j^* is a level 2 error with standard deviation σ_{u^*}. In fact,

$$\Pr(Y_{ij} \leq y_c) = \Pr(Y_{ij}^* \leq \alpha_c^*) = \Pr(e_{ij}^* \leq \alpha_c^* - \mathbf{x}_{ij}'\boldsymbol{\beta}^* - u_j^*) = g^{-1}(\alpha_c - \mathbf{x}_{ij}'\boldsymbol{\beta} - u_j).$$

Therefore, the underlying linear model (19.5) with thresholds α_c^* and level 1 error e_{ij}^* having distribution function g^{-1} is equivalent to the cumulative model (19.4) with link function g. The relationship between a parameter of the cumulative model θ and the corresponding parameter of the underlying model θ^* is $\theta = \theta^* \sigma_g / \sigma_{e^*}$, where σ_g is the standard deviation of the distribution associated to the link function (e.g., $\sigma_g = 1$ for probit and $\sigma_g = \pi/\sqrt{3} \simeq 1.81$ for logit).

When we specify the link function for the cumulative model, we implicitly specify the distribution function of the level 1 error and, consequently, we fix the standard deviation of the level 1 error to a conventional value: the probit link corresponds to a standard normal error so the standard deviation is fixed to 1, whereas the logit link corresponds to a standard logistic error so the standard deviation is fixed to $\pi/\sqrt{3} \simeq 1.81$. Indeed, the measurement unit of the underlying model is undefined since $\Pr(Y_{ij}^* \leq \alpha_c^*) = \Pr(kY_{ij}^* \leq k\alpha_c^*)$ for any constant k, thus the standard deviation σ_{e^*} is not identifiable. This indeterminacy is solved in the cumulative model (19.4) since its parameters are measured on a conventional scale defined by the link (the level 1 standard deviation does not appear as parameter). The change of scale is the reason why replacing probit with logit causes an expansion of the estimated slopes of about 1.81. The model specification requires some care in the case of level 1 heteroscedasticity, for example when σ_{e^*} changes across strata (Grilli and Rampichini, 2002).

The representation via an underlying linear model makes clear why the estimated slopes from a cumulative model are approximately invariant to merging of the categories.

19.3.2   Marginal versus conditional effects

The slopes $\boldsymbol{\beta}$ of the random intercept cumulative model (19.4) represent *conditional* or *cluster-specific* effects: they summarize the relationship between the covariates $\mathbf{x}_{ij}$ and the conditional cumulative probabilities $\gamma_{cij} = \Pr(Y_{ij} \leq y_c | \mathbf{x}_{ij}, u_j)$, which are conditional on the random effect and thus refer to a specific cluster of the population. On the other hand, the slopes $\boldsymbol{\beta}^{\circ}$ of the standard cumulative model (19.1) represent *marginal* or *population-averaged* effects: they summarize the relationship between the covariates $\mathbf{x}_{ij}$ and the marginal cumulative probabilities $\gamma_{cij}^{\circ} = \Pr(Y_{ij} \leq y_c | \mathbf{x}_{ij})$, which are marginal with respect to the random effect and thus refer to the whole population.

Marginal effects are smaller in absolute value than conditional effects: $|\beta_m^{\circ}| \leq |\beta_m|$ for every covariate. Such attenuation can be shown using the representation with the underlying linear model. Indeed, in Section 19.3.1 we showed that the mth slope of the random intercept cumulative model (19.4) is $\beta_m = \beta_m^* \sigma_g / \sigma_{e^*}$; on the other hand, if the random effect u_j^* is omitted, the underlying linear model (19.5) has a composite level 1 error $d_{ij}^* = u_j^* + e_{ij}^*$ with standard deviation $\sigma_{d^*} = \sqrt{\sigma_{u^*}^2 + \sigma_{e^*}^2}$, thus the corresponding slope of the single-level cumulative model (19.1) is $\beta_m^{\circ} = \beta_m^* \sigma_g / \sigma_{d^*}$. Since $\sigma_{d^*} \geq \sigma_{e^*}$ it follows that $|\beta_m^{\circ}| \leq |\beta_m|$ for every covariate. Clearly, the attenuation is stronger the larger is the level 2 variance compared to the level 1 variance, that is, the higher is the unobserved heterogeneity due to the clustering of the units. Under the standard assumption of a normal random effect, the analytical development outlined above is exact in the probit case (since a random intercept probit model implies a marginal probit model) and approximate in general (e.g., a random intercept logit model does not imply a marginal logit model).

Marginal and conditional slopes are population parameters: regardless of the estimation methods, a model without random effects has marginal slopes, while a model with random effects has conditional slopes. In most applications, conditional slopes are of interest as they refer to cluster-specific effects, which are more informative about causal processes.

Finally, note that if the responses are correlated within the clusters, random effects models yield correct standard errors, while marginal models, without random effects, yield incorrect standard errors (usually underestimated). Thus, if one is interested in marginal effects in the presence of correlated data, two alternatives are possible: either fit a random effects model and then recover the marginal effects, or fit a marginal model using a correction for the standard errors, such as the generalized estimating equation method or a robust estimator of the standard errors (Agresti and Natarajan, 2001).

19.3.3   Summarizing the cluster-level unobserved heterogeneity

In a linear random intercept model such as (19.5) the level of unobserved heterogeneity due to the clustering of the units is summarized by the intraclass correlation coefficient

(ICC), $\rho = \sigma_{u^*}^2/(\sigma_{u^*}^2 + \sigma_{e^*}^2)$. In a linear model the ICC is both the between-cluster variance as a proportion of the total variance and the correlation between the responses of two units of the same cluster, namely $\rho = \mathrm{Corr}(Y_{ij}^*, Y_{i'j}^*|\mathbf{x}_{ij}, \mathbf{x}_{i'j})$. Such a correlation does not depend on the covariates (it is homogeneous), so the ICC is an exhaustive indicator of the degree of correlation. Unfortunately, this property does not hold in models for categorical responses such as the random intercept cumulative model (19.4) since $\mathrm{Corr}(Y_{ij}, Y_{i'j}|\mathbf{x}_{ij}, \mathbf{x}_{i'j})$ actually depends on the covariates. An appealing solution is to summarize the degree of within-cluster correlation using the ICC for the underlying linear model, which can be easily computed using the cluster variance σ_u^2 of the cumulative model; from the relationship $\sigma_u = \sigma_{u^*}\sigma_g/\sigma_{e^*}$ of Section 19.3.1 it follows that $\rho = \sigma_u^2/(\sigma_u^2 + \sigma_g^2)$, where σ_g^2 is the variance of the distribution associated with the link function. For example, $\rho = \sigma_u^2/(\sigma_u^2 + 1)$ for probit and $\rho = \sigma_u^2/(\sigma_u^2 + \pi^2/3)$ for logit. However, the ICC for the underlying linear model is misleading if one attempts to compare it with the values usually obtained in linear models for observed continuous responses: the ICC for the underlying linear model is much lower and often gives the impression of a negligible within-cluster correlation. For example, a value $\rho = 0.01$ is negligible for an observed response but not for an underlying response.

A simple and effective way of summarizing the within-cluster correlation in models for categorical responses is to compute the probabilities under several scenarios defined by fixing the random effect u_j to a set of percentiles of its estimated distribution. Denoting by $u_{[p]}$ the percentile p, if the random intercept cumulative model (19.4) has a normally distributed u_j with estimated standard deviation $\hat{\sigma}_u$, then three scenarios could be defined by fixing the random effect to the estimated percentiles $\hat{u}_{[2.5]} = -1.96\hat{\sigma}_u$, $\hat{u}_{[50]} = 0$ and $\hat{u}_{[97.5]} = +1.96\hat{\sigma}_u$. Once the covariates have been fixed to a set of values $\mathbf{x}^0$, the cumulative probability up to category c in the scenario corresponding to percentile p is defined as $\mathrm{Pr}(Y \leq y_c|\mathbf{x}^0, \hat{u}_{[p]})$ and it is computed by replacing the model parameters with their estimates.

19.3.4   Consequences of adding a covariate

The representation of a cumulative model for ordinal responses as an underlying linear model with thresholds (Section 19.3.1) shows that the estimable parameters are scaled by the underlying level 1 standard deviation σ_{e^*}, for example $\beta_m = \beta_m^*\sigma_g/\sigma_{e^*}$. If it were possible to observe Y_{ij}^* and fit the underlying linear model (19.5), the addition of a covariate would reduce σ_{e^*}. However, in a cumulative model for the observable ordinal response Y_{ij} the level 1 standard deviation σ_{e^*} cannot change, so the effect on σ_{e^*} is dumped on the other parameters. This phenomenon can be easily seen in the hypothetical case of a new covariate w_{ij} with the following features: w_{ij} is independent of the other covariates, so its inclusion does not alter the slopes $\boldsymbol{\beta}^*$ of the other covariates; w_{ij} has no between-cluster variation, so its inclusion does not alter the level 2 standard deviation σ_{u^*}; and w_{ij} has some within-cluster variation, so its inclusion reduces the level 1 standard deviation to $k\sigma_{e^*}$ with $k < 1$. It follows that the addition of the new covariate w_{ij} in the cumulative model (19.4) inflates all the parameters by $1/k$, for example $\beta_{m,\mathrm{new}} = \frac{\beta_m^*\sigma_g}{k\sigma_{e^*}} = \frac{1}{k}\beta_{m,\mathrm{old}}$. Note that the cluster variance also increases, a phenomenon that may appear surprising since the added covariate has no between-cluster variation.

The simple pattern outlined for the hypothetical covariate w_{ij} does not hold in general, but it is clear that the change of scale induced by the new covariate hinders a direct comparison of the parameters before and after its inclusion. This issue (which also affects models for binary responses) is considered by Winship and Mare (1984) in the case of single-level models and by Fielding (2004) and Bauer (2009) in the case of multilevel models.

19.3.5   Predicted probabilities

In multilevel models for categorical responses there are three types of predicted probability (Skrondal and Rabe-Hesketh, 2009): *conditional probability* (a unit in a hypothetical cluster); *population-averaged probability* (a unit in a new, randomly sampled cluster); and *cluster-averaged probability* (a unit in a specific cluster of the sample). All these types of predicted probability require the parameters to be replaced by their estimates, and the covariates by arbitrary values. The three types of probability differ in the way the random effect u_j is handled: in conditional probability, the random effect u_j is fixed at an arbitrary value (usually chosen as a percentile of its estimated distribution in the population; see Section 19.3.3); in the other instances, the random effect u_j is averaged out using its estimated distribution in the whole population (population-averaged type) or using its estimated distribution for the jth cluster of the sample (cluster-average type). Skrondal and Rabe-Hesketh (2009) give guidelines on computation and interpretation.

Predicted probabilities are essential for an effective and intelligible report of the model results. In the random intercept cumulative model (19.4) the effects of the covariates are summarized by the vector of slopes $\boldsymbol{\beta}$. Unfortunately, the interpretation of $\boldsymbol{\beta}$ is not straightforward as it refers to a transformation g of the cumulative probabilities. In consequence, as in any model for categorical responses, the change in a given probability due to a unit increase in a covariate depends on the value of such a probability (the closer the probability to 0 or 1, the smaller the change). It is therefore important to express the effects of the covariates in terms of changes in the predicted probabilities with reference to some relevant scenarios. A popular strategy in a model with M covariates is to compute $M + 1$ sets of predicted probabilities of the categories $\widehat{\pi}_1^{(m)}, \widehat{\pi}_2^{(m)}, \ldots, \widehat{\pi}_C^{(m)}$ ($m = 0, 1, \ldots, M$), where the set $m = 0$ refers to a hypothetical baseline subject and the other sets consider a unit increase in the mth covariate.

19.3.6   Cluster-level covariates and contextual effects

As in any multilevel model, the covariates $\mathbf{x}_{ij}$ of the random intercept cumulative model (19.4) can include cluster-level covariates and cross-level interactions. The multilevel analysis with an ordinal response follows the basic principles explained in textbooks with reference to linear models for continuous responses (e.g., Raudenbush and Bryk, 2002), even if there are complications. For example, we noted in Section 19.3.1 that the level 1 and level 2 variances cannot be estimated separately since only their ratio is identified. As a consequence, the level 2 variance of the cumulative model may increase after the inclusion of a covariate with no cluster-level variation (see Section 19.3.4).

Another complication with categorical responses relates to the assessment of a contextual effect, which is a key quantity in education and sociology (Raudenbush and Bryk, 2002). In a linear model for a continuous response, the *contextual effect* of a covariate z_{ij} is the coefficient δ of its cluster mean $\bar{z}_j$ when both z_{ij} and $\bar{z}_j$ enter as covariates. Thus δ is the change in the expectation of the response following a unit increase in the cluster mean $\bar{z}_j$ while keeping the individual-level covariate z_{ij} constant. In a linear model, the change in the expectation of the response does not depend on the values of $\bar{z}_j$ and z_{ij}, so the contextual effect is a unique value denoted by the parameter δ. However, in a model for categorical responses, δ is just the contextual effect on the scale of the linear predictor: in order to assess the contextual effect on the probabilities, it is necessary to compute the predicted probabilities under several scenarios and draw plots. This approach is illustrated in Skrondal and Rabe-Hesketh (2009).

19.3.7   Estimation of model parameters

The estimation of the parameters of the random intercept cumulative model (19.4) is usually based on maximum likelihood (ML), yielding unbiased estimates under the missing at random (MAR) assumption (Rubin, 1976); that is, the missingness mechanism may depend on both model covariates and observed responses. Under mild regularity conditions, ML estimators have good asymptotic properties – consistency, normality and efficiency. In this framework the asymptotic theory requires increasing the number of clusters (increasing the cluster sizes is not enough), so the number of clusters J is the key quantity for asymptotics.

Let $\mathbf{Y}_j$ be the vector of n_j ordinal responses of the jth cluster and let $\mathbf{X}_j$ be the covariate matrix of cluster j with rows $\mathbf{x}'_{ij}$. Moreover, let $\boldsymbol{\theta} = (\boldsymbol{\alpha}, \boldsymbol{\beta}, \sigma_u)'$ be the vector of model parameters, where $\boldsymbol{\alpha}' = (\alpha_1, \ldots, \alpha_{C-1})$. The likelihood of $\mathbf{Y}_j$ conditional on u_j is equal to the product of the conditional probabilities of the responses

$$L_j(\mathbf{Y}_j \mid u_j, \mathbf{X}_j; \boldsymbol{\alpha}, \boldsymbol{\beta}) = \prod_{i=1}^{n_j} \prod_{c=1}^{C} \pi_{cij}^{d_{cij}} = \prod_{i=1}^{n_j} \prod_{c=1}^{C} \left(\gamma_{cij} - \gamma_{c-1,ij} \right)^{d_{cij}}$$

where d_{cij} is the indicator of $\{Y_{ij} = y_c\}$, $\gamma_{0ij} = 0$ and $\gamma_{Cij} = 1$. The previous likelihood is obtained by integrating out the (unobservable) random effect u_j,

$$L_j(\mathbf{Y}_j \mid \mathbf{X}_j; \boldsymbol{\theta}) = \int_{u_j} L_j(\mathbf{Y}_j \mid u_j, \mathbf{X}_j; \boldsymbol{\alpha}, \boldsymbol{\beta}) f(u_j; \sigma_u) du_j, \qquad (19.6)$$

where $f(u_j; \sigma_u)$ is the density of u_j, usually assumed to be a normal density with zero mean and standard deviation σ_u. Given independence across clusters, the log-likelihood for the J clusters is

$$\log L = \sum_{j=1}^{J} \log L_j(\mathbf{Y}_j \mid \mathbf{X}_j; \boldsymbol{\theta}),$$

which is maximized to obtain ML estimates of the model parameters.

In general, the integral in the likelihood (19.6) is not in closed form, thus some type of approximation is needed. Various approaches have been proposed in the literature, including Gaussian quadrature, Laplace approximation and Monte Carlo integration. Reviews are given by Skrondal and Rabe-Hesketh (2004) and Hedeker (2008). Each technique has advantages and drawbacks in terms of precision and computational burden. Different techniques usually yield slightly different parameter estimates, especially for the variance–covariance parameters of the random effects.

The most widely used technique for numerical integration is Gaussian quadrature, which can be ordinary, adaptive or spherical. The precision of the estimates and the computational burden depend heavily on the number of quadrature points: increasing the latter leads to higher accuracy and longer computation time. The computational burden, which may be prohibitive in some cases, also depends on factors such as the number of observations and the number of random effects (dimensionality of the integral). Simulations show that the adaptive version of

Gaussian quadrature performs well in a wide variety of situations as long as the dimensionality of the integral does not exceed 5 or 6 (Rabe-Hesketh *et al.*, 2005).

The sixth-order Laplace approximation by Raudenbush *et al.* (2000) appears to be very efficient and sufficiently accurate in many situations (Joe, 2008). Muthén's limited information approach is an excellent alternative for models with multivariate normal random effects when the cluster sizes are nearly constant and there are few missing data (Muthén and Satorra, 1995).

Quasi-likelihood methods, such as marginal quasi-likelihood (MQL) and penalized quasi-likelihood (PQL), are based on first- or second-order Taylor approximation of the likelihood. They are computationally efficient but, in some situations, underestimate the cluster variance and thus yield attenuated slopes (Mealli and Rampichini, 1999; Rodríguez and Goldman, 1995). MQL is faster but more biased than PQL. Both methods have the drawback of preventing likelihood-based inference since the likelihood function is not evaluated.

Standard likelihood-based methods require a parametric continuous distribution for the random effects to be specified, typically the normal distribution. Alternatively, it is possible to specify an arbitrary discrete distribution and estimate both the locations and masses (Aitkin, 1999). The model with discrete random effects is also called 'finite mixture' or 'latent class'. The latent class is a set of clusters, which is latent because the class membership of clusters is unobservable. Each class is characterized by a (prior) probability and a location for the random effect. For a fixed number of mass points, the estimation is straightforward since the likelihood is a finite mixture and no integration is involved. However, choosing the number of mass points is a difficult task, since the comparison of models with different numbers of mass points cannot be done with standard likelihood-based tests. A practical solution is to compare the models using fit indices such as the Akaike information criterion and the Bayesian information criterion (with many variations) or the Gâteaux derivative method (Rabe-Hesketh *et al.*, 2003). The resulting estimator is known as non-parametric maximum likelihood (Lindsay *et al.*, 1991).

In the Bayesian approach to multilevel models, both fixed and random effects are considered to be random variables with a given prior distribution and inference is based on their joint posterior distribution. This approach is more demanding than ML since it requires specification of the prior distribution of the model parameters and use of computationally intensive MCMC algorithms. The effort may be worthwhile in highly complex models since the Bayesian approach is better than ML in assessing the uncertainty of the estimates, a feature that may have considerable consequences for the coverage of confidence intervals (Browne and Draper, 2006). Moreover, Bayesian methods do not rely on asymptotics, thus outperforming ML in small samples (Austin, 2010).

19.3.8   Inference on model parameters

Standard large-sample inference procedures are applicable when the model is estimated via ML methods. Hypothesis testing for the fixed-effects parameters (i.e., α and β) can be conducted in the usual way, using the Wald test or the likelihood ratio test (LRT).

Inference about the cluster variance requires some care. In fact, unless the number of clusters is very large, the Wald test should not be used since the sampling distribution of the estimator of the cluster variance is highly asymmetric. The LRT is preferable. However, the null hypothesis of main interest, $\sigma_u^2 = 0$, is on the boundary of the parameter space and thus standard asymptotic results do not hold for the test statistics, including LRT. Indeed, the asymptotic distribution of the LRT statistic for $\sigma_u^2 = 0$ is not a chi-square with 1 degree of freedom, but rather a 50:50 mixture of a mass point at 0 and a chi-square with 1 degree of freedom (Berkhof and Snijders, 2001; Stram and Lee, 1994). A practical solution is to perform the usual LRT and then halve the p-value (otherwise the test is conservative, i.e. the actual probability of Type I error is lower than the nominal level). Alternatively, Verbeke and Molenberghs (2003) derived general one-sided score tests for variance components in models with several random effects.

19.3.9   Prediction of random effects

In many cases, it is useful to assign values to the random effects. Predicted random effects can be used for inference regarding clusters, for example to assess effectiveness of schools, universities, hospitals or firms (Grilli and Rampichini, 2009). Moreover, predicted random effects are useful quantities in model diagnostics, for example to check for violations of the normality assumption for random effects or to find outliers (Snijders and Berkhof, 2008).

The u_j are usually predicted using empirical Bayes (EB) methods (Skrondal and Rabe-Hesketh, 2009). In this setting the population distribution of the random effects is termed *prior*, whereas the distribution of the random effects conditional on the data of a given cluster is termed *posterior*. The EB prediction is the mean of the posterior distribution with parameter estimates plugged in, combining data information (likelihood) with population information (prior),

$$\hat{u}_j^{\text{EB}} = E(u_j \mid \mathbf{Y}_j, \mathbf{X}_j; \widehat{\boldsymbol{\theta}}) = \int u_j h(u_j \mid \mathbf{Y}_j, \mathbf{X}_j; \widehat{\boldsymbol{\theta}}) du_j, \tag{19.7}$$

where $h(\cdot)$ is the empirical posterior distribution of u_j. The mean of the posterior distribution is a value between 0 (the mean of the prior) and the mode of the likelihood of the jth cluster: the prediction that would be obtained using only the cluster-specific likelihood is thus shrunken, with a stronger shrinkage for small clusters. The EB predictor is conditionally biased towards zero and unconditionally unbiased (Skrondal and Rabe-Hesketh, 2009).

In the multilevel ordinal model (19.4) the EB predictions (19.7) do not have closed form, thus numerical or simulation-based integration methods must be used.

An alternative way to assign values to the random effects uses the posterior mode of the random effects. EB modal predictions do not require numerical integration.

Note that EB predictions are a by-product of the ML estimation algorithms relying on adaptive quadrature. For example, the *gllamm* procedure of Stata yields posterior means (Rabe-Hesketh *et al.*, 2005), while the routines implemented in R yield posterior modes (Pinheiro and Bates, 1995).

There are two kinds of standard errors of the EB predictions (19.7), depending on their use. *Comparative standard errors* are used for inference regarding the *true* values of u_j for specific clusters – for example, for making comparisons between clusters (Snijders and Bosker, 1999). On the other hand, the *diagnostic standard errors* are useful for model diagnostics – for example, for finding outliers (Snijders and Berkhof, 2008).

The comparative standard error is the square root of the posterior variance,

$$\text{Var}(u_j \mid \mathbf{Y}_j, \mathbf{X}_j; \widehat{\boldsymbol{\theta}}) = \int (u_j - \hat{u}_j^{\text{EB}})^2 h(u_j \mid \mathbf{Y}_j, \mathbf{X}_j; \widehat{\boldsymbol{\theta}}) du_j, \tag{19.8}$$

which has no closed form and thus the integral in (19.8) must be approximated.

For model diagnostics it is useful to consider the marginal sampling variance of the EB predictor, that is, the variance of the prediction under repeated sampling of the responses from their marginal distribution, keeping the covariates fixed and plugging in parameter estimates. There is no closed form expression for the marginal sampling variance in the ordinal multilevel model. Skrondal and Rabe-Hesketh (2009) suggest the approximation

$$\text{Var}(\hat{u}_j^{\text{EB}} \mid \mathbf{X}_j; \widehat{\boldsymbol{\theta}}) \approx \hat{\sigma}_u^2 - \text{Var}(u_j \mid \mathbf{Y}_j, \mathbf{X}_j; \widehat{\boldsymbol{\theta}}).$$

Both the posterior standard deviation (comparative standard error) and the sampling standard deviation (diagnostic standard error) of the EB prediction are lower than the estimated prior standard deviation $\hat{\sigma}_u$. The posterior standard deviation tends to decrease as the cluster size n_j increases, reflecting the increasing accuracy in the prediction of u_j. In contrast, the sampling standard deviation increases with n_j because the EB prediction is less shrunken.

On the basis of a simulation study, Skrondal and Rabe-Hesketh (2009) recommend using the posterior standard deviation as comparative standard error, while they find that the sampling distribution of the empirical Bayes predictions is often too discrete and non-normal for the diagnostic standard error to be used in the usual way for identifying outliers.

19.3.10   Software

Multilevel models for ordinal data can be fitted with ML or Bayesian methods using procedures in general purpose statistical packages (e.g., R, SAS and Stata), specialized software for multilevel analysis (e.g., MLwiN and HLM) or specialized software for latent variable models (e.g., Mplus and Latent GOLD). Multilevel modelling software reviews are available at the website of the Centre for Multilevel Modelling in Bristol. The programs are different in many respects. In particular, it is important to bear in mind that the parameter estimates may change with the method used to numerically evaluate the likelihood.

In the following we give a list with special emphasis on those programs implementing ML via numerical integration, giving references for more details. The list is not complete and relies mainly on our personal experience.

Software for ML estimation

Multilevel ordinal models can be fitted with ML by several programs. Most programs perform ML estimation via numerical integration, often using some form of quadrature.

The *ordinal* package of R (Christensen, 2010) fits cumulative link mixed models for ordinal data, though it is limited to random intercept models. The package includes the proportional odds model but it also allows for general regression structures for the location and the scale of the latent distribution (additive and multiplicative structures, structured thresholds and flexible link functions). Furthermore, several estimation procedures and auxiliary functions are implemented.

SAS PROC NLMIXED (SAS, 2009) is a general routine for non-linear mixed models. Multilevel ordinal models can be estimated by writing down the model likelihood using SAS statements. The procedure offers a wide choice of integral approximations and optimization techniques.

The *gllamm* command (Rabe-Hesketh and Skrondal, 2008) of Stata provides tools for analysing multilevel ordinal data. This procedure fits models of the generalized linear latent and mixed models (GLLAMM) class by ML with several kinds of quadrature. Moreover, it allows the parallel regression assumption (see section 19.2.1) to be relaxed by specifying a model for the thresholds or by using a scaled probit link.

The freeware program MIXOR provides ML estimates for mixed effects ordinal regression models (Hedeker and Gibbons, 1996). The commercial version is implemented in the program SUPERMIX (Hedeker and Gibbons, 2008).

Another freeware program for mixed effects ordinal models is aML, which is a general program for multilevel, multiprocess models (Lillard and Panis, 2003).

ML estimates of multilevel ordinal models via numerical integration are also provided by programs for latent variable models, such as Latent GOLD (Vermunt and Magidson, 2005) and Mplus (Muthén and Muthén, 2010).

Particular estimation techniques are available in the specialized multilevel software HLM (Raudenbush et al., 2004), which uses a combination of a fully multivariate Taylor expansion and a Laplace approximation, and MLwiN (Rasbash et al., 2005), which implements quasi-likelihood algorithms (MQL and PQL). Finally, the econometric program LIMDEP (Greene, 2007) uses simulated maximum likelihood.

Software for Bayesian estimation

Bayesian Markov chain Monte Carlo (MCMC) algorithms are available in Mplus (Muthén and Muthén, 2010) and MLwiN (Rasbash et al., 2005). MCMC algorithms can be also implemented using the freeware BUGS (Spiegelhalter et al., 1997) and its Windows version WinBUGS (Lunn et al., 2000). Marshall and Spiegelhalter (2001) provide an example of multilevel modelling using BUGS, including some syntax and discussion of the program.

19.4   Multilevel models for ordinal data in practice: An application to student ratings

In this section we present an application of multilevel models for ordinal responses to data on student satisfaction with university courses. We give guidelines on model specification, estimation and interpretation. The analysis is carried out with the R package *ordinal* which yields ML estimates using adaptive Gaussian quadrature (Christensen, 2010). The data set and the R script can be downloaded from the book website.

Student ratings are an old and widely recognized instrument for evaluating university courses. The statistical analysis of student ratings calls for special techniques which take into account the ordinal nature of the response and the hierarchical structure of the phenomenon (ratings are nested in courses which are nested in schools). Moreover, in order to use the student ratings to measure course quality, it should be recognized that student satisfaction depends not only on the characteristics of the course (lecture hall, clarity of the teacher, textbook, and so on), but also on the traits and expectations of the student. Therefore, a fair comparison among courses requires the calculation of net measures adjusting for individual characteristics. To this end, multilevel modelling is a well-suited technique (Grilli and Rampichini, 2009).

For this application we use the data of Rampichini et al. (2004), taken from a survey carried out by the University of Florence in the second semester of the academic year 1999/2000. The data set is restricted to courses with at least five respondents taken during the first year in the School of Engineering. The number of courses evaluated is 30 and the number of questionnaires is 767, while the number of questionnaires per course varies from 5 to 60. The items of the questionnaire require a response on the following four-point ordinal scale, where 1 means 'decidedly no', 2 'more no than yes', 3 'more yes than no', and 4 'decidedly yes'.

The main goal of the analysis is to identify 'good' and 'bad' courses on the basis of student overall satisfaction with the course (satisfaction) while adjusting for student characteristics. In particular, we consider a binary variable for the full-time status (fulltime) and three self-assessed individual characteristics measured on the ordinal four-point scale already mentioned: attendance with the intention of taking the exam in the first examination session (exam), previous knowledge of the subject (knowledge), and interest in the subject (interest).

The ordinal response satisfaction is studied via the random intercept cumulative model (19.4) using the logit link and C=4 categories:

$$\text{logit}(\gamma_{cij}) = \alpha_c - (\mathbf{x}'_{ij}\boldsymbol{\beta} + u_j), \quad c = 1, 2, 3,$$

where $j = 1, 2, \ldots, 30$ is the course index and i is the student index, while γ_{cij} is the cumulative probability up to the cth category for student i in course j. The covariate vector $\mathbf{x}_{ij}$ includes the student characteristics, whereas the term u_j is a random effect representing unobserved factors at the course level interpretable as 'perceived quality'.

The analysis begins with the random intercept model without covariates (null model). This model is a benchmark for subsequent models and provides a cluster variance $\hat{\sigma}_u^2 = 0.8800$ (the standard deviation is $\hat{\sigma}_u = 0.9381$). To test whether the cluster variance is statistically significant, we compare the models with and without random effects. The LRT statistic is 98.12 with 1 degree of freedom and a tiny p-value so that the null hypothesis is rejected (as noted in Section 19.3.8, the p-value must be halved, even if in this case the result of the test is unchanged). Therefore, there is evidence of unobserved heterogeneity at course level: as expected, the courses have different levels of satisfaction.

The sample frequencies of the response $(0.12, 0.27, 0.41, 0.20)$ are equal to the estimated probabilities from the single-level model (19.1) without covariates, which can be computed using the estimated thresholds ($\hat{\alpha}_1^\circ = -1.9685$, $\hat{\alpha}_2^\circ = -0.4480$, $\hat{\alpha}_3^\circ = 1.3572$). For example, the marginal probability that a student responds 'more yes than no' ($Y_{ij} = 3$) is

$$\hat{\pi}_3^\circ = \hat{\gamma}_3^\circ - \hat{\gamma}_2^\circ = \frac{1}{1 + e^{-\hat{\alpha}_3^\circ}} - \frac{1}{1 + e^{-\hat{\alpha}_2^\circ}} = \frac{1}{1 + e^{-1.3572}} - \frac{1}{1 + e^{0.4480}} = 0.41.$$

A similar computation with the random intercept null model gives the conditional probabilities, that is, the probabilities for a course with a hypothetical value of the random effect (see Section 19.3.5). For example, given the estimated thresholds ($\hat{\alpha}_1 = -2.2397, \hat{\alpha}_2 = -0.5379, \hat{\alpha}_3 = 1.5624$), the probability that a student responds 'more yes than no' ($Y_{ij} = 3$) for a course with a mean level of satisfaction ($u_j = 0$) is

$$\hat{\pi}_3 = \hat{\gamma}_3 - \hat{\gamma}_2 = \frac{1}{1 + e^{-\hat{\alpha}_3 + u_j}} - \frac{1}{1 + e^{-\hat{\alpha}_2 + u_j}} = \frac{1}{1 + e^{-1.5624}} - \frac{1}{1 + e^{0.5379}} = 0.46.$$

The amount of course-level unobserved heterogeneity is summarized by the ICC $\hat{\rho} = 0.8800/(0.8800 + \pi^2/3) = 0.21$ (see Section 19.3.3): this means that about one fifth of the total variability in the underlying satisfaction is at the course level. This is best appreciated by comparing some conditional probabilities as explained in Section 19.3.3, for example $\Pr(Y_{ij} \geq 3 \mid u_j = -1.96 \times 0.9381) = 0.21$ and $\Pr(Y_{ij} \geq 3 \mid u_j = +1.96 \times 0.9381) = 0.92$; thus, the probability that a student rates a course positively ('more yes than no' or 'decidedly yes')

Table 19.1   Estimates, standard errors and predicted probabilities $\hat{\pi}_c$ for the random intercept proportional odds model for overall satisfaction with the course (University of Florence, School of Engineering, academic year 1999/2000)

	Estimate	Std. Error	$\hat{\pi}_1$	$\hat{\pi}_2$	$\hat{\pi}_3$	$\hat{\pi}_4$
Thresholds						
First	−3.2567	0.2851				
Second	−0.9063	0.2556				
Third	1.9603	0.2664				
Baseline			0.04	0.25	0.59	0.12
Slopes						
Fulltime	0.3740	0.1808	0.03	0.19	0.61	0.17
Exam	0.4530	0.0901	0.02	0.18	0.61	0.18
Knowledge	0.5344	0.0882	0.02	0.17	0.61	0.19
Interest	1.2309	0.0966	0.01	0.09	0.57	0.33
Random effects						
Course-level σ_u	0.9477					
'Bad' course ($-1.96\sigma_u$)			0.20	0.53	0.26	0.02
'Good' course ($+1.96\sigma_u$)			0.01	0.05	0.47	0.47

ranges from 0.21 for a 'bad' course (at the 2.5th percentile) to 0.92 for a 'good' course (at the 97.5th percentile).

The analysis goes on by adding the covariates representing the characteristics of the students. Each of the variables measured on a four-point ordinal scale (exam, knowledge, interest) is tried in two alternative codings: a set of three binary indicators, and a single numerical covariate with values −2, −1, 0 and 1 (the third category is thus taken as the baseline). The second coding, which is more parsimonious and easier to interpret, is chosen on the basis of the LRT.

The estimates for the random intercept cumulative model are reported in Table 19.1, along with the predicted conditional probabilities discussed in Sections 19.3.3 and 19.3.5. In particular, the baseline student is a student who is not full-time and who responds 'more yes than no' to exam, knowledge and interest, while the baseline course has an average level of satisfaction, $u_j = 0$. The table shows how the baseline predicted probabilities change for a unit increase in each covariate and for a 'bad' course ($u_j = -1.96\sigma_u$) and a 'good' course ($u_j = 1.96\sigma_u$).

The effects of the student characteristics on the level of satisfaction are in the expected direction, that is, the probability of being satisfied is higher for full-time students, for students intending to take the exam in the first examination session, for students with good background knowledge, and for students interested in the subject. The latter feature has the largest effect, even if its estimate may be biased by endogeneity due to reverse causality.

The random effects represent the course-level unobserved heterogeneity in the ratings after controlling for the student characteristics: therefore, they may be interpreted as net measures of satisfaction for the course. The last two lines of Table 19.1 make clear that the courses are quite different in terms of satisfaction and that the features of the course have an overall effect on the ratings higher than any of the features of the students (e.g., the baseline probability 0.12

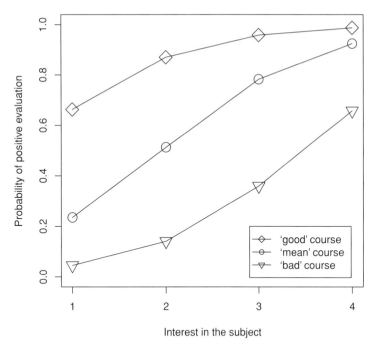

Figure 19.2   Probability of positive evaluation ('more yes than no' or 'decidedly yes') versus interest in the subject, for a full-time student responding 'more yes than no' to the questions on exam and previous knowledge; the probabilities are conditional on three hypothetical values of the random effect ('good' course, $u_j = +1.96\sigma_u$; 'mean' course, $u_j = 0$; 'bad' course, $u_j = -1.96\sigma_u$).

of being very satisfied becomes 0.19 for a student with fully adequate previous knowledge and 0.47 for a student attending a 'good' course).

For effective communication of the results, it is helpful to draw graphs of the predicted probabilities such as that in Figure 19.2, where the probability of positive evaluation ('more yes than no' or 'decidedly yes') is plotted against interest in the subject, for a full-time student responding 'more yes than no' to the questions on exam and previous knowledge. The probabilities are computed for three hypothetical courses defined by fixing the random effect at 0 and $\pm 1.96\sigma_u$. The graph shows that the effect of interest in the subject is weak for good courses, which receive favourable evaluations anyway.

Compared with the null model, the course-level standard deviation is nearly unchanged: this means that in the linear model for the underlying satisfaction (19.5), the reduction of the level 2 variance due to the covariates is similar to the reduction of the level 1 variance (see Section 19.3.4). The course-level variance could be reduced by course-level covariates, such as the subject of the course or features of the teacher. However, the data set does not include course-level covariates, so the regression model can adjust the evaluations for the student characteristics, but it cannot explain why the adjusted evaluations are different among courses.

An effective way to report the course evaluations adjusted for student characteristics is the 'caterpillar' graph in Figure 19.3, where the EB predicted random effects are plotted in ascending order along with 95% confidence intervals based on comparative standard errors

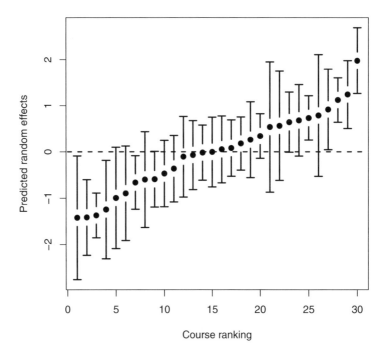

Figure 19.3   Empirical Bayes predictions of random effects with 95% confidence intervals.

(see Section 19.3.9). Each confidence interval has length inversely related to the number of collected ratings for the course and can be used to test whether the random effect of the corresponding course is significantly different from zero, which is the population mean: therefore, a course whose interval is entirely above (below) zero has an adjusted satisfaction significantly higher (lower) than the mean. In this application, it turns out that 10 courses have an adjusted satisfaction significantly different from the population mean (five higher and five lower): such courses should be inspected to establish good and bad practices and to plan interventions for increasing the overall quality.

References

Agresti, A. (2010) *Analysis of Ordinal Categorical Data*, 2nd edn. Hoboken, NJ: John Wiley & Sons, Inc.

Agresti, A. and Natarajan, R. (2001) Modeling clustered ordered categorical data: A survey. *International Statistical Review*, 69, 345–371.

Austin, P.C. (2010) Estimating multilevel logistic regression models when the number of clusters is low: A comparison of different statistical software procedures. *International Journal of Biostatistics*, 6(1), article 16.

Aitkin, M. (1999) A general maximum likelihood analysis of variance components in generalized linear models. *Biometrics*, 55, 117–128.

Bauer, D.J. (2009) A note on comparing the estimates of models for cluster-correlated or longitudinal data with binary or ordinal outcomes. *Psychometrika*, 74, 97–105.

Berkhof, J. and Snijders, T.A.B. (2001) Variance component testing in multilevel models. *Journal of Educational and Behavioral Statistics*, 26, 133–152.

Brant, R. (1990) Assessing proportionality in the proportional odds model for ordinal logistic regression. *Biometrics*, 46, 1171–1178.

Breen, R. and Luijkx, R. (2010) Mixture Models for ordinal data. *Sociological Methods & Research*, 39, 3–24.

Browne, W.J. and Draper, D. (2006) A comparison of Bayesian and likelihood-based methods for fitting multilevel models. *Bayesian Analysis*, 1, 473–550.

Christensen, R.H.B. (2010) *ordinal: Regression models for ordinal data*. Sofware and manual downloadable from http://cran.r-project.org/web/packages/ordinal/index.html

Cox, C. (1995) Location-scale cumulative odds models for ordinal data: A generalized non-linear model approach. *Statistics in Medicine*, 14, 1191–1203.

Fielding, A. (1997) On scoring ordered classifications. *British Journal of Mathematical and Statistical Psychology*, 50, 285–307.

Fielding, A. (2004) Scaling for residual variance components of ordered category responses in generalised linear mixed multilevel models. *Quality and Quantity*, 38, 425–433.

Greene, W.H. (2007) *LIMDEP 9.0, Econometric Modeling Guide*. Plainview, NY: Econometric Software Inc.

Greene, W.H. and Hensher, D.A. (2010). *Modeling Ordered Choices: A Primer*. Cambridge: Cambridge University Press.

Grilli, L. and Rampichini, C. (2002) Specification issues in stratified variance component ordinal response models. *Statistical Modelling*, 2, 251–264.

Grilli, L. and Rampichini, C. (2009) Multilevel models for the evaluation of educational institutions: A review. In P. Monari, M. Bini, D. Piccolo and L. Salmaso (eds), *Statistical Methods for the Evaluation of Educational Services and Quality of Products*, pp. 61–80. Heidelberg: Physica-Verlag.

Hedeker, D. (2008) Multilevel models for ordinal and nominal variables. In J. de Leeuw and E. Meijer (eds), *Handbook of Multilevel Analysis*, pp. 237–274. New York: Springer.

Hedeker, D. and Gibbons, R.D. (1996) MIXOR: A computer program for mixed-effects ordinal regression analysis. *Computer Methods and Programs in Biomedicine*, 49, 157–176.

Hedeker, D. and Gibbons, R.D. (2008) *Supermix – Mixed Effects Models*. Chicago: Scientific Software International.

Hilbe, M.H. (2009) *Logistic Regression Models*. Boca Raton, FL: Chapman & Hall/CRC.

Johnson, T.R. (2003) On the use of heterogeneous thresholds ordinal regression models to account for individual differences in response style. *Psychometrika*, 68, 563–583.

Johnson, V.E. and Albert, J.H. (1999) *Ordinal Data Modeling*. New York: Springer

Joe, H. (2008) Accuracy of Laplace approximation for discrete response mixed models. *Computational Statistics & Data Analysis*, 52, 5066–5074.

Lillard, L.A. and Panis, C.W.A. (2003) *aML Multilevel Multiprocess Statistical Software, Version 2.0*. Los Angeles: EconWare.

Lindsay, B.G., Clogg, C.C. and Grego, J. (1991) Semiparametric estimation in the Rasch model and related exponential response models, including a simple latent class model for item analysis. *Journal of the American Statistical Association*, 86, 96–107.

Lunn, D.J., Thomas, A., Best, N. and Spiegelhalter, D. (2000) WinBUGS, a Bayesian modelling framework: Concepts, structure, and extensibility. *Statistics and Computing*, 10, 325–337.

Marshall, E.C. and Spiegelhalter, D. (2001) Institutional performance. In A.H. Leyland and H. Goldstein (eds), *Multilevel Modelling of Health Statistics*, pp. 127–142. New York: John Wiley & Sons, Inc.

Masters, G.N. (1982) A Rasch model for partial credit scoring. *Psychometrika*, 47, 149–174.

McCullagh, P. (1980) Regression models for ordinal data. *Journal of the Royal Statistical Society, Series B*, 42, 109–142.

McKelvey, R.D. and Zavoina, W. (1975) A statistical model for the analysis of ordinal level dependent variables. *Journal of Mathematical Sociology*, 4, 103–120.

Mealli, F. and Rampichini, C. (1999) Estimating binary multilevel models through indirect inference, *Computational Statistics & Data Analysis*, 29, 313–324.

Moineddin, R., Matheson, F.I. and Glazier, R.H. (2007) A simulation study of sample size for multilevel logistic regression models. *BMC Medical Research Methodology*, 7, 34.

Muthén, B.O. and Kaplan, D. (1985) A comparison of some methodologies for the factor analysis of non-normal Likert variables. *British Journal of Mathematical and Statistical Psychology*, 38, 171–189.

Muthén, B.O. and Satorra, A. (1995) Technical aspects of Muthns LISCOMP approach to estimation of latent variables relations with a comprehensive measurement model. *Psychometrika*, 60, 489–503.

Muthén, L.K. and Muthén, B.O. (2010) *Mplus Users Guide. Sixth Edition*. Los Angeles: Muthén & Muthén.

O'Connell, A.A. (2006) *Logistic Regression Models for Ordinal Response Variables*. Thousand Oaks, CA: Sage.

Olsson, U. (1979) On the robustness of factor analysis against crude classification of the observations. *Multivariate Behavioral Research*, 14, 485–500.

Peterson, B. and Harrell, F.E. (1990) Partial proportional odds models for ordinal response variables. *Applied Statistics*, 39, 205–217.

Pinheiro, J.C. and Bates, D.M. (1995) Approximations to the log-likelihood function in the nonlinear mixed effects model. *Journal of Computational and Graphical Statistics*, 4, 12–35.

Rabe-Hesketh, S. and Skrondal, A. (2008) *Multilevel and Longitudinal Modeling using Stata*, 2nd edn. College Station, TX: Stata Press.

Rabe-Hesketh, S., Pickles, A. and Skrondal, A. (2003) Correcting for covariate measurement error in logistic regression using nonparametric maximum likelihood estimation. *Statistical Modelling*, 3, 215–232.

Rabe-Hesketh, S., Skrondal, A. and Pickles, A. (2005) Maximum likelihood estimation of limited and discrete dependent variable models with nested random effects. *Journal of Econometrics*, 128, 301–323.

Rampichini, C., Grilli, L. and Petrucci, A. (2004) Analysis of university course evaluations: From descriptive measures to multilevel models. *Statistical Methods & Applications*, 13, 357–373.

Rasbash, J., Steele, F., Browne, W.J. and Prosser, B. (2005) *A Users Guide to MLwiN. Version 2.0*. Bristol: Centre for Multilevel Modelling, University of Bristol

Raudenbush, S.W. and Bryk, A.S. (2002) *Hierarchical Linear Models: Applications and Data Analysis Methods*, 2nd edn. Thousand Oaks, CA: Sage.

Raudenbush, S.W., Bryk, A.S., Cheong, Y.F. and Congdon, R. (2004) *HLM 6: Hierarchical Linear and Nonlinear Modeling*. Chicago: Scientific Software International.

Raudenbush, S.W., Yang, M.L. and Yosef, M. (2000) Maximum likelihood for generalized linear models with nested random effects via high-order, multivariate Laplace approximation. *Journal of Computational and Graphical Statistics*, 9, 141–157.

Rodríguez, G. and Goldman, N. (1995) An assessment of estimation procedures for multilevel models with binary responses. *Journal of the Royal Statistical Society, Series A*, 158, 73–89.

Rubin, D.B. (1976) Inference and missing data. *Biometrika*, 63, 581–592.

Samejima, F. (1969) Estimation of latent trait ability using a response pattern of graded scores. *Psychometric Monograph*, 17. Bowling Green, OH: Psychometric Society.

SAS Institute (2009) *SAS/STAT(R) 9.2 User's Guide*, 2nd edn. Cary, NC: SAS Institute.

Skrondal, A. and Rabe-Hesketh, S. (2004) *Generalized Latent Variable Modeling: Multilevel, Longitudinal, and Structural Equation Models*. Boca Raton, FL: Chapman & Hall/CRC.

Skrondal, A. and Rabe-Hesketh, S. (2009) Prediction in multilevel generalized linear models. *Journal of the Royal Statistical Society, Series A*, 172, 659–687.

Snijders, T.A.B. and Berkhof, J. (2008) Diagnostic checks for multilevel models. In J. de Leeuw and E. Meijer (eds), *Handbook of Multilevel Analysis*, pp. 141–175. New York: Springer.

Snijders, T.A.B. and Bosker, R.J. (1999) *Multilevel Analysis. An Introduction to Basic and advanced Multilevel Modeling*. London: Sage.

Spiegelhalter, D.J., Thomas, A., Best, N.G. and Gilks, W. (1997) *BUGS: Bayesian inference Using Gibbs Sampling (Version 0.60)*. Cambridge: Medical Research Council, Biostatistics Unit.

Stram, D.O. and Lee, J.W. (1994) Variance component testing in the longitudinal mixed effects model. *Biometrics*, 50, 1171–1177.

Thissen, D. and Steinberg, L. (1986) A taxonomy of item response models. *Psychometrika*, 51, 567–577.

Verbeke, G. and Molenberghs, G. (2003) The use of score tests for inference on variance components. *Biometrics*, 59, 254–262.

Vermunt, J.K. and Magidson, J. (2005) *Technical Guide for Latent GOLD 4.0: Basic and Advanced*. Belmont, MA: Statistical Innovations Inc.

Winship, C. and Mare, R.D. (1984) Regression models with ordinal variables. *American Sociological Review*, 49, 512–525.

20

Quality standards and control charts applied to customer surveys

Ron S. Kenett, Laura Deldossi and Diego Zappa

This chapter presents an application of methods and standards used in quality management and quality control to the analysis of customer satisfaction surveys. It covers, in detail, the ISO 10004 standard on guidelines for monitoring and measuring customer satisfaction, and other ISO quality standards with relevance to customer satisfaction surveys. Control charts are then introduced and described with the corresponding ISO 7870 guidelines. It discusses how standard control charts (p, c, and u charts) can be used to monitor, over time, the number or proportion of satisfied or unsatisfied customers. Additional sections describe some non-standard analysis for handling items with different importance levels, customer responses as multivariate ordinal variables, respondent patterns, prior information in a Bayesian approach, and dependence structures, when longitudinal panel data is available. Next, it presents an application of the M-test for testing how well the response sample represents the full customer population. The M-test is particularly effective in customer satisfaction surveys conducted over the internet. In such surveys, emails are typically sent to all customers so that these are, in fact, an attempted census with non-response effects. Applying the M-test assesses the representativeness of the responses and helps determine if the analysis requires weighting of the responses.

20.1   Quality standards and customer satisfaction

According to International Organization for Standardization (ISO) standard ISO 10004:2010, customer satisfaction is the 'customer's perception of the degree to which the customer's requirements have been fulfilled'. It is 'determined by the gap between the customer's expectations and the customer's perception of the product as delivered by the organization'.

Modern Analysis of Customer Surveys: with applications using R, First Edition. Edited by Ron S. Kenett and Silvia Salini.
© 2012 John Wiley & Sons, Ltd. Published 2012 by John Wiley & Sons, Ltd.

'ISO is a non-governmental organization that forms a bridge between the public and private sectors. Standards ensure desirable characteristics of products and services such as quality, environmental friendliness, safety, reliability, efficiency and interchangeability – and at an economical cost.' (www.iso.org). ISO's programme of standards ranges from traditional activities such as agriculture and construction, through mechanical engineering, manufacturing and distribution, to transport, medical devices, information and communication technologies, as well as standards for good management practice and for services. Its primary aim is to share concepts, definitions and tools to guarantee that products and services meet expectations. When standards are absent, products may turn out to be of poor quality, might be incompatible with available equipment, unreliable or even dangerous.

The goals and objectives of customer satisfaction surveys are described in ISO 10004 as follows: 'The information obtained from monitoring and measuring customer satisfaction can help identify opportunities for improvement of the organization's strategies, products, processes and characteristics that are valued by customers, and serve the organization's objectives. Such improvements can strengthen customer confidence and result in commercial and other benefits.' This suggests that it is necessary to study the evolution of customer satisfaction indices over time in order to monitor customer loyalty and satisfaction with products and services. We show here that in order to achieve this goal, one can apply tools from industrial statistics and statistical quality control such as control charts. Before describing how control charts are used in analysing customer satisfaction surveys, we begin with a comprehensive description of the ISO 10004 standard.

20.2   ISO 10004 guidelines for monitoring and measuring customer satisfaction

The rationale of the ISO 10004 standard – as reported in Clause 1 – is to provide 'guidance in defining and implementing processes to monitor and measure customer satisfaction'. It is intended for use 'by organizations regardless of type, size or product provided' but it is related only 'to customers external to the organization'.

One common definition of customer satisfaction is the gap between the customer's expectations and perception of a product or service quality (see Zeithaml *et al.*, 1990; Chapter 1, 7 and 14, this volume). Given customer expectations of an organization that delivers a product or service, customer perceptions can be measured by semi-structured or structured surveys. The smaller the gap between expectation and perception, the higher the customer satisfaction. The ISO approach outlines three phases in the processes of measuring and monitoring customer satisfaction: planning (Clause 6); monitoring and measuring (Clause 7); and maintenance and improvement (Clause 8).

The planning phase refers to the definition of the purposes and objectives of measuring customer satisfaction and the determination of 'the frequency of data gathering, which can be on a regular basis, on an occasional basis, or both, as dictated by business needs or specific events' (Clause 6.2). For example, an organization might be interested in investigating reasons for customer complaints after the release of a new product or for the loss of market share. Alternatively it might want to regularly compare its position relative to other organizations. Moreover, 'information regarding customer satisfaction might be obtained indirectly from the organization's internal processes (e.g. customer complaints handling) or from external sources (e.g. reports in the media) . . . [or] . . . directly from customers' (Clause 6.3).

The operation phase represents the core of the standard and introduces the operational steps an organization should follow in order to meet the requirements of ISO 10004. These steps are to:

(a) identify the customers (current or potential) and their expectations;

(b) gather customer satisfaction data directly from customers by a survey and/or indirectly examining existing sources of information, after having identified the main characteristics related to customer satisfaction (product, delivery, or organizational characteristic);

(c) analyse customer satisfaction data after having chosen the appropriate method of analysis;

(d) communicate customer satisfaction information;

(e) monitor customer satisfaction at defined intervals, making sure that 'the customer satisfaction information is consistent with, or validated by, other relevant business performance indicators' (Clause 7.6.5).

The maintenance and improvement phase includes the review, evaluation, and continual improvement of processes for periodically monitoring and measuring customer satisfaction.

Statistical issues mentioned in ISO 10004 relate to the number of customers to be surveyed (sample size), the method of sampling (Clause 7.3.3.3 and Annexes C.3.1 and C3.2), and the choice of measurement scale (Clause 7.3.3.4 and Annex C.4). All these issues are discussed in Chapters 3 and 4 in this book.

Control charts are mentioned in Table D.1 of the ISO 10004 standard as one of the methods for monitoring a customer satisfaction index. They are described in detail in the next sections. For completeness, we review below a list of other ISO standards that are of relevance to customer satisfaction surveys.

ISO 9001:2008, Quality management systems – Requirements. This is the standard in the ISO 9000 family against whose requirements a quality management system can be certified by an external body. The standard recognizes that the term 'product' applies to services, processed material, hardware, and software intended for your customer. There are five sections in the standard that specify activities that need to be considered in implementing a quality system: (1) overall requirements for the quality management system and documentation; (2) management responsibility, focus, policy, planning and objectives; (3) resource management and allocation; (4) product realization and process management; (5) measurement, monitoring, analysis and improvement.

ISO 9004:2009, Managing for the sustained success of an organization – A quality management approach. This standard gives guidance on a range of objectives of a quality management system that is broader than ISO 9001, particularly in managing for the long-term success of an organization. ISO 9004 is recommended as a guide for organizations whose top management wishes to extend the benefits of ISO 9001 in pursuit of systematic and continual improvement of the organization's overall performance. However, it is not intended for certification or contractual purposes.

ISO 10001:2007, Quality management – Customer satisfaction – Guidelines for codes of conduct for organizations. This document provides guidance for planning, designing, developing, implementing, maintaining, and improving customer satisfaction codes of conduct. ISO 10001 applies to product-related codes containing promises made to customers by an organization concerning its behaviour. Such promises and related provisions are aimed at enhanced customer satisfaction. Annex A provides simplified examples of codes of conduct for different types of organizations. It is intended for use by organizations regardless of type, size and product provided. Annex C gives guidance specifically for small businesses.

ISO 10002:2004, Quality management – Customer satisfaction – Guidelines for complaints handling in organizations. This standard provides guidance on the process of complaints handling related to products within an organization, including planning, design, operation, maintenance and improvement. The complaints-handling process described is suitable for use within an overall quality management system. It does not apply to disputes referred for resolution outside the organization or for employment-related disputes.

ISO 10003:2007, Quality management – Customer satisfaction – Guidelines for dispute resolution external to organizations. This standard provides guidance for an organization to plan, design, develop, operate, maintain and improve an effective and efficient dispute-resolution process for complaints that have not been resolved by the organization. It applies to: (1) complaints relating to the organization's products intended for, or required by, customers, the complaints-handling process or dispute-resolution process; (2) resolution of disputes arising from domestic or cross-border business activities, including those arising from electronic commerce.

ISO/TR 10017:2003, Guidance on statistical techniques for ISO 9001:2000. This guide provides guidance on the selection of appropriate statistical techniques that may be useful to an organization in developing, implementing, maintaining and improving a quality management system in compliance with ISO 9001. The document examines requirements of ISO 9001 that involve the use of quantitative data and then identifies and describes statistical techniques that can be useful when applied to such data.

ISO/TR 13425:2006, Guidelines for the selection of statistical methods in standardization and specification. This gives guidance on the selection of all the referenced standards, guides, technical reports and draft international standards (DIS) developed by ISO/TC 69 from a user prospective. It also gives two descriptions of the content of the standards by two sets of abstracts: non-technical abstracts and technical abstracts. Each abstract presents a brief survey of the content of the standard or DIS with some indications of the use of the document in different areas.

ISO 7870-1: 2007 Control Charts – Part 1: General guidelines. This document presents key elements and philosophy of the control chart approach, and lists a wide variety of control charts, including an overview of basic principles and concepts.

Other standards on control charts are:

- ISO/CD 7870-2, Control charts – Part 2: Shewhart control charts;

- ISO/DIS 7870-3, Control charts – Part 3: Acceptance control charts;

- ISO 8258:1991, Shewhart control charts;

- ISO 11462-1:2001, Guidelines for implementation of statistical process control (SPC) – Part 1: Elements of SPC;

- ISO 11462-2:2010, Guidelines for implementation of statistical process control (SPC) – Part 2: Catalogue of tools and techniques.

20.3   Control Charts and ISO 7870

Assessing customer satisfaction with the quality of services or products provides critical information on changes over time. With such information, management can determine which initiatives proved effective and which activities had no impact on customer satisfaction levels. Some organizations only monitor the number of complaints over time. These firms focus their attention on a lagging indicator that provides retrospective information which reflects past decisions. The goal should obviously be proactive and forward-looking, not relying only on a rear-view mirror for driving the organization.

Perceived quality, satisfaction levels and customer complaints can be effectively managed with statistical process control (SPC) techniques. SPC methods were originally developed in the 1920s to improve the quality of products. For a general perspective on the evolution of industrial statistics see Kenett and Zacks (1998, Chapter 1). The ISO 7870-1:2007 standard 'presents key elements and philosophy of the control chart approach, identifies a wide variety of control charts . . . and illustrates the relationship among various control chart approaches to aid in the selection of the most appropriate standard for given circumstances'.

The application of SPC methods generally assumes a process that works continuously or at discrete time intervals, and quality factors that can be clearly identified. Examples of quality factors are process yield, concentration of substances, assembly errors, and percentage of satisfied customers. In order to determine whether a process is under control, data is collected at fixed or random time intervals. A sample that presents local process variability is called a 'rational sample'. Such samples represent variability due to common (or chronic) causes that reflect the process design and normal performance. Summarized data from such samples is tracked over a chart with control limits typically set at three standard deviations above and below the overall mean or central line. When special (or sporadic) causes affect the process and change its normal behaviour, the control chart will signal an alarm. A control chart is therefore a graphical display of a quality characteristic, computed from a sample drawn from the process at different time instances. When the sample size n is equal to 1, we have an *individual control chart* and individual measurements are plotted over time.

Since natural variability due to common causes is always present in a process, we expect some variability on the control chart. Special cause events, sometimes labelled assignable causes, result in excessive variability, beyond the process natural variability. When only common causes of variation are operating, the process is said to be stable or in statistical control (Kenett and Zacks, 1998).

Basic control chart principles are illustrated in Figure 20.1. We used the R library qcc (Scrucca, 2004) to generate the figure with the commands:

```
library(qcc)
set.seed(123)
data <- rnorm(100, mean = .5, sd = .05)
```

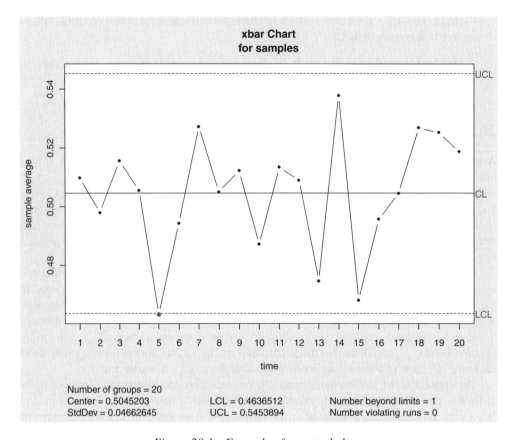

Figure 20.1   Example of a control chart

```
times <- rep(5,20)
samplefreq <- rep(1:20,times)
samples <- qcc.groups(data, samplefreq)
q <- qcc(samples, type="xbar", confidence.level=0.95)
plot(q, xlab="time", ylab="sample average")
```

For details on the R language, see the Appendix of this book and R Development Core Team (2010). The centre line (CL) represents the location of the process, usually an average, determined by a process capability analysis study. The upper and lower control limits (UCL and LCL, respectively) are lines parallel to the centre line. The control limits are positioned so that the quality characteristic being tracked is expected to fall between them with high probability $(1 - \alpha)$, if the process is in control. The UCL and LCL values are computed in accordance to the probability distribution of the statistic used in the chart.

At any given time, the points on the chart show an estimate of the position of the process. Because of natural variability due to uncontrollable random factors (the common causes), we expect the data points to fall between the upper and lower control limits. When this occurs, we are confident that the process is in statistical control and no intervention is necessary. On the other hand, when a point falls outside the control limits, we suspect that some special cause of variation has affected the process and an investigation is required. Sometimes the special cause has a negative effect and a corrective action is required. In other cases, the special cause can have a positive effect and the organization has an opportunity to learn what works well and document it as best practice. If the point is within the control limits, the process is declared to be in control, if the point is outside the control limits, then the process is declared to be out of control, with α representing the probability of a false alert. More sophisticated rules have been designed to improve the performance of a control chart. For more on this, see Kenett and Zacks (1998).

Control charts are generally classified into two groups. If the quality characteristic is measured on a continuous scale, we have a *control chart for variables*. When the quality characteristic is classified as conforming or not conforming on the basis of whether or not it possesses certain attributes, then *control charts for attributes* are used. Whatever the control chart, for a variable or for an attribute, a process under control produces independent successive samples from a stable distribution.

In using control charts for analysing customer satisfaction survey data, we exploit the relationship between control charts and hypothesis testing. At any time, control charts may be thought of as a graphical representation of the following test:

$$\begin{cases} H_0 : \text{process is in control,} \\ H_1 : \text{process is out of control.} \end{cases}$$

When the statistic being tracked is within the control limits, this is equivalent to not rejecting the null hypothesis, H_0, and continuing to monitor the process. Conversely, if a point falls outside the control limits, we have evidence for rejecting the hypothesis that the process is under statistical control. Interpretation of control charts as tests of hypothesis provides a measure of performance in detecting a shift at a given Type I error probability, α (i.e., concluding that the process is out of control when it is not). Computing the operating characteristic (OC) curve of the control chart, we can see how the probability of a Type II error, β (i.e., concluding that the process is in control when it is not), varies for different parameters under H_1 (see Kenett and Zacks, 1998). It should be noted that, unlike the use of hypothesis testing for decision making, control charts are used for process monitoring. In this context, an out-of-control indication typically triggers a corrective action to readjust the process to its original condition.

In the case of customer satisfaction surveys, we can use control charts to identify a shift from previous surveys or investigate the achievement of preset targets. In general, we test the hypothesis

$$\begin{cases} H_0 : \theta = \theta_0 \\ H_1 : \theta \neq \theta_0 \end{cases}$$

where θ can be the mean, the standard error, or a proportion, depending on the particular kind and scope of the control chart (i.e., for variables or for attributes).

All the above details also hold when we are interested in testing a specific shift of the parameter such as $\theta > \theta_0$ or $\theta < \theta_0$. In these cases, only one control limit, either UCL or LCL, is reported on the control chart.

20.4   Control charts and customer surveys: Standard assumptions

20.4.1   Introduction

Principles and instruments of SPC can be effectively implemented in the analysis of customer satisfaction survey data. For example, control charts can be used to monitor the temporal evolution of satisfaction with respect to a product or service. In such a context, given the time interval between samples (typically a week, a month or a year), the purpose of control charts is to distinguish between random variation due to common causes and significant changes due to assignable causes. In such applications, it is natural to use control charts for attributes tracking the percentage of positive or negative responses. For an application of a continuous variable control chart to customer satisfaction surveys, see Spiring (2008).

Customer satisfaction surveys, such as the ABC questionnaire presented in Chapter 2, usually include:

(1) lists of demographic or contextual alternatives. For example, respondents can be asked to mark areas where they experienced a problem.

(2) questions (items) related to specific aspects of a product or service. Respondents are typically asked to declare their degree of satisfaction on an ordinal scale from 1 ('very low') to 5 ('very high').

(3) general questions related to the overall satisfaction, intention to repurchase or intent to recommend to others.

Referring to the ABC survey, customers were asked to express, for a variety of topics, their level of satisfaction or degree of agreement with detailed statements on a five-point scale. Questions 4 and 5 in the questionnaire are related to loyalty and repurchasing intentions (see Chapter 2).

In a longitudinal panel, the dependence in the evaluation given by the same customer over time can be taken into account with appropriate control chart for autocorrelated data. The ABC survey data covers only one year and therefore does not allow us to monitor responses over time. The modelling of dependence in the data is discussed in Section 20.5.6.

20.4.2   Standard control charts

The standard control charts for attributes are the p chart, the c chart, and the u chart.

The p chart is used, for example, to monitor the percentage of respondents who answered '5' to a question on overall satisfaction. In general, the quality index monitored at time t is $p_t = D_t/n_t$, where n_t is the sample size and D_t is the number of respondents who gave a specific response. The distribution of D_t is typically hypergeometric; for large population, this can be approximated by the binomial distribution. Then $D_t \sim \text{Bin}(n_t, \pi)$, so that $E(D_t) = n_t \pi$ and $\text{var}(D_t) = n_t \pi (1 - \pi)$. The use of p_t provides a direct interpretation of the index with $E(p_t) = \pi$ and $\text{var}(p_t) = \pi(1 - \pi)/n_t$ with the advantage that the expected value of the quality index will not vary when different sample sizes are used over time.

The c chart is used to monitor the number of specific responses to a questionnaire from a given sample of customers. The quality index is $C_t = d_{1,t} + d_{2,t} + \ldots + d_{n,t}$ where $d_{i,t}$ is the number of specific responses given by subject i to the items of the questionnaire used for the survey run at time t. C_t is generally assumed to be Poisson distributed, $C_t \sim P(\lambda)$. A property of the Poisson distribution is that the mean and the variance are the same, $E(C_t) = \mathrm{var}(C_t) = \lambda$. This implies that the variance increases with the mean. The latter condition is observed when the data is collected under homogeneous conditions and no mixtures of Poisson distributions are involved. This assumption is often verified for industrial processes. When dealing with survey respondents, proper stratification is necessary to guarantee that the sample of customers is 'homogeneous' (i.e., they behave in almost the same way). If the latter assumption is violated, it may happen that the variance is larger than the mean. This gives rise to overdispersion, which can be modelled by distributions different from the Poisson such as the negative binomial or COM-Poisson. For further information, refer to Sellers and Shmueli (2010), Kushler and Radka (2010), Kenett and Zacks (1998) or Montgomery (2005).

The u chart is used to monitor the number of specific responses to a questionnaire, when the counts are adjusted for the size of the survey sample size. In this case, the quality index is $U_t = C_t / n_t$ and most of the properties for C_t also apply to U_t. The main difference is that $E(U_t) = \lambda / n_t = u_t$ and $\mathrm{var}(U_t) = u_t / n_t$ and thus they depend on sample sizes used over time.

To plot control charts, we need to set the centre line and control limits. In customer satisfaction surveys, target values for such values can be set by management. Control charts can then be used to test whether the targets are reached by testing a null hypothesis based on the target value ($H_0 : \pi = \pi_0$ or $\lambda = \lambda_0$).

The parameters of control charts are estimated in a pre-sampling phase known in statistical quality control as phase I. Suppose we collected m preliminary samples of *rational subgroups*, that is, samples from homogenous groups. The control charts parameters are estimated as follows. For the p chart,

$$\bar{p} = \frac{\sum_1^m d_t}{\sum_1^m n_t};$$

if $n_t = n$ for all t, this reduces to

$$\bar{p} = \frac{\sum_1^m d_t}{m \cdot n}.$$

For the c chart,

$$\bar{c} = \frac{\sum_1^m c_t}{m}.$$

For the u chart,

$$\bar{u} = \frac{\sum_1^m c_t}{\sum_1^m n_t};$$

if $n_t = n$ for all t, this reduces to

$$\bar{u} = \frac{\sum_1^m c_t}{m \cdot n}.$$

In spite of the asymmetry that characterizes the distributions of the random variables used to design the p, c and u charts, the control limits (UCL and LCL) are commonly set equidistantly from the centre line (CL) using the normal distribution approximation. The control limits are given by:

$$\bar{p} \pm k\sqrt{\bar{p}(1 - \bar{p})/n_t} \qquad (20.1)$$

for the p chart;

$$\bar{c} \pm k\sqrt{\bar{c}} \qquad (20.2)$$

for the c chart; and

$$\bar{u} \pm k\sqrt{\bar{u}/n_t} \qquad (20.3)$$

for the u chart. In (20.1)–(20.3) k is coefficient representing a percentile of the distribution of the centre line estimator. The typical justification for this relies on the central limit theorem or theorems related to maximum likelihood estimation. Then, in general, assuming that the estimators of π and λ are asymptotically normally distributed, the customary choice of $k = 3$ ensures a probability coverage of $1 - \alpha = 0.9973$. Software such as Minitab or JMP performs exact calculations for small sample sizes.

Data from the m preliminary samples is plotted on the control chart. If some points are outside the control limits and an assignable cause is determined, the out-of-control data is removed and new control limits are calculated.

If the parameter is known from a benchmark or best practice study, control limits for the above charts are obtained by the same formulas, substituting for the average values of p, c, and u the corresponding standard value p_0, c_0, and u_0.

The control limits depend on the sample size so that if $n_t \neq n$ for some t, the control lines will vary over t. The more respondents, the narrower the control limits.

Example 1

Consider the ABC questionnaire presented in Chapter 2 and focus on customers' overall satisfaction with ABC as measured by responses to question 1. In this example, we design a control chart for monitoring the proportion of customers expressing a low or very low level of satisfaction (i.e., responding '1' or '2').

The ABC annual customer satisfaction survey data is available only for one year, so that we cannot perform a trend analysis investigating changes in the proportion of low-satisfaction customers over time. Consider the single ($m = 1$) sample of 266 respondents. We observe that the proportion of customers whose overall satisfaction with ABC is very low or low (labelled 'BOT1+2') is $36/266 = \bar{p} = 0.135$. Following (20.1), the control limits for the implementation of a p chart are

$$(\text{UCL, LCL}) = 0.135 \pm 3\sqrt{0.135 \cdot (1 - 0.135)/266} = 0.135 \pm 0.063,$$

so that LCL $= 0.072$ and UCL $= 0.198$. Now, assume that on the basis of 10 monthly surveys ($t = 1, 2, \ldots, 10$) with a fixed sample size of $n = 266$ monthly respondents, the estimated proportion of unsatisfied customers is reported in Table 20.1.

Table 20.1 Proportion of unsatisfied customers, by month, over 10 months, with $n = 266$ every month

Month	1	2	3	4	5	6	7	8	9	10
BOT1+2	0.116	0.080	0.149	0.166	0.129	0.153	0.092	0.110	0.156	0.219

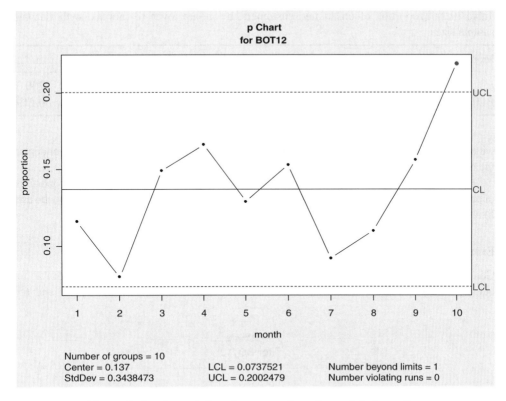

Number of groups = 10
Center = 0.137   LCL = 0.0737521   Number beyond limits = 1
StdDev = 0.3438473   UCL = 0.2002479   Number violating runs = 0

Figure 20.2   Control chart for proportion of unsatisfied suppliers

Plot $\hat{p}_t$, using the qcc library in R and the following commands:

```
n <- rep(266,10)
proportion <- c(0.116, 0.080, 0.149, 0.166, 0.129, 0.153,
0.092, 0.110, 0.156, 0.219)
BOT12 <- proportion*n
q <- qcc(BOT12, sizes=n, type="p",plot=FALSE, nsigmas=3)
plot(q, xlab="month", ylab="proportion")
```

The R code above produces the control chart presented in Figure 20.2 where all points are located within the control limits, except for point 10 that is above the upper control limit. Since the proportion of BOT1+2 is greater than 0.198, we conclude that, after 10 months, the proportion of suppliers not satisfied with ABC has increased to a new level.

If the number of respondents to surveys in subsequent months is not constant, as in Table 20.2, the control lines vary over time. Applying the R commands

```
n <- c(266,266,266,320,320,320,320,200,200,200)
fraction <- c(0.116, 0.080, 0.149, 0.166, 0.129, 0.153, 0.092,
0.110, 0.156, 0.219)
BOT12 <- fraction*n
q <- qcc(BOT12, sizes=n, type="p",plot=FALSE,nsigmas=3)
plot(q, xlab="month", ylab="proportion")
```

Table 20.2   Proportion of unsatisfied customers, by month, over 10 months, with varying sample sizes

Month	1	2	3	4	5	6	7	8	9	10
n_t	266	266	266	320	320	320	320	200	200	200
BOT1+2	0.116	0.080	0.149	0.166	0.129	0.153	0.092	0.110	0.156	0.219

we obtain the chart in Figure 20.3. As expected from the mathematics, the more respondents, the narrower the control limits.

Following a similar procedure, we can produce a p control chart to monitor the proportion of customers with repurchasing intentions (question 5) or the proportion of customers who are likely to recommend ABC to others (question 4).

Example 2

Consider again the ABC questionnaire and focus on customers responding '1' or '2' to questions 6–66 with the exception of items 11, 17, 25, 31, 38, 42, 43, 49, 57, and 65

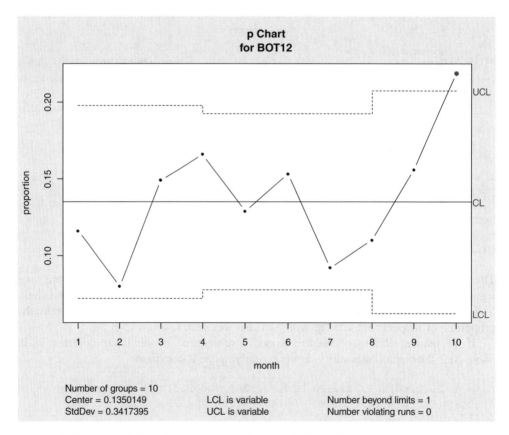

Figure 20.3   Control chart for proportion of unsatisfied suppliers with variable sample size

(measuring overall satisfaction) and questions 39, 40 and 44 (yes/no responses). For these 48 items, the average *number* of '1' or '2' responses, for the 266 respondents, is equal to 6.75. To construct a *u* chart we use the control limits described in (20.3), i.e. $6.75 \pm 3\sqrt{6.75/266} = 6.75 \pm 0.48$, so that LCL $= 6.27$ and UCL $= 7.23$.

Suppose we wish to test whether 6.75 is consistent with an average number of $6.48 = 0.135 \cdot 48$, obtained assuming that the proportion $\bar{p} = 0.135$ of low-satisfaction respondents to question 1 (i.e. customers who responded '1' or '2' to the overall satisfaction question) responded '1' or '2' also in all the previously mentioned 48 questions. Is 6.75 statistically different from 6.48? Control charts can help answer this question.

The use of control limits from a *u* chart, in this context, is equivalent to performing a hypothesis test (see Section 20.3). The control limits are $6.48 \pm 3\sqrt{6.48/266} = 6.48 \pm 0.47$, so that LCL $= 6.01$ and UCL $= 6.95$. Since 6.75 is within the LCL and UCL control limits, we conclude that low satisfaction in the 48 questions is not statistically different from the low overall satisfaction. This consistency between low satisfaction levels in the individual questions (items) and low overall satisfaction indicates that the questionnaire adequately covered the areas of concern of the customers.

If one computes control limits from the Poisson or binomial distribution, and not from a normal approximation, one can get non-equidistant control limits with respect to the centreline. This can be justified in several cases:

- When we monitor negative responses such as BOT1+2, there is a difference between 'out of control' signals above the upper control limit or below the lower limit. From a practical perspective, points above the upper control limit indicate trouble, while points below the lower control limit indicate improvement. Thus, it makes sense to consider the two directions separately, and possibly use asymmetric control limits. Similar considerations apply to TOP5, the number of responses with very high satisfaction level.

- In customer satisfaction surveys, the Poisson and binomial distributions are more appropriate than the normal distribution approximations commonly used in setting up control limits. Apart from some specific values of the parameter space, these distributions are typically asymmetric. A more exact, but computationally demanding approach, is to set exact control limits, e.g., at the 0.135th and 99.865th percentiles, which will turn out to be asymmetric with respect to CL. In general one can set

$$LCL = F^{-1}(\alpha/2) \quad \text{and} \quad UCL = F^{-1}(1 - \alpha/2),$$

where $F^{-1}(\cdot)$ is the inverse of the cumulative distribution function for the Poisson or binomial distribution and α is a percentile level. Moderate computational issues may arise because these distributions are discrete and exact solution may not exist. As mentioned, this is how programs such JMP or Minitab set up control limits, when computationally possible. Another approach is to apply Monte Carlo simulation. This requires four major steps:

0. Define $\bar{p}$ as the probability of a negative response (BOT1+2) to an item within a sample of n_t respondents.

1. Generate, by simulation, m sets of n_t random numbers from $D_t \sim \text{Bin}(1, \bar{p})$.

2. For each set, compute a pseudo-estimate of $\bar{p}$.

3. Using the cumulative distribution function obtained by the m pseudo-estimates of $\bar{p}$, find the $\alpha/2$ and $(1 - \alpha/2)$ percentiles.

For very large m the solutions in step 3 are consistent estimates of the percentiles of the estimator of $\bar{p}$. Yet another approach is to bootstrap the data in order to derive empirical bootstrapped control limits. Analytical approximations to the exact solutions and bootstrapping solutions are provided in Kenett and Zacks (1998).

- If standard control limits are used, it may happen that the limits are outside the parametric domain and that $0 < \bar{p} \pm k\sqrt{\bar{p}(1 - \bar{p})/n_t} < 1$ is not satisfied. To avoid this, if $\bar{p}$ is known from a preliminary survey, we need a large enough sample size so that $n_t > \frac{1-\bar{p}}{\bar{p}}k^2$. This suggests that the smaller the value of $\bar{p}$, the larger n_t must be. Analogous procedures can be applied to the c and u charts.

The c and u charts are known also as Shewhart control charts for counts. Their major drawback is that they are not sensitive to small parameter shifts. For this reason, more sensitive charts such as the cumulative sum (CUSUM) control charts (see Section 20.5.3), Shiryayev–Roberts charts or exponentially weighted moving average (EWMA) charts are often recommended to monitor a process with Poisson counts. For more on this topic, see Kenett and Pollak (1996), Kenett and Zacks (1998) and Ryan and Woodall (2010).

20.5   Control charts and customer surveys: Non-standard methods

The previous sections have presented straightforward applications of control charts. The literature has also focused on aspects of customer surveys that require techniques based on non-standard assumptions. In this section, we present some advanced methods that have proved effective in customer satisfaction survey data analysis. We discuss how to weight counts, to monitor responses to items with k different categories, to cumulate the proportion of negative responses (BOT1+2), to prepare a control chart for several items, to apply Bayesian methods, and to consider time-dependent data.

20.5.1   Weights on counts: Another application of the c chart

It is fairly common to consider each item within the questionnaire as *a priori* equally important. This supports the application of the traditional c chart. From a different point of view, it is not unusual to consider negative responses to specific items as more important than negative responses to generic items. To deal with this assumption, suppose the questionnaire is structured with l different items or l different groups of items. The overall number of negative responses, given by subject i at time t, can be computed by weighting each item or group of items according to the relevance that an item receives within the questionnaire. In the ABC questionnaire, weights are chosen by the respondents on the basis of a pre-specified importance scale from 1 for 'low importance' to 3 for 'high importance'.

Let $_wd_{i,t} = l(w_1q_{i,t,1} + w_2q_{i,t,2} + \ldots + w_lq_{i,t,l})$ be the weighted number of negative responses, where $\sum w_j = 1$ and $q_{i,t,j}$, for $j = 1, 2, \ldots, l$, are the negative responses to each item or group of items. Consider a sample of n respondents. The statistic to be used is $_w c_t = \sum_{i=1}^{n} w d_{i,t}$. If $w_j = 1/l$, for all j, we have the same statistic used in the standard c chart. Assuming that at time t, $\sum_{i=1}^{n} q_{i,t,j} = Q_j \sim P(\lambda_j)$, using m preliminary samples, λ_j can be estimated by $\bar{q}_j = \sum_{i=1}^{n} \sum_{t=1}^{m} q_{i,t,j}/m$. The centre line of this control chart is $_w\bar{c} = l \sum_{j=1}^{l} w_j \bar{q}_j$ and the variance is $l^2 \sum_{j=1}^{l} w_j^2 \bar{q}_j$.

A similar approach can be applied to reflect the different degree of satisfaction in answers to the same item. This may happen, for example, when we wish to balance differently a very negative evaluation as opposed to a mild one.

20.5.2   The χ^2 chart

In Section 20.3 we show the connection between control charts, tests of hypothesis, and the performance of p charts in monitoring dichotomous categories. In many cases items are structured so that respondents are allowed to choose one category from k alternatives. If one is interested in taking into account the joint proportions of respondents in the different categories, the multinomial chart should be considered.

By analogy with the p chart, we start by assuming that the 'in control' probabilities, corresponding to the K categories, are either known or estimated from a preliminary sample. Let $\pi_{1j}, \pi_{2j}, \ldots, \pi_{Kj}$ be the 'target' values for item j, n_j the sample size of respondents to item j and $n_{1j}, n_{2j}, \ldots, n_{Kj}$ the number of respondents in each category, so that $\sum_{i=1}^{K} n_{ij} = n_j$. A statistic that measures the distance of $n_{1j}, n_{2j}, \ldots, n_{Kj}$ from $\pi_{1j}, \pi_{2j}, \ldots, \pi_{Kj}$ is Pearson's goodness-of-fit statistic

$$\chi_j^2 = \sum_{i=1}^{K} \frac{(n_{ij} - n_j \pi_{ij})^2}{n_j \pi_{ij}}. \tag{20.4}$$

If n_j is sufficiently large and the process is in control, (20.4) is approximately chi-square distributed with $K - 1$ degrees of freedom. The chart will monitor (20.4) by checking whether the observed vector value is not larger than UCL $= \chi_{K-1,1-\alpha}^2$, the $1 - \alpha$ percentile of a χ_{K-1}^2 distribution.

A drawback of the multinomial-based control chart is that it signals when the current sample significantly differs from the in-control probabilities for the K categories, but does not indicate which categories are causing the disturbance. For a fixed total, counts are negatively correlated: as one count increases, the sum of the others decreases. The multinomial chart is therefore useful when increments are in all but one category (see Marcucci, 1985). Duran and Albin (2009) present a method, called the p-tree method, that offers a way to interpret an out-of-control signal, indicating which categories are causing the problem. The authors show that their method has a performance similar to Marcucci (1985). Both the p-tree and multinomial chart approach monitor either nominal or ordinal categorical data. A method for ordinal categorical data is proposed by Tucker *et al.* (2002) whose sensitivity is better than Marcucci's method when the underlying continuous distribution of the ordinal characteristic is supposed to be known. Fuchs and Kenett (1980) propose a method to identify changes in specific cells in the multinomial case. For a general treatment of multivariate quality control in the continuous case, see Fuchs and Kenett (1998).

The multinomial chart is derived within a much broader methodological framework that makes use of the likelihood ratio statistic to discriminate between hypotheses on the proportion of responses. Samimi *et al.* (2010) discuss how to monitor customer loyalty by assuming independence from or dependence on customer satisfaction. In the first case they suggest the application of standard control charts, and in the second case they treat loyalty as an ordinal random variable. Using ordinal logistic regression to link loyalty to customer satisfaction, they show, by an example, the efficacy of the likelihood ratio statistic $-2\ln(L_{\text{reduced}}/L_{\text{full}})$, where L_{reduced} and L_{full} are the likelihoods without and after fitting the ordinal logistic model, respectively. They conclude that the likelihood ratio statistic is informative in monitoring customer loyalty.

20.5.3 Sequential probability ratio tests

Special sequential methods can be used for effectively monitoring changes over time in the proportion of 'nonconformities' (responses) to an item administered in periodical surveys. These methods can improve the performance of the standard p chart. The basic sequential method is the sequential probability ratio test (SPRT) or its alternative representation by CUSUM charts (see Wald, 1947; Kenett and Zacks, 1998; Montgomery, 2005). For the many optimality properties of SPRT and CUSUM statistics, see Siegmund (1985).

Exploiting the duality of control charts and hypothesis tests, suppose you are interested in testing

$$\begin{cases} H_0 : \pi = \pi_0, \\ H_1 : \pi = \pi_1. \end{cases}$$

At time $j = 1, 2, \ldots, t$ a survey is run reporting the estimate $p_j = d_j/n$. For the sake of simplicity, we suppose that $n_j = n$ for all j (i.e., the sample size does not change in time).

Let $D_j \sim \text{Bin}(n, \pi)$. Applying the Neyman–Pearson lemma, at time j we accept H_0 if the probability ratio test

$$\frac{l_{H_1,j}}{l_{H_0,j}} \equiv \frac{l_{1,j}}{l_{0,j}} \geq C_\alpha,$$

where $l_{i,j}$ is the likelihood of the sample under H_i ($i=0,1$), and C_α is a critical value such that the Type I error probability is α. The test is said to be *sequential* if, at time j, a rule is defined such that either H_0 or H_1 is accepted or an additional sample must be extracted because not enough information is available to discriminate between H_0 or H_1 and the decision is postponed to step $j+1$ (called the continuation step).

Suppose that samples are gathered up to time t and that between-sample (survey) dependence does not exist. The sequential probability ratio test statistic is

$$\frac{l_{1,t}}{l_{0,t}} = \prod_{j=1}^{t} \frac{l_{1,j}}{l_{0,j}}.$$

Let

$$A = \frac{\beta}{1-\alpha} \quad \text{and} \quad B = \frac{1-\beta}{\alpha},$$

where α and β are the Type I and Type II error probabilities. Using Wald's theorems on sequential tests it has been shown that at time t the decision will be

$$\begin{cases} H_0 & \text{is accepted if } S(p_j) \leq a_0 + \delta t, \\ H_1 & \text{is accepted if } S(p_j) \geq a_1 + \delta t, \\ & \text{otherwise continue} \end{cases}$$

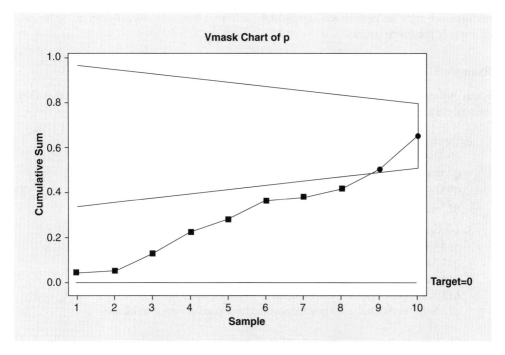

Figure 20.4   V-mask for data in Example 20.1

where

$$S(p_j) = \sum \frac{d_j}{n}.$$

From a practitioner point of view, the efficacy of SPRT is evident if one considers that the two decisions are graphically displayed by two parallel straight lines. When $H_1 : \pi \neq \pi_0$ a much more efficient chart is the CUSUM chart with the so-called V-mask. Using data in Table 20.1 and Minitab^{TM} version 16 we derive the CUSUM chart presented in Figure 20.4.

A process in control corresponds to points within the (V-mask) region whose boundaries are the straight lines. Once a sample is available, the V-mask is positioned so that the sample estimate is at the centre of the vertical line of the V-mask. When the cumulative statistic crosses either the upper or lower line an out-of-control signal is declared and we may also estimate at what time the process has started to be out of control. In the example, we register a signal at time 10 but the out-of-control condition started at 9. The graphical representation is useful for monitoring the direction of the cumulative average value of the proportion of satisfied or unsatisfied customers.

20.5.4   Control chart over items: A non-standard application of SPC methods

It is not so unusual to find control charts monitoring the proportion of top (or bottom) evaluation ratings over items in order to quickly highlight the characteristic of the product/service whose performance may be considered significantly different from the overall average value. An example is presented below.

Example 3

Focus on the questions on equipment and sales support in the ABC questionnaire. The questions we analyse are:

Equipment
    q6 The equipment's features and capabilities meet your needs.
    q7 Improvements and upgrades provide value.
    q8 Output quality meets or exceeds expectations.
    q9 Uptime is acceptable.

Sales Support
    q12 Verbal promises have been honoured.
    q13 Sales personnel communicate frequently enough with you.
    q14 Sales personnel respond promptly to requests.
    q15 Sales personnel are knowledgeable about equipment.
    q16 Sales personnel are knowledgeable about market opportunities.

For each question, we compute the number of responses with very high satisfaction level (5) which we label 'TOP5'. Table 20.3 presents this data.

A p chart of the TOP5 statistics for the equipment questions, computed with Minitab™ version 16, is presented in Figure 20.5. The chart shows an average TOP5 proportion of 14.4%. Because of the small number of questions, the UCL and LCL were positioned at 2 standard deviations above and below the average (CL). Question 9 on 'uptime' is showing up with a TOP5 proportion significantly higher than the average, indicating that 'uptime' stands out as an area of excellence from the customer's point of view.

Figure 20.6 shows both the TOP5 proportions for the equipment and sales support questions. The sales support average TOP5 proportion is 18.2% with question 14, relative to the prompt response by sales personnel, significantly high. For more on this approach to customer survey data analysis, see Kenett and Zacks (1998) and Kenett (2004).

Table 20.3   TOP5 responses in ABC survey

Question	Topic	n	TOP5
6	Equipment	262	32
7	Equipment	262	40
8	Equipment	263	30
9	Equipment	260	49
12	Support	261	39
13	Support	261	50
14	Support	261	60
15	Support	260	43
16	Support	259	45

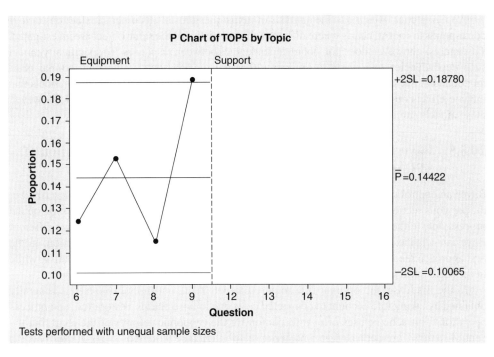

Figure 20.5   p chart of proportion of TOP5 ratings in questions on equipment

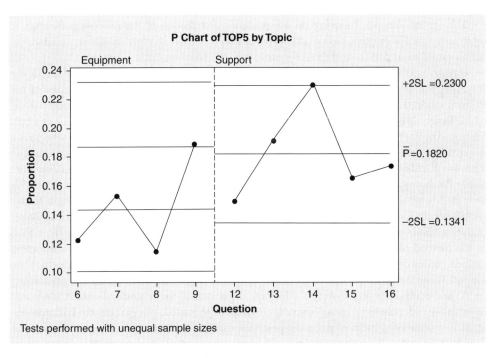

Figure 20.6   p chart of proportion of TOP5 ratings in questions on equipment and support

Applying control charts to the analysis of items/questions in customer satisfaction surveys corresponds to control charts where different characteristics of the same object are investigated. This analysis, however, does not model dependencies between responses. To specifically model such dependencies, we needs to apply multivariate models such as the multinomial chart mentioned in Section 20.5.2, log-linear models presented in Chapter 12, or Rasch models that include effects due to the responder and to the items as discussed in Chapter 14. However, these models can be limited by computational constraints.

20.5.5   Bayesian control chart for attributes: A modern application of SPC methods

Bayesian control charts differ from the classic approach for process monitoring described in the previous sections. In this context, Bayes' theorem is used to update the information on the process state (either 'in control' or 'out of control'). Following Colosimo (2007), two different Bayesian schemes can be adopted, depending upon the use of the posterior function. In the first approach the focus is on the posterior probability that the process is out of control, while in the second the posterior distribution of the process parameters is considered.

In the first case, the optimal rule required to detect the out-of-control state is generally obtained by taking into account the expected costs due to false alarms, restoration, and process operation. This scheme uses prior information for the economic settings of the chart, but the control statistics are not based on a posterior distribution (see Stoumbos et al., 2000). There is a large literature on economic models of this kind, but they are not suitable for the customer satisfaction survey framework, and we do not consider them either in the traditional or in the Bayesian approach.

The second scheme, focusing on the posterior distribution of the process parameter, is more attractive for customer satisfaction surveys. In this context a few results are available in the literature related to control chart for attributes. We mention three proposals.

Bayarri and Garcia-Donato (2005) propose an alternative to the usual u chart in the Bayesian framework. First of all, the authors overcome the restrictive assumptions of the Poisson model dealing with overdispersion (or extra variance) typically present in the context of customer satisfaction survey due to non-homogeneity of the sample of customers (see Section 20.4.2). They introduce a hierarchical model in which the rates λ of people who gave a negative evaluation in the questionnaire are allowed to vary from survey to survey according to a gamma distribution, $Ga(\alpha, \beta)$. The final discrete distribution, known in the Bayesian literature as the Poisson–gamma distribution, reduces to the negative binomial for integer values of α and, in this sense, may be considered an extension of the proposal of Kushler and Radka (2010) recalled in Section 20.4.2. Bayarri and Garcia-Donato exploit the empirical Bayes (naïve method) approach to estimate the hyperparameters (α, β) by maximum likelihood or the method of moments, and they consequently obtain the control limits by computing the percentiles of the resulting Poisson–gamma distribution by Monte Carlo methods. Then the full Bayesian approach is implemented assuming a prior distribution on the parameters (α, β) and using both objective (no information external to the monitored data is used) and informative (information coming from expert knowledge is used) priors. This kind of approach leads to a sequential method since the posterior distribution at time $t-1$ becomes the prior distribution for the observations at time t. The inclusion of prior information has some relevant advantages and typically results in control limits that are narrower than those obtained when no previous information is incorporated. Furthermore, the last sequential approach does not need the implementation of a phase I control to set up the chart, so representing a positive point in the customer survey context.

The extension of the multinomial charts (see Section 20.5.2) in the Bayesian framework is proposed by Laviolette (1995) and Shiau *et al.* (2005). They add the assumption that the probability parameters of the multinomial model vary according to a prior distribution, typically the Dirichlet distribution because of its flexibility in modelling very different assumptions on the hyperparameter and for computational ease in using conjugate families. In particular, Shiau *et al.* (2005) assume that the Dirichlet prior distribution has unknown hyperparameters and they utilize the process data to estimate them (empirical Bayes method). A pseudo-maximum likelihood and the method of moment estimators are considered and compared. It is shown that the latter performs slightly better than the former.

The same distributional assumptions on the outcome (multinomial random variable) and on the prior distribution (Dirichlet random variable) have also been explored in the Bayesian network framework (see Chapter 11) by Colosimo and Semeraro (2002). They propose a control chart to assess the overall satisfaction item and to find the nodes in the network more responsible for the overall out-of-control signal, if it was observed.

20.5.6   Control chart for correlated Poisson counts: When things become fairly complicated

A basic assumption usually made in constructing a control chart is that the process behaves independently in time. For a discussion of cases when this assumption is not met, see Psarakis and Papaleonida (2007). When the same customers respond repeatedly to a survey, we have a repeated measure condition. In the case of continuous variables, autocorrelation in time series can be modelled by autoregressive integrated moving average (ARIMA) models. A similar approach has been explored also for count processes. Weiss (2007) showed that these counts can be modelled by a member of the integer-valued ARMA family (INARMA). In particular, assuming that the process of counts is stationary INAR(1) with Poisson (Weiss, 2007, 2009b, 2009d, 2010; Weiss and Testik, 2009) or binomial (Weiss, 2009a, 2009c; Weiss and Atzmüller, 2010) distributed marginals in the state of statistical control, the authors propose a number of charts (Shewhart, CUSUM and EWMA) to monitor a process over time and investigate the ability of the charts to detect several types of out-of-control situations. No general results on best performance have been obtained since, depending on the type of out-of-control situation, specific charts perform better.

20.6   The *M*-test for assessing sample representation

Once questionnaires have been returned, an analysis of possible bias in response distribution is necessary. Such bias can be evaluated using the *M*-test (Fuchs and Kenett, 1980; Kenett, 1991; Kenett and Zacks, 1998). In simple terms, the *M*-test compares the number of expected returns, by demographic strata, conditioned on the total number of returns. By standardizing the difference between actual and expected returns and using a Bonferroni bound, we can determine whether there are strata with under- (or over-)representation. If bias is identified, a follow-up analysis is carried out to determine whether there are significant differences in satisfaction levels between the various strata with a bias in returns. If such differences are demonstrated, weighted estimators are used; if not, unweighted estimators are computed (for a related discussion; see Chapter 3).

The application of the M-test is a prime example of the *plan–do–check–act* (PDCA) cycle originally developed by Shewhart at Bell Laboratories who proposed in 1924 the use of control charts to monitor production processes (Kenett and Zacks, 1998). Later, the PDCA cycle was made famous by Deming, who used it as a key component of quality improvement initiatives (Kenett and Zacks, 1998).

The M-test is designed to check whether the responses to a customer satisfaction survey match the underlying distribution of the customers being surveyed.

Assume our customers are classified into K categories (strata) with p_i the proportion of customers in category i, for $i = 1, \ldots, K$. The K categories can represent customers from different geographical areas, customers of various types of products or customers of various ranks representing their strategic value to the company.

In the *plan* phase we derive expected returns by category given no bias, and perform the following steps:

1. Compute the expected number of returns in category i, $E_i = Np_i$, $i = 1, \ldots, K$.

2. Compute, for each category i, $S_i = \sqrt{Np_i(1 - p_i)}$, $i = 1, \ldots, K$.

3. Compute the adjusted residuals for each category i, $Z_i = (n_i - E_i)/S_i$, $i = 1, \ldots, K$.

In the *do* phase we compare actual returns to expected values with the following step:

4. Determine the critical value C of Z_i from the M-test table (Table 20.4).

The *check* and *act* phases involve comparing Z_i to C and determining follow-up actions accordingly:

5a. If all adjusted residuals Z_i are smaller, in absolute value, than C, no significant bias is declared and we can proceed to analyse the return data without weighting for bias correction.

5b. If some adjusted residuals Z_i are larger, in absolute value, than C, a potential significant bias is declared and we proceed to analyse differences between categories in key

Table 20.4   Critical values for M-test with K categories at 10%, 5% and 1% significance levels

K	10%	5%	1%
4	1.95	2.24	2.81
5	2.05	2.32	2.88
6	2.12	2.39	2.93
7	2.18	2.44	2.99
8	2.23	2.49	3.04
9	2.28	2.53	3.07
10	2.32	2.57	3.1
20	2.57	2.81	3.3
30	2.71	2.94	3.46

Table 20.5 *M*-test of returns for ABC data

Region	Population	P_i	Expected	Returns	Z	Significance
Benelux	64	0.109215	29	26	−0.58973	OK
France	61	0.104096	28	15	−2.61009	5% significance
Germany	215	0.366894	98	112	−1.78106	OK
Israel	73	0.124573	33	23	−1.85668	OK
Italy	78	0.133106	35	39	0.722001	OK
UK	95	0.162116	43	51	1.330896	OK
Total	586	1		266		

questions using the return data. If we find no significant differences we proceed without weighting for bias correction; alternatively, we weight responses by number of returns to derive overall performance measures.

Table 20.5 shows an application of the *M*-test to the ABC data. The 266 returned questionnaires represent a 45% return rate for the population of 566 customers who were asked to respond to the survey. The 64 Benelux customers represent 10.9% of the install base population. With 266 responses we expect 29 to be from Benelux. The *M*-test shows that the 26 returns from Benelux represent well that group proportion ($Z = -0.59$). In France, however, we have slightly fewer responses than expected (15), which is significant at the 5% level but not at 1% (see Table 20.4, with $K = 6$). This was eventually considered not significant and the data was analysed without proportional weighting.

20.7 Summary

This chapter provides a review of international standards related to customer satisfaction surveys and focuses on a non-standard application of control charts in their analysis.

Modern methods for analysing customer surveys focus on proportions of extremes in the survey outcome. This, for example, leads to an analysis using control charts for attributes of the proportion of customer responding 'very low' or 'low' (1 or 2) or 'very high' (5) to a question on overall satisfaction. We label these proportions BOT1+2 and TOP5, respectively. Such an analysis identifies areas of excellence (significantly high TOP5) and problem areas (significantly high BOT1+2).

When surveys include questions on item relative importance, one can combine importance with the identification of excellence and problem area items to identify strengths and weaknesses. In the ABC survey, questions on equipment and system (6–10), sales support (12–16), technical support (18–24), training (26–30), supplies (32–37), software add-on solutions and customer website (41, 45–48), purchasing support (50–56), contracts and pricing (58–64), and system installation (67) all include a rating of importance of the item to the customer. This declared importance can be evaluated, for example, by the proportion of '3' responses, representing a high level of importance. A strength is an area where the TOP5 is significantly high and the relative declared importance is high. A weakness is a problem area with significantly high BOT1+2 and high declared importance.

In analysing data over time, control charts can be used in a traditional way, for identifying trends and pointing out areas of improvement or deterioration. In a longitudinal study of consumer satisfaction, with several surveys performed over a period, several issues related to missing data can occur:

1. Some of the customers may refuse to participate in the study. This type of missing data is a *unit non-response*, and we presented in Section 20.6 a test for verifying proper representation of the returns relative to the population frame of a specific survey (see also Chapter 3).

2. Some customers may participate in initial studies but drop out of the study afterwards. In this attrition case with dropouts, the missing data is still considered a *unit non-response* but only for the late surveys in which they do not participate.

3. In each particular survey, respondents may decide not to respond to specific question, for example when questions are sensitive, or the respondent ends the interview before its completion. The resulting data thus contains *item non-responses*. Item non-response may also occur when respondents have no opinion or knowledge on issues (see also Chapter 3).

In many cases, the transformation of data with missing observations into a complete data set precedes the analyses aimed at assessing the parameters of the data. The two basic methods for transforming the data set into a rectangular array for analysis are to discard the incomplete cases and then analyse the units with complete information (complete-case analysis), or to impute estimated values instead of the missing ones and then analyse the filled-in data. For more information on missing data, see Fuchs and Kenett (2007) and Chapter 8 of this book.

In the longitudinal study of consumer satisfaction example, for each survey separately, missing data have to be considered in the analysis, usually by properly weighting the observed items. Alternatively, incomplete data can be analysed by more sophisticated methods which do not require a complete data set. In some cases, the common practice is to perform a basic comparison between the characteristics of the units (say, respondents) with complete observations and the characteristics of the units with incomplete data.

The chapter indicates in Section 20.5.2 how to handle multinomial data, that is, a combined set of responses and data with correlation over time (Section 20.5.6) and an analysis in the Bayesian context (Section 20.5.5).

The topics covered by this chapter represent state-of-the-art methods in the context of customer satisfaction surveys and are active areas of further research and investigation.

References

Bayarri, M.J. and Garcia-Donato, G. (2005) A Bayesian sequential look at u-control charts. *Technometrics*, 47(2), 142–151.

Colosimo, B.M. (2007) Bayesian control charts. In F. Ruggeri, R.S. Kenett and F. Faltin (ed.), *Encyclopaedia of Statistics in Quality and Reliability*, pp. 169–174. Chichester: John Wiley & Sons, Ltd.

Colosimo, B.M. and Semeraro, Q. (2002) A Bayesian control chart for service quality control. Paper presented to the Quality and Productivity Research Conference, Tenye, AZ, 5–7 June.

Duran, R.I. and Albin, S.L. (2009) Monitoring and accurately interpreting service processes with transactions that are classified in multiple categories. *IIE Transactions*, 42(2), 136–145.

Fuchs, C. and Kenett, R.S. (1980) A test for detecting outlying cells in the multinomial distribution and two-way contingency tables. *Journal of the American Statistical Association*, 75, 395–398.

Fuchs, C. and Kenett, R.S. (1998) *Multivariate Quality Control: Theory and Applications*. New York: Marcel Dekker.

Fuchs, C. and Kenett, R.S. (2007) Missing data and imputation. In F. Ruggeri, R.S. Kenett and F. Faltin (ed.), *Encyclopaedia of Statistics in Quality and Reliability*. Chichester: John Wiley & Sons, Ltd.

Kenett, R.S. (1991) Two methods for comparing Pareto charts. *Journal of Quality Technology*, 23, 27–31.

Kenett, R.S. (2004) The integrated model, customer satisfaction surveys and Six Sigma. In *Proceedings of the First International Six Sigma Conference*, Center for Advanced Manufacturing Technologies, Wrocław University of Technology, Wrocław, Poland.

Kenett, R.S. and Pollak, M. (1996) Data-analytic aspects of the Shiryayev–Roberts control charts: Surveillance of a non-homogenous Poisson process. *Journal of Applied Statistics* 23, 125–137.

Kenett, R.S. and Zacks, S. (1998) *Modern Industrial Statistics: Design and Control of Quality and Reliability*. Pacific Grove, CA: Duxbury Press.

Kushler, R. and Radka, G. (2010) *Control Charts for Customer Satisfaction Surveys*. RDA Group, www.rdagroup.com (accessed 21 May 2011).

Laviolette, M. (1995) Bayesian monitoring of multinomial processes. *Journal of the Industrial Mathematics Society*, 45, 41–49.

Marcucci, M. (1985) Monitoring multinomial processes. *Journal of Quality Technology*, 17, 86–91.

Montgomery, D.C. (2005) *Introduction to Statistical Quality Control*, 5th edn. Hoboken, NJ: John Wiley & Sons, Inc.

Psarakis, S. and Papaleonida, G.E.A. (2007) SPC procedures for monitoring autocorrelated processes. *Quality Technology and Quantitative Management*, 4, 501–540.

R Development Core Team (2010) *R: A Language and Environment for Statistical Computing*. Vienna: R Foundation for Statistical Computing.

Ryan, A.G. and Woodall, W.H. (2010) Control charts for Poisson count data with varying sample sizes. *Journal of Quality Technology*, 42(3), 260–275.

Samimi, Y., Aghaie, A. and Tarokh, M.J. (2010) Analysis of ordered categorical data to develop control charts for monitoring customer loyalty. *Applied Stochastic Models in Business and Industry*, 26, 668–688.

Scrucca, L. (2004) qcc: An R package for quality control charting and statistical process control. *R News*, 4/1, 11–17.

Sellers, K.F. and Shmueli, G. (2010) A flexible regression model for count data. *Annals of Applied Statistics*, 4(2), 943–961.

Shiau, J.J., Chen, C. and Feltz, C.J. (2005) An empirical Bayes process monitoring technique for polytomous data. *Quality and Reliability Engineering International*, 21, 13–28.

Siegmund, D. (1985) *Sequential Analysis: Tests and Confidence Intervals*. New York: Springer.

Spiring, F. (2008) A process capability/customer satisfaction approach to short-run processes. *Quality and Reliability Engineering International*, 24, 467–483.

Stoumbos, Z.G., Reynolds, M.R., Ryan, T.P. and Woodall, W.H. (2000) The state of statistical process control as we proceed into the 21st century. *Journal of the American Statistical Association*, 95, 992–998.

Tucker, G.R., Woodall, W.H. and Tsui, K.L. (2002) A control chart for ordinal data. *American Journal of Mathematical and Management Sciences*, 22, 31–48.

Wald, A. (1947) *Sequential Analysis*. New York: John Wiley and Sons, Inc.

Weiss, C.H. (2007) Controlling correlated processes of Poisson counts. *Quality Reliability Engineering International*, 23(6), 741–754.

Weiss, C.H. (2009a) Controlling correlated processes with binomial marginals. *Journal of Applied Statistics*, 36(4), 399–414.

Weiss, C.H. (2009b) Controlling jumps in correlated processes of Poisson counts. *Applied Stochastic Models in Business and Industry*, 25(5), 551–564.

Weiss, C.H. (2009c) Jumps in binomial AR(1) processes. *Statistics and Probability Letters*, 79(19), 2012–2019.

Weiss, C.H. (2009d) EWMA monitoring of correlated processes of Poisson counts. *Quality Technology and Quantitative Management*, 6(2), 137–153.

Weiss, C.H. (2010) The INARCH(1) model for overdispersed time series of counts. *Communications in Statistics – Simulation and Computation*, 39(6), 1269–1291.

Weiss, C.H. and Atzmüller, M. (2010) EWMA control charts for monitoring binary processes with applications to medical diagnosis data. *Quality and Reliability Engineering International*, 26(8), 795–805.

Weiss, C.H. and Testik, M.C. (2009) CUSUM monitoring of first-order integer-valued autoregressive processes of Poisson counts. *Journal of Quality Technology*, 41(4), 389–400.

Zeithaml, V.A., Parasuraman, A. and Berry, L.L. (1990) *Delivering Quality Service Balancing Customer Perceptions and Expectations*. New York: Free Press.

21

Fuzzy Methods and Satisfaction Indices

Sergio Zani, Maria Adele Milioli and Isabella Morlini

This chapter develops a framework that uses fuzzy set theory in order to measure customer satisfaction, starting from a survey with several questions. The basic concepts of the theory of the fuzzy numbers are briefly described. A criterion based on the sampling cumulative function, which assigns values to the membership function with reference to each quantitative, ordinal and binary variable, is suggested. Weighting and aggregation operators for the variables are considered. An application to ABC 2010 annual customer satisfaction survey data shows the usefulness of the fuzzy set approach: the gradual transition from very dissatisfied to really satisfied customers is captured by fuzzy composite indices. The comparison with the classical methods for the measurement of customer satisfaction highlights the advantages of the suggested criterion from both the theoretical and operational points of view.

21.1 Introduction

Fuzzy set theory was introduced by Zadeh (1965), and Zadeh describes it in the foreword to Zimmermann (2001) as basically 'a theory in which everything is a matter of degree or, to put it figuratively, everything has elasticity'.

In classical (crisp) set theory, an element either belongs to a set, or not, and in conventional dual logic, a statement can either be true or false. By contrast, in fuzzy set theory, an element presents a membership function value to a fuzzy set in the closed interval [0, 1]. The membership function may also be considered as the grade of compatibility or the degree of truth. Therefore, fuzzy set theory provides a mathematical framework in which vague concepts or linguistic variables can be precisely and rigorously studied. For a complete and formal description of the theory we refer to the volumes of Dubois and Prade (1980), Klir and Folger (1988), Zimmermann (2001), Smithson and Verkuilen (2006), Viertl (2011), for application

---

Modern Analysis of Customer Surveys: with applications using R, First Edition. Edited by Ron S. Kenett and Silvia Salini.
© 2012 John Wiley & Sons, Ltd. Published 2012 by John Wiley & Sons, Ltd.

in statistical analysis to Coppi *et al.* (2006), and for recent advances to the journal *Fuzzy Sets and Systems*. The application of fuzzy set theory to several statistical techniques is presented in the review of Zimmermann (2010).

In the social sciences, fuzzy sets have been applied, for example, to the measurement of poverty (Cerioli and Zani, 1990; Lemmi and Betti, 2006), well-being (Chiappero Martinetti, 2000; Baliamoune-Lutz, 2006), quality of life (Lazim and Osman, 2009), science and technology indices (Moon and Lee, 2005), and market share and preference prediction (Turksen and Wilson, 1995). Customer satisfaction can also be considered as a vague concept and may be defined, for example, as the degree of happiness that a customer experiences with a company's product or service and it is a function of the gap between expected and perceived quality (see Chapter 1 in this volume for a complete description). The gradual transition from very dissatisfied to really satisfied customer can be captured by means of fuzzy indices (Chien and Tsai, 2000; Ammar *et al.*, 2008; Darestami and Jahromi, 2009; Zani *et al.*, 2010).

This chapter is organized as follows. Section 21.2 introduces the basic concept of fuzzy set theory. Section 21.3 presents the definitions of triangular and trapezoidal fuzzy numbers and their use with reference to the categories of an ordinal variable. Section 21.4 suggests a criterion, based on the sampling cumulative function, for the fuzzy transformation of quantitative, ordinal and binary variables related to customer satisfaction. Section 21.5 considers a few weighting criteria for the aggregation of the variables, in order to obtain a fuzzy composite indicator of customer satisfaction. Section 21.6 presents the application of the suggested approach to ABC 2010 annual customer satisfaction survey data, described in Chapter 2 of this volume. Fuzzy satisfaction indices are computed for different subsets of the questions of the survey. Finally, these results are compared with those obtained by simple crisp criteria for variables and with other methods presented in the previous chapters for the evaluation of customer satisfaction.

21.2   Basic definitions and operations

Let X be a set of elements $x \in X$.

Definition 21.2.1   *A fuzzy subset A of X is a set of ordered pairs,*

$$[x, \mu_A(x)], \qquad \forall x \in X, \tag{21.1}$$

where $\mu_A(x)$ is the membership function (m.f.) of x *in A in the closed interval* [0, 1].

If $\mu_A(x) = 0$ then x does not belong to A, while if $\mu_A(x) = 1$ then x completely belongs to A. If $0 < \mu_A(x) < 1$ then x partially belongs to A and its membership in A increases according to the values of $\mu_A(x)$. For example, if X is the set of the customer of a company, A is the fuzzy subset of the satisfied customers. A customer x with $\mu_A(x) = 1$ is totally satisfied, whereas $\mu_A(x) = 0$ corresponds to a very dissatisfied customer.

Definition 21.2.2   *The complement C of a fuzzy set A is a fuzzy set with m.f.*

$$\mu_C(x) = 1 - \mu_A(x), \qquad \forall x \in X. \tag{21.2}$$

This m.f. value is interpreted as the membership level of the element x in the fuzzy set C representing the negation of the concept expressed by A. Thus, if A is the fuzzy set of satisfied customers, its complement is the fuzzy set of unsatisfied customers.

Definition 21.2.3 *If A and B are two fuzzy sets with m.f. μ_A and μ_B, respectively, then the union $A \cup B$ is a fuzzy set and its membership function is defined by*

$$\mu_{A \cup B} = \max[\mu_A(x), \mu_B(x)] \tag{21.3}$$

Definition 21.2.4 *If A and B are two fuzzy sets with m.f. μ_A and μ_B, respectively, then the intersection $A \cap B$ is a fuzzy set and its membership function is defined by*

$$\mu_{A \cap B} = \min[\mu_A(x), \mu_B(x)]. \tag{21.4}$$

The previous definitions are the simplest (standard) definitions of complement, union and intersection for fuzzy sets. When the m.f. values are restricted to 0 and 1, these functions perform precisely as the corresponding operators for crisp sets. However, for each of these three fuzzy operators, several different classes of functions, which possess appropriate axiomatic properties, have been proposed. They are presented and discussed, for example, in Klir and Folger (1988, pp. 37–58) and Zimmermann (2010).

21.3 Fuzzy numbers

In some papers on fuzzy sets, a variable whose values are words of sentences is called a 'linguistic variable' (Zadeh, 1975; Zimmermann, 2010). For example, 'Agreement' is a linguistic variable whose values are words such as very strongly agree, strongly agree, agree, undecided, disagree, strongly disagree, and very strongly disagree. These are called terms or labels of the linguistic variable 'Agreement' (Lazim and Osman, 2009).

In the statistical literature we define such variables as categorical or qualitative variables, and they may be nominal or ordinal. An example is a variable with ordinal categories low, medium, high (in the ABC 2010 annual customer satisfaction survey, the importance level of each item is defined in this way). Variables of this type can be fuzzified.

Definition 21.3.1 *A triangular fuzzy number is a fuzzy set defined by the triplet a_1, a_2, a_3 where $a_1 \leq a_2 \leq a_3$, whose m.f. is the following:*

$$\mu_A(x) = \begin{cases} \dfrac{x - a_1}{a_2 - a_1}, & a_1 \leq x \leq a_2, \\[2ex] \dfrac{x - a_3}{a_2 - a_3}, & a_2 \leq x \leq a_3, \\[2ex] 0, & \text{otherwise,} \end{cases} \tag{21.5}$$

where x is a real number and $\mu_A(x) = 1$ if and only if $x = a_2$.

Definition 21.3.2 *A trapezoidal fuzzy number is a fuzzy set defined by the quadruplet a_1, a_2, a_3, a_4 where $a_1 \leq a_2 \leq a_3, \leq a_4$, whose m.f. is the following:*

$$\mu_A(x) = \begin{cases} \dfrac{x - a_1}{a_2 - a_1}, & a_1 \leq x \leq a_2, \\[2ex] 1, & a_2 \leq x \leq a_3, \\[2ex] \dfrac{x - a_4}{a_3 - a_4}, & a_3 \leq x \leq a_4, \\[2ex] 0, & \text{otherwise.} \end{cases} \tag{21.6}$$

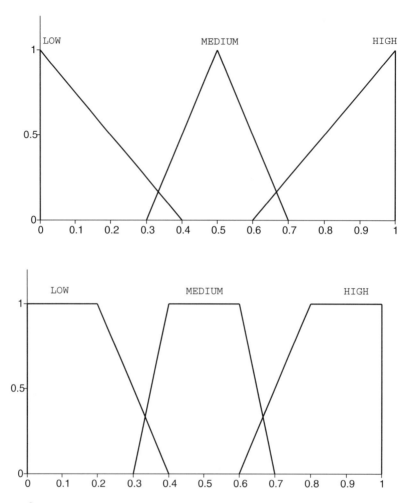

Figure 21.1   An example of triangular and trapezoidal fuzzy numbers corresponding to the three ordinal categories of a variable.

The previous definitions may be considered particular cases of a more general definition of fuzzy data suggested in D'Urso (2007, p. 161).

Triangular and trapezoidal fuzzy numbers are appropriate for quantifying linguistic or ordinal variables. Usually, for comparison purposes, all the categories of a variable are normalized in the interval [0, 1]. The previously mentioned variable with levels low, medium, high, can be fuzzified as shown in Figure 21.1. The values defining triangular and trapezoidal fuzzy numbers corresponding to each category are given in Table 21.1. Obviously, other triplets or quadruplets of values could be chosen to define the fuzzy numbers corresponding to each label of the variables. For example, the category 'medium' could correspond to the triangular fuzzy number defined by the triplet (0.25, 0.5, 0.75) or to the trapezoidal fuzzy number defined by the quadruplet (0.4, 0.45, 0.55, 0.6).

Table 21.1   Triangular and trapezoidal fuzzy numbers corresponding to each label of the variable shown in Figure 21.1.

	Triangular	Trapezoidal
low	(0, 0, 0.4)	(0, 0, 0.2, 0.4)
medium	(0.3, 0.5, 0.7)	(0.3, 0.4, 0.6, 0.7)
high	(0.6, 1, 1)	(0.6, 0.8, 1, 1)

21.4   A criterion for fuzzy transformation of variables

In a survey about customer satisfaction, there are several questions reflecting the different aspects of this latent phenomenon. Each question corresponds to a statistical variable, which may be binary (in the ABC 2010 annual customer satisfaction survey the question 'Is ABC your best supplier?'), ordinal (e.g., on a Likert scale, as in the case of the question on overall satisfaction in the ABC survey) or quantitative (e.g., the total amount of sales in dollars).

Each ordinal variable can be transformed into a set of fuzzy numbers, as shown in the previous section. Then, for each customer, we can apply union, intersection or other aggregation operators of the fuzzy sets corresponding to the relevant categories of the questions of the survey. However, this approach has two main drawbacks:

(i) The lower and upper values defining the fuzzy set corresponding to each category of a variable are *a priori* defined, irrespective of the observed frequency distribution of the labels, and several choices are possible, as previously highlighted.

(ii) For each customer we end up with a fuzzy set and not a crisp number. Therefore the ranking of the customers entails the comparison of fuzzy sets and it is less clear than the ranking of crisp numbers.

Therefore, in this chapter we use a fuzzy transformation criterion suggested in a paper on poverty evaluation (Cerioli and Zani, 1990), based on the frequency distribution of each variable. In a general framework, consider a set of n elements (customers) e_i $(i - 1, 2, \ldots, n)$ and p variables X_s $(s = 1, 2, \ldots, p)$ reflecting customer satisfaction. Without loss of generality, let us assume that each variable is positively related to customer satisfaction, that is, it satisfies the property 'the larger the better'. If a quantitative variable X_s shows negative correlation, we substitute it with a simple decreasing function transformation, such as $f(x_{si}) = \max x(x_{si}) - x_{si}$; in the case of an ordinal variable, we consider it in reverse order.

With reference to each variable X_s, we can obtain the fuzzy subset A_s of satisfied customers. In order to define the m.f. for each variable it is necessary to identify the extreme situations such that $\mu_A(x) = 0$ (non-membership) and $\mu_A(x) = 1$ (full membership), and to define a criterion for assigning m.f. values to the intermediate values of the variable. Let us assume that X_s is a quantitative variable and, for simplicity of notation, we omit index s. For that variable X we choose a lower threshold l and an upper threshold u and define the m.f. as follows:

$$\begin{cases} \mu_A(x_i) = 0, & x_i \leq l, \\ \mu_A(x_i) = \dfrac{x_i - l}{u - l}, & l < x_i < u, \\ \mu_A(x_i) = 1, & x_i \geq u. \end{cases} \qquad (21.7)$$

This m.f. is a linear function between the values of the two thresholds. Alternatively, we can arrange the values x_i in non-decreasing order according to i and define the following m.f.:

$$\begin{cases} \mu_A(x_i) = 0, & x_i \leq l, \\ \mu_A(x_i) = \mu_A(x_{i-1}) + \dfrac{F(x_i) - F(x_{i-1})}{1 - F[x_{i(l)}]}, & l < x_i < u, \\ \mu_A(x_i) = 1, & x_i \geq u, \end{cases} \quad (21.8)$$

where $F(x_i)$ is the sampling cumulative function of the variable X and $x_{i(l)}$ is the highest value $x_i \leq l$. If $l = x_1 = \min(x_i)$ and $u = x_n = \max(x_i)$, formula (21.8) corresponds to the 'totally fuzzy and relative approach' suggested by Cheli and Lemmi (1995). However, in the literature, other specifications have been considered. In this chapter we are trying to measure the degree of customer satisfaction. Therefore, the distance $d(x)$ between the value x and the goal (i.e., a totally satisfied customer) is an indicator of the success in achieving the target. If $d(x) = 0$ there is full membership in A, and then $\mu_A(x) = 1$. If $d(x) > 0$ then $\mu_A(x) < 1$. So we can write

$$\mu_A(x) = \frac{1}{1 + d(x)}. \quad (21.9)$$

In general, as highlighted by Zimmerman (2001), the relationship between physical measures and perception takes an exponential form. The distance $d(x)$ can be expressed as

$$d(x) = e^{-a(x-b)}, \quad (21.10)$$

so that the m.f. is defined as

$$\mu_A(x) = \frac{1}{1 + e^{-a(x-b)}}. \quad (21.11)$$

It is worth noting that the parameter a represents the extent of vagueness and the parameter b may be viewed as the point in which the tendency of the subject's attitude changes from rather positive to rather negative. Baliamoune-Lutz (2006) uses m.f. (21.11) in the problem of measuring the human well-being with a fuzzy approach.

If X_s is ordinal with k categories, a suitable choice is the following: the m.f. values of the x_i up to the threshold l are equal to 0 (no customer satisfaction) and those of the $x_i \geq u$ are equal to 1 (complete satisfaction); the intermediate m.f. values are defined by formula (21.8). For example, consider the scores of the variable 'Overall satisfaction' in the ABC survey for 208 respondents shown in Table 21.2. Two choices may be considered for the lower threshold:

Table 21.2 Membership function for the subset of satisfied customers with reference to ABC 2010 annual customer satisfaction survey data.

Categories	x_i	n_i	$F(x_i)$	$m.f._A(x_i)$	$m.f._B(x_i)$
very low	1	10	0.05	0.00	0.00
low	2	21	0.15	0.00	0.11
medium	3	54	0.41	0.31	0.38
high	4	90	0.84	0.81	0.83
very high	5	33	1.00	1.00	1.00
Total		208			

$l = low = 2$ or $l = very\ low = 1$. In the first case, a customer with 'low' satisfaction does not belong to the fuzzy subset A of satisfied customers, while with the second choice, he belongs to A with m.f. value equal to 0.11. The upper threshold is always $u = very\ high = 5$. The m.f. values for the two lower levels are presented, respectively, in the last two columns of the table. For example, a customer with 'medium' satisfaction belongs to A with m.f. value equal to 0.31 with the first choice, and equal to 0.38 with the second choice. If variable X_s is binary, one of the categories can be considered as a symptom of customer satisfaction. Therefore the m.f. is a crisp function assuming only values equal to 0 (absence) and 1 (presence). However, in general we consider a set of q ($q \leq p$) binary variables reflecting several aspects of the phenomenon; in this situation the m.f. can be defined as follows:

$$\mu_A(x_i) = \frac{1}{q} \sum_{s=1}^{q} z_{si}, \tag{21.12}$$

where $z_{si} = 1$ if the corresponding x_{si} denotes presence of the symptom of satisfaction and $z_{si} = 0$ otherwise. Definition (21.12) is consistent with interpreting membership values as the proportion of subjects who rate the ith unit as an actual member of the fuzzy set A (Cerioli and Zani, 1990).

21.5 Aggregation and weighting of variables

In the analysis of the data from a survey about customer satisfaction, two very difficult steps are the weighting and aggregation of the variables, in order to obtain a composite indicator of overall satisfaction. In Section 21.2 we presented the union and intersection of two fuzzy sets, and these operators may be easily extended to p fuzzy sets. For the ith subject we have the intersection

$$_I\mu_A(i) = \min[\mu_A(x_{1i}), \mu_A(x_{2i}), \ldots, \mu_A(x_{pi})], \tag{21.13}$$

and the union

$$_U\mu_A(i) = \max[\mu_A(x_{1i}), \mu_A(x_{2i}), \ldots, \mu_A(x_{pi})]. \tag{21.14}$$

A general aggregation function is the weighted generalized mean (Klir and Folger, 1988, p. 61)

$$\mu_A(i) = \sum_{s=1}^{p} [\mu_A(x_{si})]^\alpha \cdot w_s^{1/\alpha}, \tag{21.15}$$

where $w_s > 0$ is the normalized weight that expresses the relative importance of variable X_s, and $\sum_{s=1}^{p} w_s = 1$.

For fixed arguments and weights, function (21.15) is monotonically increasing with α; if $\alpha \to -\infty$, then formula (21.15) becomes the intersection; if $\alpha \to +\infty$, then (21.15) is equal to the union. For $\alpha \to 0$ formula (21.15) tends to the weighted geometric mean.

The weighting criteria in (21.15) may be:

- equal weights, which imply a careful selection of the items in order to ensure a balance of the different aspects of the latent phenomenon of customer satisfaction;

- factor loadings, obtained by principal components analysis (PCA) for quantitative variables or by nonlinear PCA for ordinal variables (see Chapter 17 in this volume) – this method of weighting is valid if the first component accounts for a high percentage of the total variance and the weights (loadings) of the variables are proportional to their correlation with the first component (factor) reflecting the underlying concept of customer satisfaction;

- obtained from expert judgements, for example using focus groups;

- determined by an analytic hierarchy process (Saaty, 1980; Kwong and Bai, 2002).

We suggest a criterion for the determination of the weights considering for each variable X_s the fuzzy proportion $g(X_s)$ of the achievement of the target (i.e., customer satisfaction):

$$g(X_s) = \frac{1}{n} \sum_{i=1}^{n} \mu(x_{si}).$$

(21.16)

If X_s is binary, formula (21.16) coincides with the crisp proportion and, in general, may be seen as an index of the proportion of the units having (totally or partially) the latent phenomenon (Cheli and Lemmi, 1995). The normalized weights may be determined as an inverse function of $g(X_s)$, in order to give higher importance to the rare features in the n subjects. To avoid excessive weights on the variables with low value of $g(X_s)$ we choose (Cerioli and Zani, 1990):

$$w_s = \frac{\ln [1/g(X_s)]}{\sum_{s=1}^{p} \ln [1/g(X_s)]}.$$

(21.17)

Using (21.17), it is possible to attach to each variable a weight sensitive to the fuzzy membership of the units to A. This may be perceived as an advantage in the weighting step from a logical point of view. In practice, we suggest comparing the solutions obtained by different weighting criteria, in order to gain insights into the stability of the pattern highlighted by the different methods.

21.6   Application to the ABC customer satisfaction survey data

21.6.1   The input matrices

A number of analytic tools have been presented in the previous chapters of this volume in order to evaluate customer satisfaction. The fuzzy approach has the advantage of pointing out the different degree of membership of each subject in the subset of satisfied customers. In the literature, a few papers present applications of these methods in different fields of customer satisfaction evaluation: service quality (Chien and Tsai, 2000); e-commerce (Fasanghari and Roudsari, 2008); and university education (Crocetta and Del Vecchio, 2007). In a previous paper, we applied fuzzy methods to the results of a survey on the satisfaction of the users of a contact centre; see Zani et al. (2010) and Zani and Berzieri (2008) for a complete description of the survey.

In this section, we obtain fuzzy composite indicators of customer satisfaction using the ABC 2010 annual customer satisfaction survey data. ABC is a global supplier of software, hardware and telecommunications services. Its customers include media companies, network operators and service providers. In 2001, it launched its first annual customer satisfaction survey (ACSS), which has been running ever since. The ACSS questionnaire consists of 81 questions. The first part includes questions on overall satisfaction, repurchasing intentions and willingness to recommend ABC. Other sections of the questionnaire cover topics such as equipment and system, sales support, technical support, and training. Each topic is covered by specific items and by an overall satisfaction question for this aspect (see Chapter 2 of this volume for further information about the company and the survey).

As a pre-processing stage, we eliminate both observations and variables with many missing values. We first consider variables related to overall satisfaction, both global and for the topics previously mentioned, and we select the following:

X_1: Overall satisfaction level with ABC.

X_2: Overall satisfaction level with ABC's improvements during the last year.

X_3: Is ABC your best supplier?

X_4: Would you recommend ABC to other companies?

X_5: If you were in the market to buy a product, how likely would it be for you to purchase an ABC product again?

X_{11}: Overall satisfaction level with the equipment and system.

X_{17}: Overall satisfaction level with sales support.

X_{25}: Overall satisfaction level with technical support.

X_{31}: Overall satisfaction level with ABC training.

X_{38}: Overall satisfaction level with ABC supplies and orders.

X_{57}: Overall satisfaction level with purchasing support.

X_{65}: Overall satisfaction level with contracts and pricing.

Considering observations with no missing values in these items, we obtain a 137×12 data matrix. In the following, we will refer to this data set as the *overall*$_1$ data. Eliminating variable X_{31}, which has a large number of missing values, we obtain a 208×11 data matrix. We will refer to this second data set as the *overall*$_2$ data.

Both in the *overall*$_1$ and *overall*$_2$ data, all variables but X_3 are measured on a Likert scale with categories very low (1), low (2), medium (3), high (4), and very high (5). Item X_3 is binary, with responses 'yes' or 'no'.

We conduct two analyses, by first considering the lower threshold $l = 2$ (and thus assigning to 'very low' and 'low' an m.f. equal to 0) and then considering $l = 1$ (and thus assigning only to 'very low' ab m.f. value equal to 0 and to 'low' an m.f. value according to the cumulative function of the corresponding variable). The value of the higher threshold u is always 'very high'. In X_3 an m.f. value equal to 1 is assigned to the label 'yes' and equal to 0 to the label 'no'.

We then consider the single variables related to the evaluation of the specific aspects of satisfaction, deleting the variables with several missing values: equipment and system (4 variables), sales support (5 variables), technical support (7 variables), supplies (2 variables), purchasing support (5 variables), contracts and pricing (5 variables). All these variables are measured on Likert scale with five categories. Considering observations with no missing values or a missing value in only one of these items, we obtain a 151×28 data matrix. We replace the few remaining missing values with the median. In the following, we will refer to this data set as the *specific* data (see Chapter 8 in this volume for a complete description of the treatment of missing values).

21.6.2   Main results

For each variable in the three data sets, we compute the fuzzy m.f. values using formula (21.8). For each customer, the corresponding value shows its degree of membership in the fuzzy subset A of satisfied customers based on a single item.

Then we compute the fuzzy composite indices with the two lower thresholds mentioned above and with two weighting criteria:

- equal weights W_1 for each variable;

- normalized weights W_2, as an inverse function of the fuzzy proportion of each variable according to formula (21.17).

Normalized factor loadings obtained with PCA on the τ rank correlation matrix are not used in this application, since the first component in each data set explains less than 30% of the total variability.

Tables 21.3 and 21.4 show the frequency distributions of the fuzzy composite indices in the *overall*$_1$ and *overall*$_2$ data sets, respectively. In both tables, the last column gives the frequency distribution of the crisp composite index obtained as the average value of the numerical codes assigned to the categories. In order to have variables with the same range, in item X_3 a numerical code equal to 1 is assigned to the category 'no' and a numerical code equal to 5 to the category 'yes'. In both data sets, the pairwise rank correlations among the different indices are very high (Spearman's correlation coefficient is always higher than 0.95). Indeed, the rankings of the customers based on different data sets and different aggregation criteria are very similar. This provides evidence that the latent concept of customer satisfaction really is present and

Table 21.3   Frequency distribution of the fuzzy composite indicators and the normalized average score (crisp indicator) in the *overall*$_1$ data set

Classes	$l = 1$ W_1	W_2	$l = 2$ W_1	W_2	Crisp indicator
0.0 ⊢ 0.1	0	0	1	1	1
0.1 ⊢ 0.2	1	1	3	7	0
0.2 ⊢ 0.3	6	7	8	4	6
0.3 ⊢ 0.4	5	11	13	18	8
0.4 ⊢ 0.5	16	18	12	16	18
0.5 ⊢ 0.6	17	22	18	22	20
0.6 ⊢ 0.7	23	19	26	14	25
0.7 ⊢ 0.8	19	17	13	16	22
0.8 ⊢ 0.9	18	16	16	16	13
0.9 ⊢ 1.0	20	13	17	12	16
1	12	13	10	11	8
Tot	137	137	137	137	137

Table 21.4 Frequency distribution of the fuzzy composite indicators and the normalized average score (crisp indicator) in the *overall₂* data set

Classes	$l = 1$		$l = 2$		Crisp indicator
	W_1	W_2	W_1	W_2	
0.0 ⊢ 0.1	0	0	2	2	1
0.1 ⊢ 0.2	3	4	8	10	2
0.2 ⊢ 0.3	10	12	10	9	9
0.3 ⊢ 0.4	9	17	18	24	9
0.4 ⊢ 0.5	21	19	18	18	21
0.5 ⊢ 0.6	22	32	28	35	24
0.6 ⊢ 0.7	38	25	28	19	38
0.7 ⊢ 0.8	26	26	26	23	36
0.8 ⊢ 0.9	24	22	27	29	31
0.9 ⊢ 1.0	32	27	26	18	27
1	23	24	17	21	10
Tot	208	208	208	208	208

is not an artificial phenomenon driven by a single composite indicator. Increasing the lower threshold leads to a decrease in the average and median values of the distribution of the fuzzy index. The effect of decreasing the mean and the median is also driven by imposing different weights on the variables in the construction of the composite indicators. The fuzzy indices show a larger dispersion, as measured by the interquartile range, than the crisp indicator.

Table 21.5 shows the frequency distributions of the composite indicators in the *specific* data set. Using the variables related to specific questions, none of the respondents can be

Table 21.5 Frequency distribution of the fuzzy composite indicators and the normalized average score (crisp indicator) in the *specific* data set

Classes	$l = 1$		$l = 2$		Crisp indicator
	W_1	W_2	W_1	W_2	
0.0 ⊢ 0.1	0	0	0	0	0
0.1 ⊢ 0.2	1	1	1	1	0
0.2 ⊢ 0.3	6	9	2	2	2
0.3 ⊢ 0.4	7	6	7	7	3
0.4 ⊢ 0.5	12	15	7	9	6
0.5 ⊢ 0.6	24	23	17	15	14
0.6 ⊢ 0.7	33	36	32	33	35
0.7 ⊢ 0.8	26	19	33	31	42
0.8 ⊢ 0.9	20	24	20	22	30
0.9 ⊢ 1.0	18	15	26	25	16
1	4	3	6	6	3
Tot	151	151	151	151	151

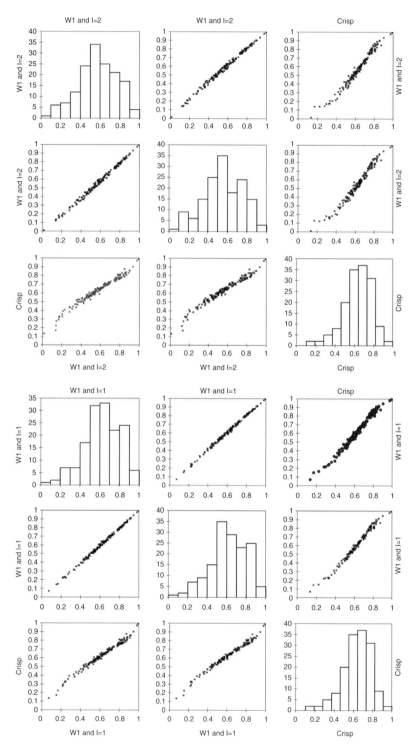

Figure 21.2   Scatter plot matrices of the fuzzy indicators and of the crisp indicator in the specific *data set, with different thresholds and weights.*

regarded as completely unsatisfied. The majority of the respondents belongs to the set of satisfied customers, with m.f. values higher than 0.5. Figure 21.2 displays the scatter plot matrices of the fuzzy indicators and of the crisp indicator in the *specific* data set. Here again, the fuzzy and crisp indices show high pairwise correlations. With $l = 1$ the distribution of the values is more symmetric.

In order to test whether the responses to the specific questions (contained in the *specific* data set) are related to the overall scores, we have computed the correlation coefficient between the values of the indicators obtained in the *overall*₁ (or *overall*₂) data set and the values of the indicators obtained in the *specific* data set (considering the subsample of observations for which both sets of indices have been computed). Table 21.6 shows correlation coefficients among fuzzy indicators obtained in the *overall*₁ and *specific* data sets. The correlations between the pairs of variables in the same data set are very high (always greater than 0.98): different thresholds for fuzzy m.f. and different weighting criteria have poor influence on the evaluation of the satisfaction of each customer. The correlation coefficients with reference to pairs of variables in the two data sets are also high (always greater than 0.83) and indicate strong relations between the level of satisfaction based on global questions and on specific items.

In order to verify the coherence between the overall satisfaction level with ABC measured by X_1 and the degree of satisfaction recorded for specific aspects, we have considered the empirical distributions of the different composite indices for each value of X_1. Figure 21.3 shows the empirical distribution of the fuzzy indicators obtained with W_2 and $l = 2$, for each category of X_1. The median and average values increase with the ordinal categories of the overall satisfaction with ABC, as we would expect. In all three data sets (*specific*, *overall*₂ and *overall*₂), the values of the fuzzy indicators show a bigger dispersion for the 'high' category. Indeed, there are many respondents declaring high overall satisfaction but indicating a medium or small satisfaction with specific items. In contrast, much coherence is present for the respondents declaring very low, low or a very high overall satisfaction. Figure 21.4 shows the boxplots of the values of the fuzzy indicators (computed with W_2 and $l = 2$) and of the

Table 21.6 Pairwise correlations among the fuzzy composite indicators and crisp indicators obtained in the *overall*₁ and *specific* data sets.

	*overall*₁ data set					*specific* data set				
	$W_1,$ $l=2$	$W_2,$ $l=2$	$W_1,$ $l=1$	$W_2,$ $l=1$	Crisp	$W_1,$ $l=2$	$W_2,$ $l=2$	$W_1,$ $l=1$	$W_2,$ $l=1$	Crisp
$W_1, l=2$	1	0.995	0.998	0.996	0.990	0.871	0.851	0.868	0.831	0.862
$W_2, l=2$	0.995	1	0.992	0.996	0.988	0.863	0.844	0.860	0.822	0.855
$W_1, l=1$	0.998	0.992	1	0.997	0.992	0.874	0.854	0.871	0.835	0.867
$W_2, l=1$	0.996	0.996	0.997	1	0.993	0.867	0.847	0.864	0.827	0.860
Crisp	0.990	0.988	0.992	0.993	1	0.861	0.843	0.860	0.823	0.859
$W_1, l=2$	0.871	0.863	0.874	0.867	0.861	1	0.994	0.999	0.986	0.991
$W_2, l=2$	0.851	0.844	0.854	0.847	0.843	0.994	1	0.995	0.997	0.993
$W_1, l=2$	0.868	0.860	0.871	0.864	0.860	0.999	0.995	1	0.990	0.994
$W_2, l=1$	0.831	0.822	0.835	0.827	0.823	0.986	0.997	0.990	1	0.991
Crisp	0.862	0.855	0.867	0.860	0.859	0.991	0.993	0.994	0.991	1

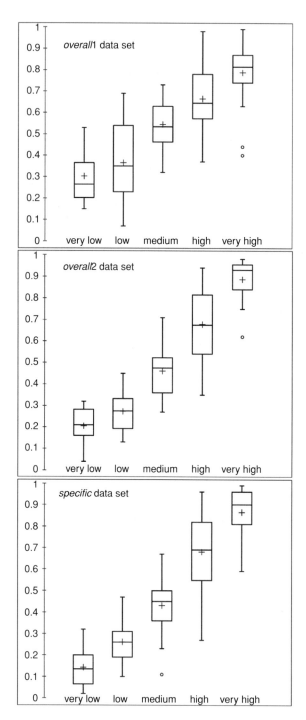

Figure 21.3   Boxplots of the values of the fuzzy composite indicator computed with weights W_2 *and* $l = 1$ *in the* overall$_1$, overall$_2$ *and* specific *data sets, for different categories of the variable* X_1 = *overall satisfaction with ABC.*

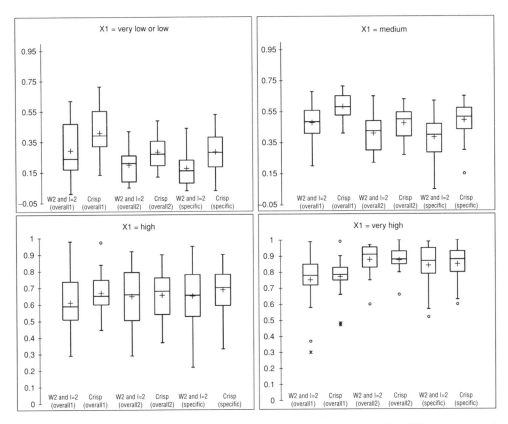

Figure 21.4   Boxplots of the values of the fuzzy composite indicators for different values of the variable $X_1 =$ *overall satisfaction with ABC*

crisp indicator in each data set and for each value of X_1. Here, the first two values, $X_1 = 1$ and $X_1 = 2$, have been aggregated, since for some indices they pertain to only few respondents. Figure 21.5 shows boxplots of the values of the fuzzy composite indicators computed with variables X18–X24 for different values of X25 = overall satisfaction with technical support.

21.7   Summary

Customer satisfaction may be considered as a latent phenomenon that can be measured by several questions (manifest variables) in a survey. The simple traditional methods based on a crisp composite indicator, using the average of the numbers 1, 2, ... representing the ordered categories of the items, does not take proper account of the ordinal nature of the variables: the assumption of equal distance between consecutive labels of a variable is not realistic.

In this chapter, we suggest using fuzzy set theory to evaluate the fuzzy subset A of satisfied customers. For each (ordinal) variable in the survey, we suggest computing the m.f. in A based on its cumulative function in the sample of respondents. Each subject presents a m.f. value to A, according to the relevant category of that variable.

Next we obtain fuzzy composite indicators with reference to all or to a subset of the questions of the survey, using convenient criteria for weighting and aggregating the variables.

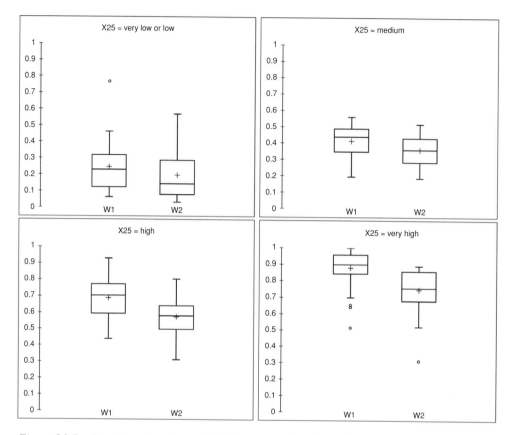

Figure 21.5 Question related to technical support: boxplots of the values of the fuzzy composite indicators computed with variables X_{18}–X_{24} *for different values of* X_{25} = *overall satisfaction with technical support*

This procedure has many advantages from both theoretical and operational points of view. Customer satisfaction is not a dichotomous phenomenon, but there are many successive levels of satisfaction that can be captured by m.f. values, with reference first to single items and then to fuzzy composite indicators.

We obtain rankings of the subjects, from very dissatisfied to completely satisfied. These rankings could be compared with those computed by other multivariate methods, such as using the scores of nonlinear principal components analysis (see Chapter 17 in this volume). The application of this approach to the ABC 2010 annual customer satisfaction survey data highlights its practical usefulness for customer satisfaction evaluation.

For each of the different subsets of the questions of the survey, we obtain the distribution of the respondents according to the value of their fuzzy composite indicator. The results for different subsets of variables are highly correlated, showing that the latent concept of customer satisfaction really is present in the data set. We also compute the values of the fuzzy composite indicators for each category of the first question on overall satisfaction with ABC. The boxplots show an increase in the median of the values of the fuzzy indicators, according to the ordinal categories of overall satisfaction, and a larger dispersion for the label 'high'.

It is interesting to note that this comparison reveals a few non-coherent respondents declaring 'very high' overall satisfaction, but with small values of the composite indices based on several items (or vice versa). These subjects may be considered as multiple ordinal outliers.

References

Ammar, S., Moore, D. and Wright, R. (2008) Analysing customer satisfaction surveys using a fuzzy rule-based decision support system: Enhancing customer relatioship management. *Journal of Database Marketing and Customer Strategy Management*, 15, 2, 91–105.

Baliamoune-Lutz. M. (2006) On the measurement of human well-being: Fuzzy set theory and Sen's capability approach. In M. McGillivray and M. Clarke (eds), *Understanding Human Well-being*. New York: United Nations University Press.

Cerioli, A. and Zani, S. (1990) A fuzzy approach to the measurement of poverty. In C. Dagum and M. Zenga (eds), *Income and Wealth Distribution, Inequality and Poverty*, pp. 272–284. Berlin: Springer.

Cheli, B. and Lemmi, A. (1995) A totally fuzzy and relative approach to the multidimensional analysis of poverty. *Economic Notes*, 24(1), 115–134.

Chiappero Martinetti, E. (2000) A multidimensional assessment of well-being based on Sen's functioning approach. *Rivista Internazionale di Scienze Sociali*, 108, 2, 207–239.

Chien, C.J. and Tsai, H.H. (2000) Using fuzzy numbers to evaluate perceived service quality. *Fuzzy Sets and Systems*, 116, 289–300.

Coppi, R., Gil, M.A. and Kiers, H.A.L. (2006) A fuzzy approach to statistical analysis. *Computational Statistics & Data Analysis*, 51, 1–14.

Crocetta, C. and Delvecchio, G. (2007) A fuzzy measure of satisfaction for university education as a key for employment. In L. Fabbris (ed.), *Effectivness of University Education in Italy*, pp. 11–27. Heidelberg: Physica-Verlag.

Darestani, A.Y. and Jahromi, A.E. (2009) Measuring customer satisfaction using a fuzzy inference system. *Journal of Applied Sciences*, 9(3), 469–478

Dubois, D. and Prade, H. (1980) *Fuzzy Sets and Systems: Theory and Applications*. New York: Academic Press.

D'Urso, P. (2007) Fuzzy clustering of fuzzy data. In J.V. De Oliveira and W. Pedrycz (eds), *Advances in Fuzzy Clustering and its Applications*, pp. 155–192. Chichester: John Wiley and Sons, Ltd.

Fasanghari, M. and Roudsari, F.H. (2008) The fuzzy evaluation of e-commerce customer satisfaction. *World Applied Sciences Journal*, 4(2), 164–168.

Klir, G.J. and Folger, T.A. (1988) *Fuzzy Sets, Uncertainity and Information*. London: Prentice Hall International.

Kwong, C.K. and Bai, H. (2002) A fuzzy AHP approach to the determination of importance weights of customer requirements in quality function deployment. *Journal of Intelligent Manufacturing*, 13, 367–377.

Lazim, M.A. and Osman, M.T.A. (2009) A new Malaysian quality of life index based on fuzzy sets and hierarchical needs. *Social Indicators Research*, 94, 3, 499–508.

Lemmi, A and Betti, G (eds) (2006) *Fuzzy Set Approach to Multidimensional Poverty Measurement*. New York: Springer.

Moon, H.S. and Lee, J.D. (2005) A fuzzy set theory approach to national composite S&T indices. *Scientometrics*, 64(1), 67–83.

Saaty, T.L. (1980) *The Analytic Hierarchy Process*. New York: McGraw-Hill.

Smithson, M. and Verkuilen, J. (2006) *Fuzzy Set Theory: Applications in the Social Sciences*. Thousand Oaks, CA: Sage.

Turksen, I.B. and Wilson, I.A. (1995) A fuzzy set model for market share and preference prediction. *European Journal of Operational Research*, 82, 39–52.

Viertl, R. (2011) *Statistical Methods for Fuzzy Data*. Chichester: John Wiley & Sons, Ltd.

Zadeh, L.A. (1965) Fuzzy sets. *Information and Control*, 8, 338–353.

Zadeh, L.A. (1975) The concept of a linguistic variable and its application to approximate reasoning, Parts 1 and 2. *Information Sciences*. 8, 199–249, 301–357.

Zani, S. and Berzieri, L. (2008) Measuring customer satisfaction using ordinal variables: An application in a survey on a contact center. *Statistica Applicata – Italian Journal of Applied Statistics*, 20(3–4), 331–351.

Zani, S., Milioli, M.A. and Morlini, I. (2010) Fuzzy composite indicators: An application for measuring customer satisfaction. In *Proceedings of the Meeting of Italian Statistical Society*, Padua [CD].

Zimmermann, H.J. (2001) *Fuzzy Sets Theory and its Applications*, 4th edn. Boston: Kluwer.

Zimmermann, H.J. (2010) Fuzzy sets theory. *Wiley Interdisciplinary Review: Computational Statistics*, 2, 317–332.

Appendix

An introduction to R

Stefano Maria Iacus

This chapter reviews the basics of the R statistical environment. The topics discussed here include the R programming language, the R graphics and some elementary statistical analysis.

A.1 Introduction

R is, at first glance, user unfriendly in the sense that it is a real programming language rather than point-and-click software. A typical statistical analysis in R is an interactive session where the user types commands on the R console. Indeed, the standard installation of R has a very limited graphical user interface (GUI) which can be enhanced using the package Rcmdr. There are also integrated environments like the free software RStudio (http://www.rstudio.org) which is a spreadsheet-like tool with R as a back end, or the commercial product Revolution R (http://www.revolutionanalytics.com). The following pages are intended as a quick start guide to the R language so that the reader of this book can immediately start to replicate the analyses of the previous chapters. Here we will also mention additional tools that may be valuable in more advanced analyses and production environments. The reader is also invited to read the quick guide called *An Introduction to R* that comes with every installed version of R (and can be found in the 'manual' folder) or the introductory book on the R environment by Dalgaard (2008).

A.2 How to obtain R

R is open source and free software available in ready-to-install binary form for many platforms. Although the home page of the software is http://www.R-Project.org the main repository for downloading R is called CRAN– the Comprehensive R Archive Network – and its main address is http://cran.r-project.org/. Three main sections are available on that page: 'Linux', 'MacOS X' and 'Windows'. Inside each section, one should check for the 'base' installation of R. For example, the base installation of R for Windows is available at

Modern Analysis of Customer Surveys: with applications using R, First Edition. Edited by Ron S. Kenett and Silvia Salini.
© 2012 John Wiley & Sons, Ltd. Published 2012 by John Wiley & Sons, Ltd.

http://cran.r-project.org/bin/windows/base while the basic installation for Mac OS X is available at http://cran.r-project.org/bin/macosx. For Linux users there are different subdirectories that contain binary files for debian, redhat, suse and ubuntu, but these users can always install R from the source code. Expert users can try to install the 64-bit version of R which allows for the manipulation of larger data bases in memory at the cost of some external packages possibly not being available on some platforms.

A.3   Type rather than 'point and click'

All the commands are given as inputs to R after the prompt > (written below as R>) and are analysed by the R parser after the user presses the return/enter key (or a newline character is encountered in the case of a script file).

```
R> cat("help me!")

help me!
```

R inputs can be multiline; hence, if the R parser thinks that the user did not complete some command (because of unbalanced parentheses or quotation marks), on the next line a + symbol will appear instead of a prompt.

```
R> cat("help me!"
+
```

This can be quite frustrating for novice users, so it is better to know how to exit from this impasse. Depending on the implementation of R, usually pressing CTRL+C or ESC on the keyboard helps. Otherwise, for GUI versions of R, clicking on the stop button of the R console will exit the parser. Of course, another solution is to complete the command (with a ')' in our example).

A.3.1   The workspace

Almost every command in R creates an object and not just text output, and objects live in the workspace. The workspace, or groups of objects, can be saved and loaded into R with the save.image (or save) and load commands, respectively. The user is prompted about saving workspace when exiting from R. This workspace is saved in the current directory as a hidden (on some operating systems) file named .RData and reloaded automatically the next time R is started.

A.3.2   Graphics

Usually R graphics are displayed on a device that corresponds to a window for a GUI version of R (e.g., under MS Windows, X11, or Mac OS X). Otherwise a Postscript file Rplots.ps is generated in the current working directory. Sometimes, in interactive uses of R, it is useful to use par(ask=TRUE) to pause R at each new plot or par(ask=FALSE) to avoid such pauses. We do not discuss the multiple R graphic systems here, but the reader can refer to Murrel (2005) and Deepayan (2008).

A.3.3   Getting help

The casual user will find it very hard to get started without prior knowledge of which command is needed to perform a particular task. The help system is not that useful either. But the R system is constructed in such a way that every command has its help page and documentation always matches the actual implementation of the command. To get information about a particular command one should use `help`, for example `help(load)` or `?help`. For some special operators, the user should specify the argument like this: `?"for"`, `?"+"`, etc. When the documentation contains an 'Examples' section with R code inside, the code from that page can be executed automatically with `example(topic)`, where `topic` is the corresponding R command of interest – try, for example, `example(plot)`.

If one wishes to execute a fuzzy search on the help system, one can use the command `help.search("topic")` and R will return several options which partly match the word `topic` – try, for example, `help.search("regression")`. It is also possible to extend the search for a term or for more complicated queries to the web using `RSiteSearch`, – try, for example, `RSiteSearch("nonlinear regression")`. The search will be extended to all documentation pages for packages in the R repository and to all the pertinent mailing lists.[1] The website for R mailing lists and related projects is http://stat.ethz.ch/mailman/listinfo.

There is a rich repository of quick guides and electronic books on R and its use in different disciplines which can be found in the 'Documentation/Contributed' section on CRAN. The direct link to the page is http://cran.r-project.org/other-docs.html. Finally, we mention the 'Task Views', which are collections of R packages organized by macro areas – for example, 'Multivariate'. These are again hosted on CRAN and the direct link is http://cran.r-project.org/web/views.

A.3.4   Installing packages

R is a layered software suite consisting of a core of basic functionalities which serves as the glue for additional capabilities grouped as a collection of functions in the so-called packages. The notion of *package* in R corresponds to that of *tool* or *module* in other software. There are more 3000 additional packages available for R as of May 2011.

This book, like many others, requires several add-on packages which are not distributed with the basic R system. The main repository for R packages is CRAN. To install a package in the R system, one should use the command `install.packages` with a package name as argument – for example, `install.packages("pls")` to install the package `pls` for partial least squares regression. R GUIs usually offer some option to get the list of all packages at the repository and install those selected by point-and-click actions.

Another important source of R packages is the `R-Forge` repository. This is a repository mainly for developers but where users can also find pre-releases of developer versions of packages already on CRAN or even packages not necessarily hosted on CRAN. The home page of `R-Forge` is http://r-forge.r-project.org. For example, to install the package `permute` (for generating restricted permutations of data) from `R-Forge` one can use a command like

```
R> install.packages("permute",repos="http://R-Forge.R-project.org")
```

---

[1] If you have a question, have a look at the mailing list archives first.

where we have specified for install.packages the argument repos with a proper web address.

In order to install the complete suite of packages from a particular Task View one first needs to install the ctv package and load it into R

```
R> install.packages("ctv")
R> library("ctv")
```

and then install the packages from a Task View, say 'Multivariate', with

```
R> install.views("Multivariate")
```

A.4   Objects

As mentioned, most functions in R return objects rather than text output. Clearly objects can be created anew with commands. We now describe how to create, inspect and manipulate objects.

A.4.1   Assignments

To create an object it is necessary to use the operator '<-' which has the meaning 'assign the right-hand side to the left-hand side', or use the more common operator '=', as in the following lines in which we create an object named x and assign the number 4 to it:

```
R> x <- 4
R> x = 4
```

Similarly, one can use the operator '->' which assigns the left-hand side to the right-hand side, i.e. x -> 4. The following command creates a more interesting vector y containing the numbers 2, 7, 4, and 1 concatenated in a single object using the function c():

```
R> y <- c(2, 7, 4, 1)
R> y
```

```
[1] 2 7 4 1
```

A matrix can be created using the matrix command

```
R> z <- matrix(1:30, 5, 6)
R> z
```

```
     [,1] [,2] [,3] [,4] [,5] [,6]
[1,]    1    6   11   16   21   26
[2,]    2    7   12   17   22   27
[3,]    3    8   13   18   23   28
[4,]    4    9   14   19   24   29
[5,]    5   10   15   20   25   30
```

where 1:30 produces a sequence from 1 to 30 in unit steps.

```
R> 1:30
```

```
 [1]  1  2  3  4  5  6  7  8  9 10 11 12 13 14 15 16 17 18 19 20 21 22
[23] 23 24 25 26 27 28 29 30
```

The command `matrix` requires at least three arguments, where the second and third are the number of rows and columns and the first one is an object which is used recursively to fill the elements of the matrix. Of course, we can create an empty matrix with

```
R> matrix(, 5, 6)
```

```
     [,1] [,2] [,3] [,4] [,5] [,6]
[1,]   NA   NA   NA   NA   NA   NA
[2,]   NA   NA   NA   NA   NA   NA
[3,]   NA   NA   NA   NA   NA   NA
[4,]   NA   NA   NA   NA   NA   NA
[5,]   NA   NA   NA   NA   NA   NA
```

where NA is the R symbol for missing values, or empty numerical vectors with `numeric`

```
R> numeric(4)
```

```
[1] 0 0 0 0
```

The command `ls()` shows the current content of the workspace

```
R> ls()
```

```
[1] "x" "y" "z"
```

Notice that objects that have been created but not assigned are not kept in the workspace. Like `numeric`, there are several command to allocate objects for the different data types available in R, that is, we have functions like `integer`, `character`, etc. or use the function `vector` as follows:

```
R> vector(mode = "numeric", 4)
```

```
[1] 0 0 0 0
```

is equivalent to `numeric(4)`. Objects can also have length zero

```
R> w1 <- numeric(0)
R> w1
```

```
numeric(0)
```

or can be initialized as NULL

```
R> w2 <- NULL
R> w2
```

```
NULL
```

which is useful if one is intending to enlarge these objects later. In the above, w1 and w2 are objects of different types and, in particular, w1 is an object of class numeric while w2 is not. It is also possible to use the command assign to create objects, and this is sometimes useful when the name of the object has to be created dynamically in the R code. The following is an example of use in which O1 is created as before and O2 is created via assign.

```
R> O1 <- 1:4
R> O1

[1] 1 2 3 4

R> ls()

[1] "O1" "w1" "w2" "x"   "y"   "z"

R> assign("O2", 5:8)
R> ls()

[1] "O1" "O2" "w1" "w2" "x"   "y"   "z"

R> O2

[1] 5 6 7 8
```

A.4.2   Basic object types

Classes of objects can be created from scratch in the R language, and this is usually the case for many R packages, but the basic classes are integer, numeric, complex, character, etc. which can be aggregated in vectors, matrixes, arrays or lists. While vectors, arrays and lists contain elements all of the same type, lists are more general and can contain objects of different size and type and can also be nested. For example, the following code loads a data set, estimates a linear model via lm and assigns the result to an object mod. The statistical analysis *per se* is not relevant here, we simply wish to observe that an estimated regression model in R is not just an output of coefficients with their significance, but an object.

```
R> data(cars)
R> mod <- lm(dist ~  speed, data = cars)
R> mod

Call:
lm(formula = dist ~  speed, data = cars)

Coefficients:
(Intercept)        speed
   -17.579        3.932
```

We now look at the structure of the object created by the linear regression using the command str which inspects the structure of the object.

```
R> str(mod)

List of 12
 $ coefficients : Named num [1:2] -17.58 3.93
  ..- attr(*, "names")= chr [1:2] "(Intercept)" "speed"
 $ residuals    : Named num [1:50] 3.85 11.85 -5.95 12.05 2.12 ...
  ..- attr(*, "names")= chr [1:50] "1" "2" "3" "4" ...
 $ effects      : Named num [1:50] -303.914 145.552 -
    8.115 9.885 0.194 ...
  ..- attr(*, "names")= chr [1:50] "(Intercept)" "speed" "" "" ...
 $ rank         : int 2
 $ fitted.values: Named num [1:50] -1.85 -1.85 9.95 9.95 13.88 ...
  ..- attr(*, "names")= chr [1:50] "1" "2" "3" "4" ...
 $ assign       : int [1:2] 0 1
 $ qr           :List of 5
  ..$ qr   : num [1:50, 1:2] -7.071 0.141 0.141 0.141 0.141 ...
  .. ..- attr(*, "dimnames")=List of 2
  .. .. ..$ : chr [1:50] "1" "2" "3" "4" ...
  .. .. ..$ : chr [1:2] "(Intercept)" "speed"
  .. ..- attr(*, "assign")= int [1:2] 0 1
  ..$ qraux: num [1:2] 1.14 1.27
  ..$ pivot: int [1:2] 1 2
  ..$ tol  : num 1e-07
  ..$ rank : int 2
  ..- attr(*, "class")= chr "qr"
 $ df.residual  : int 48
 $ xlevels      : Named list()
 $ call         : language lm(formula = dist ~  speed, data = cars)
 $ terms        :Classes 'terms', 'formula' length 3 dist ~  speed
  .. ..- attr(*, "variables")= language list(dist, speed)
  .. ..- attr(*, "factors")= int [1:2, 1] 0 1
  .. .. ..- attr(*, "dimnames")=List of 2
  .. .. .. ..$ : chr [1:2] "dist" "speed"
  .. .. .. ..$ : chr "speed"
  .. ..- attr(*, "term.labels")= chr "speed"
  .. ..- attr(*, "order")= int 1
  .. ..- attr(*, "intercept")= int 1
  .. ..- attr(*, "response")= int 1
  .. ..- attr(*, ".Environment")=<environment: R_GlobalEnv>
  .. ..- attr(*, "predvars")= language list(dist, speed)
  .. ..- attr(*, "dataClasses")= Named chr [1:2] "numeric" "numeric"
  .. .. ..- attr(*, "names")= chr [1:2] "dist" "speed"
 $ model        :'data.frame': 50 obs. of  2 variables:
  ..$ dist : num [1:50] 2 10 4 22 16 10 18 26 34 17 ...
  ..$ speed: num [1:50] 4 4 7 7 8 9 10 10 10 11 ...
  ..- attr(*, "terms")=Classes 'terms', 'for-
mula' length 3 dist   speed
  .. .. ..- attr(*, "variables")= language list(dist, speed)
  .. .. ..- attr(*, "factors")= int [1:2, 1] 0 1
  .. .. .. ..- attr(*, "dimnames")=List of 2
  .. .. .. .. ..$ : chr [1:2] "dist" "speed"
```

```
.. .. .. .. ..$ : chr "speed"
.. .. ..- attr(*, "term.labels")= chr "speed"
.. .. ..- attr(*, "order")= int 1
.. .. ..- attr(*, "intercept")= int 1
.. .. ..- attr(*, "response")= int 1
.. .. ..- attr(*, ".Environment")=<environment: R_GlobalEnv>
.. .. ..- attr(*, "predvars")= language list(dist, speed)
.. .. ..- attr(*, "dataClasses")= Named chr [1:2] "numeric"
"numeric"
.. .. .. ..- attr(*, "names")= chr [1:2] "dist" "speed"
- attr(*, "class")= chr "lm"
```

We see that `mod` is essentially a `list` object of 12 elements and it is of class `lm` (for linear models). For example, the first one is called `coefficients` and can be accessed using the symbol $ as follows:

```
R> mod$coefficients
```

```
(Intercept)        speed
 -17.579095     3.932409
```

```
R> str(mod$coefficients)
```

```
 Named num [1:2] -17.58 3.93
 - attr(*, "names")= chr [1:2] "(Intercept)" "speed"
```

The vector `coefficients` is a 'named vector'. One can obtain or change the names of the elements of a vector with

```
R> names(mod$coefficients)
```

```
[1] "(Intercept)" "speed"
```

or change them with

```
R> names(mod$coefficients) <- c("alpha", "beta")
R> mod$coefficients
```

```
     alpha         beta
-17.579095    3.932409
```

Similarly, one can assign or get the names of the rows or the columns of an R matrix:

```
R> z
```

```
     [,1] [,2] [,3] [,4] [,5] [,6]
[1,]    1    6   11   16   21   26
```

```
[2,]    2    7   12   17   22   27
[3,]    3    8   13   18   23   28
[4,]    4    9   14   19   24   29
[5,]    5   10   15   20   25   30

R> rownames(z) <- c("a", "b", "c", "d", "e")
R> colnames(z) <- c("A", "B", "C", "D", "E", "F")
R> z

  A  B  C  D  E  F
a 1  6 11 16 21 26
b 2  7 12 17 22 27
c 3  8 13 18 23 28
d 4  9 14 19 24 29
e 5 10 15 20 25 30
```

As anticipated, lists can be nested. For example, the object model inside mod is itself a list

```
R> str(mod$model)

'data.frame':          50 obs. of  2 variables:
 $ dist : num  2 10 4 22 16 10 18 26 34 17 ...
 $ speed: num  4 4 7 7 8 9 10 10 10 11 ...
 - attr(*, "terms")=Classes 'terms', 'formula' length 3 dist ~  speed
  .. ..- attr(*, "variables")= language list(dist, speed)
  .. ..- attr(*, "factors")= int [1:2, 1] 0 1
  .. .. ..- attr(*, "dimnames")=List of 2
  .. .. .. ..$ : chr [1:2] "dist" "speed"
  .. .. .. ..$ : chr "speed"
  .. ..- attr(*, "term.labels")= chr "speed"
  .. ..- attr(*, "order")= int 1
  .. ..- attr(*, "intercept")= int 1
  .. ..- attr(*, "response")= int 1
  .. ..- attr(*, ".Environment")=<environment: R_GlobalEnv>
  .. ..- attr(*, "predvars")= language list(dist, speed)
  .. ..- attr(*, "dataClasses")= Named chr [1:2] "numeric" "numeric"
  .. .. ..- attr(*, "names")= chr [1:2] "dist" "speed"
```

or, more precisely, a data.frame which is essentially a list with the property that all the elements have the same length. The data.frame object is used to store data sets, such as the cars data set

```
R> str(cars)

'data.frame':          50 obs. of  2 variables:
 $ speed: num  4 4 7 7 8 9 10 10 10 11 ...
 $ dist : num  2 10 4 22 16 10 18 26 34 17 ...
```

and it is assumed that the elements of a data.frame correspond to variables, while the length of each object is the same as the sample size.

A.4.3   Accessing objects and subsetting

We have seen that $ can be used to access the elements of a list and hence of a data.frame, but R also offer operators for enhanced subsetting. The first one is [which returns an object of the same type as the original object

```
R> y

[1] 2 7 4 1

R> y[2:3]

[1] 7 4

R> str(y)

 num [1:4] 2 7 4 1

R> str(y[2:3])

 num [1:2] 7 4
```

or, for matrix-like objects,

```
R> z

   A  B  C  D  E  F
a  1  6 11 16 21 26
b  2  7 12 17 22 27
c  3  8 13 18 23 28
d  4  9 14 19 24 29
e  5 10 15 20 25 30

R> z[1:2, 5:6]

   E  F
a 21 26
b 22 27
```

and subsetting can occur also on non-consecutive indices

```
R> z[1:2, c(1, 3, 6)]

   A  C  F
a  1 11 26
b  2 12 27
```

or in different order

```
R> z[1:2, c(6, 5, 4)]

   F  E  D
a 26 21 16
b 27 22 17
```

One can subset objects also using names, for example

```
R> z[c("a", "c"), "D"]

 a  c
16 18
```

We can also use a syntax like

```
R> z["c", ]

 A  B  C  D  E  F
 3  8 13 18 23 28
```

leaving one argument out to mean 'run all the elements' for that index. Further, R allows for negative indices which are used to exclude indices

```
R> z[c(-1, -3), ]

  A  B  C  D  E  F
b 2  7 12 17 22 27
d 4  9 14 19 24 29
e 5 10 15 20 25 30
```

but positive and negative indexes cannot be mixed.

The subsetting operator [also works for lists

```
R> a <- mod[1:2]
R> str(a)

List of 2
 $ coefficients: Named num [1:2] -17.58 3.93
  ..- attr(*, "names")= chr [1:2] "alpha" "beta"
 $ residuals   : Named num [1:50] 3.85 11.85 -5.95 12.05 2.12 ...
  ..- attr(*, "names")= chr [1:50] "1" "2" "3" "4" ...
```

where we have extracted the first two elements of the list mod using mod[1:2]. We can use names as well and the commands below return the same objects:

```
R> str(mod["coefficients"])

List of 1
 $ coefficients: Named num [1:2] -17.58 3.93
  ..- attr(*, "names")= chr [1:2] "alpha" "beta"
```

```
R> str(mod[1])
```

```
List of 1
 $ coefficients: Named num [1:2] -17.58 3.93
  ..- attr(*, "names")= chr [1:2] "alpha" "beta"
```

Notice that mod[1] returns a list with one element but not just the element inside the list. For this purpose one should use the subsetting operator [[. The next group of commands returns the element inside the list:

```
R> str(mod[[1]])
```

```
 Named num [1:2] -17.58 3.93
 - attr(*, "names")= chr [1:2] "alpha" "beta"
```

```
R> str(mod[["coefficients"]])
```

```
 Named num [1:2] -17.58 3.93
 - attr(*, "names")= chr [1:2] "alpha" "beta"
```

```
R> str(mod$coefficients)
```

```
 Named num [1:2] -17.58 3.93
 - attr(*, "names")= chr [1:2] "alpha" "beta"
```

We have mentioned that a data.frame looks like a particular list, but with more structure and used to store data sets. The latter are always thought of as matrices, and indeed it is possible to access the elements of a data.frame using the subsetting rules for matrices:

```
R> cars[, 1]
```

```
 [1]  4  4  7  7  8  9 10 10 10 11 11 12 12 12 12 13 13 13 13 14 14 14
[23] 14 15 15 15 16 16 17 17 17 18 18 18 18 19 19 19 20 20 20 20 20 22
[45] 23 24 24 24 24 25
```

which is equivalent to the following:

```
R> cars$speed
```

```
 [1]  4  4  7  7  8  9 10 10 10 11 11 12 12 12 12 13 13 13 13 14 14 14
[23] 14 15 15 15 16 16 17 17 17 18 18 18 18 19 19 19 20 20 20 20 20 22
[45] 23 24 24 24 24 25
```

```
R> cars[[1]]
```

```
 [1]  4  4  7  7  8  9 10 10 10 11 11 12 12 12 12 13 13 13 13 14 14 14
[23] 14 15 15 15 16 16 17 17 17 18 18 18 18 19 19 19 20 20 20 20 20 22
[45] 23 24 24 24 24 25
```

Notice that only the output is not a `data.frame` while[2]

```
R> str(cars[1])

'data.frame':       50 obs. of  1 variable:
 $ speed: num   4 4 7 7 8 9 10 10 10 11 ...

R> head(cars[1])

  speed
1     4
2     4
3     7
4     7
5     8
6     9
```

is a proper (sub-) `data.frame` although the matrix-like subsetting operator behaves differ-ently if used on columns or rows: `cars[,1]` returns the element but, for example,

```
R> cars[1:3, ]

  speed dist
1     4    2
2     4   10
3     7    4
```

returns a `data.frame` with the selected number of rows and all columns.

A.4.4   Coercion between data types

Functions such as `names` and `colnames`, but also `levels` and `attributes`, are used to retrieve and set properties of objects and are called accessor functions. Objects can be transformed from one type into another using functions with names as `.*`. For example, `as.integer` transforms an object into an integer whenever possible or returns a missing value:

```
R> pi

[1] 3.141593

R> as.integer(pi)

[1] 3

R> as.integer("3.14")

[1] 3
```

---

[2] The commands `head` and `tail` show the first and last rows of a `data.frame` respectively.

```
R> as.integer("a")
```

```
[1] NA
```

Other examples are as.data.frame to transform matrix objects into true data.frame objects and vice versa. For more complex classes one can also try the generic function as.

A.5   S4 objects

We have several times used the term 'class' for R objects. This is because each object in R belongs to some class and for each class there exist generic functions called methods which perform some task on that object. For example, the function summary provide summary statistics which are appropriate for a given object:

```
R> summary(cars)
```

```
     speed              dist
 Min.   : 4.0   Min.   :  2.00
 1st Qu.:12.0   1st Qu.: 26.00
 Median :15.0   Median : 36.00
 Mean   :15.4   Mean   : 42.98
 3rd Qu.:19.0   3rd Qu.: 56.00
 Max.   :25.0   Max.   :120.00
```

```
R> summary(mod)
```

```
Call:
lm(formula = dist ~ speed, data = cars)

Residuals:
    Min      1Q  Median      3Q     Max
-29.069  -9.525  -2.272   9.215  43.201

Coefficients:
      Estimate Std. Error t value Pr(>|t|)
alpha -17.5791     6.7584  -2.601   0.0123 *
beta    3.9324     0.4155   9.464 1.49e-12 ***
---
Signif. codes:  0 '***' 0.001 '**' 0.01 '*' 0.05 '.' 0.1 ' ' 1

Residual standard error: 15.38 on 48 degrees of freedom
Multiple R-squared: 0.6511,        Adjusted R-squared: 0.6438
F-statistic: 89.57 on 1 and 48 DF,  p-value: 1.490e-12
```

The standard set of classes and methods in R is called S3. In this framework, a method for an object of some class is simply an R function named method.class – for example, summary.lm is the function which is called by R when the function summary is called with an argument which is an object of class lm. In R methods like summary are very generic and the function methods provides a list of specific methods (which apply to specific types of objects) for some particular method. For example

```
R> methods(summary)
```

```
 [1] summary.aov            summary.aovlist
 [3] summary.aspell*        summary.connection
 [5] summary.data.frame     summary.Date
 [7] summary.default        summary.ecdf*
 [9] summary.factor         summary.glm
[11] summary.infl           summary.lm
[13] summary.loess*         summary.manova
[15] summary.matrix         summary.mlm
[17] summary.nls*           summary.packageStatus*
[19] summary.POSIXct        summary.POSIXlt
[21] summary.ppr*           summary.prcomp*
[23] summary.princomp*      summary.srcfile
[25] summary.srcref         summary.stepfun
[27] summary.stl*           summary.table
[29] summary.tukeysmooth*

   Non-visible functions are asterisked
```

The dot '.' naming convention is quite unfortunate because one can artificially create functions which are not proper methods – for example, the t.test function is not the method t for objects of class test but it is just an R function which performs an ordinary two-sample *t* test. The new system of classes and methods which is now fully implemented in R is called S4. Objects of class S4 apparently behave like all other objects in R but they possess properties called 'slots', which can be accessed differently from other R objects. The next code estimates the maximum likelihood estimator for the mean of a Gaussian law. It uses the function mle from the package stats4, which is an S4 package as the name suggests. Again, we are interested in the statistical part of this example.

```
R> require(stats4)
R> set.seed(123)
R> y <- rnorm(100, mean = 1.5)
R> f <- function(theta = 0) -sum(dnorm(x = y, mean = theta,
+      log = TRUE))
R> fit <- mle(f)
R> fit

Call:
mle(minuslogl = f)

Coefficients:
   theta
1.590406
```

We now take a look at the object fit returned by the mle function

```
R> str(fit)

Formal class 'mle' [package "stats4"] with 8 slots
  ..@ call      : language mle(minuslogl = f)
  ..@ coef      : Named num 1.59
  .. ..- attr(*, "names")= chr "theta"
  ..@ fullcoef : Named num 1.59
  .. ..- attr(*, "names")= chr "theta"
  ..@ vcov      : num [1, 1] 0.01
  .. ..- attr(*, "dimnames")=List of 2
  .. .. ..$ : chr "theta"
  .. .. ..$ : chr "theta"
  ..@ min       : num 133
  ..@ details  :List of 6
  .. ..$ par         : Named num 1.59
  .. .. ..- attr(*, "names")= chr "theta"
  .. ..$ value       : num 133
  .. ..$ counts      : Named int [1:2] 6 3
  .. .. ..- attr(*, "names")= chr [1:2] "function" "gradient"
  .. ..$ convergence: int 0
  .. ..$ message     : NULL
  .. ..$ hessian     : num [1, 1] 100
  .. .. ..- attr(*, "dimnames")=List of 2
  .. .. .. ..$ : chr "theta"
  .. .. .. ..$ : chr "theta"
  ..@ minuslogl:function (theta = 0)
  ..@ method    : chr "BFGS"
```

We see that this is an S4 objects with slots that, as the structure suggests, can be accessed using the symbol @ instead of $. For example,

```
R> fit@coef

   theta
1.590406
```

To get a list of methods for S4 objects one should use the function showMethods:

```
R> showMethods(summary)

Function: summary (package base)
object="ANY"
object="mle"
```

A.6   Functions

In the previous section we created a new function called f to define the log-likelihood of the data. In R functions are created with the command function followed by a list of arguments, and the body of the function (if longer than one line) has to be contained within '{' and '}' as in the next example in which we define the payoff function of a call option:

```
R> g <- function(x, K = 110) {
+     max(x - K, 0)
+ }
```

The function returns the last calculation unless the command `return` is used. By default, in the function g we have set the strike price K=100 and x is the argument which represents the price of the underlying asset.

```
R> g(120)
```

```
[1] 10
```

```
R> g(99)
```

```
[1] 0
```

```
R> g(115, 120)
```

```
[1] 0
```

In R arguments are always named, so the function can be called with arguments in any order if named, for example

```
R> g(150, 120)
```

```
[1] 30
```

```
R> g(K = 120, x = 150)
```

```
[1] 30
```

In the definition of g we have fixed a default value of 100 for the argument K, so if it is missing in a call, it is replaced by R with its default value. The argument x cannot be omitted, therefore a call such as g(K=120) will produce an error.

A.7 Vectorization

Most R functions are vectorized, which means that if a vector is passed to a function, the function is applied to each element of the function and a vector of results is returned as in the next example:

```
R> set.seed(123)
R> x <- runif(5, 90, 150)
R> x
```

```
[1] 107.2547 137.2983 114.5386 142.9810 146.4280
```

```
R> sin(x)
```

```
[1]   0.4263927 -0.8026760   0.9916244 -0.9992559   0.9414204
```

But functions should be prepared to be vectorized. For example, our function g is not vector-ized:

```
R> g(x)
```

```
[1] 36.42804
```

Indeed, in the body of g the function max is used and it operates as follows. First x-K is calculated:

```
R> x - 100
```

```
[1]   7.254651 37.298308 14.538615 42.981044 46.428037
```

Then max calculates the maximum of the vector c(x-100, 0). To vectorize it we can use the function sapply as follows:

```
R> g1 <- function(x, K = 110) {
+      sapply(x, function(x) max(x - K, 0))
+ }
R> x
```

```
[1] 107.2547 137.2983 114.5386 142.9810 146.4280
```

```
R> g1(x)
```

```
[1]   0.000000 27.298308   4.538615 32.981044 36.428037
```

We get five different payoffs. The functions of class *apply are designed to work iteratively on different objects. The function sapply iterates the vector in the first argument and applies the functions in the second argument. The function apply works on arrays (e.g. matrices), lapply iterates over lists, etc.

The usual for and while constructs exist in R as well, but their use should be limited to real iterative tasks which cannot be parallelized as in our example. Here is a for version of the function g:

```
R> g2 <- function(x, K = 110) {
+      n <- length(x)
+      val <- numeric(n)
+      for (i in 1:n) val[i] <- max(x[i] - K, 0)
+      val
+ }
```

or, in a more R-like fashion,

```
R> g3 <- function(x, K = 110) {
+       val <- NULL
+       for (u in x) val <- c(val, max(u - K, 0))
+       val
+ }
R> g1(x)

[1]   0.000000 27.298308   4.538615 32.981044 36.428037

R> g2(x)

[1]   0.000000 27.298308   4.538615 32.981044 36.428037

R> g3(x)

[1]   0.000000 27.298308   4.538615 32.981044 36.428037
```

The vectorized versions are usually faster then those iterated using `for` loops:

```
R> y <- runif(10000, 90, 150)
R> system.time(g1(y))

   user   system elapsed
  0.030    0.002   0.032

R> system.time(g2(y))

   user   system elapsed
  0.044    0.003   0.047

R> system.time(g3(y))

   user   system elapsed
  0.218    0.009   0.226
```

Notice that the function `g3` is particularly inefficient because instead of allocating and assigning the results, it grows the vector `val` dynamically.

A.8   Importing data from different sources

R offers facilities to import data from the most common file formats via the package `foreign` which is included in the standard distribution of R. The `foreign` library can manage the following file formats:

- Stata. R can read and write the `.dta` data format from Stata versions 5–11. The two commands are `read.dta` and `write.dta`.

- EpiInfo. EpiInfo is a public domain database and statistics package produced by the US Centers for Disease Control, and EpiData is a freely available data entry and validation system. R can read `.REC` files using the function `read.epiinfo`. Version 2000 of EpiInfo uses the Microsoft Access file format to store data. This may be read with the RODBC or RDCOM packages.

- Minitab portable worksheet. R is able to read files in this particular format but, while in general the output of these input functions is a `data.frame`, in this case the returned object is a `list`. The input function is `read.mtp`.

- SAS Transport (XPORT). This is the format used by the SAS system to export data. The function `read.xport` reads a file as an SAS XPORT format library and returns a list of `data.frames`. With the exception of the Mac OS X platform, if SAS is installed in the system, it is possible to use the wrapper function `read.ssd`, which executes an SAS script that transforms the SAS files with extension `.ssd` (or `.ssd7bdat`) into the XPORT format and then calls the function `read.xport`.

- SPSS. The function `read.spss` can read files stored using the commands 'save' and 'export' in SPSS.

- S. It is possible to read the old file formats of the Unix and Windows versions of S-PLUS (3.x, 4.x or 2000 32-bit). The function is called `read.S`. It is also possible to use the function `data.restore` to read S-PLUS files created with the S-PLUS command `data.dump` using option `oldStyle=T`.

- DBF. The function `read.dbf` reads a DBF file into a data frame, converting character fields into factors, and if possible respecting NULL fields.

- Systat. The function `read.systat` reads a rectangular data file stored by the Systat SAVE command as (legacy) `*.sys` or more recently `*.syd` files.

- WEKA. WEKA attribute-relation file format (ARFF) files can be read and saved using the commands `read.arff` and `write.arff`.

- Octave. The `read.octave` imports a file in the Octave text data format into a list.

Other packages may exist to read files from other commercial or free software.

A.9   Interacting with databases

With R it is possible to interact with database management systems (DBMSs) and their relational counterparts (RDBMSs). The strengths of DBMSs and RDBMSs are that they:

- provide fast access to selected parts of large databases;

- provide powerful ways to summarize and cross-tabulate columns in databases;

- store data in more organized ways than the rectangular grid model of spreadsheets and R data frames;

- allow concurrent access from multiple clients running on multiple hosts while enforcing security constraints on access to the data;

- have the ability to act as a server to a wide range of clients.

The sort of statistical applications for which DBMSs might be used are to extract a 10% sample of the data, to cross-tabulate data to produce a multi-dimensional contingency table, and to extract data group by group from a database for separate analysis.

Most R packages are designed to work with a specific database server. For example, ROracle, RPostgreSQL and RSQLite. The RODBC package provides access to databases in a broader sense (including Microsoft Access and Microsoft SQL Server) through an ODBC interface. A typical use of the RODBC package is as follows:

```
R> library(RODBC)
R> myconn <- odbcConnect("mydsn", uid = "Othello", pwd =
"2beornot2be")
R> crimedat <- sqlFetch(myconn, Crime)
R> pundat <- sqlQuery(myconn, "select * from Punishment")
R> close(myconn)
```

The package DBI acts as a generic front-end to some of the specific packages mentioned above. The package RJDBC uses Java and can connect to any DBMS that has a JDBC driver, while RpgSQL is a specialist interface to PostgreSQL built on top of RJBDC.

A.10   Simple graphics manipulation

R offers complete control over its graphics at the cost of learning many small bits of information. By default, the standard R graphics tend to be simple but effective. In some cases, labels on the axis may be dropped to avoid overlaps, or the bins in a histogram may be less than that specified by the user (R considers this number as a suggestion only), and other similar little things which drive the user crazy at first. We briefly recall here some very basic graphics parameters which can be applied to almost all plots in R. We go through the different options using examples. We begin by plotting a scatterplot and two regression lines with the command abline with two different line widths: lwd=1 (the default value) and lwd=4.

```
R> data(cars)
R> par(mfrow = c(1, 3))
R> plot(cars)
R> plot(cars, main = "lwd=1")
R> abline(lm(dist ~ speed, data = cars))
R> plot(cars, main = "lwd=4")
R> abline(lm(dist ~  speed, data = cars), lwd = 4)
```

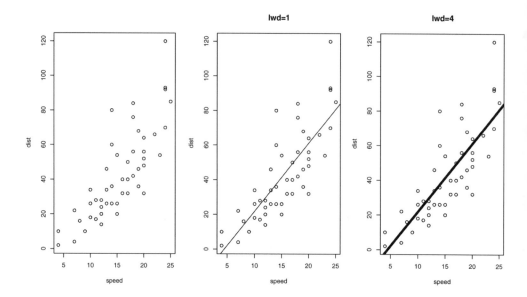

Here we have split the graphical area into one row and three columns and specified a different name at the top of the graphics via the argument `main` in the command `plot`.

Line types can either be specified as an integer (0=blank, 1=solid (default), 2=dashed, 3=dotted, 4=dotdash, 5=longdash, 6=twodash) or as one of the character strings `blank`, `solid`, `dashed`, `dotted`, `dotdash`, `longdash`, or `twodash`, where `blank` uses invisible lines (i.e., does not draw them).

```
R> plot(1, type = "n")
R> abline(h = 0.6, lty = 2)
R> abline(h = 0.8, lty = 3)
R> abline(h = 0.9, lty = 4)
R> abline(h = 1.2, lty = 5)
R> abline(h = 1.4, lty = 6)
```

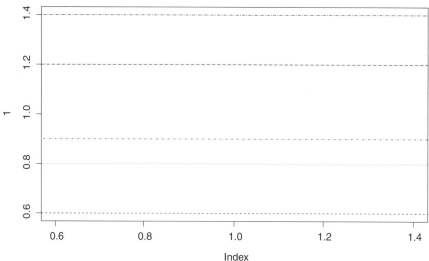

We can also specify colours using the argument col="red", for example, if we want to draw the plot in red. Colours can be chosen by name or from the palette

```
R> str(colors())

 chr [1:657] "white" "aliceblue" "antiquewhite" ...

R> palette()

[1] "black"   "red"      "green3"  "blue"     "cyan"     "magenta"
[7] "yellow"  "gray"
```

It is possible to modify title, subtitle, axis labels and tick values. For example:

```
R> par(mfrow = c(1, 2))
R> x <- c(1, 2, 5, 9, 10, 11)
R> y <- c(1, 7, 5, 4, 3, 1)
R> plot(x, y, main = "Title", sub = "subtitle", ylab =
"values of y")
R> plot(x, y, main = "Title\n under the title", sub = "subtitle",
+      axes = FALSE, xlab = "numbers", type = "b")
R> axis(2)
R> axis(1, x, c("one", "two", "five", "nine", "ten", "eleven"))
```

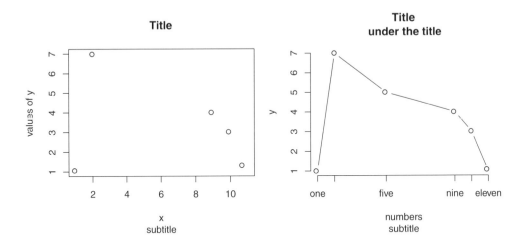

Here we have also used the argument AXES=FALSE because we wish to draw the axes ourselves. To this end, we use axes(2) to draw the vertical axis and axis(1,...) to draw the values of X at given positions and with given labels.

It is also possible to add text and formulas to the graph, making use of the expression command. The text is added using the text command and the corresponding horizontal and vertical lines.

```
R> plot(cars)
R> abline(lm(dist ~ speed, data = cars), lty = 3, lwd = 2)
```

```
R> text(10, 100, expression(hat(beta) == (X^t * X)^{
+      -1
+ } * X^t * y))
R> text(10, 80, "the regression line")
```

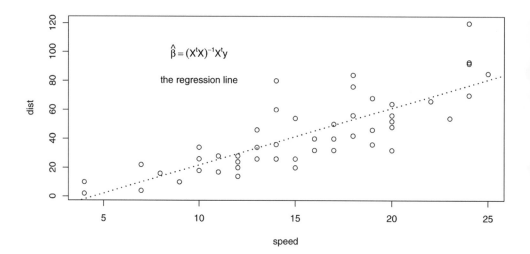

The text can further be aligned in different ways:

```
R> plot(1, main = "Text alignment")
R> text(1, 0.8, "hello", adj = 0)
R> text(1, 0.9, "hello", adj = 0.5)
R> text(1, 1.1, "hello", adj = 1)
R> abline(v = 1, lty = 3)
```

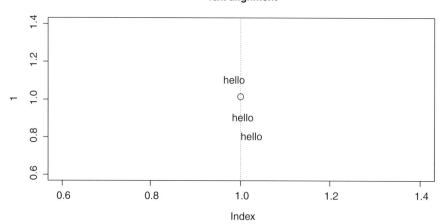

A.11   Basic analysis of the ABC data

Now that we have some knowledge of the basic concepts of the language itself, we show how to produce some elementary statistical analysis in R. As a working example, we consider the data from Chapter 2 on the ABC annual customer satisfaction survey. These data were available to us as an Excel file. The quick way to import this data into R is to save the data as a csv file and load it into R with

```
R> dat <- read.csv("ABCdata.csv")
```

We can check the structure of this data set:

```
R> str(dat, list.len = 15)

'data.frame':        266 obs. of  135 variables:
 $ ID         : int  1 2 3 4 5 6 7 8 9 10 ...
 $ q1         : int  1 3 4 3 5 4 4 3 3 3 ...
 $ q4         : int  1 4 3 4 4 4 4 4 4 2 ...
 $ q5         : int  1 4 2 4 4 4 3 4 4 2 ...
 $ q6         : int  2 4 4 3 5 5 0 3 4 4 ...
 $ qi6        : int  3 NA 3 3 3 3 3 3 3 3 ...
 $ q7         : int  3 1 2 2 5 5 4 4 3 3 ...
 $ qi7        : int  3 NA 3 2 3 3 1 3 2 1 ...
 $ q8         : int  2 3 3 2 4 3 0 4 4 3 ...
 $ qi8        : int  3 NA 3 3 3 3 3 3 3 3 ...
 $ q9         : int  3 4 5 2 4 4 0 4 3 3 ...
 $ qi9        : int  3 NA 3 3 3 3 3 3 3 2 ...
 $ q10        : int  0 NA NA 0 4 0 0 4 0 0 ...
 $ qi10       : int  NA NA NA NA 3 NA 3 3 NA NA ...
 $ q11        : int  1 4 4 3 4 4 4 4 3 3 ...
   [list output truncated]
```

We see that the data frame consists of 266 observations and 135 variables. We have truncated the output to the first 15 entries. A quick look at the data can be provided using the summary command which can be very lengthy for this data set. Purely as an exercise, we reduce the number of columns in the data matrix to those we will use in the subsequent analysis. In order to do this we just subset the data frame using column names:

```
R> dat <- dat[, c("q1", "q4", "q5", "q11", "q17", "q25",
+      "q31", "q38", "country", "cseniority", "var4", "var6",
+      "var9", "var11")]
R> str(dat, vec.len = 2)

'data.frame':        266 obs. of  14 variables:
 $ q1         : int  1 3 4 3 5 ...
 $ q4         : int  1 4 3 4 4 ...
 $ q5         : int  1 4 2 4 4 ...
 $ q11        : int  1 4 4 3 4 ...
 $ q17        : int  1 3 3 4 5 ...
 $ q25        : int  1 3 4 2 4 ...
```

```
$ q31        : int  NA NA 4 3 NA ...
$ q38        : int  2 3 4 3 3 ...
$ country    : Factor w/ 6 levels "Benelux","France",..: 1 1 6 3 3 ...
$ cseniority: int  2 2 4 3 2 ...
$ var4       : int  3 2 3 1 3 ...
$ var6       : int  2 2 4 3 2 ...
$ var9       : int  2 2 1 2 1 ...
$ var11      : int  1 6 5 2 1 ...
```

R> summary(dat)

```
      q1                q4                q5                q11
Min.   :1.000    Min.   :1.000    Min.   :1.000    Min.   :1.000
1st Qu.:3.000    1st Qu.:3.000    1st Qu.:3.000    1st Qu.:3.000
Median :4.000    Median :4.000    Median :4.000    Median :4.000
Mean   :3.561    Mean   :3.716    Mean   :3.779    Mean   :3.572
3rd Qu.:4.000    3rd Qu.:5.000    3rd Qu.:5.000    3rd Qu.:4.000
Max.   :5.000    Max.   :5.000    Max.   :5.000    Max.   :5.000
NA's   :4.000    NA's   :5.000    NA's   :4.000    NA's   :9.000
      q17               q25               q31               q38
Min.   : 1.00    Min.   :1.000    Min.   : 1.000    Min.   : 1.000
1st Qu.: 2.75    1st Qu.:3.000    1st Qu.: 3.000    1st Qu.: 3.000
Median : 3.00    Median :4.000    Median : 4.000    Median : 3.000
Mean   : 3.29    Mean   :3.746    Mean   : 3.841    Mean   : 3.409
3rd Qu.: 4.00    3rd Qu.:5.000    3rd Qu.: 4.000    3rd Qu.: 4.000
Max.   : 5.00    Max.   :5.000    Max.   : 5.000    Max.   : 5.000
NA's   :18.00    NA's   :2.000    NA's   :90.000    NA's   :14.000
    country          cseniority          var4              var6
Benelux: 26    Min.   : 1.000    Min.   : 1.00    Min.   : 1.000
France : 15    1st Qu.: 2.000    1st Qu.: 1.00    1st Qu.: 2.000
Germany:112    Median : 3.000    Median : 2.00    Median : 3.000
Israel : 23    Mean   : 3.466    Mean   :17.62    Mean   : 3.466
Italy  : 39    3rd Qu.: 5.000    3rd Qu.: 4.00    3rd Qu.: 5.000
UK     : 51    Max.   :99.000    Max.   :99.00    Max.   :99.000

      var9             var11
Min.   : 1.00    Min.   :1.000
1st Qu.: 1.00    1st Qu.:1.000
Median : 2.00    Median :1.000
Mean   :11.04    Mean   :2.083
3rd Qu.: 3.00    3rd Qu.:3.000
Max.   :99.00    Max.   :7.000
```

where q1 is 'Overall satisfaction level with ABC'; q4 is 'Would you recommend ABC to other companies?'; q5 is 'If you were in the market to buy a product, how likely would it be for you to purchase an ABC product again?'; country is the country; cseniority is the customer seniority (in years); var4 is segmentation; var6 is the age of ABC's equipment; var9 is profitability; and var11 is position within the customer's company.

We can now quickly reproduce Table 2.1, which is the distribution of respondents by country, with the command summary or table

```
R> summary(dat$country)
```

```
Benelux  France Germany  Israel   Italy      UK
     26      15     112      23      39      51
```

```
R> table(dat$country)
```

```
Benelux  France Germany  Israel   Italy      UK
     26      15     112      23      39      51
```

which in this case produce the same result. Now we focus on variable `var4` and we need to transform it into a factor variable and recode the values 1, 2, 3, 4 and 99 into "Other", "Silver", "Gold", "Platinum" and "Not allocated". We first proceed by coercing a `numeric` variable into a `factor`:

```
R> table(dat$var4)
```

```
  1    2    3    4   99
112   42   43   26   43
```

```
R> sum(table(dat$var4))
```

```
[1] 266
```

```
R> dat$var4 <- factor(dat$var4)
```

Then we access the `levels` of the factor and change them:

```
R> levels(dat$var4)
```

```
[1] "1"  "2"  "3"  "4"  "99"
```

```
R> levels(dat$var4) <- c("Other", "Silver", "Gold", "Platinum",
+     "Not allocated")
```

Now we are ready to reproduce Table 2.2:

```
R> summary(dat$var4)
```

```
     Other         Silver          Gold      Platinum Not allocated
       112             42            43            26            43
```

```
R> table(dat$var4)
```

```
     Other         Silver          Gold      Platinum Not allocated
       112             42            43            26            43
```

Next, we need to recode `var6` in a similar way. First we look at the data

```
R> table(dat$var6)
```

```
 1  2  3  4  5 99
46 65 44 35 75  1
```

```
R> summary(dat$var6)
```

```
   Min. 1st Qu.  Median    Mean 3rd Qu.    Max.
  1.000   2.000   3.000   3.466   5.000  99.000
```

from which we understand that variable var6 is stored as a numeric value while we need a factor here.

```
R> dat$var6[which(dat$var6 == 99)] <- NA
R> dat$var6 <- factor(dat$var6)
R> levels(dat$var6)
```

```
[1] "1" "2" "3" "4" "5"
```

```
R> levels(dat$var6) <- c("less than 1", "1-2", "2-3", "3-4",
+      "More than 4")
```

Now we can generate Table 2.3:

```
R> table(dat$var6)
```

```
less than 1            1-2          2-3          3-4 More than 4
         46             65           44           35          75
```

Similar work is needed for variable var9 in order to generate Table 2.4:

```
R> table(dat$var9)
```

```
  1   2   3  99
 78 105  58  25
```

```
R> summary(dat$var9)
```

```
   Min. 1st Qu.  Median    Mean 3rd Qu.    Max.
   1.00    1.00    2.00   11.04    3.00   99.00
```

```
R> dat$var9[which(dat$var9 == 99)] <- NA
R> dat$var9 <- factor(dat$var9)
R> levels(dat$var9)
```

```
[1] "1" "2" "3"
```

```
 R> levels(dat$var9) <- c("Profitable", "Break-Even", "Below Break-
Even")
```

```
R> table(dat$var9)
```

```
     Profitable          Break-Even Below Break-Even
             78                 105                 58
```

In order to reproduce Table 2.5 we need to recode the values "99" into missing values:

```
R> dat$cseniority[which(dat$cseniority == 99)] <- NA
R> table(dat$cseniority)
```

```
 1  2  3  4  5
46 65 44 35 75
```

Again, to reproduce Table 2.6 we need the following code:

```
R> table(dat$var11)
```

```
  1   2   3   4   5   6   7
142  49  30  15  20   6   4
```

```
R> dat$var11 <- factor(dat$var11)
R> levels(dat$var11)
```

```
[1] "1" "2" "3" "4" "5" "6" "7"
```

```
R> levels(dat$var11) <- c("Owner", "Management", "Technical Manage-
ment",
+      "Technical Staff", "Operator", "Administrator", "Other")
R> table(dat$var11)
```

```
            Owner        Management Technical Management
              142                49                   30
  Technical Staff          Operator        Administrator
               15                20                    6
            Other
                4
```

Variables q1, q4 and q5 are coded with numbers 1 to 5 denoting, respectively "very low", "low", "average", "high" and "Very high". So we proceed with the recoding of these variables:

```
R> dat$q1 <- factor(dat$q1)
R> levels(dat$q1)
```

```
[1] "1" "2" "3" "4" "5"
```

```
R> levels(dat$q1) <- c("Very Low", "Low", "Average", "High",
+      "Very High")
R> dat$q4 <- factor(dat$q4)
R> levels(dat$q4)
```

```
[1] "1" "2" "3" "4" "5"

R> levels(dat$q4) <- c("Very Low", "Low", "Average", "High",
+     "Very High")
R> dat$q5 <- factor(dat$q5)
R> levels(dat$q5)

[1] "1" "2" "3" "4" "5"

R> levels(dat$q5) <- c("Very Low", "Low", "Average", "High",
+     "Very High")
```

We can ow easily reproduce Table 2.7 by the simple command

```
R> table(dat$q1)
```

```
Very Low      Low   Average      High Very High
      11       25        70       118        38
```

and a version of what appears in Figure 2.1 using the barplot command. First of all, we construct the relative frequency distribution for both q4 and q5 by applying prop.table to the output of table:

```
R> prop.table(table(dat$q4))
```

```
  Very Low        Low    Average       High  Very High
0.04980843 0.08812261 0.22988506 0.36015326 0.27203065
```

Then we rescale by 100 and round to one decimal digit:

```
R> tq4 <- round(prop.table(table(dat$q4)) * 100, 1)
R> tq5 <- round(prop.table(table(dat$q5)) * 100, 1)
R> tq4
```

```
Very Low      Low   Average      High Very High
     5.0      8.8      23.0      36.0      27.2
```

```
R> tq5
```

```
Very Low      Low   Average      High Very High
     5.0      9.9      20.2      32.1      32.8
```

Now we prepare the plot. To this end, we divide the plotting area into two. This is done in R using the par command with argument mfrow=c(1,2), which informs R that we wish to plot different graphs on the same graphical devices by partitioning the whole area as a matrix of one row and two columns; this is followed by the command barplot.

```
R> par(mfrow = c(1, 2))
R> barplot(tq4)
R> barplot(tq5)
```

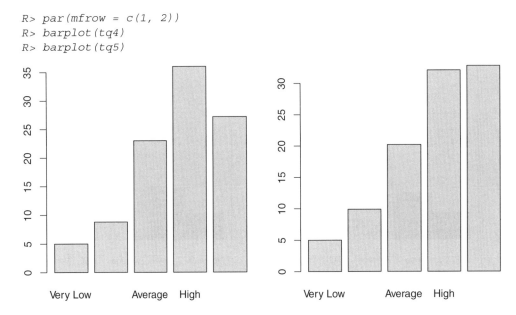

As such, this graphic display is not particularly nice. So we can tune it by changing some arguments to the `barplot` command in order to make the plot look more similar to what is presented in Figure 2.1. We can add a title to each plot using the `main` argument; we can reduce the size of the labels by setting the rescale parameter `cex` to be a fraction of the current font size; we can decide to extend the vertical axis slightly using `ylim` so that this axis goes up to 40; decide to plot the axis ourselves using `axes=FALSE` inside `barplot` and then use the command `axis`; finally, we can add the percentage on top of each bar using the command `text` to place the text in the proper positions:

```
R> par(mfrow = c(1, 2))
R> mycex <- 0.8
R> barplot(tq4, cex.names = mycex, space = 0, col = gray(seq(0,
+     1, len = 5)), main = "Likelihood to Recommend", ylim = c(0,
+     40), axes = FALSE)
R> axis(2, seq(0, 40, len = 5), sprintf("%d%%", seq(0, 40,
+     len = 5)), cex.axis = mycex)
R> text(-0.5 + 1:5, tq4 + 2, sprintf("%.1f%%", tq4), col = "red",
+     cex = mycex)
R> barplot(tq5, cex.names = mycex, space = 0, col = gray(seq(0,
+     1, len = 5)), main = "Likelihood to Repurchase",
+     ylim = c(0, 40), axes = FALSE)
R> axis(2, seq(0, 40, len = 5), sprintf("%d%%", seq(0, 40,
+     len = 5)), cex.axis = mycex)
R> text(-0.5 + 1:5, tq5 + 2, sprintf("%.1f%%", tq5), col = "red",
+     cex = mycex)
```

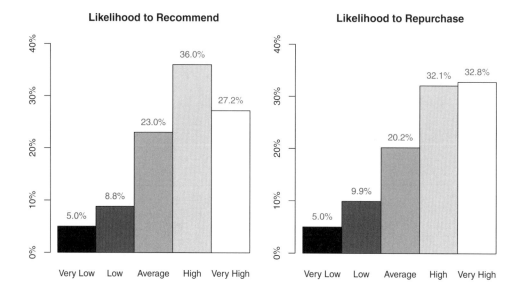

As described in Chapter 2, from the responses on overall satisfaction q1, recommendation q4 and repurchasing intention q5 one can generate three indices that assess customer loyalty. The customer risk index (CRI) is the percentage of customers who gave a rating of '1' or '2' to any of these three questions. Such customers are considered detractors who, in a sense, work for the competition. The customer loyalty index (CLI) is the percentage of customers who gave a rating of '5' or '4' to overall satisfaction, '5' to recommendation and '5' to repurchasing intentions. Finally, the net CLI is the difference between CLI and CRI. We now try to reproduce Table 2.10. We calculate the CLI and CRI using indicator variables. To this end, we create two variables CLI and CRI for each customer which, by default, are set equal to 0.

```
R> dat$CRI <- 0
R> dat$CLI <- 0
```

now we check the CRI by selecting the indexes (idex.cri) of customers such that their rating is in the set "Very Low" or "Low" in any of the three questions using the or operator "|"

```
R> idx.cri <- which((dat$q5 %in% c("Very Low", "Low")) |
+      (dat$q4 %in% c("Very Low", "Low")) | (dat$q1 %in%
+      c("Very Low", "Low")))
R> idx.cri

  [1]   1   3  10  12  14  25  26  33  34  37  39  41  42  48  50  51
 [17]  54  56  60  62  63  64  65  71  79  83  84  86  87  90  96  98
 [33] 101 113 114 121 123 127 128 134 139 142 145 150 157 166 167 168
 [49] 173 178 187 201 205 229 238 248 256

R> dat$CRI[idx.cri] <- 1
```

Then we select the indices (idex.cli) of the customers who give a very high rating on both q4 and q5:

```
R> idx.cli <- which((dat$q5 == "Very High") & (dat$q4 ==
+     "Very High"))
R> idx.cli
```

```
 [1]   17  66  73  77  88  99 100 102 103 104 108 109 110 117 118 119
[17] 120 122 138 143 151 152 155 158 159 160 161 175 176 182 183 186
[33] 188 190 197 198 206 207 208 209 210 211 212 213 214 215 216 217
[49] 218 219 220 223 224 228 231 234 235 237 239 250 252 259 263 265
```

```
R> dat$CLI[idx.cli] <- 1
```

Finally, to take missing values into account, we identify missing values and drop the corresponding observations from the data set:

```
R> idx.na <- which(is.na(dat$q1) | is.na(dat$q4) | is.na(dat$q5))
R> idx.na
```

```
[1]   19  29  36 105 204
```

Now we create a simple data frame, called dat1, which contains only the variable of interest without missing data:

```
R> dat1 <- na.omit(dat[-idx.na, c("CLI", "CRI", "country")])
R> n <- NROW(dat1)
```

We have added the variable country because we wish to calculate the corresponding CRI, CLI and net CLI country *by* country and thus we use the R function by. The function by requires a set of indices, a group variable (country in our case) and a function to be applied. So we define a function h which, taken a subset of indexes, evaluates the corresponding statistics:

```
R> h <- function(x) {
+      cri <- sum(dat1$CRI[x])
+      cli <- sum(dat1$CLI[x])
+      tmp <- round(c(cli - cri, cli, cri)/length(x) * 100,
+          1)
+      tmp <- c(tmp, length(x))
+      names(tmp) <- c("Net CLI", "CLI", "CRI", "n")
+      tmp
+ }
```

Now we can iterate the function h on the groups generated by country. The results are as in Table 2.10.

```
R> by(1:n, dat1$country, h)
```

```
dat1$country: Benelux
Net CLI     CLI      CRI       n
    -56       0       56      25
----------------------------------------------------
dat1$country: France
```

```
Net CLI     CLI      CRI        n
  -21.4     14.3     35.7     14.0
----------------------------------------------------
dat1$country: Germany
Net CLI     CLI      CRI        n
   32.4     43.2     10.8    111.0
----------------------------------------------------
dat1$country: Israel
Net CLI     CLI      CRI        n
   17.4     26.1      8.7     23.0
----------------------------------------------------
dat1$country: Italy
Net CLI     CLI      CRI        n
  -10.3     12.8     23.1     39.0
----------------------------------------------------
dat1$country: UK
Net CLI     CLI      CRI        n
  -24.5      6.1     30.6     49.0
```

However, we can create a more sophisticated version of the above to obtain a real table ready to insert in a publication:

```
R> countries <- unique(dat$country)
R> tab <- NULL
R> for (cnt in countries) {
+      x <- which(dat1$country == cnt)
+      cri <- sum(dat1$CRI[x])
+      cli <- sum(dat1$CLI[x])
+      tmp <- round(c(cli - cri, cli, cri)/length(x) * 100,
+         1)
+      tmp <- c(tmp, length(x))
+      tab <- rbind(tab, tmp)
+ }
R> rownames(tab) <- countries
R> colnames(tab) <- c("Net CLI", "CLI", "CRI", "n")
R> ABC <- c(sum(dat1$CLI), sum(dat1$CRI))
R> ABC <- c(ABC[1] - ABC[2], ABC)/n * 100
R> ABC <- c(round(ABC, 1), n)
R> rbind(ABC, tab)
```

```
          Net CLI  CLI   CRI    n
ABC           2.7 24.5  21.8  261
Benelux     -56.0  0.0  56.0   25
UK          -24.5  6.1  30.6   49
Germany      32.4 43.2  10.8  111
France      -21.4 14.3  35.7   14
Israel       17.4 26.1   8.7   23
Italy       -10.3 12.8  23.1   39
```

Here we have used a `for` loop instead of the `by` command.

To end this section, we try to reproduce the graph in Figure 2.2a because it implies the drawing of two different scales on the same plot. As before, we need a function to build the necessary statistics. The plot of Figure 2.2a shows two indicators: the proportion of very low and low ratings, labelled Bot1+2, and the proportion of very high ratings, labelled Top5. We define the function g which transforms all `factor` objects into `numeric` ones.

```
R> g <- function(x) {
+       x <- as.numeric(x)
+       n <- sum(!is.na(x))
+       tmp <- round(c(sum(x == 5, na.rm = TRUE), sum(x %in%
+           1:2, na.rm = TRUE))/n * 100, 1)
+       tmp <- c(tmp, round(mean(x, na.rm = TRUE), 2))
+       c(tmp, n)
+ }
```

If we apply this to overall satisfaction, we get

```
R> abc <- g(dat$q1)
R> abc
```

```
[1]   14.50   13.70   3.56 262.00
```

We can now apply this function to the different sections of the questionnaire: equipment and system (q11), sales support (q17), technical support (q25), training (q31), and supplies and orders (q38):

```
R> equip <- g(dat$q11)
R> equip
```

```
[1]    6.60    9.30   3.57 257.00
```

```
R> sales <- g(dat$q17)
R> sales
```

```
[1]   15.70   25.00   3.29 248.00
```

```
R> tech <- g(dat$q25)
R> tech
```

```
[1]   28.00   17.00   3.75 264.00
```

```
R> train <- g(dat$q31)
R> train
```

```
[1]   17.60    5.10   3.84 176.00
```

```
R> supp <- g(dat$q38)
R> supp
```

```
[1]    3.60    8.70   3.41 252.00
```

From these we create a table:

```
R> tab <- rbind(abc, equip, sales, tech, train, supp)
R> rownames(tab) <- c("ABC", "Equipment", "Sales", "Tech Supp",
+    "Training", "Supplies")
R> colnames(tab) <- c("Top5", "BOT1+2", "Average", "n")
R> tab
```

```
           Top5 BOT1+2 Average    n
ABC        14.5   13.7    3.56  262
Equipment   6.6    9.3    3.57  257
Sales      15.7   25.0    3.29  248
Tech Supp  28.0   17.0    3.75  264
Training   17.6    5.1    3.84  176
Supplies    3.6    8.7    3.41  252
```

Then we make a plot using `barplot` with option `beside=TRUE` and adding some default legend. In addition to this, we need to specify `par(new=TRUE)` to indicate that a new plot will be added to the previous one. The second plot adds some triangles on top of the bar plot using a different scale on the vertical axis. For this reason, we need to use the function `axis(4)` to specify the fourth axis. Along with the triangles, we also add the average values of the rating for each sector using the `text` function. The result looks very similar to Figure 2.2a.

```
R> barplot(t(tab[, 1:2]), beside = TRUE, legend = c("Top5",
+    "BOT1+2"), col = c("gray", "black"))
R> avg <- tab[, "Average"]
R> par(new = TRUE)
R> plot(1:6, avg, type = "p", pch = 17, xaxt = "n", yaxt = "n",
+    xlab = "", ylab = "", ylim = c(1, 4.5), col = "red")
R> text(1:6, avg, avg, col = "red", pos = 3)
R> axis(4)
```

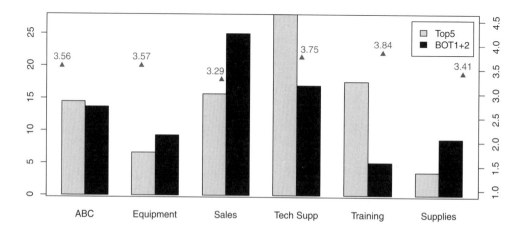

Similar code can be used to plot Figure 2.2b.

We can now try to construct contigency tables to test the association between the variables q1 and q4 and between q1 and q5. This is extremely simple to do with a `table` command:

```
R> tabQ1Q4 <- table(dat$q1, dat$q4)
R> tabQ1Q4
```

	Very Low	Low	Average	High	Very High
Very Low	7	4	0	0	0
Low	5	11	9	0	0
Average	1	5	37	26	1
High	0	3	13	63	38
Very High	0	0	1	5	32

We can add the marginal distributions by row and column with `addmargins`:

```
R> addmargins(tabQ1Q4)
```

	Very Low	Low	Average	High	Very High	Sum
Very Low	7	4	0	0	0	11
Low	5	11	9	0	0	25
Average	1	5	37	26	1	70
High	0	3	13	63	38	117
Very High	0	0	1	5	32	38
Sum	13	23	60	94	71	261

We can obtain relative frequencies applying the function `prop.table` to the output of `table`:

```
R> prop.table(tabQ1Q4)
```

	Very Low	Low	Average	High
Very Low	0.026819923	0.015325670	0.000000000	0.000000000
Low	0.019157088	0.042145594	0.034482759	0.000000000
Average	0.003831418	0.019157088	0.141762452	0.099616858
High	0.000000000	0.011494253	0.049808429	0.241379310
Very High	0.000000000	0.000000000	0.003831418	0.019157088

	Very High
Very Low	0.000000000
Low	0.000000000
Average	0.003831418
High	0.145593870
Very High	0.122605364

We can also reduce the output to something more familiar by applying the command `round`:

```
R> round(prop.table(tabQ1Q4), 2)
```

```
          Very Low  Low Average High Very High
Very Low      0.03 0.02    0.00 0.00      0.00
Low           0.02 0.04    0.03 0.00      0.00
Average       0.00 0.02    0.14 0.10      0.00
High          0.00 0.01    0.05 0.24      0.15
Very High     0.00 0.00    0.00 0.02      0.12
```

`R> addmargins(round(prop.table(tabQ1Q4), 2))`

```
          Very Low  Low Average High Very High  Sum
Very Low      0.03 0.02    0.00 0.00      0.00 0.05
Low           0.02 0.04    0.03 0.00      0.00 0.09
Average       0.00 0.02    0.14 0.10      0.00 0.26
High          0.00 0.01    0.05 0.24      0.15 0.45
Very High     0.00 0.00    0.00 0.02      0.12 0.14
Sum           0.05 0.09    0.22 0.36      0.27 0.99
```

Similarly, we can obtain the conditional distributions by row and column using the argument `margin`, specifying the dimension along which the calculation should be done:

`R> round(prop.table(tabQ1Q4, margin = 1), 2)`

```
          Very Low  Low Average High Very High
Very Low      0.64 0.36    0.00 0.00      0.00
Low           0.20 0.44    0.36 0.00      0.00
Average       0.01 0.07    0.53 0.37      0.01
High          0.00 0.03    0.11 0.54      0.32
Very High     0.00 0.00    0.03 0.13      0.84
```

`R> addmargins(round(prop.table(tabQ1Q4, margin = 1), 2))`

```
          Very Low  Low Average High Very High  Sum
Very Low      0.64 0.36    0.00 0.00      0.00 1.00
Low           0.20 0.44    0.36 0.00      0.00 1.00
Average       0.01 0.07    0.53 0.37      0.01 0.99
High          0.00 0.03    0.11 0.54      0.32 1.00
Very High     0.00 0.00    0.03 0.13      0.84 1.00
Sum           0.85 0.90    1.03 1.04      1.17 4.99
```

and, for the columns,

`R> round(prop.table(tabQ1Q4, 2), 2)`

```
          Very Low  Low Average High Very High
Very Low      0.54 0.17    0.00 0.00      0.00
Low           0.38 0.48    0.15 0.00      0.00
Average       0.08 0.22    0.62 0.28      0.01
High          0.00 0.13    0.22 0.67      0.54
Very High     0.00 0.00    0.02 0.05      0.45
```

`R> addmargins(round(prop.table(tabQ1Q4, 2), 2))`

	Very Low	Low	Average	High	Very High	Sum
Very Low	0.54	0.17	0.00	0.00	0.00	0.71
Low	0.38	0.48	0.15	0.00	0.00	1.01
Average	0.08	0.22	0.62	0.28	0.01	1.21
High	0.00	0.13	0.22	0.67	0.54	1.56
Very High	0.00	0.00	0.02	0.05	0.45	0.52
Sum	1.00	1.00	1.01	1.00	1.00	5.01

R provides a particular plot for `table` objects, which is useful to obtain a visual impression of the conditional distributions:

```
R> plot(tabQ1Q4)
```

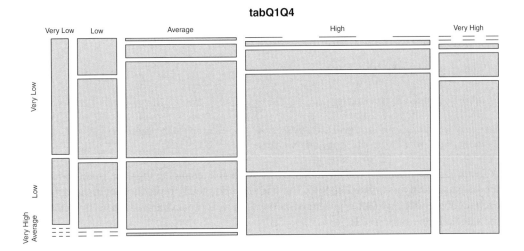

From the above plot we can see that there is probably some association between high levels of satisfaction for this set of customers. This association can be test using the standard χ^2 test simply using the `summary` command:

```
R> summary(tabQ1Q4)
```

```
Number of cases in table: 261
Number of factors: 2
Test for independence of all factors:
    Chisq = 298.03, df = 16, p-value = 6.521e-54
    Chi-squared approximation may be incorrect
```

This analysis confirms that the test of independence between q1 and q4 is rejected with an extremely small *p*-value. This is clearly just a trivial example. A more in-depth analysis of association can be found in the main body of this book, and in Chapter 11 in particular.

A.12   About this document

The present chapter was entirely written by mixing R and LaTeX using the Sweave approach. This means that the source of this text is a plain LaTeX file with special R tags. R reads and processes the file producing the correct LaTeX code, graphics, tables and all the R output. This new file is passed to a standard LaTeX compiler, and the result is the present text. The next two lines are an example of typical processing of the file from a batch shell:

```
$ R CMD Sweave appendixR
Writing to file appendixR.tex
Processing code chunks ...
 1 : term verbatim

( some output suppressed )

91 : echo term verbatim eps pdf

You can now run LaTeX on 'appendixR.tex'

$ R CMD pdflatex appendixR

( some output suppressed )

Output written on appendixR.pdf (34 pages, 423555 bytes).
Transcript written on appendixR.log.
```

A Sweave document is a live document,in the sense that changing the data frame updates all the statistical analysis in Section A.11. But it is also completely reproducible, in the sense that each number, table and plot is generated in this Sweave process and Sweave is included in the standard distribution of R.

A.13   Bibliographical notes

There are many basic books apart form the one mentioned earlier (Dalgaard, 2008), such as Crawley (2007), which cover the basic functionalities of the R language. A simple search with the keyword R in on-line book stores will return hundreds of titles. For advanced programming techniques on the standard S language we should mention Chambers (2004) and Venables and Ripley (2000). For S4 programming some recent references are Chambers (2008) and Gentleman (2008). For advanced graphics one should not overlook the books by Murrel (2005) and Deepayan (2008).

References

Chambers, J. (2004) *Programming with Data: A Guide to the S Language* (revised version). New York: Springer.

Chambers, J. (2008) *Software for Data Analysis: Programming with R*. New York: Springer.

Crawley, M. (2007) *The R Book*. Chichester: John Wiley & Sons, Ltd.

Dalgaard, P. 2008 *Introductory Statistics with R*, (2nd edn). New York: Springer

Deepayan, S. (2008) *Lattice: Multivariate Data Visualization with R*. New York: Springer.

Gentleman, R. (2008) *R Programming for Bioinformatics*. Boca Raton, FL: Chapman & Hall/CRC.

Murrel, P. (2005) *R Graphics*. Boca Raton, FL: Chapman & Hall/CRC.

Venables, W.N. and Ripley, B.D. (2000) *S Programming*. New York: Springer.

Index

Modern Analysis of Customer Surveys: with applications using R, First Edition. Edited by Ron S. Kenett and Silvia Salini.
© 2012 John Wiley & Sons, Ltd. Published 2012 by John Wiley & Sons, Ltd.